WITHDRAWN

PRESTRESSED CONCRETE

McGRAW-HILL CIVIL ENGINEERING SERIES

HARMER E. DAVIS, *Consulting Editor*

BABBITT, DOLAND, AND CLEASBY · Water Supply Engineering
BENJAMIN · Statically Indeterminate Structures
CHOW · Open-channel Hydraulics
DAVIS, TROXELL, AND WISKOCIL · The Testing and Inspection of Engineering Materials
DUNHAM · Advanced Reinforced Concrete
DUNHAM · Foundations of Structures
DUNHAM · The Theory and Practice of Reinforced Concrete
DUNHAM AND YOUNG · Contracts, Specifications, and Law for Engineers
HALLERT · Photogrammetry
HENNES AND EKSE · Fundamentals of Transportation Engineering
KHACHATURIAN AND GURFINKEL · Prestressed Concrete
KRYNINE AND JUDD · Principles of Engineering Geology and Geotechnics
LEONARDS · Foundation Engineering
LINSLEY, KOHLER, AND PAULHUS · Applied Hydrology
LINSLEY, KOHLER, AND PAULHUS · Hydrology for Engineers
LUEDER · Aerial Photographic Interpretation
MATSON, SMITH, AND HURD · Traffic Engineering
MEAD, MEAD, AND AKERMAN · Contracts, Specifications, and Engineering Relations
NORRIS, HANSEN, HOLLEY, BIGGS, NAMYET, AND MINAMI · Structural Design for Dynamic Loads
PEURIFOY · Construction Planning, Equipment, and Methods
PEURIFOY · Estimating Construction Costs
TROXELL, DAVIS, AND KELLY · Composition and Properties of Concrete
TSCHEBOTARIOFF · Soil Mechanics, Foundations, and Earth Structures
WANG AND ECKEL · Elementary Theory of Structures
WINTER, URQUHART, O'ROURKE, AND NILSON · Design of Concrete Structures

PRESTRESSED CONCRETE

Narbey Khachaturian
Professor of Civil Engineering
University of Illinois

German Gurfinkel
Associate Professor of Civil Engineering
University of Illinois

McGraw-Hill Book Company
New York   St. Louis   San Francisco   London
Sydney   Toronto   Mexico   Panama

PRESTRESSED CONCRETE

Copyright © 1969 by McGraw-Hill, Inc. All rights reserved. Printed in the United States of America. No part of this publication may be reproduced, stored in a retrieval system, or transmitted, in any form or by any means, electronic, mechanical, photocopying, recording, or otherwise, without the prior written permission of the publisher.

Library of Congress Catalog Card Number 69-14488

34465

1 2 3 4 5 6 7 8 9 - MAMM - 1 0 9 8 7

Preface

The present book represents a first course in prestressed concrete for advanced undergraduate students in civil engineering as well as professional engineers and architects. The material is based upon a course on the subject which was initiated by the senior author and later continued jointly by both authors.

Emphasis is placed upon topics which have been the subject of considerable research during recent years. The approach used is novel to structural concrete, in that the emphasis is placed upon the fundamentals, with minimum dependence upon codes and approximations.

The text represents analysis and design of basic structures, presented as accurately as current knowledge permits. A large number of illustrative examples are included to supplement the text. Considerable space is given to each selected topic in order to discuss it in depth. The selection of the topics was made carefully to stimulate the interest of the student for further research and study beyond the scope of this book. Some areas that were excluded to reduce the size of the text will hopefully be presented in another book at some future date.

Chapter 1 is a brief background on the development of prestressed concrete, to familiarize the reader with the basic concepts of prestressing. Chapter 2 describes various methods of prestressing used in practice.

Chapters 3 to 6 are on behavior and analysis of beams. The reader should follow the book in the order presented. Chapter 3 discusses the flexural behavior of the beam and includes a method of analysis which leads to moment-curvature and moment-deflection relations for the entire range of load. This chapter constitutes a fundamental approach and is intended to familiarize the student with the basic problems that arise in analysis of a beam. Chapter 4 covers the behavior of the beam in the region of combined stresses. The difficulties associated with a theoretical study of the problem are discussed, and the thinking behind the present methods is exposed. Chapter 5 contains problems concerning bond and anchorage. The stress-transfer bond, flexural bond, and the stresses at the anchorage zone are discussed in detail. Chapter 6 is on time effects and includes discussion of losses in prestress as well as calculation of deflections due to long-time loads.

The emphasis in Chapters 3 through 6 is on the knowledge that has been accumulated from research. Students at this stage are not burdened by specifications with all their usual approximations and shortcuts. This has been done intentionally in order not to confuse the student, for whom it is important to know the fundamental principles first before worrying about approximations. This does not mean that specifications and approximations are unimportant or that they should be neglected. Rather, with this approach it is possible to understand and evaluate them for various situations.

Chapters 7 and 8 present the working-stress and ultimate methods of design of simply supported beams, respectively. Chapter 8 shows for the first time a logical approach to design of beams on the basis of strength and ductility. Chapter 7 is a detailed presentation of the working-stress method for those readers who may prefer to use this traditional method of design. Specifications and, particularly, the provisions of the ACI code are extensively discussed in these chapters.

Chapters 9 and 10 present the analysis and design of composite and continuous beams, with specialized applications to buildings and bridges.

Chapter 11 is on prestressed-concrete columns. It shows for the first time the application of the ultimate interaction diagram in the analysis of the behavior of eccentrically loaded prestressed-concrete columns. The effects of initial prestrain in the steel on the behavior of the column are also discussed thoroughly.

The authors are grateful to all the manufacturers for supplying the information concerning the methods of prestressing used, for which full acknowledgment is given in the text.

Narbey Khachaturian
German Gurfinkel

Contents

Preface v

CHAPTER 1 INTRODUCTION   1

    1-1 *The Basic Idea of Prestressing* 1
    1-2 *Prestressed Concrete—Historical Background* 5
    1-3 *Prestressed Concrete—An Example* 7
    1-4 *Application of Prestressed Concrete* 14
         References 15

CHAPTER 2 METHODS OF PRESTRESSING   16

    2-1 *Introduction* 16
    2-2 *Pretensioning Systems* 17
    2-3 *Post-tensioning Systems* 22
    2-4 *Summary* 39

CHAPTER 3 FLEXURAL BEHAVIOR AND STRENGTH OF PRESTRESSED–CONCRETE BEAMS   41

3-1 *Introduction*   41
3-2 *Behavior of Prestressed-concrete Beams*   42
3-3 *Strain Distribution with Depth*   44
3-4 *The Relationship between Strain in Prestressed Steel and Strain in Concrete*   45
3-5 *The Relationship between Strain in Non-prestressed Steel and Strain in Concrete*   52
3-6 *The Stress-Strain Diagrams for the Materials Used in Prestressed Concrete*   53
3-7 *Analysis*   59
3-8 *Solution of Special Problems*   64
3-9 *Approximate Concrete Stress-Strain Diagram*   67
3-10 *Curvature*   70
3-11 *Illustrative Problem 3-1*   71
3-12 *Calculation of Deflection*   85
3-13 *Moment-deflection Relationship for Illustrative Problem 3-1*   90
3-14 *Illustrative Problem 3-2*   91
3-15 *Illustrative Problem 3-3*   105
3-16 *Influence of the Variables on the Behavior of a Prestressed-concrete Beam*   107
3-17 *Illustrative Problem 3-4*   108
3-18 *Flanged Sections*   112
3-19 *Illustrative Problem 3-5*   121
Problems   121
References   124

CHAPTER 4 BEHAVIOR OF PRESTRESSED–CONCRETE BEAMS IN THE REGION OF COMBINED STRESSES   126

4-1 *Introduction*   126
4-2 *Behavior of Beams before Cracking*   128
4-3 *Cracking of Beams*   132
4-4 *Behavior of Beams after Cracking*   134
4-5 *Shear at Inclined Cracking—Shear-compression Failure*   137
4-6 *Shear at Inclined Cracking—Web-distress Failure*   140
4-7 *Beams with Web Reinforcement*   143
4-8 *Illustrative Example 4-1*   145
4-9 *Illustrative Problem 4-2*   149
Problems   154
References   155

CHAPTER 5 BOND AND ANCHORAGE 156

5-1 *Introduction* 156
5-2 *Stress-transfer Bond in Pretensioned Concrete Beams* 156
5-3 *Flexural Bond in Pretensioned Concrete Beams* 161
5-4 *Anchorage-zone Stresses* 163
5-5 *Determination of Transverse Reinforcement* 164
5-6 *Illustrative Problem 5-1* 169
Problems 172
References 172

CHAPTER 6 LOSSES AND LONG-TIME DEFLECTIONS 173

6-1 *Introduction* 173
6-2 *Loss of Prestress due to Friction in Post-tensioned Beams* 174
6-3 *Illustrative Example 6-1* 176
6-4 *Loss of Prestress due to Elastic Shortening* 177
6-5 *Illustrative Problem 6-2* 178
6-6 *Loss of Prestress due to Anchorage Set* 179
6-7 *Long-time Effects* 180
6-8 *Shrinkage* 180
6-9 *Creep* 181
6-10 *Relaxation* 182
6-11 *Determination of Effectiveness* 182
6-12 *Long-time Deflections* 184
6-13 *Calculation of Long-time Deflections* 189
6-14 *Illustrative Example 6-3* 191
Problems 196
References 196

CHAPTER 7 WORKING-STRESS DESIGN OF SIMPLY SUPPORTED PRESTRESSED-CONCRETE BEAMS 197

7-1 *Introduction* 197
7-2 *Conditions of Loading* 199
7-3 *Allowable Stresses and Stress Coefficients* 201
7-4 *The Four Basic Requirements* 204
7-5 *Problems in Design* 206

DESIGN USING STANDARD SECTIONS

7-6 *Determination of Prestressing Force and Eccentricity* 207
7-7 *Illustrative Problem 7-1* 208
7-8 *Variation of Eccentricity with the Prestressing Force* 210

LEAST-WEIGHT DESIGN

7-9   The Dimensionless Variables   212
7-10 Relations among the Dimensionless Unknowns   215
7-11 Design for Economy   216
7-12 Applicability of the Least-Weight Design Criteria   222
7-13 Relationship among the Dimensionless Variables   223
7-14 The Idealized Sections   232
7-15 Relationship among ρ, Δ, and the Section Properties of Idealized Sections   234
7-16 Determination of ω, Δ, and ρ 240
7-17 The Allowable Stress in Prestressing Steel   242
7-18 Illustrative Problem 7-2   243
7-19 Illustrative Problem 7-3   248
7-20 Stresses at the Ends   252
7-21 The Profile of the Prestressing Steel   255

PRESTRESSING WITH SUPERIMPOSED DEAD LOAD

7-22 The Four Requirements   258
7-23 Illustrative Problem 7-4   260
      Problems   263
      References   264

CHAPTER 8 ULTIMATE DESIGN OF SIMPLY SUPPORTED PRESTRESSED–CONCRETE BEAMS   266

8-1   Introduction   266
8-2   Ultimate Moment   267
8-3   Illustrative Problem 8-1   269
8-4   Simplified Methods for the Determination of Ultimate Moment   270
8-5   Illustrative Problem 8-2   274
8-6   Approximate Methods for the Determination of Ultimate Moment in Flanged Sections   276
8-7   Provisions of ACI Code 318-63 for the Ultimate Flexural Strength   278
8-8   Analysis of the Beam for Ultimate Design   282
8-9   Ultimate Design   283
8-10 Illustrative Example 8-3   287
8-11 Illustrative Example 8-4   292
8-12 Comparison of the Three Solutions   293
8-13 Provisions of ACI Code 318-63 for Shear   294
8-14 Illustrative Example 8-5   296
      Problems   299
      References   302

CHAPTER 9 DESIGN OF COMPOSITE PRESTRESSED–CONCRETE BEAMS   303

9-1 *Introduction* 303
9-2 *Conditions of Loading* 306
9-3 *The Four Basic Requirements* 307
9-4 *The Effective Slab Area* 309
9-5 *The Standard Sections* 310
9-6 *Illustrative Problem 9-1* 312
9-7 *Design of a Beam for Composite Construction* 320
9-8 *Relations among the Dimensionless Variables* 323
9-9 *Design for Least Area of Concrete* 324
9-10 *The Idealized Sections* 326
9-11 *Illustrative Problem 9-2* 331
9-12 *Ultimate Design* 337
9-13 *Illustrative Problem 9-3* 339
Problems 341
References 341

CHAPTER 10 ANALYSIS AND DESIGN OF CONTINUOUS PRESTRESSED–CONCRETE BEAMS   342

10-1 *Introduction* 342
10-2 *Loss in the Prestressing Force Due to Friction* 344
10-3 *Reversal of Sign of the Live-load Moment* 347
10-4 *Design of Continuous Prestressed-concrete Beams* 347
10-5 *Elastic Analysis of Continuous Beams* 348
10-6 *The Variation of Eccentricity* 351
10-7 *The Fixed-end Moments* 359
10-8 *Graphical Representation of Fixed-end Moments* 365
10-9 *Illustrative Problem 10-1* 368
10-10 *The Equivalent-load Method* 371
10-11 *Numerical Analysis* 380
10-12 *Illustrative Problems 10-2 and 10-3* 382
10-13 *Equations for Design* 383
10-14 *Illustrative Problem 10-4* 387
Problems 397
References 399

CHAPTER 11 PRESTRESSED–CONCRETE COLUMNS   400

11-1 *Introduction* 400
11-2 *Determination of Ultimate Interaction Diagram* 403
11-3 *Illustrative Example 11-1* 406

11-4 *Determination of Zero-tensile-strain Interaction Diagram* 412
11-5 *Illustrative Example 11-2* 414
      Problems 419
      References 419

APPENDIX A 421

A-1 *Derivation of Eqs. (7-5) to (7-8)* 421
A-2 *Derivation of Eqs. (7-10) to (7-14)* 423
A-3 *Derivation of Eq. (9-9)* 425
A-4 *Derivation of Eq. (9-10)* 426
A-5 *Fixed-end Moments* 427

APPENDIX B. PHYSICAL AND GEOMETRIC PROPERTIES OF PRESTRESSING SYSTEMS 430

B-1 *Freyssinet* 430
B-2 *Stressteel* 435
B-3 *Roebling* 439
B-4 *BBRV* 442
B-5 *Ryerson* 445
B-6 *Prescon* 446
B-7 *CCL* 452

Index 457

1
Introduction

1-1 THE BASIC IDEA OF PRESTRESSING

In prestressing a structural member permanent internal stresses are induced in the member in order to neutralize, to a desired degree, the stresses of opposite sign caused by the acting loads. For example, if an axially precompressed bar is subjected to an axial tensile force, the preexisting compressive stress will neutralize the tensile stress to a certain extent. Figure 1-1 shows a short precompressed member subjected to an axial tensile force of T. Before the member is subjected to the tensile force of T, it carries a permanent compressive force of P resulting in permanent compressive stress, as shown in Fig. 1-1a. The effect of the tensile force T alone is shown in Fig. 1-1b. If the compressive force P is equal to the tensile force T, these two forces will completely eliminate each other and there will be no stress in the member, as shown in Fig. 1-1c. If the force P is greater than T, the superposition of the two forces of opposite direction will result in some compressive stress in the section, as shown in Fig. 1-1d.

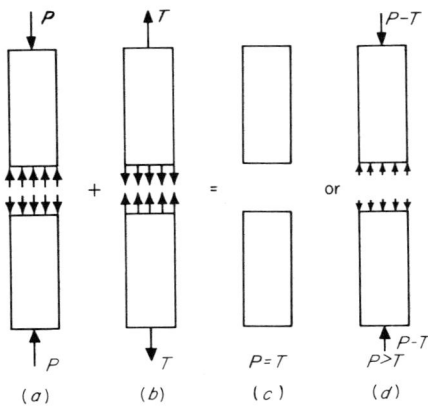

Fig. 1-1 Precompressed bar subjected to tensile force.

Consequently, if the material of this member has a small tensile strength while its compressive strength is fairly high, it will be possible to permit an axial tensile force in the member by first applying a compressive force which is at least equal to the tensile force. In order to design such a member there are two requirements that must be met. First, the compressive force P should be small enough that the resulting compressive stress does not exceed the allowable compressive stress in the material. Second, the load P should be large enough that after the load T is applied, the tensile stresses will be small. If the two requirements are not met simultaneously, the cross-sectional area of the member should be increased. *The force P is generally referred to as the prestressing force.*

One of the early applications of prestressing principles was made

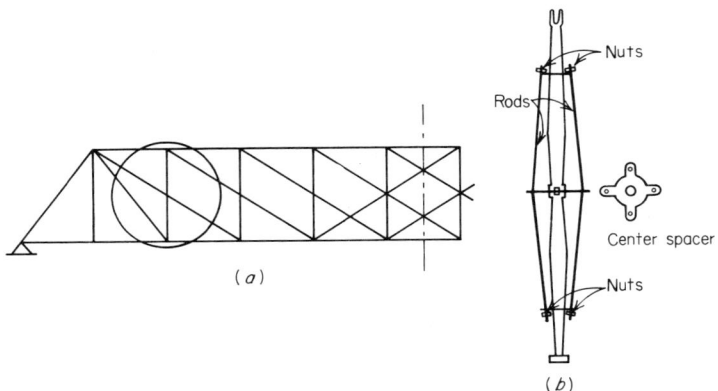

Fig. 1-2 The Whipple truss; the elevation is shown at (a) and a detail of cast-iron verticals is shown at (b).

INTRODUCTION

in 1837 by Squire Whipple, an American instrument maker from Troy, New York. He developed what is known as the *Whipple* or *double-intersection* truss in which the ties or the web system extended over two panels, as shown in Fig. 1-2. The top chord and the verticals were made of cast iron and the lower chord was composed of wrought-iron links. Since cast iron has a small tensile strength, he introduced axial compression in the verticals of the truss by means of four rods which were tightened by nuts, as shown in Fig. 1-2b. These rods introduced initial compression in the verticals, thus prestressing them against possible tensile stress.[1]

The principle of prestressing can be and has been applied to beams. A beam subjected to downward loads will deflect, introducing compressive stress at the top fiber and tensile stress at the bottom fiber. If the beam were somehow initially subjected to permanent loads, causing an upward deflection, the effect of downward loads would be counteracted to a certain extent. Figure 1-3a shows a simply supported beam subjected to a permanent upward load of P and the resulting moment diagram. Subsequently this beam is subjected to a uniformly distributed load. The load and the moment diagram are shown in Fig. 1-3b. If it is assumed that the effect of the downward load is greater than that of the upward load, the superposition of the two loads will result in the moment diagram shown in Fig. 1-3c. It can be seen in this case that the moments due to downward loads are greatly reduced by prestressing.

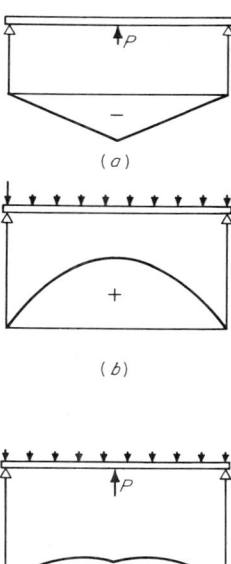

Fig. 1-3 Prestressing of a beam by means of a vertical load.

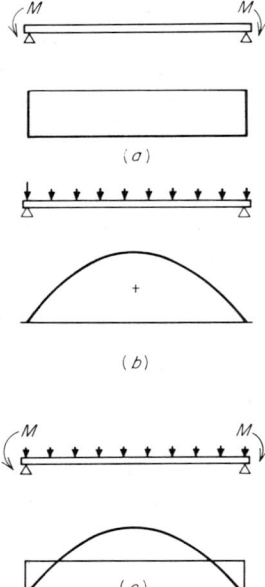

Fig. 1-4 Prestressing of a beam by means of moments.

A somewhat similar problem is shown in Fig. 1-4. In this case the beam is somehow subjected permanently to equal negative moments acting at the ends. The beam and the moment diagram are shown in Fig. 1-4a. As before, the beam is subjected to a uniformly distributed load acting down, and the combined effect is shown in Fig. 1-4c. It should be noted that in this case after prestressing, the negative moment due to prestress is equal to the final positive moment at midspan. Although in this solution the positive moment at the center of the beam has been reduced to one-half by prestressing, negative moments of the same magnitude exist in the ends.

Beams may also be prestressed by direct axial compression. However, this method is not efficient.

In prestressed beams, as in axially loaded members, there are two sets of requirements that must be met. First, the stress at any point in the beam due to prestressing and the weight of the beam, before the application of service loads, must not exceed the allowable stresses for the material. Second, the stresses after the application of acting loads must not exceed the allowable stresses.

One of the best-known applications of prestressing in beams is the king-post truss or the single-strut trussed beam. This structure consists of a timber beam which is reinforced by rods passing over a

INTRODUCTION

timber strut. If the pull in the rod is known, the structure is statically determinate and the upward vertical force applied by the strut tends to neutralize the effect of acting loads. Figure 1-5a shows diagrammatically a king-post truss. A queen-post truss, shown in Fig. 1-5b, is a variation of the same idea.

A common application of prestressing which often is not recognized as such is the utilization of guys for stability of towers. The tension in the guys introduces compression in the tower where the guy is attached to the tower. In this way the tower is capable of resisting a considerable lateral force. This example is only one of the variety of problems to which the idea of prestressing has been or is being applied.

The principles of prestressing have been applied to steel, timber, concrete, and other materials. Both axially loaded and flexural members can be prestressed. Prestressing has a varied and versatile use in engineering. Its principles are used in repairing old structures, in designing new ones, and in waterproofing some.

Though prestressing has had numerous applications, in this text our attention will be directed to prestressed concrete, which is one of the most important applications. Specifically, we shall discuss behavior and design of prestressed-concrete beams for various types of construction.

1-2 PRESTRESSED CONCRETE—HISTORICAL BACKGROUND

The most efficient and extensive application of prestressing is related to concrete. Since concrete is weak in tension and it can resist considerable compression, it is particularly suitable for prestressing.

The first application of prestressing to concrete is generally attributed to the American engineer P. A. Jackson of San Francisco, who patented a novel scheme for construction of arches and vaults in 1872.[2]

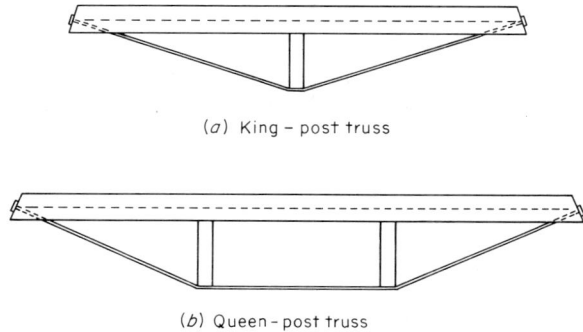

(a) King-post truss

(b) Queen-post truss

Fig. 1-5 Examples of trussed beams. The prestressing is accomplished by means of tie rods passing under struts.

In this method, Jackson passed iron tie rods through masonry or concrete blocks and tightened them by means of nuts. A sketch of this scheme is shown in Fig. 1-6.

During the five decades following Jackson's novel construction method, comparatively small progress has been recorded in prestressed concrete. Several American and European engineers applied the same idea to different structures, with some variations. Among the engineers prominent in the pioneering period are the Norwegian engineer J. G. F. Lund and the American engineer G. R. Steiner.

Lund in 1907 initiated the fabrication of prestressed vaults made up of concrete blocks joined by mortar. In this method the prestressing was accomplished by means of iron tie rods, and the compression was transmitted to the blocks by bearing plates at the ends while bond was destroyed by stretching. In 1908 a similar method was initiated by G. R. Steiner, who proposed first to tighten the prestressing rods against the wet concrete in order to destroy bond and then to increase the tension after the hardening of the concrete.[3]

There is no evidence that in any of the early prestressed-concrete members the losses in the prestress due to shrinkage and creep of concrete were taken into account. If the losses were recognized, certainly no adequate remedy was devised to counteract their effect. Since the stress in iron or steel rods used was low, it can be assumed that the early prestressing ideas were not satisfactory from a practical point of view.

To R. E. Dill of Alexandria, Nebraska, goes the credit of first recognizing the significance of shrinkage and creep. The prestressing process in Dill's method was carried out after most of the shrinkage in concrete had taken place. To compensate for the effect of creep the nuts were tightened occasionally. In Dill's system, bonding was prevented by coating the steel with a plastic substance.[4]

In 1922, W. H. Hewett of Minneapolis, Minnesota, successfully applied prestressing to concrete tanks for the primary purpose of developing a crackless or waterproof concrete. Hewett similarly recognized the significance of shrinkage in concrete.[3] Horizontal hoops with three turnbuckles each placed around the tank were used. Turnbuckles were tightened by hand, and an outer layer of concrete 3 in. thick was cast to cover the steel. Occasionally the concrete was gunited.

In 1928 a French engineer, E. Freyssinet, introduced a significant

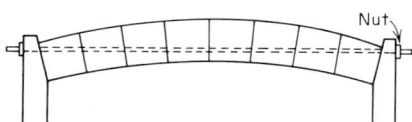

Fig. 1-6 Jackson's prestressing method for masonry and concrete floors.

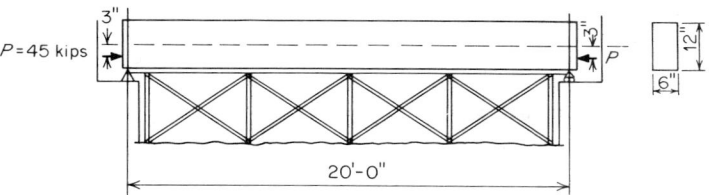

Fig. 1-7 Prestressing of a plain concrete beam on falsework.

innovation by using high-strength steel for prestressing. This not only resulted in a considerable saving in steel but allowed such high prestress that even after losses the remaining tensioning force was sufficient to exert large compressive stresses in the beam. In his beams, he bonded steel with concrete, creating a homogeneous material. In addition, Freyssinet clearly showed the effect of creep in concrete, and by use of high-strength steel he demonstrated that the greater part of the prestress can be retained.[5]

This is considered to be the beginning of prestressed concrete as we know it today. There are many variations of the original Freyssinet method; however, the basic idea remains the same.

1-3 PRESTRESSED CONCRETE—AN EXAMPLE

In order to understand the effect of prestressing on concrete, let us think of a plain concrete member resting on some sort of shoring. Let us assume that the 28-day strength of concrete is 3500 psi, and that the member has a rectangular cross section 12 in. deep and 6 in. wide. This member, which is 22 ft long, is to be utilized as a beam carrying a load of 300 lb per ft on a span of 20 ft.

Let us apply a horizontal prestressing force of 45 kips 3 in. from the bottom at each end by means of jacks bearing against rigidly anchored walls which have practically no horizontal deflection under the load. This arrangement is shown in Fig. 1-7.

As the horizontal prestressing force of 45 kips is applied, since the point of action of the prestressing force is 3 in. below the neutral axis, end moments are created and the beam begins to bend up. As soon as this occurs, the self-weight of the beam becomes a downward load acting on the beam.

The stresses at the extreme top and bottom fibers along the beam due to the prestressing force only can be expressed as follows:

$$-\frac{P}{A} + \frac{Pey}{I}$$

and
$$-\frac{P}{A} - \frac{Pey}{I}$$

respectively, where a negative sign indicates compressive stress and where P = prestressing force
A = cross-sectional area of beam
e = eccentricity of prestressing force
y = distance between extreme fiber and centroidal axis
I = moment of inertia of section about horizontal axis

In this problem the section properties of the beam are as follows:

$A = 12 \times 6 = 72$ in.^{2}

$I = \frac{1}{12}6(12)^3 = 864$ in.^{4}

$\frac{I}{y} = \frac{864}{6} = 144$ in.^{3}

In addition, the prestressing force and eccentricity are known:

$P = 45$ kips

$e = 3$ in.

The extreme top fiber stress in any section along the beam due to the prestressing force only is

$$-\frac{P}{A} + \frac{Pey}{I} = -\frac{45}{72} + \frac{45 \times 3}{144} = -0.625 + 0.938$$
$$= 0.313 \text{ ksi (tensile)}$$

The stress at the extreme bottom fiber is

$$-\frac{P}{A} - \frac{Pey}{I} = -\frac{45}{72} - \frac{45 \times 3}{144} = -0.625 - 0.938$$
$$= -1.563 \text{ ksi (compressive)}$$

Figure 1-8a shows the stress distribution due to the prestressing force. The diagram at the left shows the stress distribution through the depth of the beam due to the axial load, P. The middle diagram shows the stress distribution due to moment, Pe. The diagram at the right shows the stress distribution at any section along the beam due to superposition of the axial and bending effects of the prestressing force.

As mentioned previously, the prestressing force does not act alone. As soon as the beam bends upward leaving the form, the weight of the beam starts to act. The weight of the beam, however, has no effect at the ends of the beam since the bending moment due to the weight of the beam at the ends is zero. Hence, the stresses at the ends of the beam will be the same as those shown in Fig. 1-8a. At midspan, how-

INTRODUCTION

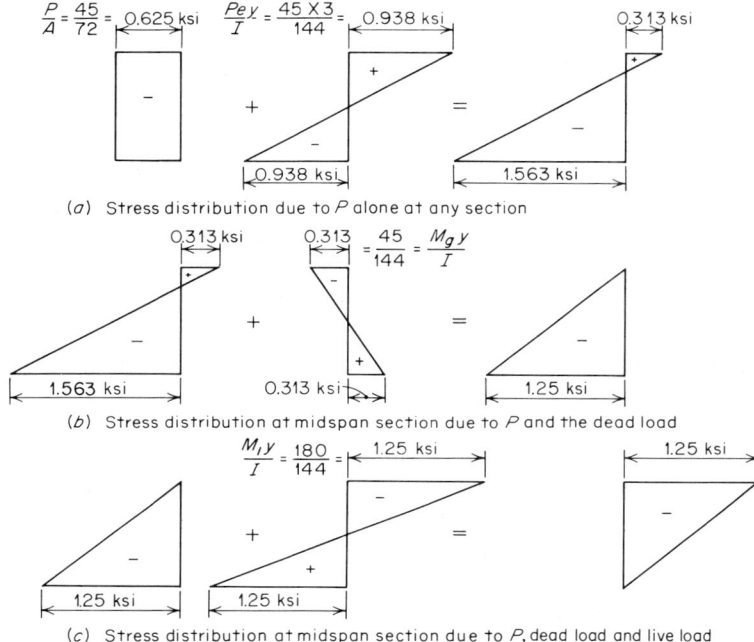

Fig. 1-8 Stress distribution in a prestressed-concrete beam.

ever, the stresses due to the prestressing force and the weight of the beam at the top and bottom fibers will be

$$-\frac{P}{A} + \frac{Pey}{I} - \frac{M_g y}{I}$$

and

$$-\frac{P}{A} - \frac{Pey}{I} + \frac{M_g y}{I}$$

respectively, where M_g is the moment at midspan due to the weight of the beam.

For this problem we have

$$M_g = 72\,\frac{150}{144}\,\frac{1}{1000}\,\frac{(20)^2}{8}\,12 = 45 \text{ in.-kips}$$

We are assuming that the unit weight of the concrete is 150 pcf.

The extreme top fiber stress at midspan due to the prestressing force and dead load is the following:

$$-\frac{P}{A} + \frac{Pey}{I} - \frac{M_g y}{I} = 0.313 - \tfrac{45}{144} = 0$$

The extreme bottom fiber stress at midspan due to the prestressing force and dead load is the following:

$$-\frac{P}{A} - \frac{Pey}{I} + \frac{M_o y}{I} = -1.563 + 0.313$$
$$= -1.250 \text{ ksi (compressive)}$$

It can be seen that due to the prestressing force and the weight of the beam there is a compressive stress of 1.25 ksi at the bottom fiber, and zero stress at the top fiber. Figure 1-8b shows the stress distribution at midspan due to the dead load and the prestressing force.

Now we shall apply a uniformly distributed load of 300 plf on the beam. The extreme top fiber stress at midspan due to the prestressing force, dead load, and applied load is the following:

$$-\frac{P}{A} + \frac{Pey}{I} - \frac{M_o y}{I} - \frac{M_a y}{I}$$

where M_a is the moment at midspan due to the applied load.

The extreme bottom fiber stress due to the prestressing force, dead load, and applied load will be the following:

$$-\frac{P}{A} - \frac{Pey}{I} + \frac{M_o y}{I} + \frac{M_a y}{I}$$

For this problem we have

$$M_a = 0.3 \frac{(20)^2}{8} 12 = 180 \text{ in.-kips}$$

and the total stresses at the top and bottom fibers are

$$-\frac{P}{A} + \frac{Pey}{I} - \frac{M_o y}{I} - \frac{M_a y}{I} = 0 - \tfrac{180}{144} = -1.25 \text{ ksi (compressive)}$$

and

$$-\frac{P}{A} - \frac{Pey}{I} + \frac{M_o y}{I} + \frac{M_a y}{I} = -1.25 + \tfrac{180}{144} = 0$$

Figure 1-8c shows both the stress distribution at midspan due to the applied load, and total load.

Figure 1-9a shows the variation of the top fiber stress along the span. The dashed line shows the variation of the top fiber stress due to the prestressing force and dead load. The solid line shows the variation of the top fiber stress due to all the loads acting on the beam. Figure 1-9b shows the variation of stresses along the span for the bottom fiber.

A study of magnitude of stress along the span indicates that all the stresses are within the usual allowable limits. The tensile stress

INTRODUCTION

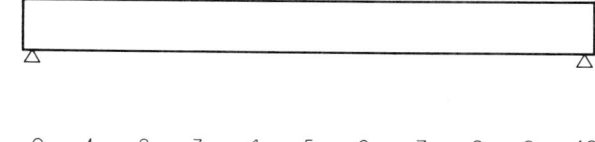

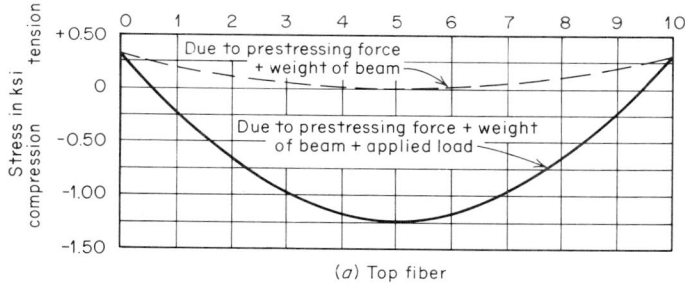

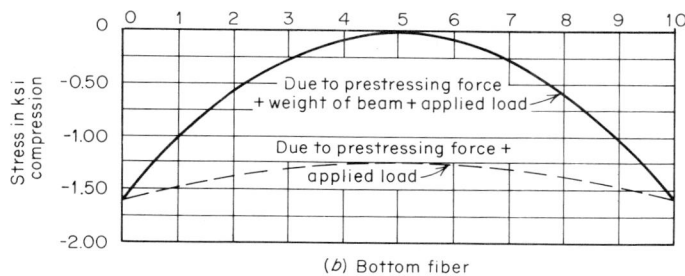

Fig. 1-9 Variation of stress along the beam.

of 0.313 ksi at the top fiber at the end is probably less than the flexural tensile strength of concrete.

If it were possible to apply the prestressing force of 45 kips permanently against unyielding anchor walls, the beam could satisfactorily carry the load. However, practically it is not possible to ensure rigid abutments, and a comparatively small movement will eliminate the prestress.

Now if we place two No. 9 ordinary reinforcement bars 3 in. from the bottom fiber and stretch them to a stress of 22.5 ksi each, passing through the preformed holes and anchoring them at the ends, we shall have a total force of 45 kips. The properties of the section will change slightly because of the holes; however, we can neglect this change as being small. It is clear that this scheme would provide the same prestressing as did the jacks.

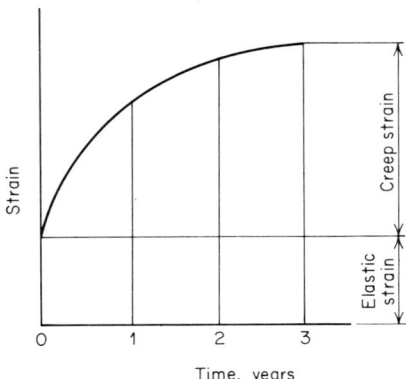

Fig. 1-10 Typical time–creep strain relationship.

Immediately after concrete is cast, it will start to shrink, and eventually it will reach a definite limit. The total shrinkage strain varies between 0.0004 to about 0.0002, depending upon the particular mix and strength of concrete. For this problem we shall assume a total shrinkage strain of 0.0003. Let us assume that by the time concrete is set and the beam is ready for prestressing about one-third of the shrinkage has taken place. Hence, only two-thirds of the total shrinkage will be effective in reducing the prestressing force. Total shortening of the beam due to shrinkage will be about

$$0.0003 \times \tfrac{2}{3} \times 20 \times 12 = 0.048 \text{ in.}$$

Since concrete in the beam is subjected permanently to dead load and prestress, there will be creep in concrete due to these loads. The creep phenomenon in concrete occurs in loaded specimens, and it may be assumed that virtually all the creep takes place in 3 years; however, about 75 to 85 percent of it occurs within 1 year. The amount of creep depends on many variables. The important variables seem to be the level of stress, water-cement ratio, and type of cement. The total creep strain is usually between 1.5 and 2.5 times the elastic strain. Figure 1-10 shows a typical time–creep strain relationship.

In this problem we shall assume that creep strain is approximately twice the elastic strain. The total shortening of concrete in the beam at the level of steel will be

$$2 \frac{f}{E_c} 20 \times 12$$

where f = average stress in concrete at level of steel due to prestressing force and dead load
E_c = modulus of elasticity of concrete

INTRODUCTION

The stress at the level of steel due to prestress is

$$0.625 + 0.938 \times \tfrac{3}{6} = 1.094 \text{ ksi}$$

and this stress is a constant quantity along the beam. The average stress due to dead load at the level of the steel may be taken as

$$\tfrac{2}{3} \times 0.313 \times \tfrac{3}{6} = -0.104 \text{ ksi}$$

Hence, the value of f is

$$1.094 - 0.104 = 0.990 \text{ ksi}$$

For a 3500-psi concrete we may assume $E_c = 3$ million psi.
Hence, the total shortening of the beam at the level of steel will be

$$2\frac{fL}{E_c} = 2 \times \tfrac{0.99}{3000} \, 240 = 0.158 \text{ in.}$$

The total shortening effect of creep and shrinkage is

$$0.158 + 0.048 = 0.206 \text{ in.}$$

It can be seen that after about 2 or 3 years the concrete beam will be 0.206 in. shorter at the level of steel.

The elongation to be maintained in the prestressed reinforcement in order to keep the 45-kip force is the following:

$$22.5 \times \tfrac{240}{30,000} = 0.18 \text{ in.}$$

As soon as the concrete at the level of steel begins to shorten, the total elongation in the steel due to prestress becomes less, resulting in a loss in prestressing force. When the total shortening reaches the value of 0.18 in., all the prestress is lost and a plain concrete beam action results.

One way this situation can be remedied is to tighten the nuts continually during the first few years of the life of the structure until creep and shrinkage have taken place. This is not practical, for it would demand constant checking for the presence of the required amount of prestressing force.

Another more practical solution to this problem is to increase the total elongation of prestressing steel to such a level that the shortening of concrete will be comparatively small. This may be accomplished by use of high-strength steel with an ultimate strength in the neighborhood of 250,000 psi. If we use this type of steel, we can safely stress it to at least 150 ksi, in which case the total elongation in steel will be

$$150 \times \tfrac{240}{30,000} = 1.2 \text{ in.}$$

In comparison with the above elongation the 0.206-in. shortening of concrete at the level of the steel is small. After shrinkage and creep have taken place, only $\frac{0.206}{1.2} = 17$ percent of the prestress will be lost.

As discussed previously, the credit for the clear understanding of losses caused by creep and shrinkage and the counteraction of losses through use of high-strength steel goes to Freyssinet.

It should be emphasized that the above calculations for shrinkage and creep are at best approximate, giving indication only as to the order of magnitude of the deformations. Although shrinkage can be measured and determined fairly accurately and independently of creep, it is impossible to measure or calculate creep independently of shrinkage. More detailed discussion of this problem is given in Chap. 6.

In addition to the calculated loss of about 17 percent in the prestress due to creep and shrinkage in the concrete, there is loss of prestress due to relaxation of stress in the steel. Relaxation takes place when the load level of prestress is more than half the ultimate strength of steel. Generally for 150,000 psi prestress, relaxation loss may be from 5 to 10 percent.

1-4 APPLICATION OF PRESTRESSED CONCRETE

Prestressed concrete is perhaps the most important innovation in structural concrete and in the building industry in recent years. Prestressing of concrete by high-strength steel permits use of steel and concrete to a very high degree of efficiency.

The conventional reinforced concrete becomes massive and impractical for simple spans over 40 ft. In reinforced concrete, efficient use of high-strength concrete requires larger quantities of steel, and this is not necessarily economical. Furthermore, the limitation of allowable stresses in steel under working conditions for containing of cracks makes it inefficient to use high-strength steel in reinforced concrete.

No such limitations exist in prestressed concrete. Prestressed concrete may be used for spans over 100 ft. Both high-strength steel and concrete can be used with great advantage in prestressed-concrete structures.

Prestressed concrete provides many possibilities for construction and can be used in many situations with advantage. It can be used in simple spans, in continuous spans, and in composite construction, with a large variety of methods of prestressing. It is a field highly receptive to creative approach.

Prestressed concrete in combination with precasting provides an economical way to build structures.

In the subsequent chapters the problems associated with prestressing of beams are discussed in detail. The discussions concern the behavior and strength of prestressed-concrete beams under load, and include acceptable rational design methods. The discussions are limited to simply supported beams—composite and noncomposite construction—and continuous beams.

REFERENCES

1. Whipple, Squire: "A Work on Bridge Building," pp. 77–78, H. H. Curtiss, Printer, Devereux Block, Utica, N.Y., 1847.
2. Ransome, E. L., and Alexis Saurbrey: "Reinforced Concrete Buildings," pp. 32–33, McGraw-Hill Book Company, New York, 1912.
3. Abeles, P. W.: "Principles and Practice of Prestressed Concrete," pp. 14–16, Frederick Ungar Publishing Co., New York, 1948.
4. Dill, R. E.: Some Experience with Prestressed Steel in Small Concrete Units, *Proc. Am. Concrete Inst.*, vol. 38, pp. 165–168, 1942.
5. Freyssinet, E.: "Un Révolution dans les techniques du Beton," Librairie de l'Enseignement Technique, Editeur Leon Eyrolles, Paris, 1936.

2
Methods of Prestressing

2-1 INTRODUCTION

In general, there are two ways in which prestressing of concrete by steel elements can be accomplished, namely, pretensioning or post-tensioning. The main distinction between the two methods is the condition of the concrete at the time when the steel elements are stretched. In the pretensioning method the steel is stretched before the concrete is cast, while in the post-tensioning method the steel cables are stretched after the concrete has been cast and only when it is strong enough to support the stress.

    Other important differences exist between the two procedures. Pretensioning as practiced in the United States requires an elaborate industrial layout with stressing beds and prestressing equipment in addition to all other conventional facilities, and involves a substantial capital investment. Post-tensioning can also be carried out in a manufacturing plant, but considerably less equipment and facilities are required for it in comparison with pretensioning. Post-tensioning also allows the *in situ* construction of structures, such as continuous bridges and frames, building slabs, and shells that require prestressing of the concrete but cannot be fabricated in a plant. In continuous structures, where curved cables are more efficient than straight cables, post-tensioning is particularly suitable, since it is easy to obtain curved paths for the cables by use of permanent sheathing cast in the concrete. On the other hand, even when some deflecting of the strands from a straight trajectory is possible in the pretensioning systems through hold-down devices, it is always a limited and costly procedure. There can be no doubt, however, that the efficiency of prestressing, measured by the cost per pound of tensioning load, is greater in the pretensioning system. This is true because of the costs of material and labor involved in the additional sheathing, end anchorages, and grouting required by the post-tensioning system. Also the individual jacking of cables in a post-tensioned member takes more time and effort than the simultaneous jacking of all strands that is normal practice in modern pretensioning operations.

In what follows, both methods of prestressing will be studied in detail. American practice will be emphasized throughout except for some necessary background on methods of post-tensioning that originated in Europe. The pretensioning system will be discussed first, followed by the description of the most important methods of post-tensioning available at present.

2-2 PRETENSIONING SYSTEMS

In pretensioning systems used in the United States the prestressing elements consist of several seven-wire strands. This system of prestressing takes its name from the fact that the steel strands are stretched before the concrete is cast. Considering how the prestressed steel is held until it is released to the concrete, we note that there are two ways in which pretensioning can be accomplished.

The first method, which is not often used in this country, consists in stretching the strands and anchoring them directly to the metal forms before casting the concrete. After the concrete has attained sufficient strength, the prestress is released. In this method the metal forms must be made strong enough to resist the buckling stresses created by the steel strands. This provision increases the cost of this method.

The second method for pretensioning is used predominantly in the United States because of its suitability for mass production at precasting plants. In a precasting yard, a stressing bed or bench is set which consists of a long reinforced-concrete slab on ground, vertical steel anchor walls called uprights at the ends, and stressing equipment. The steel strands are stretched and anchored at the vertical uprights, which

Fig. 2-1 A top view of the 130-ft casting bed readied for two single T beams at Iowa Falls Prestressed Company. A Drott Travelift straddles the bucket for filling the form. (*Concrete Industries Yearbook.*)

are very stiff, and are usually made of wide-flanged steel sections on reinforced-concrete foundations. Uprights must be designed to carry the eccentric forces created by the prestressing steel. This technique lends itself to efficient mass production, since the stressing beds are made long enough to permit fabrication of a number of similar members simultaneously with only one common jacking operation.

The first pretensioning plant was established in the United States in 1950 in Pottstown, Pennsylvania. The stressing bed of this plant was only 3 ft wide and 125 ft long and was designed to use $\frac{1}{4}$-in. seven-wire strand with pressed-on copper-sleeve friction anchors. In modern plants, stressing beds have an average 350-ft length, with several up to 600 ft in length, and a capacity of 1000 kips of pretensioning force. Figure 2-1 shows a modern pretensioning plant. The $\frac{3}{8}$-in. seven-wire strand is the type of steel most often used. However, the present tendency is toward increased use of the $\frac{1}{2}$-in. strand to reduce the number of strands required in a given element. Temporary anchorage of the strand to the stressing equipment is obtained by frictional-type split-cone wedges or by quick-release grips. Examples of the latter are the three-wedge steel strandvises fabricated by Reliable Electric Company and shown in Fig. 2-2 and the British Europa four-wedge grip. Seven-wire strand is supplied in reels which range between 6000 ft for $\frac{1}{2}$-in. strand to 25,000 ft for $\frac{1}{4}$-in. strand. To prevent the

METHODS OF PRESTRESSING

formation of harmful scaling rust in the strands, the reels are stored at the plant in dry, clean platforms. Contact with the soil is avoided at all times.

Modern plants are equipped with stressing equipment which may stretch each strand individually or all the strands at one time. The main component of the stressing system is the hydraulic jack, which should have a stroke of at least 48 in. (which includes allowance for slack strand) to be suitable to stretch the strands in a one-step operation. In order to release the strands gradually and apply the prestressing force to the concrete without impact, short-stroke (6-in.) large jacks are used at the releasing end in beds where stressing is done for each strand separately. If large jacks are used, releasing of prestress can be effected from the stressing end by the same jacks. It is generally recognized that the complete jacking operation is by far one of the most inexpensive procedures in a pretensioning plant.

The stressing beds in some plants are provided with hold-down devices for deflecting the strands. By means of these devices it is possible to provide the desired profile for the strands. Figure 2-3 shows one type of hold-down device specifically intended for strands. The stressing bed has to be provided with enough reinforcement to resist

Fig. 2-2 A typical quick-release-grip strandvise used for gripping strands in pretensioning.

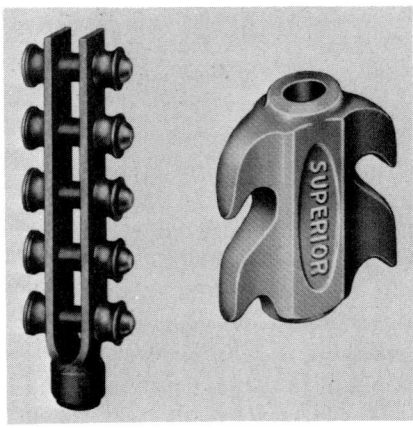

Fig. 2-3 Typical hold-down devices for use in deflecting the strands to provide the desired profile. (*Superior Concrete Accessories.*)

the vertical forces applied by the hold-down devices. As shown in Fig. 2-4, a high-strength bolt projects through the bottom of the form to hold the device down. Special care must be exercised in releasing the connection between the hold-down device and the stressing bed before the strands are released; otherwise, the longitudinal motion of the member can jam the device and may prevent its release. On the other hand, the designer should take into account the effect of the concentrated upward vertical forces on the beam caused by the release of the hold-down devices. It may be necessary to prestress the member partially by

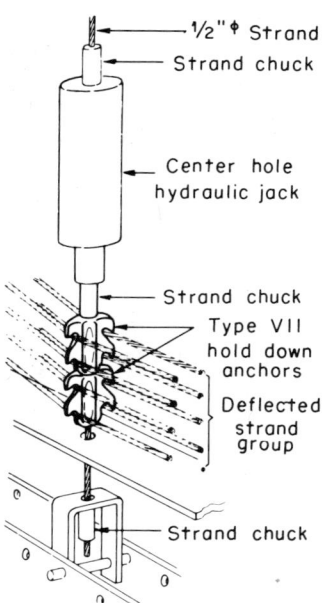

Fig. 2-4 Typical installation of a hold-down device to deflect a group of eight strands simultaneously. (*Superior Concrete Accessories.*)

METHODS OF PRESTRESSING

releasing some of the strands before releasing the hold-down devices in order to prevent cracks in the top.

After tensioning of the high-strength steel has occurred, fabrication of the member proceeds as in conventional precast construction, with only some additional precautions necessary to prevent damage to the strands. For example, the strands should be protected from exposure to the excessive heat or the electric arc created by welding of the reinforcing cage of the member in their vicinity.

Generally, for members that are produced in large numbers, steel forms will be available in precasting yards (see Fig. 2-5). Several advantages of steel forms over wood forms more than offset their higher initial cost. Steel forms last longer, can be manufactured with a high degree of precision, are easy to handle, may be adjusted with ease for minor variations in member shape, and finally are convenient for form vibration for the concrete. Forms are not removed until curing of the concrete takes place. The best curing process consists in wet heat in the form of live steam applied under a protective cover 2 or 3 hr after the concrete is cast, and then continued for 12 to 14 hr.

Fig. 2-5 Details of a two-piece steel end bulkhead with a slotted top section. Openings around the strands are closed with cardboard. Strand in bottom section was threaded through bulkhead holes before the top section was set in place. (*Concrete Industries Yearbook*.)

The final operation on the member while still on the stressing bed consists in releasing the strands from the end anchorages. This occurs after test cylinders have shown that the concrete in the member has attained the strength specified by the designer. Transportation and erection of the finished members follow immediately after cutting of the strands. There is no doubt that efficient handling and shipping are very important in keeping production costs competitive. Pretensioning industries must have, therefore, all necessary facilities for efficient in-plant handling and storage of the finished products. In addition, plants must be located in such a way as to be able to take full advantage of the highway and railway systems. This usually allows products to move from plant to jobsite economically.

2-3 POST-TENSIONING SYSTEMS

Post-tensioning systems are those in which the prestressing of steel cables is carried out after the concrete has been cast and cured and is capable of resisting the imposed prestress. To accomplish this, permanent-type sheathing, which follows the intended profile for the steel cables, is placed in the forms. Thus, after casting of the concrete takes place, ducts are formed in the concrete for the passage of the steel cables. In most cases threading of the steel through the ducts is eliminated by placing the cables with the sheathing as an assembled unit. This is necessary because of the large dimensions of the integral anchoring device of the cables, which would otherwise require a large-diameter duct throughout the beam for proper passage. Some systems like Freyssinet and Stresssteel do not require the cables to be placed before concreting of the member, as anchorages in both these systems are assembled later to the ends of the cables and are not integral parts of the unit.

After post-tensioning, it is the usual practice to inject cement mortar to fill the space between the cable and its sheathing. This operation, known as grouting, protects the steel from corrosion and bonds the cables to the concrete, thus considerably increasing the flexural strength and ductility of the member. It is possible to use galvanized-steel cables and eliminate grouting entirely, leaving the cables unbonded to the concrete. However, this practice is not desirable, because it results in beams with low flexural strength and ductility.

Several post-tensioning systems are available in the United States, all of them being fundamentally the same. It is in the details of end anchorage and type of cables, however, that these systems differ. In the following paragraphs some of the important systems of post-tensioning will be described fully. These systems use various types of cables, namely, the cable with parallel wires, the high-strength steel bars, and

METHODS OF PRESTRESSING

the monostrand and multiple-strand cables. Anchorages for these cables are based on either grip friction or direct bearing. A system of post-tensioning is characterized by the particular device used at the anchorage.

It is the usual practice in the United States for designers not to select any particular method of post-tensioning. The design may be finished by showing the minimum amount of steel and prestress required and the profile of the center of gravity of the steel cables. By making use of the information on physical and geometric properties of the available systems, the engineer may furnish a finished design that can be reasonably post-tensioned by any of the existing methods. The final cost of post-tensioning will be based upon the competition among the various systems.

A description of all systems of post-tensioning available in the United States is beyond the scope of this work. An effort is made to present those systems which typify the cable and anchorage devices frequently used. The systems of post-tensioning described are Freyssinet, Stressteel, Roebling, BBRV, Ryerson, Prescon, and CCL. The latest available information on these systems is compiled in Appendix B. The engineer is advised, however, to contact the various companies directly for information on any subsequent changes or improvements.

Freyssinet system One of the pioneers of prestressed concrete, Eugene Freyssinet was also the creator of an ingenious method for post-tensioning concrete which bears his name. Originally, the Freyssinet system was used for parallel-wire cables only, and three sizes were available, namely, 18-0.196, 12-0.196, and 12-0.276. The first figure indicates the number of wires per cable, and the second indicates the diameter of the wire in inches. The wires are inserted into flexible metal sheathing, stiff tubing, or preformed ducts and are anchored at each end with a special Freyssinet anchorage, which is shown in Fig. 2-6. The anchorage consists of a simple set of male and female concrete cones that act as a wedge in anchoring all wires of the cable simultaneously and preventing slip. The female part is a heavily reinforced concrete cylinder with a central conical hole lined with closely wound helical wire, and the male part is a fluted concrete plug which spaces the wires evenly around its perimeter and wedges them against the inside of the female cone. At the stressing end the male cone is inserted by the jack with great force after prestressing of the wires has occurred. At the anchor end the male cone is pushed initially into the female cone with enough force to just grip the wires. As jacking proceeds and tension is developed, final seating of the male cone takes place by itself.

To meet the post-tensioning market requirements for an anchorage that could be used with strands, the Freyssinet method developed an anchorage unit made of forged, high-strength alloy steel that is composed of a tapered cone and fluted plug with 12 grooves (see Fig. 2-7). This wedge-type anchorage creates a gripping action that is specifically designed to post-tension ASTM grade or type 270K strand. Appendix B-1 contains detailed information concerning these strands. Each ASTM grade strand is $\frac{1}{2}$ in. in diameter, has seven uncoated wires, and is stress-relieved in accordance with ASTM Specification 416-59T. Type 270K also meets the above characteristics and requirements but is approximately 15 percent stronger.

The Freyssinet strand cable is composed of 6, 8, 9, or 12 strands of ASTM grade or type 270K with a special strand anchorage at the end. A 12-strand $\frac{1}{2}$-in. Freyssinet cable in ASTM grade is designated as 12/500. For type 270K, the designation is followed by a K. For example, a 12-strand $\frac{1}{2}$-in. Freyssinet cable of type 270K grade is designated as 12/500K. The strand anchorage can also be used with $\frac{1}{2}$-in. galvanized prestressed-concrete strand, except that values of strength 10 percent lower than the corresponding strand size of ASTM grade should be

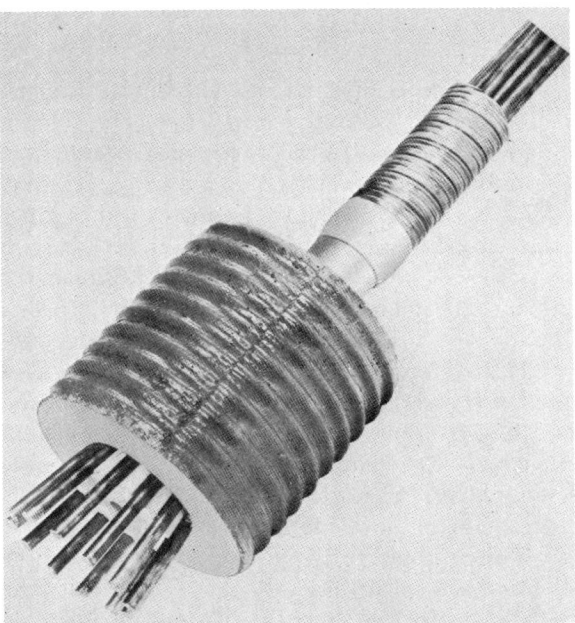

Fig. 2-6 Assembled Freyssinet wire anchorage consisting of a simple set of male and female cones. *(Freyssinet Co., Inc.)*

METHODS OF PRESTRESSING

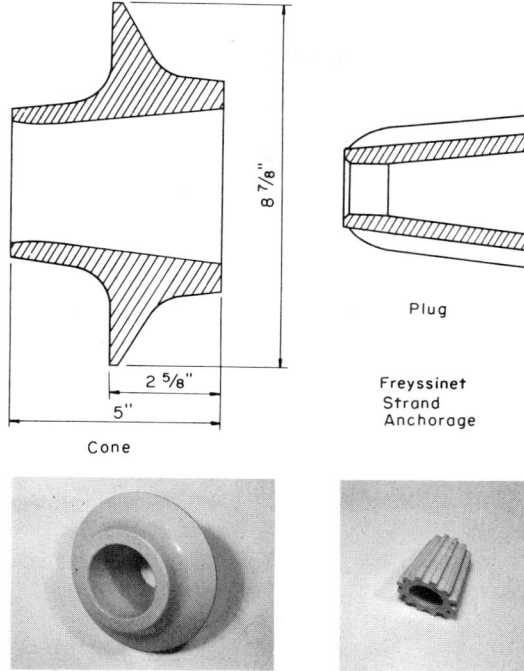

Fig. 2-7 Freyssinet anchorage for strands.

taken. Figure 2-8 shows the schematic layout of the internal and external Freyssinet-type anchorages for strand cables.

Grouting of the cables in the Freyssinet system can be accomplished through a small hole in the male plug. It may be convenient to plug the openings which exist between the wires or strands and the male and female cones to prevent the escape of a substantial amount of grout.

Stressteel system The Stressteel system of post-tensioning has been used extensively in the United States since 1952, with the introduction of high-strength steel bars ranging from $\frac{1}{2}$ to $1\frac{3}{8}$ in. in diameter. The bars may be anchored with a choice of wedge or grip-nut anchorages, as shown in Figs. 2-9 and 2-10, which are specially designed to anchor unthreaded bars, and any variety of size and slope of anchorage plate. The designer may select from a wide range of post-tensioning working forces, starting with individual bars from 16 kips for $\frac{1}{2}$-in. bars up to 142 kips for $1\frac{3}{8}$-in.

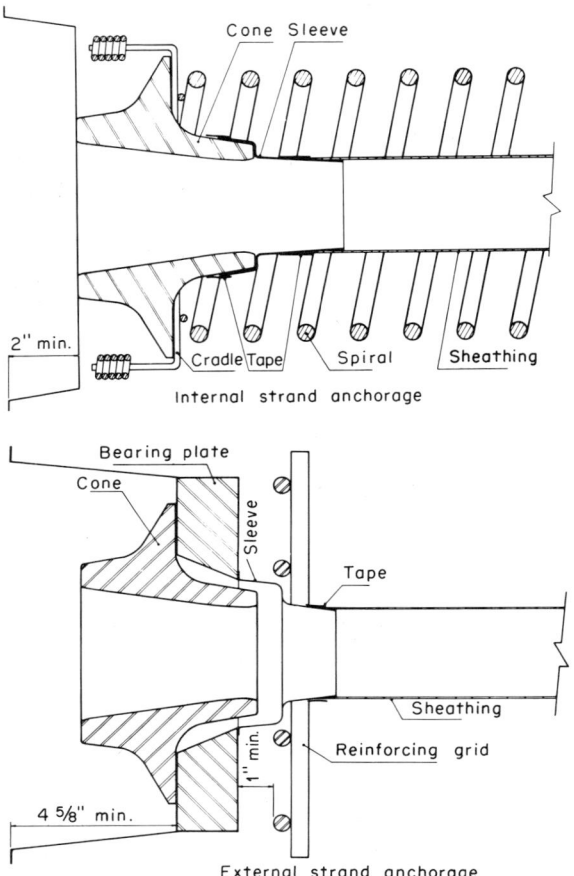

Fig. 2-8 Schematic layouts for internal and external use of Freyssinet strand anchorages.

bars. In multiple-bar arrangements, anchorages have been built with a working force as large as 7000 kips.

The Stressteel bars are manufactured from hot-rolled alloy steels, and are first cold-stretched to uniformly cold-work the cross section and develop a high yield strength. The bars are then stress-relieved in a gas-fired furnace for adequate ductility and uniform stress-strain characteristics. As a result of cold-stretching, bars are proof-stressed to the minimum guaranteed yield stress, and bars with surface imperfections or metallurgical defects are eliminated.

METHODS OF PRESTRESSING

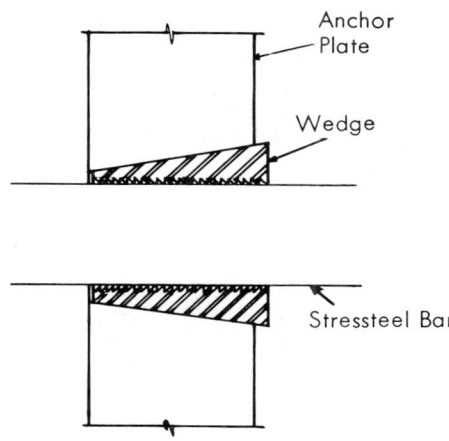

Fig. 2-9 Stressteel wedge-type anchorage of high-strength steel bars. (*Stressteel Corporation.*)

Wedge Anchor

The use of bars instead of wire cables or strands has some advantages that compensate for their lower ratio of prestressing force per pound of steel. A lower friction and wobble coefficient, as shown in Table 6-1, may be used with high-strength bars than with cables or galvanized strands. From a construction point of view, bars require less frequent tying or securing in the forms. Also, devices are available which will anchor the bar at any point where it emerges from the form, thereby making unnecessary the determination and specification of an exact bar length. Finally, the existence of easily attached couplers, as shown in Fig. 2-11, permits the efficient use of bars in cases where construction advances in stages requiring either partial or full tensioning of segments; where precast elements are tied together to become an integrated prestressed structure; and where long, continuous bars are required.

The system also allows for the bars to adopt a curved profile, as shown in Fig. 2-12. The bars are placed in flexible metal tubing which is tied or otherwise supported to prevent vertical or horizontal displacement during casting of the concrete. The post-tensioning procedure is very simple, and anchorage is accomplished by the jack itself by ramming the seat wedge in place. The jacking procedure requires a minimum outside bar length ranging from 21 in. for 60-kip jacks to 30 in. for 200-kip jacks. The protruding length of bar after tensioning may be flame-cut to within $\frac{1}{2}$ in. of the anchorage assembly.

Stressteel has lately introduced the SEEE strand system developed in France by Societé d'Études et d'Équipements d'Entreprises. The

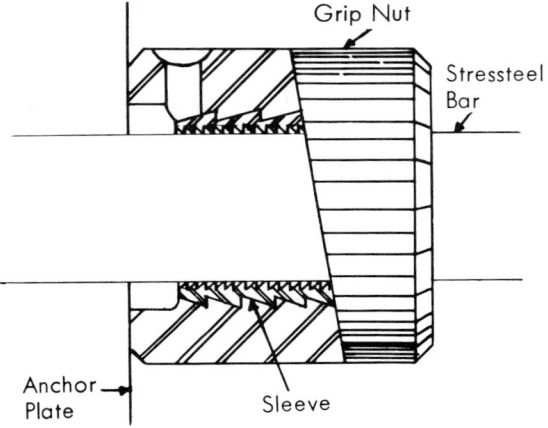

Grip Nut Assembly

Fig. 2-10 Stressteel grip-nut anchorage of high-strength steel bars.

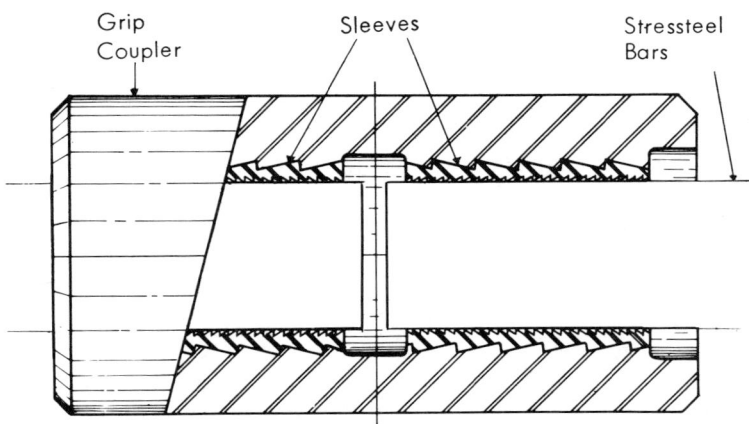

Grip Coupler

Fig. 2-11 Grip coupler device for high-strength steel bars in Stressteel system.

METHODS OF PRESTRESSING

Fig. 2-12 Stressteel high-strength bars, encased in flexible tubing, are positioned in the forms of a precast girder. Cast-in-place plates are attached to forms normal to the slope of the bar at the proper trajectory.

method uses cables composed of parallel seven-wire stress-relieved strands, available in units of one, four, seven, or nineteen elements. Using 270K-type strand, units of the SEEE strand system are manufactured in a range of sizes, from the smallest $\frac{1}{2}$-in. monostrand with a working force of 24.8 kips to the largest unit composed of $\frac{1}{2}$-in. strands with a working force of 470 kips. The tensioning procedure of SEEE strands requires a pulling bar that is threaded to the anchor sleeve of the strand and to the pulling nut of the jack. When the required post-tensioning has been applied, the anchoring nut is turned tightly against the bearing plate, after which the jacking equipment and the pulling nut are removed. The pulling bar is finally removed by unthreading. Details of strands and anchorages of both Stressteel systems may be found in Appendix B-2.

Roebling system The Roebling system of post-tensioning was one of the first to appear in the United States. It employs galvanized strand which is machine-fabricated in the same manner as strands used in suspension bridges. The strands are fabricated from hot-dip galvanized wire, which guarantees complete protection against corrosion without further treatment.

Roebling cables are fabricated in various sizes, starting with the smallest, 0.6 in. in diameter and 26 kips design prestressing force, to the largest cable, which has a diameter of $1\frac{11}{16}$ in. and may be prestressed to 208 kips design load.

The strands are anchored by spreading the wires and burying them with molten metal in a cast-steel tube. To allow stretching of the strand by the hydraulic jack, the outer end of the anchoring steel tube is threaded in the inside so that a pulling rod can be attached to it and to the jack. For safety reasons while the cable is being stretched, the anchoring nut is kept close to the bearing plate, using the threaded outside portion of the cast-steel tube. As soon as the required force has been attained, the anchoring nut is set tight against the bearing plate and the jack is released. A tensioning operation is shown in Fig. 2-13. This anchorage procedure requires that Roebling cables be ordered to the desired length, which is specified as the length out-to-out of bearing plates. The cable is then fabricated to the dimensions specified by the designer, and flexible metal hose is assembled on the strand as required. As shown in Fig. 2-14, the tendons are shipped as separate coils. Details of available strands, fittings, and bearing-plate assemblies are given in Appendix B-3.

BBRV, Ryerson, and Prescon systems The BBRV system of post-tensioning was developed in Switzerland by Birkenmaier, Brandestini, Ros, and Vogt in 1949. The main feature of this system, which employs cables made of a number of parallel uncoated stress-relieved wires, is the way in which the wires are fixed to the anchorage devices. A button-head is made at the end of each component wire by slowly squashing a small protruding end of wire against the gripping face of the head-forming machine. The head has a diameter approximately 40 percent larger than the actual diameter of the wire. Special precautions, such as careful measuring under controlled conditions, are taken to guarantee that all wires of a given cable are of the same length.

Fig. 2-13 Tensioning a $1\frac{9}{16}$-in.-diameter Roebling galvanized strand on pier A at Hoboken, New Jersey. A single long-stroke center-hole ram and threaded connections are used to cut jacking time to a minimum. Power pumps can be used to further save time. (*Colorado Fuel and Iron Corporation.*)

METHODS OF PRESTRESSING

Fig. 2-14 Roebling galvanized strands of $1\tfrac{9}{16}$ in. diameter assembled for shipment. Each strand is encased in flexible metal hose with fittings attached.

The BBRV buttonhead anchorage facilitates the simultaneous tensioning and anchoring of a large number of wires. The system is capable, therefore, of furnishing very large cables, an example being the C220, which is made out of fifty-five 7-mm wires and can develop 488 kips of permanent tension force. Several types of BBRV anchorages are available as series B, J, C, S, and E; data on these can be found in Appendix B-4.

Type B anchors are shown in Fig. 2-15. Four standard sizes are available, namely, B32, B64, B100, and B138, the last number indicating

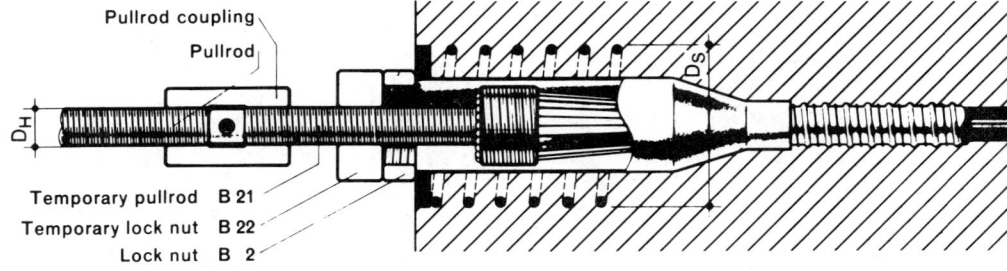

Fig. 2-15 BBRV type B movable anchor for prestressing cables can be efficiently used for post-tensioning in stages. (Courtesy American BBR Inc.)

the capacity of the cable in metric tons. Three sizes of wires—5, 6, and 7 mm—can be used to fabricate these standard cables, the required number varying with the size of the wire. Also, by leaving away single wires in the standard cables any desired forces may be obtained. The B-type anchors can be used for post-tensioning in stages, and can be temporarily anchored as shown in Fig. 2-15 at the end of each pre-stressing stage. Final injection of the grout is effected through the center hole of the anchor head. The J-type anchor is the most frequently used movable cable anchorage in the BBRV system, particularly with short cables, and has the practical advantage that no part of the anchor projects beyond the member, as shown in Fig. 2-16. With the J-type anchor, however, the full prestressing force is available only at the end of the spiral, at approximately 12 in. from the ends. For this reason it is not recommended where limited bearing conditions of the member require full prestressing immediately at the face of the support. The standard cables available in the J series are the same as in the B series. For prestress forces larger than 304 kips, the BBRV system fabricates

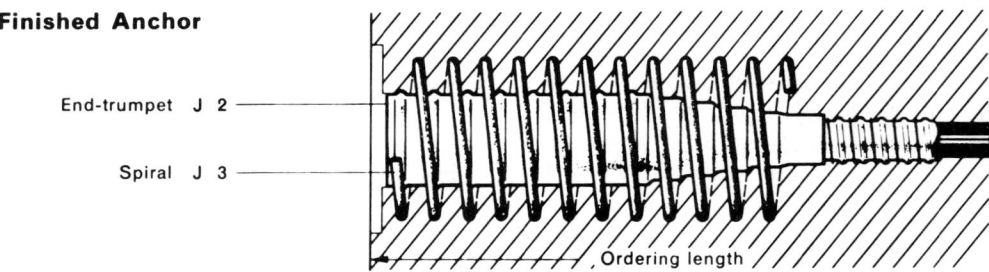

Fig. 2-16 BBRV type J, the most frequently used movable anchor for prestressing cables in the series.

Grouting

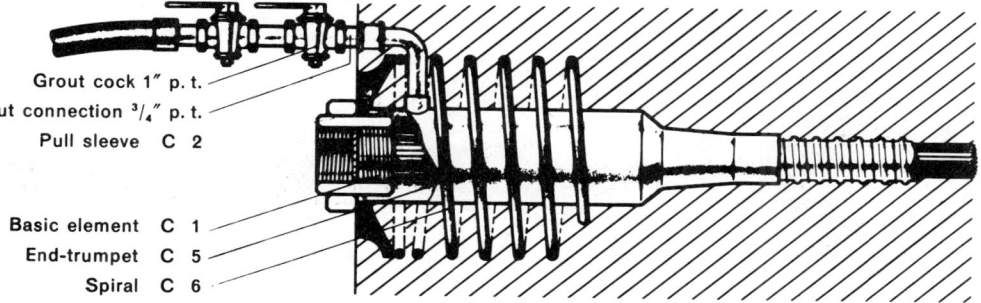

Fig. 2-17 BBRV type C movable anchor for large prestressing cables.

the type C series, which is characterized by a high concentration of 7-mm wires around the cable axis. The corresponding type C anchors can also be used for prestressing in stages, as they can be temporarily anchored at the end of each stage. Final injection of the cement grout takes place through a lateral pipe connection, as shown in Fig. 2-17.

For cables smaller than 165 ft and for longer cables with small

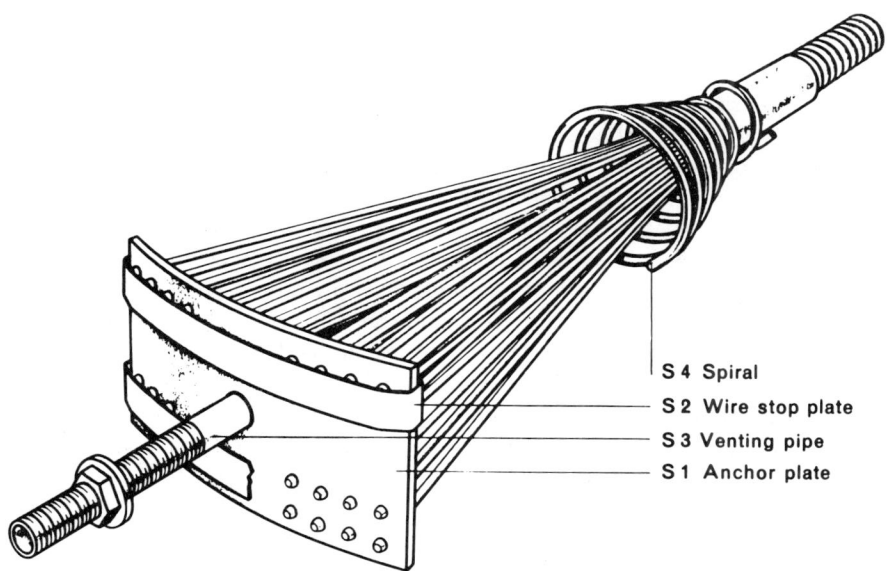

Fig. 2-18 BBRV type S fixed anchor for prestressing cables used opposite to movable types series B and J.

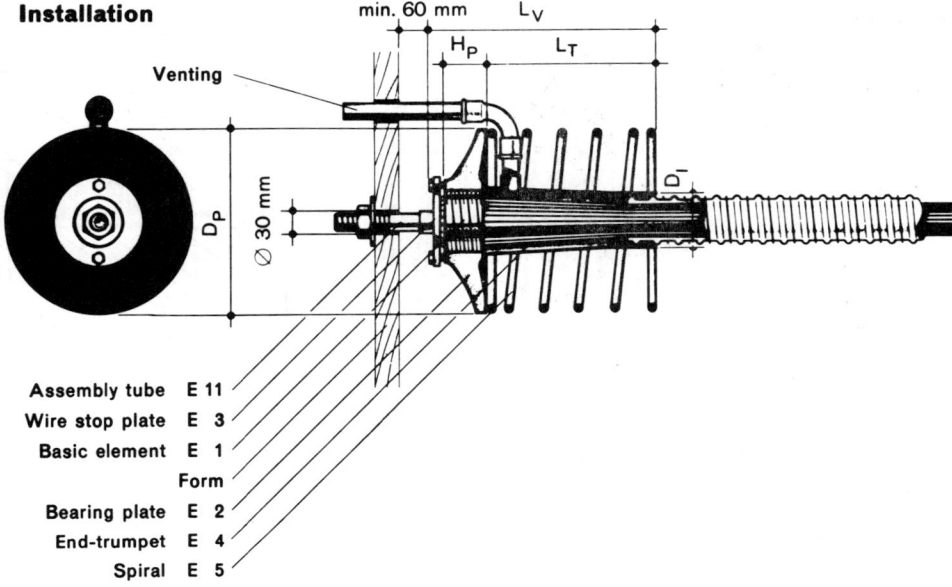

Fig. 2-19 BBRV type E fixed anchor for prestressing cables, having a larger capacity than the type S series used opposite to movable type series C.

curvature in their profiles, it may be possible to jack from only one end of the member without causing large losses due to friction. If the designer determines that one-end jacking is admissible, a simpler fixed-type anchor can be used at the other end of the cable. The BBRV system provides the fixed-anchor series type S, as shown in Fig. 2-18, for prestressing forces up to 304 kips to be used opposite to movable types series B and J. To distribute the frictional loss of prestress uniformly along the member it may be advisable to place one half of the fixed S-type anchors at one end and the other half at the other end and jack half the cables from each end. Fixed anchors for the larger forces of the movable type C series are also available as type E series, shown in Fig. 2-19.

In the United States, the Ryerson and Prescon systems of post-tensioning also use the BBRV principle of buttonhead anchorage for the component wires of the cables. However, different types of anchor heads have been developed in both systems.

The Ryerson system provides two types of cables: those that are bonded to the concrete by grouting and those left unbonded by greasing or paper wrapping. Both types are shown in Figs. 2-20 and 2-21 in the movable- and fixed-end versions. The movable-end anchor for the bonded cable is held in its final position by means of a locknut screwed

METHODS OF PRESTRESSING

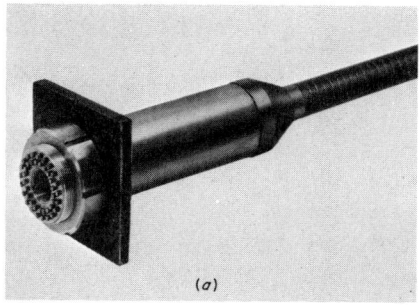
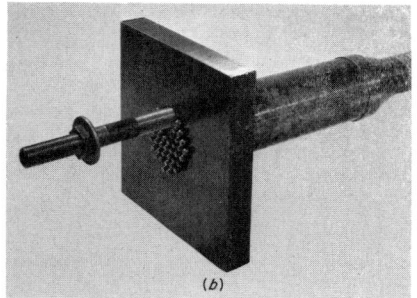

Fig. 2-20 Ryerson (a) movable- and (b) fixed-end anchors for grouted cables. (*Joseph T. Ryerson & Sons, Inc.*)

against the bearing plate. Also the center hole, after removal of the pull rod, is used as a grout inlet. The unbonded cable uses a different anchor system at the movable end. Shims of a predetermined height, which depend on the elongation of the cable, are inserted through the wires between the bearing plate and the anchor head of the wires. After the jack is released and removed, the protruding anchor must be enclosed with concrete or otherwise protected from rust and fire. For both types of cables the fixed-end anchor is very simple, consisting only of a bearing plate to which the buttonhead wires are attached. The only difference between the fixed-end anchor types is the presence of a holding-grout pipe required for the injection of cement mortar and subsequent bonding of the cable. Dimension and stressing data for bonded and unbonded cables are given in Appendix B-5.

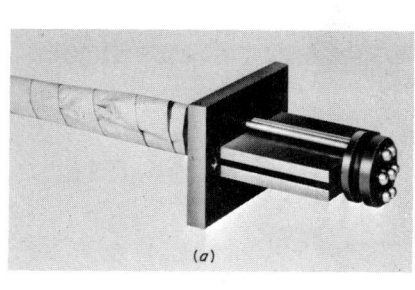
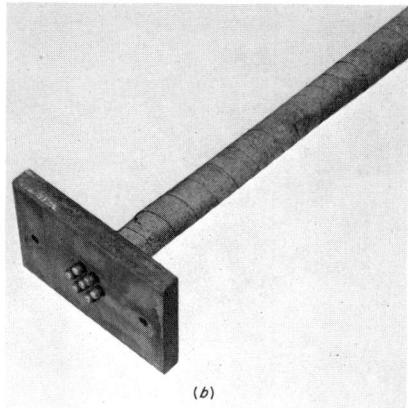

Fig. 2-21 Ryerson (a) movable- and (b) fixed-end anchors for greased and wrapped (unbonded) cables.

PRESTRESSED CONCRETE

Fig. 2-22 Standard Prescon cable arrangement showing buttonhead wires, stressing washer, shims, and the bearing plate. (*The Prescon Corporation.*)

The Prescon system is also available in both bonded and unbonded cables. Grouting cables are enclosed in flexible metal tubing, fully interlocked and mortar-tight. Coated cables are spirally wrapped with waterproof, fiberglass-reinforced kraft paper. The standard Prescon end anchorage (see Fig. 2-22) consists of three components, namely, a drilled stressing washer, precut shims to maintain the elongation of the stretched cable, and a bearing plate. After the completion of the jacking operation in those jobs where bonded cables are required, grout fittings are attached to the protruding hardware to permit grout injection through

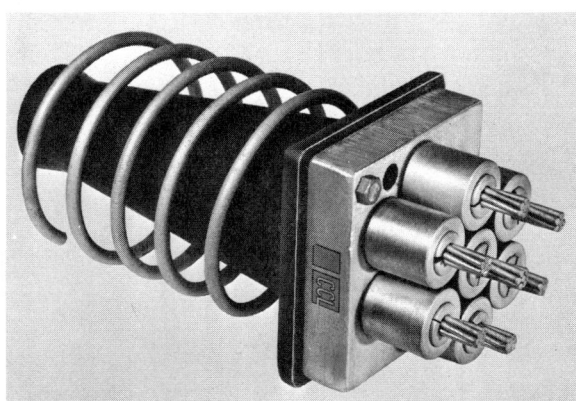

Fig. 2-23 CCL anchorage for seven-strand cables. (*Cable Covers Limited.*)

METHODS OF PRESTRESSING 37

the end anchor and conduit. These attachments are furnished by the Prescon system in the required sizes. Details of end hardware and the cable size chart on the Prescon system can be found in Appendix B-6.

Cable Covers Ltd. system The CCL system of post-tensioning originated in Great Britain several years ago, and later the Gifford-Udall and Gifford-Burrow systems were incorporated with it. The method uses cables made up of a combination of seven or twelve strands. Figures 2-23 and 2-24 show the anchorages for the seven-strand system and the

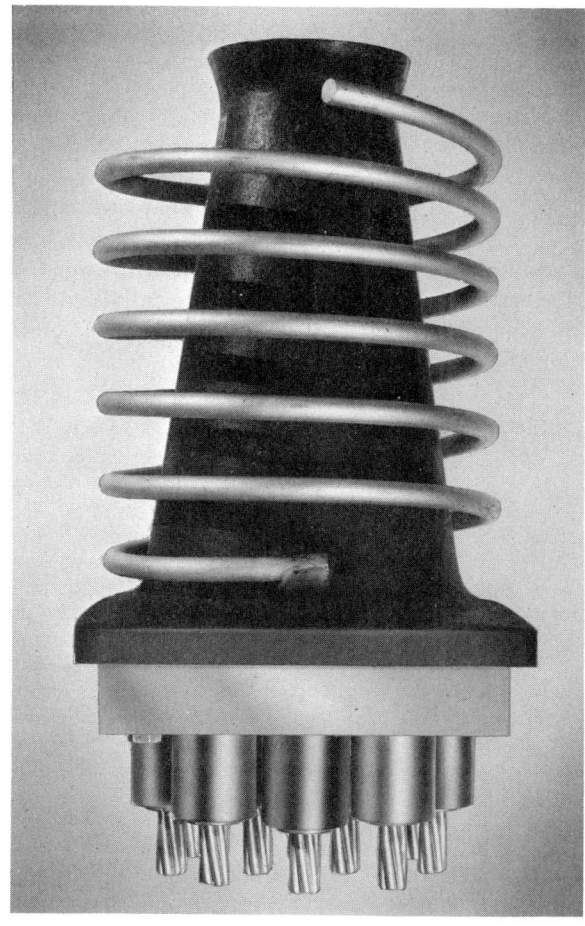

Fig. 2-24 CCL anchorage for twelve-strand cables.

twelve-strand system respectively. Both systems can be used for 0.5-, 0.6-, and 0.7-in.-diameter strands. In each case the cable consists of seven or twelve strands, depending on the system, each strand being stressed separately.

For steel profiles in which there is no reversal of curvature, none of the seven-strand systems requires the use of spacers, nor do the strands need to be made up into cables. The strands are threaded through the duct by a special pulling device, so that the exact location of each strand is secured. Separate stressing proceeds in the specified order, by means of lightweight power-operated jacks, and each strand is anchored mechanically in CCL grips. Should the force required be smaller than that given by seven strands, the number of strands can be reduced accordingly.

An idea of the forces that can be developed by the cables is given in the following example. Each 0.5-in. strand of a seven-strand cable can be individually stressed to a load of 25.9 kips for a total force of 181 kips per cable, within a duct of 2 in. internal diameter. The compactness of the CCL cable can be realized by noting that this force is 60 percent more than that given by a cable with twelve 0.276-in. wires, and 40 percent more than that of a single $1\frac{1}{8}$-in.-diameter strand.

The following are advantages of the CCL system. A small number of cables are required to produce a given prestress in a member, because of the high initial prestressing force provided by the cables. As a consequence, the total cost of sheathing and anchorages is reduced. No spacers are required among the strands of a cable, and as the strands are stressed separately, the application of the prestressing force to the structure takes place gradually. Light jacks may be used even when the

Fig. 2-25 CCL spiral strand anchorage for large strands.

METHODS OF PRESTRESSING

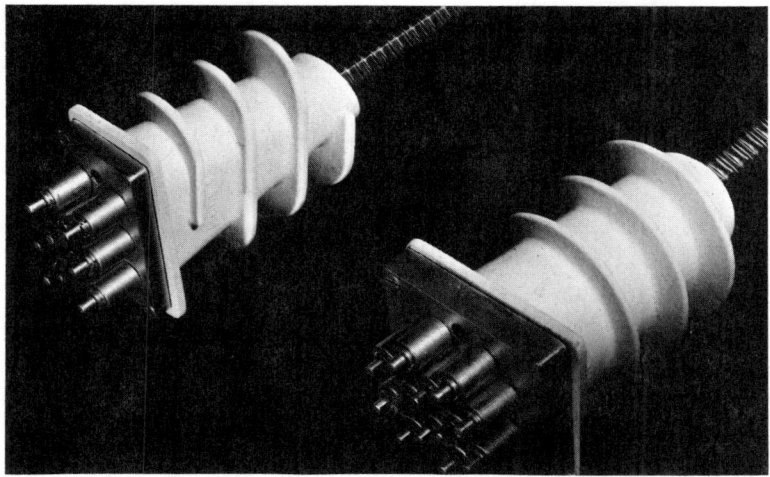

Fig. 2-26 CCL spiral anchorages for wire cables.

total force in a cable can be as large as 191 tons; and finally, since individual grips are used to anchor each strand of a given cable, the prestressing force can be checked and adjusted easily.

The CCL system has also developed a special anchorage for large strands, which is shown in Fig. 2-25. This anchor consists of a metal casting with a large external spiral rib and an internal tapered hole. The anchor is cast into the concrete and the strand is secured by a steel wedge inside the taper. No helical reinforcement is required, and grouting takes place from the front of the anchor directly into the duct. The same principle of spiral ribs and bearing area is used by CCL to provide anchorage for wire cables. The anchorages are available for 4-, 8-, and 12-wire cables, as shown in Fig. 2-26. Additional design information on the CCL system is given in Appendix B-7.

2-4 SUMMARY

Both methods of prestressing that are used currently in the United States have been discussed in the preceding sections. It is not difficult to conclude that the designer has a great variety of choice and freedom in whatever method he selects. There is a possibility that mass-produced standard structural members, manufactured by the pretensioning method, are available for use in his particular design. Use of these standard members as a rule results in substantial economy. If the job is large enough, however, the designer may even consider his design as capable of standardization and mass production with the consequent reduction in cost. Pretensioning is usually cheaper because of all the savings in sheathing, end anchorage, and grouting, and principally because it is made in the shop rather than in the field. If the beam is heavy, or the project is far from a pretensioning plant, a post-tensioning system may be used either for prestressing an entire member cast in place or to connect the various elements of a prefabricated member.

The following chapters in the book are devoted to the discussion and understanding of behavior of both types of prestressed members. Much of what has been discussed in this chapter regarding characteristics of various methods will become clearer in the reader's mind as the material of the following chapters is developed.

3
Flexural Behavior and Strength of Prestressed-concrete Beams

3-1 INTRODUCTION

The type of prestressed-concrete beam considered here is the one most commonly used in practice. It is assumed that the beam is simply supported, horizontal or almost horizontal, and subjected to downward acting loads. It is also assumed that prestressing force is applied by tensioned steel elements properly anchored at the ends. The discussions are limited to short-time static loads.

The purpose of this chapter is to present a method for the analysis of the beam for the entire range of load up to and including the load which causes the complete collapse of the beam. In addition, it is intended to develop bases for a rational design of a prestressed-concrete beam for specified loads.

A prestressed-concrete beam of the above description has the peculiarity of being loaded by the prestressing force as well as its own weight before it is subjected to the loads for which it is designed. The stresses and deformations caused in the beam by the prestressing force and the weight of the beam should be small, and adequate provisions should be made to prevent failure of the beam due to these loads before the beam is used for its intended purpose.

The stresses caused by the prestressing force are somewhat compensated by the stresses caused by the weight of the beam. The highest stresses occur just after the prestressing force is applied. This condition occurs immediately after the strands are cut in a pretensioned construction, and after the cables are pulled and seating of anchorages is effected in a post-tensioned construction. Since the prestressing force decreases with time—because of shrinkage and creep in concrete and relaxation in steel—the stresses due to the prestressing force and the weight of the beam likewise decrease with time.

The stresses caused by the prestressing force are compensated by those resulting from the weight of the beam, superimposed dead load, and live load. The highest stresses occur when the prestressing force has reached a minimum value. The stresses and deformations in the beam caused by the weight of the beam, superimposed dead load, and live load, as well as the minimum prestressing force, should be small. Adequate provisions should also be made to ensure sufficient safety in the beam when all the loads are acting.

Although the condition of the beam is critical immediately after the application of the prestressing force, this is temporary. The reduction of the prestressing force causes the stresses to be reduced. Furthermore, the application of the superimposed dead load, if any, after the structure is erected also reduces the stresses caused by the prestressing force.

The critical condition of loading when all the loads are acting—and the prestressing force is a minimum—is more lasting and hence more important than the temporary condition.

From a practical point of view it is necessary to predict how a beam which is already loaded by the prestressing force and its own weight behaves and ultimately fails under increasing downward loads. In the following paragraphs this problem is discussed in detail.

3-2 BEHAVIOR OF PRESTRESSED-CONCRETE BEAMS

In order to establish a reasonable basis for design of a prestressed-concrete beam, it is necessary to know the relationship between the load and the resulting deformations for the entire range of load.

For a given beam and type of load, the relationship between the load and a particular deformation when the acting load varies from zero to the magnitude corresponding to the complete collapse of the beam is a measure of the behavior of the beam. The term *behavior of the beam*, therefore, refers to the variation of deformations in the beam as the load is increased to the level in which at a certain point the beam fails in a particular way.

Since both loads and deformations can be observed and measured, the behavior of the beam can be studied empirically. Hence, whatever rational formulation is made of the problem, it can be readily checked by tests. The deformations considered are those at a critical section, and may be strain in concrete at the extreme fiber, strain in the prestressed steel, curvature (the ratio of strain in concrete at the extreme fiber to the depth to neutral axis), or deflection.

The behavior of prestressed-concrete beams varies with load. For low magnitudes of load the relationship between load and deformation is linear, and it is possible to predict any deformation in terms of load by a relationship simply formulated. Furthermore, for low magnitudes of load any removal or decrease of load follows the same relationship. The deformations caused by the prestressing force and the weight of the beam are small and may be assumed linearly related to the prestressing force and the weight of the beam, respectively.

As the acting load is increased beyond that corresponding to the cracking load, the properties of the beam section become a function of load, and the relationship between load and deformation becomes nonlinear.

The load corresponding to the complete collapse of the beam is the *ultimate load* or *strength* of the beam. If at failure the beam has undergone large deformations, it displays a *ductile* behavior. If, on the other hand, the beam has only slightly deformed at failure, it displays a *brittle* behavior. There is no set deformation which distinguishes between a ductile and a brittle behavior. However, such a limit may be defined for a given material and a particular purpose.

There are several ways in which a prestressed-concrete beam can fail. It can fail in the region of pure moment, or in the region of combined stresses where both shear and moment are present. It may also fail in bond, or anchorage at the ends before the beam is subjected to the acting loads, or during the prestressing operation. The discussions in this chapter are limited to the region of pure moment. Other types of failure will be discussed in subsequent chapters.

Since the failure in the region of pure moment can be controlled more accurately, and more strength and ductility may be obtained if the beam failed in that region, it seems logical to make the region of pure moment the weakest link. Necessary provisions can be made to strengthen the region of combined stresses so that failure in that region is prevented. Similarly, the beam should not be permitted to fail in bond and in the anchorage zone before the full capacity of the section of pure moment is developed.

The acting load may be measured in terms of bending moment at a given section, which is usually the section subjected to the maximum moment, corresponding to zero shear. The behavior of the beam subjected to pure bending is called the flexural behavior of the beam. The moment which corresponds to failure at the critical section in the beam is called the *ultimate moment* or the *flexural strength* of the beam.

The flexural behavior of prestressed-concrete beams lends itself to a rational study, and is comparatively simple. It is a special case of a more complicated problem of combined stresses for which a rational solution is difficult to obtain.

The most convenient measure of deformation is curvature, from which deflections may be obtained. Both curvature and deflection may be observed and measured. The moment-curvature or the moment-deflection relationship may be developed rationally if the following are known:

1. Variation of strain in concrete with the depth of the beam
2. A relationship between strain in steel and strain in concrete
3. The stress-strain relationship for all the materials, that is, the stress-strain diagram for concrete, prestressed steel, and any other type of non-prestressed reinforcement used

In the following paragraphs our knowledge of these relations is discussed briefly.

3-3 STRAIN DISTRIBUTION WITH DEPTH

Results of numerous tests on prestressed- and reinforced-concrete beams have indicated that strain varies linearly with depth in all stages of loading.[1-3] Before the flexural cracking takes place, the strain in concrete is linearly distributed in the entire depth of the beam, in the region of tensile as well as compressive strains. After the beam cracks, the strain in concrete at the tension zone becomes concentrated at the cracks, and the linearity of strains becomes meaningful only for the uncracked region of the beam. Results of tests indicate that the strain distribution in the compression zone, that is, above the neutral axis, remains linear even at high loads nearing the ultimate load.

It should be pointed out that this conclusion has been arrived at primarily by tests on beams with rectangular section and measurement of strains at the region of constant moment. Investigations are often made with a two-point loading pattern in order to produce a region of approximately uniform moment so that the strain measurements can be carried out without disturbance. Strain gages are placed in each side face of the beam at various levels. By plotting the observed strains with depth, the shape of the strain distribution is obtained.

Some question exists whether on the basis of such tests the conclusion of linearity of strains is justified. Since the strain gages measure the strains at the outer faces rather than points inside the beam, it may be argued that outer faces do not represent what happens in the beam. Furthermore, for nonrectangular sections the strains probably are not distributed linearly; disturbances occur at points where the width of beam changes abruptly. In spite of these objections the assumption of linearity of strains appears to be reasonable and simple, and the only logical one that can be made.

3-4 THE RELATIONSHIP BETWEEN STRAIN IN PRESTRESSED STEEL AND STRAIN IN CONCRETE

The relationship between the strain in concrete at a given fiber and the strain in prestressing steel may be understood by studying the variation of strain at midspan (or a section in the region of maximum moment) when the load varies from zero to ultimate. For convenience, distinction may be made among the following three stages of loading:

Stage 1. Only prestressing force acting
Stage 2. Load corresponding to zero strain in concrete at the level of steel at midspan, or at a section in the region of maximum moment
Stage 3. The ultimate load

The variation of strain in the prestressed steel from stage 1 through stage 2 can be formulated easily since in this range of loading the concrete fiber at the level of steel is compressive. In this range usually flexural cracking has not taken place, or if it has, cracks are only slightly developed. Hence in this range there is a simple relationship between strain in steel and strain in concrete at the level of steel.

The variation of strain in the prestressed steel between stages 2 and 3 is somewhat more difficult to formulate since in this range of loading the strain in concrete at the level of steel is tensile. In this range, for low tensile strains, the strain in steel may be related to the strain in concrete at the level of steel. For high tensile strains, the concrete cracks progress considerably in the beam, and the strains become concentrated at the cracks. In this case the strain in steel becomes dependent on the condition of bond and the relative position of the cracks from the section under consideration.

In the following paragraphs the variation of strain in prestressed steel is discussed briefly for each range.

Variation of strain in steel between stages 1 and 2 Figure 3-1a shows the rectangular cross section of a prestressed-concrete beam which has only straight prestressed reinforcement. The strain distribution at any section along the span due to prestressing force only is shown in Fig. 3-1b. This condition of loading is designated as stage 1. The strain in steel due to prestressing force only is called the *strain due to effective prestress*, and is designated as ϵ_{se}. Since the strain in steel after the application of

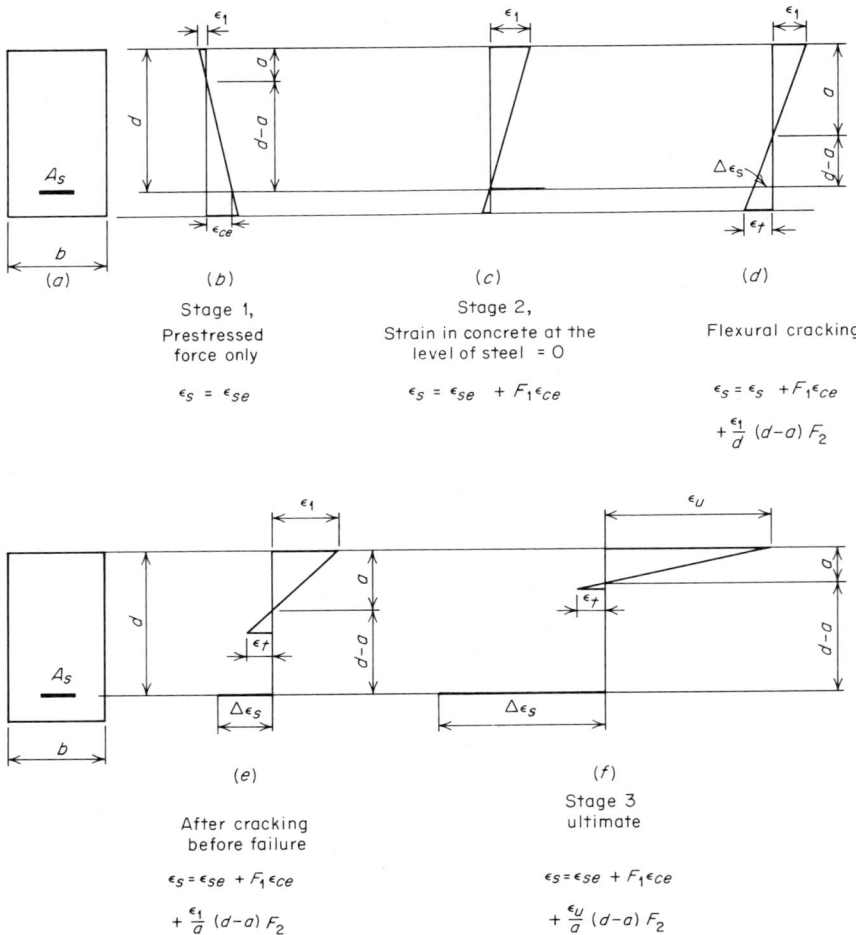

Fig. 3-1 Strain distribution in concrete at various stages of loading.

FLEXURAL BEHAVIOR AND STRENGTH OF PRESTRESSED-CONCRETE BEAMS

the prestressing force is increased somewhat because of the effect of the weight of the beam, this definition of ϵ_{se} corresponds to the fictitious case of a weightless beam. However, it is a convenient definition and serves to simplify the study of behavior of a prestressed-concrete beam. The term *effective* implies that ϵ_{se} is the strain in steel at the time of testing or analysis of the beam, and that the variation of ϵ_{se} with time is not being considered. The strain in concrete at the level of steel due to effective prestress is designated as ϵ_{ce}. Thus at stage 1, or the condition of no load on the beam, ϵ_{ce} is always a compressive strain.

As an applied load is placed on the beam, the compressive strain in concrete at the level of steel tends to decrease; that is, the fiber of concrete at the level of steel elongates. Figures 3-1c and d show the strain distribution in the section as the load is applied on the beam. At a particular load, the strain in concrete at the level of steel becomes zero. Figure 3-1c shows the strain distribution in the beam when strain in concrete at the level of steel is zero. This condition of the beam is defined as stage 2, and is important since it establishes a datum which is convenient in relating the strain in steel to strain in concrete. Before this stage is reached, the strain in concrete at the level of steel is compressive. After this stage the strain begins to become tensile.

The change in strain in steel at a given stage of loading between stages 1 and 2 may be expressed in the following general form:

$$\Delta \epsilon_s = F_1 \left[\epsilon_{ce} - \frac{\epsilon_1}{a}(d - a) \right] \tag{3-1}$$

where $\Delta \epsilon_s$ = increase in strain in steel between stage 1 and given stage of loading
ϵ_{ce} = strain in concrete at level of steel at region of maximum moment due to effective prestress
ϵ_{se} = strain in steel due to effective prestress [see (3-2)]
ϵ_1 = strain in concrete at top fiber at given stage of loading
a = depth to neutral axis, that is, distance from neutral axis to top fiber
d = effective depth
F_1 = compatibility factor

Equation (3-1) is valid when ϵ_1 changes from tension to compression as the load is increased. A point of singularity occurs when $\epsilon_1 = 0$ and $a = 0$, and Eq. (3-1) cannot be used.

Hence the strain in steel at any stage of loading between stages 1 and 2 can be expressed as follows:

$$\epsilon_s = \epsilon_{se} + F_1 \left[\epsilon_{ce} - \frac{\epsilon_1}{a}(d - a) \right] \tag{3-2}$$

where F_1 is the ratio of increase in strain in steel to the increase in strain in concrete at the level of steel at midspan or the region of maximum moment. Since in bonded beams between stages 1 and 2 the change in strain in steel is equal to the change in strain in concrete at the level of steel, F_1 is equal to 1.0.

In ideally unbonded beams where prestressing steel is not in contact with concrete (except at anchorages), the change in strain in steel is not equal to the change in strain in concrete at the level of steel. The distribution of the change in strain in steel along the span is uniform, but the distribution of the change in strain in concrete fiber at the level of steel between stages 1 and 2 varies as the moment diagram, and in the region of maximum moment the increase in strain in concrete is higher. Hence the magnitude of F_1 depends upon the shape of the moment diagram caused by the acting loads. For example, in a simply supported beam with a single concentrated load acting at midspan, the moment diagram is triangular, and the distribution of the change in strain in concrete at the level of steel is also triangular. In this case F_1 is equal to $\frac{1}{2}$, because the ratio of average to maximum strain is one-half. By the same reasoning it can be shown that for an unbonded beam subjected to a uniformly distributed load $F_1 = \frac{2}{3}$. It can be shown that for an ideally unbonded beam

$$F_1 = \frac{1}{\epsilon_{ce}L} \int_0^L \epsilon_c(x) \, dx$$

where $\epsilon_c(x)$ is the strain in concrete at the level of steel at stage 2 at any point along the span and L is the length of simple span.

In practical situations where the prestressing steel is unbonded but is in contact with concrete at certain sections, as in beams with draped reinforcement, the value of F_1 is more difficult to determine. In beams of this type, the value of F_1 is greater than that corresponding to the ideally unbonded beam, but is less than 1 since the strain in steel in the region of maximum moment is less than the strain in concrete at the level of steel. The exact value of F_1 in this case depends upon the frictional conditions between steel and concrete.

Setting $d = a$ in Eq. (3-2), we obtain the following expression for strain in steel at stage 2:

$$\epsilon_s = \epsilon_{se} + F_1 \epsilon_{ce} \tag{3-3}$$

The behavior of the beam between stages 1 and 2 is important since it corresponds to the service condition of a prestressed-concrete beam in which the beam is usually uncracked.

Variation of strain in steel between stages 2 and 3 As the applied load is increased beyond that corresponding to stage 2, the strain in concrete at the level of steel begins to become tensile. If the beam is not already cracked, the increase of the load will cause the strain in concrete at the bottom fiber to reach the cracking level, making the beam crack somewhere in the region of maximum moment. This stage of loading is shown in Fig. 3-1d. The strain in steel at flexural cracking can be expressed as follows:

$$\epsilon_s = \epsilon_{se} + F_1\epsilon_{ce} + \frac{\epsilon_1}{a}(d-a)F_2 = \epsilon_{se} + F_1\epsilon_{ce}$$
$$+ \frac{\epsilon_t}{h-a}(d-a)F_2 \quad (3\text{-}4)$$

In the above expression ϵ_t is strain corresponding to the tensile strength of concrete. The quantity F_2 is a compatibility factor similar to F_1. Equation (3-4) assumes that cracking follows stage 2, which is usually the case.

Figure 3-1e shows a stage of loading beyond cracking in which the greater part of concrete at the tension zone is cracked and can carry only a limited amount of tension. The final stage of loading, which corresponds to ultimate, is shown in Fig. 3-1f and is designated as stage 3. In this case it is assumed that ultimate moment is reached when the extreme compression fiber in concrete has reached the limiting strain ϵ_u.

The change in strain in steel at a given stage of loading between stages 2 and 3 may be expressed in the following general form:

$$\Delta\epsilon_s = \frac{\epsilon_1}{a}(d-a)F_2$$

The factor F_2 in the above equation is the same as F_1 before flexural cracking, and may be approximated as F_1 at loads considerably beyond cracking.

However, at large loads when the cracks have propagated deep in the beam, F_2 becomes dependent upon the condition of bond between concrete and steel as well as the position of the crack with respect to the given section. A study of Figs. 3-2 and 3-3 shows the erratic nature of F_2 for a given stage of loading.

Figure 3-2a shows a simply supported beam at a loading stage after the cracking has occurred but cracks are only slightly developed. The loading consists of two concentrated loads, as shown. Figure 3-2b shows the variation of strain in concrete at extreme fiber, which follows the moment diagram since the cracks are not advanced enough to cause an appreciable variation in the section properties. The variation of the position of neutral axis in the region of maximum moment is shown in Fig. 3-2c. The solid line in Fig. 3-2d corresponds to the variation of $\epsilon_1(d-a)/a$. If at a given point in the region of maximum moment the

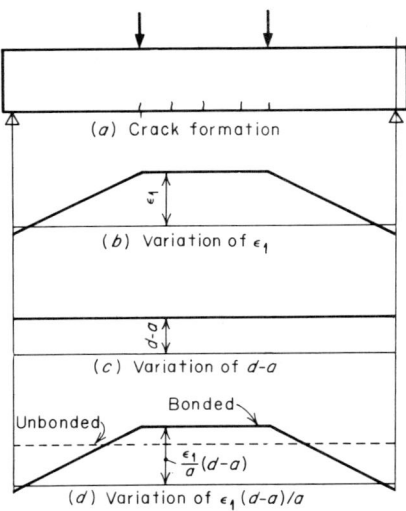

Fig. 3-2 Variation of the change in strain in concrete and increase in strain in steel for a load after cracking.

actual change in strain in steel beyond stage 2 were below this line, the quantity F_2 for that point would be less than 1.0. If, on the other hand, the actual change in strain were on or above this line, the value of F_2 would equal or be greater than 1.0. For bonded beams at this particular stage of loading it is reasonable to assume that the solid line in Fig. 3-2d represents the actual variation of change in steel strain beyond stage 2, that is, $F_2 = 1.0$. For an unbonded beam, the change in strain in steel at the middle portion of the beam is represented by the dashed line in Fig. 3-2d, and it can be seen that in this case $F_2 < 1.0$ since the dashed line is below the solid line.

Figure 3-3a through d shows the same beam and the variation of the same quantities at a high stage of loading approaching ultimate. Cracks have developed to such an extent that they influence greatly the distribution of strain at the extreme fiber, and the position of the neutral axis along the span, as shown in Fig. 3-3b and c. The solid curve in Fig. 3-3d shows the variation of change in strain in steel as defined by $\epsilon_1(d - a)/a$ at each point. For bonded beams at this particular stage of loading, F_2 may have any value, depending upon the condition of bond and the position of the point along the span. If the presence of the cracks has completely eliminated the bond between concrete and steel, the condition of the beam approaches that of an unbonded beam and $F_2 < 1.0$. The quantity F_2 is smaller for a section at the crack than for a section between the cracks. If, on the other hand, we have the other extreme condition that there is perfect bond between steel and concrete, the strain in steel will be concentrated more at the cracks, as is the case for strains in concrete. The change in strain in steel approaches $\epsilon_1(d - a)/a$ and F_1 approaches 1. For ideally unbonded beams the variation of strain in steel is uniform, as shown in Fig. 3-3d, and for the region of maximum moment $F_2 < 1.0$.

It should be pointed out that in an actual beam the bond never remains perfect; nor is it completely destroyed; nor is there an ideally unbonded beam. The exact determination of F_2 for a given section at a particular stage of loading is difficult.

Extensive experimental data are available on the variation of factor F_2, particularly at ultimate. Empirical studies have been made both on bonded and on unbonded beams at ultimate load in order to relate the variation of $F_2\epsilon_1$ with the position of the neutral axis.[4]

The strain in steel at a stage of loading between stages 2 and 3 may be expressed in the following general form:

$$\epsilon_s = \epsilon_{se} + F_1\epsilon_{ce} + \frac{\epsilon_1}{a}(d - a)F_2 \tag{3-5}$$

From the above discussion it may be concluded that when the strain in the concrete at the level of the steel is compressive, the change in strain in the concrete is $1/F_1$ times the change in strain in the steel. For bonded beams $F_1 = 1.0$. For ideally unbonded beams F_1 is the ratio of average to maximum strain in concrete at the level of steel.

When the strain in the concrete at the level of steel is tensile, the change in strain in the prestressed steel is $(\epsilon_1/a)(d - a)F_2$. For bonded beams F_2 depends on the stage of loading, but may be approximated as

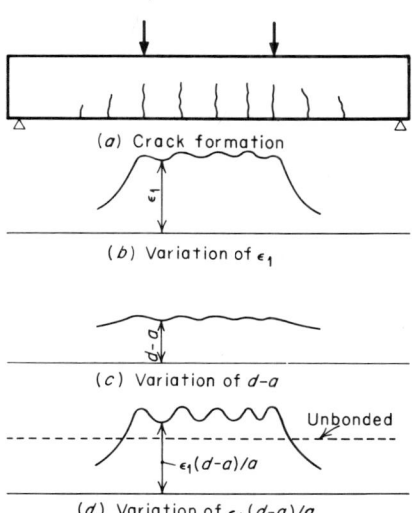

Fig. 3-3 Variation of the change in strain in concrete and increase in strain in steel for a load near ultimate.

1.0 even though it may be greater or smaller than 1.0 at the section considered. For unbonded beams $F_2 = F_1$ for low tensile strains in concrete, and decreases further with load.

3-5 THE RELATIONSHIP BETWEEN STRAIN IN NON-PRESTRESSED STEEL AND STRAIN IN CONCRETE

Relationships between strain in non-prestressed steel and strain in concrete at the level of non-prestressed steel may be obtained if such reinforcement is used in the beam. Figure 3-4 shows a beam with rectangular section which in addition to prestressed steel has non-prestressed reinforcement at the top and bottom. The area of non-prestressed compressive reinforcement is designated as A_{s1}, and that for tensile reinforcement is designated as A_{s2}.

For the non-prestressed reinforcement at the top it may be assumed that in all stages of loading the strain in concrete at the level of steel is equal to the strain in steel. From Fig. 3-4 this relationship can be formulated as follows:

$$\epsilon_{s1} = \frac{\epsilon_1}{a}(a - d_1) \tag{3-6}$$

where ϵ_{s1} = strain in non-prestressed reinforcement at top at any stage of loading
d_1 = distance from extreme top fiber to center of gravity of non-prestressed reinforcement at top

The other terms have already been defined.

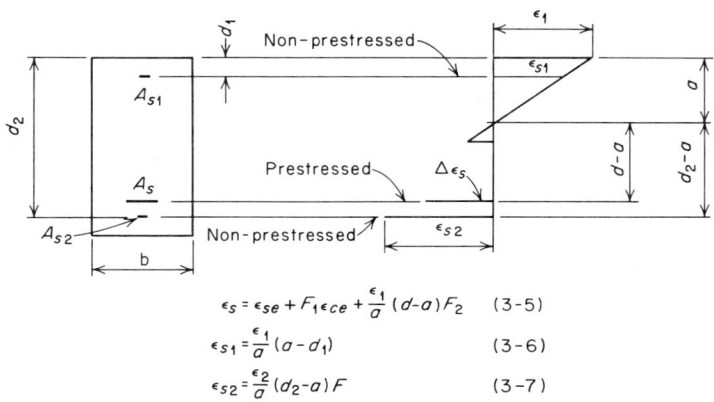

$$\epsilon_s = \epsilon_{se} + F_1\epsilon_{ce} + \frac{\epsilon_1}{a}(d-a)F_2 \tag{3-5}$$

$$\epsilon_{s1} = \frac{\epsilon_1}{a}(a-d_1) \tag{3-6}$$

$$\epsilon_{s2} = \frac{\epsilon_2}{a}(d_2-a)F \tag{3-7}$$

Fig. 3-4 Prestressed-concrete beam with non-prestressed reinforcement.

FLEXURAL BEHAVIOR AND STRENGTH OF PRESTRESSED-CONCRETE BEAMS

For the non-prestressed reinforcement at the bottom the strain in steel at any stage of loading can be expressed in the following general form:

$$\epsilon_{s2} = \frac{\epsilon_1}{a}(d_2 - a)F_3 \tag{3-7}$$

where ϵ_{s2} = strain in non-prestressed reinforcement at bottom at any stage of loading
d_2 = distance from extreme top fiber to center of gravity of non-prestressed reinforcement at bottom
F_3 = strain compatibility factor, which depends on bond and relative position of crack

The quantity F_3 is equal to 1.0 for loads below flexural cracking load, and may be approximated as 1.0 for loads beyond cracking. For loads corresponding to well-developed cracks F_3 may be more than, equal to, or less than 1.0, depending upon the condition of bond and the relative position of the section from the crack. For practical purposes F_3 may be approximated as 1.0. Figure 3-4 shows the strains in a prestressed-concrete beam with non-prestressed top and bottom reinforcement.

3-6 THE STRESS–STRAIN DIAGRAMS FOR THE MATERIALS USED IN PRESTRESSED CONCRETE

Prestressing steel The stress-strain diagram for the prestressed reinforcement may be obtained by tension tests of specimens. The prestressed steel in the beam is subjected to direct tension, and the results obtained from tension tests represent the actual condition.

The steel used in prestressing may be high-tensile-strength single-wire strands shop-fabricated from several high-strength wires and high-strength alloy bars. Table 3-1 shows a summary of the various types of steel used in the United States.[5]

The high-tensile-strength wire bears the ASTM designation A421-59T as "uncoated stress-relieved wire for prestressed concrete." The minimum tensile strength requirement ranges from 235,000 psi for 0.276-in.-diameter wires to 250,000 psi for 0.192-in.-diameter wires. These wires are generally made from high-carbon steel which is manufactured by the open-hearth or electric-furnace process. The wire is cold-drawn and stress-relieved by a controlled time-temperature treatment in order to produce the prescribed mechanical properties.

The high-strength single wires may be used for either pretensioning or post-tensioning purposes. Figure 3-5 shows a typical stress-strain diagram for a 0.199-in.-diameter high-strength single wire.

Table 3-1 A summary of the types of steel commonly used for prestressing in the United States

Name and designation	Strength (minimum), psi	Yield (minimum)	Total elongation (minimum)
High-tensile-strength wire, ASTM A421	235,000 (0.276-in. wire) 250,000 (0.192-in. wire)	At least 80 percent of strength at 1.0 percent elongation	4.0 percent (10-in. gage length)
Uncoated seven-wire strand for prestressed concrete, ASTM A416	250,000 ($\frac{3}{8}$-in. and $\frac{1}{2}$-in. strand) 247,700 ($\frac{7}{16}$-in. strand)	At least 85 percent of specified strength at 1.0 percent elongation	3.5 percent (24-in. gage length)
High-strength alloy steel bars, AISI 5160 or AISI 9260	145,000 for all diameters	At least 90 percent of specified strength at 0.2 percent of permanent strain	4.0 percent (gage length of 20 diameters)

The strands manufactured in the United States consist of 7, 19, and 37 wires. However, the seven-wire strand is the predominant type and is used mostly in pretensioned construction. The ASTM designation A416 is specifically for "uncoated seven-wire stress-relieved strand for prestressed concrete." The seven-wire strands have a center wire enclosed tightly by six helically placed outer wires with uniform pitch. They are available in five sizes from $\frac{1}{4}$ in. diameter to $\frac{1}{2}$ in. diameter in increments of $\frac{1}{16}$ in. The most common sizes are $\frac{3}{8}$- and $\frac{7}{16}$-in.-diameter

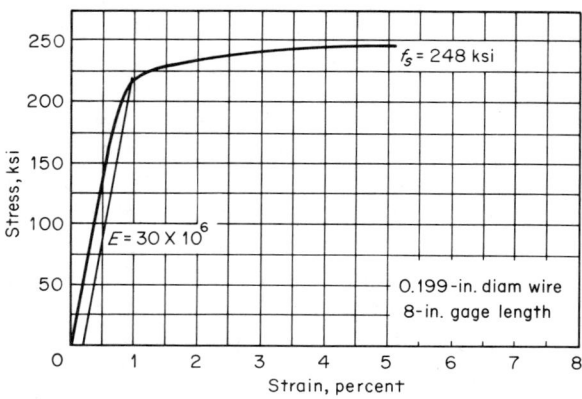

Fig. 3-5 Stress-strain relationship for a prestressing wire.

FLEXURAL BEHAVIOR AND STRENGTH OF PRESTRESSED-CONCRETE BEAMS

strands. After manufacturing, the strands are required to be stress-relieved in order to produce the prescribed mechanical properties.

The stress-strain diagram of a strand is similar to that for an individual wire.

Large-diameter high-strength strand with 19 or 37 individual wires can be used in post-tensioned construction.

Cold-stretched high-strength alloy steel bars are usually designated as AISI 5160 or AISI 9260. The bars are first hot-rolled and cold-worked or heat-treated. Each bar is then cold-stretched to a minimum of 90 percent of the specified ultimate strength. The minimum required ultimate strength for all diameters is 145,000 psi.

Non-prestressed supplementary reinforcement Prestressed-concrete beams often have non-prestressed reinforcement in addition to the prestressing steel. The non-prestressed reinforcement in the beam is subjected to axial tension or compression as the case may be, and its stress-strain diagram may be obtained from tension tests.

There are several types of steel used as non-prestressed reinforcement. In Table 3-2 the designations and properties of some of the more common types are listed for convenient reference. The steel used in non-prestressed reinforcement is the same type used in reinforced concrete.

Table 3-2 A summary of the types of non-prestressed steel used in prestressed concrete

Name and designation	Types available	Tensile strength, psi	Yield point (minimum), psi	Minimum elongation in 8 in., percent
Billet steel bars ASTM A15 Axle steel bars ASTM A160	Plain bars:			
	Structural grade	55,000 to 75,000	33,000	1,400,000/tens strength 20
	Intermediate grade	70,000 to 90,000	40,000	1,300,000/tens strength 16
	Hard grade	80,000 min	50,000	1,100,000/tens strength
	Deformed bars:			
	Structural grade	55,000 to 75,000	33,000	1,200,000/tens strength 16
	Intermediate grade	70,000 to 90,000	40,000	1,100,000/tens strength 12
	Hard grade	80,000 min	50,000	1,100,000/tens strength
Rail steel bars ASTM A16	Regular plain	80,000	50,000	1,100,000/tens strength
	Regular deformed	80,000	50,000	1,000,000/tens strength
	Special deformed	90,000	60,000	1,000,000/tens strength
High-strength billet bars, ASTM A431	Deformed bars only	100,000	75,000	Varies 5 to $7\frac{1}{2}$ depending on bar size

Figure 3-6 shows typical stress-strain diagrams for billet steel of intermediate-grade bar. In design problems often the strain-hardening is ignored and it is assumed that the stress-strain diagram consists of an inclined line—defining the elastic behavior of the bar—and a horizontal line. In this case the stress and strain at the yield point and the modulus of elasticity are sufficient to define the entire stress-strain diagram.

Concrete The stress-strain relationship in concrete in the beam, unlike that for the steel reinforcement, cannot be fully determined from the concentrically loaded cylinders.

The stress distribution in a concentrically loaded cylinder is approximately uniform, and for each measured strain, the stress may be computed by dividing the load by the cross-sectional area of the cylinder. However, in a beam the stress distribution is not uniform through the depth of the beam. Although the strain can be measured readily at any point in concrete, it is very difficult to measure the stress. Furthermore, the stress-strain relationship obtained from ordinary machines for concentrically loaded cylinders considers the range of the load from zero to maximum, and the failure takes place shortly after the maximum stress is reached. The failure takes place suddenly since the energy stored in the testing machine is released immediately after the maximum load is reached. Figure 3-7 shows typical stress-strain diagrams for a concentrically loaded cylinder. The strain recorded at the maximum load is in the neighborhood of 0.002.

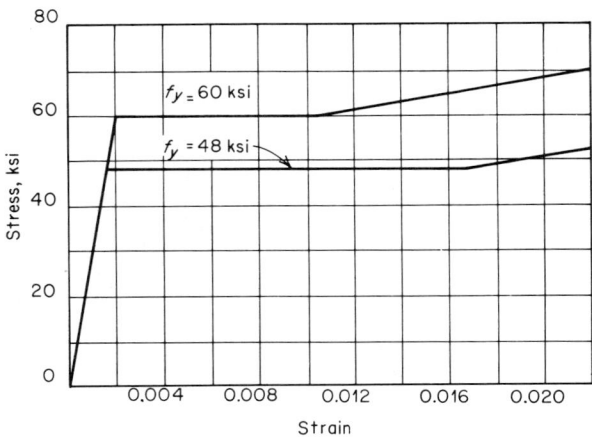

Fig. 3-6 Typical stress-strain diagrams for intermediate-grade steel.

FLEXURAL BEHAVIOR AND STRENGTH OF PRESTRESSED-CONCRETE BEAMS

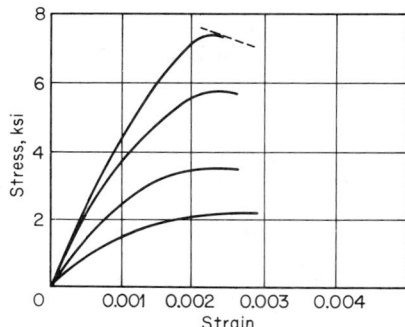

Fig. 3-7 Typical stress-strain diagrams for 28-day concrete cylinders.

The extent to which the stress-strain diagram in concrete may be developed beyond the point of maximum stress in concentrically loaded cylinders depends on the flexibility of the testing machine. In flexible machines the specimens fail at maximum load without a significant decrease in load. In very stiff machines it is possible to develop an appreciable portion of the stress-strain diagram beyond the maximum load.[6] Hence the extent to which the diagram is developed beyond the point of maximum stress is a measure of stiffness of machine.

By surrounding concrete specimens by a system of calibrated steel springs, the portion of the stress-strain diagram beyond the point of maximum stress has been obtained.[7] Figure 3-8 shows results obtained on 3- by 6-in. cylinders on such specimens.

A more complete stress-strain relationship in concrete has been obtained on flexural specimens.[1] Specimens were tested in such a way that during the entire period of testing the stress on one side was kept at zero while on the opposite face it was maximum. The strains were measured directly and the stress was computed from assumption of linearity of strains and statics. The results of these tests were compared with those obtained from the concentric compression tests on cylinders.

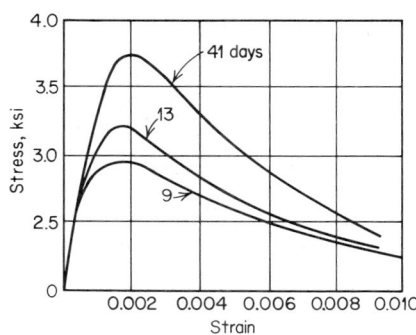

Fig. 3-8 Stress-strain diagrams for concrete cylinders tested with surrounding springs.

These tests indicated that the part of the stress-strain diagram that can be developed in compression tests of cylinders is almost identical with those in the beams. The stress-strain diagrams developed in the beams, however, were developed far beyond the strain corresponding to the maximum stress. Figure 3-9 shows results of such tests.

From tests on a great number of beams it has been found that the strain at the extreme fiber of the beam at ultimate is given from 0.003 to about 0.004.[2,4] That is, in beams the maximum strain reached is considerably greater than that in concentrically loaded cylinders.

The above findings suggest that the highly strained outer fibers in the beam yield and transfer stress to the fibers that are closer to the neutral axis and are less strained. In beams it is possible to develop a maximum strain nearly twice as much as that in concentrically loaded cylinders. Figure 3-9 shows that the maximum strain developed in the concrete in the beam is considerably larger than the strain corresponding to maximum stress.

It is seen that the exact shape of the stress-strain diagram in concrete in a beam is difficult to determine. The only quantity known to the designer of a beam is f'_c, the 28-day cylinder strength, which may not be the same as the actual strength in the beam. However, for practical purposes the stress-strain relation can be approximated by idealization. Since the stress-strain diagram for low-strength concrete is flat in the region of the maximum stress, it may be assumed elastoplastic. This simplification will be incorrect for high-strength concretes, since the stress-strain relation for high-strength concrete peaks sharply. Figure 3-10a shows idealized stress-strain diagrams that may be used for practical purposes.

Jensen idealized the stress-strain diagram in elastoplastic form, as shown in Fig. 3-10b.[8] Hognestad suggested a combination of a parabola and a straight line, as shown in Fig. 3-10c.[3]

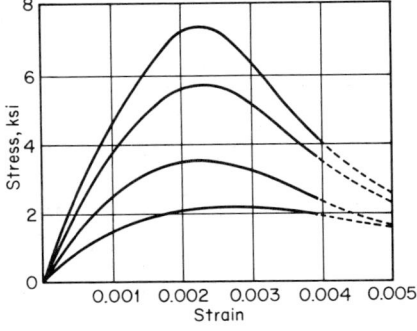

Fig. 3-9 Stress-strain diagrams for concrete at the extreme fiber of the beam.

FLEXURAL BEHAVIOR AND STRENGTH OF PRESTRESSED-CONCRETE BEAMS

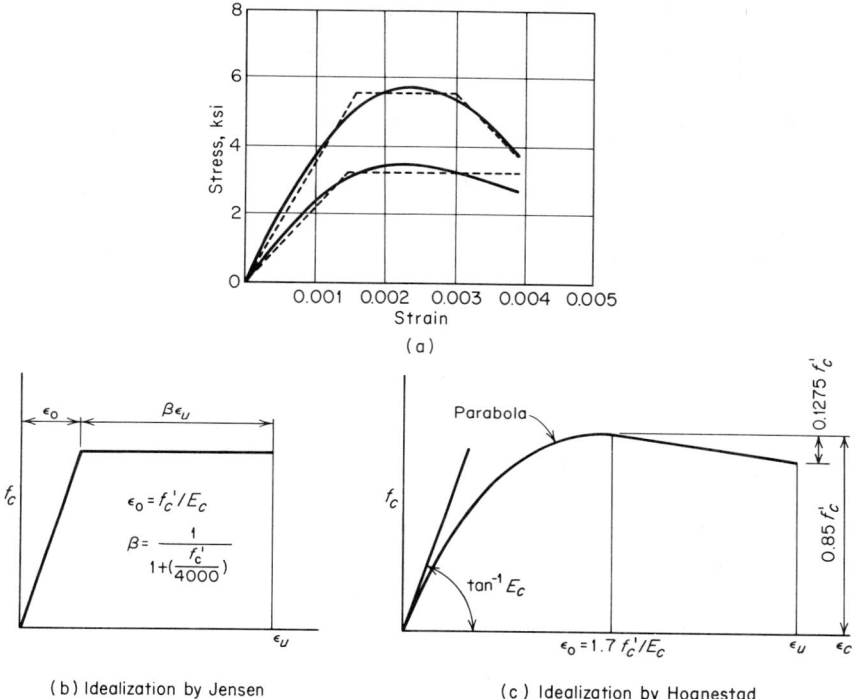

(b) Idealization by Jensen (c) Idealization by Hognestad

Fig. 3-10 Idealized stress-strain diagrams for concrete.

The stress-strain relation in tension is generally assumed to be linear. There are several empirical expressions that give the limiting tensile stress or strain. The limiting tensile stress in concrete when the stress-strain diagram is assumed to be linear is called *modulus of rupture* and is designated as f_r. Experimental work indicates that its magnitude is in the neighborhood of 10 percent of 28-day cylinder strength f'_c. The modulus of rupture has been correlated with $\sqrt{f'_c}$ in recent tests and is usually taken as $8\sqrt{f'_c}$.

Recent test data indicate that the stress-strain diagram in tension is nonlinear. In this case the stress corresponding to cracking is called tensile strength and its magnitude is usually taken as $6\sqrt{f'_c}$. The strain corresponding to tensile strength is designated as ϵ_t.

3-7 ANALYSIS

In the preceding sections the significant relations developed from the empirical study of prestressed-concrete beams were discussed briefly. This knowledge can serve as a basis for assumptions to be used in analyzing a prestressed-concrete beam. Analysis in this sense means relating the external loads with strains in concrete or steel for any given load between zero and maximum. The following assumptions can be made:

1. The strain is distributed linearly with depth in the uncracked region of the beam.
2. The compatibility factors F_1, F_2, and F_3 are known.
3. The stress-strain diagrams for all materials are known; that is, the stress-strain diagrams for the prestressing steel, supplementary non-prestressed steel, if any, and concrete are available. It is further assumed that all fibers in concrete follow the same stress-strain relation.
4. Ultimate load is defined as the load corresponding to the limiting strain in either steel or concrete, whichever occurs first.
5. The area of each type of steel is concentrated at the centroid of area, and the stress and strain considered are the average.

From these assumptions it is possible to relate the load or moment to the various stresses or strains in the beam.

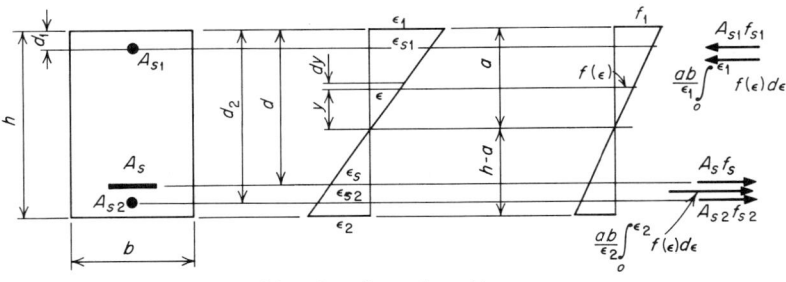

(a) Before flexural cracking

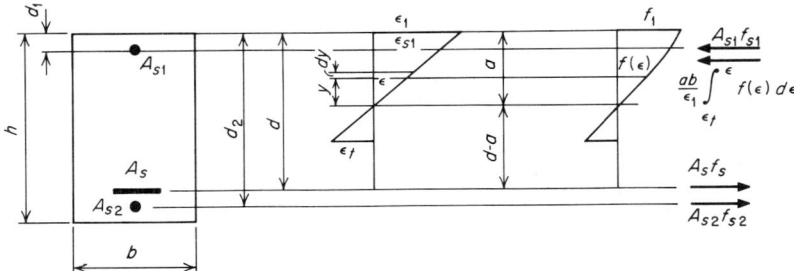

(b) After flexural cracking

Fig. 3-11 A prestressed-concrete beam with rectangular section and non-prestressed reinforcement.

FLEXURAL BEHAVIOR AND STRENGTH OF PRESTRESSED-CONCRETE BEAMS

In order to understand the fundamental relations in the analysis of a prestressed-concrete beam, a beam with a rectangular section will be considered. To analyze the beam in its general form it will be assumed that, in addition to prestressed steel, the beam has non-prestressed reinforcement of different types at top and bottom. Figure 3-11 shows a rectangular section of a prestressed-concrete beam with the three types of reinforcement. For a given load that is acting on a given beam, stress and strain in each type of steel and concrete in all the fibers are unknowns. The relations that can be used to determine these unknowns may be grouped in three categories: (1) equations of equilibrium, (2) equations of compatibility, and (3) the stress-strain relations of all the materials used. In the following paragraphs these relations are discussed in detail.

Equations of equilibrium Since we are analyzing a section of pure flexure, two equations of equilibrium are available: the equations of equilibrium of horizontal forces and those of moments.

From Fig. 3-11a equilibrium of horizontal forces can be expressed as follows:

$$b \int_{-(h-a)}^{a} f(\epsilon)\, dy + A_s f_s + A_{s1} f_{s1} + A_{s2} f_{s2} = 0$$

where A_s, f_s = area and stress in prestressed steel
A_{s1}, f_{s1} = area and stress in non-prestressed compression steel
A_{s2}, f_{s2} = area and stress in non-prestressed tensile steel

The function $f(\epsilon)$ in the above expression defines the stress-strain relation of concrete in the beam. Since it is assumed that strain varies linearly with depth in the uncracked zone, we have

$$y = \frac{a}{\epsilon_1} \epsilon$$

and

$$dy = \frac{a}{\epsilon_1} d\epsilon$$

The equation of equilibrium of horizontal forces can now be expressed as follows:

$$\frac{ba}{\epsilon_1} \int_{\epsilon_2}^{\epsilon_1} f(\epsilon)\, d\epsilon + A_s f_s + A_{s1} f_{s1} + A_{s2} f_{s2} = 0 \tag{3-8}$$

where ϵ_1 = strain at top fiber
ϵ_2 = strain at bottom fiber

The above equation applies to the uncracked beam. However, by changing the lower limit of integration ϵ_2 to ϵ_t, the strain corresponding to tensile strength, it will be applicable to the range of load corresponding to the cracked beam also.

In a similar way the equilibrium of moments yields the following equation:

$$M + \frac{ba^2}{\epsilon_1^2}\int_{\epsilon_2}^{\epsilon_1} f(\epsilon)\epsilon\, d\epsilon + (d - a)A_s f_s + (a - d_1)A_{s1}f_{s1}$$
$$+ (d_2 - a)A_{s2}f_{s2} = 0 \quad (3\text{-}9)$$

where M is the moment due to the external loads at the section.

Equation (3-9) gives the sum of moments about the neutral axis. Moment arm is taken positive when the force is above the neutral axis. Tensile stress and strain are considered positive.

Equations of compatibility If the stage of loading corresponding to the strain distribution in Fig. 3-11b is considered, the strain in prestressed steel can be expressed as follows:

$$\epsilon_s = \epsilon_{se} + F_1\epsilon_{ce} + \frac{\epsilon_1}{a}(d - a)F_2 \quad (3\text{-}5)$$

If the loading is such that the strain in concrete at the level of steel is compressive, Eq. (3-2) should be used instead of Eq. (3-5).

The strain in the non-prestressed reinforcement at the top and bottom will be the following:

$$\epsilon_{s1} = \frac{\epsilon_1}{a}(a - d_1) \quad (3\text{-}6)$$

$$\epsilon_{s2} = \frac{\epsilon_1}{a}(d_2 - a)F_3 \quad (3\text{-}7)$$

On the basis of the assumption that the strain is distributed linearly with depth, the following relationship is correct before the beam cracks:

$$a = \frac{\epsilon_1}{\epsilon_1 - \epsilon_2}h$$

The quantities ϵ_1 and ϵ_2, which are the strains in concrete at the top and bottom fibers, respectively, are taken as positive when tensile and negative when compressive. After cracking, a similar relation exists between ϵ_1 and the strain at the root of crack ϵ_t.

Stress-strain diagrams Since we have assumed that the stress-strain diagram for all the materials used is known, let the following expression define the stress-strain diagram in concrete:

$$f = f(\epsilon) \quad (3\text{-}10)$$

The above expression for stress in concrete applies to any fiber in concrete, and is valid from maximum tension to maximum compression.

FLEXURAL BEHAVIOR AND STRENGTH OF PRESTRESSED-CONCRETE BEAMS

Let the following function define the stress-strain relation in prestressed steel:

$$f_s = \phi(\epsilon_s) \tag{3-11}$$

Similarly, let the following functions define the stress-strain relation for the non-prestressed top and bottom reinforcement:

$$f_{s1} = \phi_1(\epsilon_{s1}) \tag{3-12}$$

$$f_{s2} = \phi_2(\epsilon_{s2}) \tag{3-13}$$

Usually the same type of steel is used for non-prestressed top and bottom steel. Here we have assumed them different in order to show the solution of the general problem.

The stress-strain diagrams for concrete and the various types of steel are presented by Eqs. (3-10) through (3-13) in functional form, in order to show that the method of analysis presented here is applicable to all types of stress-strain diagrams. Equation (3-10) represents stress-strain diagrams in the actual form or idealized form shown in Fig. 3-10a, b, or c. Equation (3-11) represents the stress-strain diagram in Fig. 3-5. Equations (3-12) and (3-13) represent stress-strain diagrams of the type shown in Fig. 3-6.

The above nine equations provide sufficient relations for the determination of the unknowns. The quantities b, h, d, d_1, d_2, A_s, A_{s1}, A_{s2}, ϵ_{se}, ϵ_{ce}, F_1, F_2, and M are given or are assumed known. The unknowns are the stress and strain in concrete at the top fiber (or any fiber), the stress and strain in each of the three types of steel used, and the position of the neutral axis as defined by a. That is, there are nine unknowns, namely, f_1, ϵ_1, f_s, ϵ_s, f_{s1}, ϵ_{s1}, f_{s2}, ϵ_{s2}, and a. A simultaneous solution of Eqs. (3-5) through (3-13) will give the values of these quantities.

When there is no non-prestressed tensile and compressive reinforcement, Eqs. (3-6), (3-7), (3-12), and (3-13) vanish, and in the remaining equations the following may be assumed:

$$A_{s1} = A_{s2} = 0$$

In this case there will be only the following five equations:

$$\epsilon_s = \epsilon_{se} + F_1\epsilon_{ce} + \frac{\epsilon_1}{a}(d-a)F_2 \tag{3-5}$$

$$f = f(\epsilon) \tag{3-10}$$

$$f_s = \phi(\epsilon_s) \tag{3-11}$$

$$\frac{ba}{\epsilon_1}\int_{\epsilon_2}^{\epsilon_1} f(\epsilon)\, d\epsilon + A_s f_s = 0 \tag{3-8a}$$

$$M + \frac{ba^2}{\epsilon_1^2}\int_{\epsilon_2}^{\epsilon_1} f(\epsilon)\epsilon\, d\epsilon + (d-a)A_s f_s = 0 \tag{3-9a}$$

A simultaneous solution of the above five equations will yield the unknowns f_1, ϵ_1, f_s, ϵ_s, and a. This case is the simplest in the analysis of prestressed-concrete beams.

3-8 SOLUTION OF SPECIAL PROBLEMS

The method presented in the preceding section is general, and has a wide application. In the following paragraphs some special problems will be discussed.

Reinforced-concrete beam, straight-line theory In this case both the stress-strain diagram for concrete and that for steel are assumed linear and the tension carried by concrete is neglected. It is also assumed that $F_2 = 1.0$. We have

$$\epsilon_{se} = \epsilon_{ce} = \epsilon_t = 0$$

$$F_1 = F_2 = 1.0$$

Equation (3-5) can be expressed as follows:

$$\frac{\epsilon_s}{\epsilon_1} = -\frac{d-a}{a} \tag{3-5a}$$

Since the stress-strain diagram for concrete is assumed linear, Eq. (3-10) can be written as

$$f = f(\epsilon) = E_c \epsilon \tag{3-10a}$$

where $E_c = df/d\epsilon$ is the modulus of elasticity of concrete.

Similarly, the stress-strain diagram for steel may be assumed in the following form:

$$f_s = E_s \epsilon_s \tag{3-11a}$$

where E_s is the modulus of elasticity of steel.

Substituting Eqs. (3-10a) and (3-11a) for $f(\epsilon)$ and f_s, respectively, in Eqs. (3-8) and (3-9), and taking $a = kd$, we have

$$\frac{bkd}{\epsilon_1} \int_0^{\epsilon_1} E_c \epsilon \, d\epsilon + A_s E_s \epsilon_s \equiv \frac{bkd}{2} E_c \epsilon_1 + A_s E_s \epsilon_s = 0 \tag{3-8b}$$

and

$$M + \frac{b(kd)^2}{\epsilon_1^2} \int_0^{\epsilon_1} E_c \epsilon^2 \, d\epsilon + d(1-k)A_s E_s \epsilon_s$$

$$\equiv M + \frac{b(kd)^2}{3} E_c \epsilon_1 + d(1-k)A_s E_s \epsilon_s = 0 \tag{3-9b}$$

By introducing $A_s/bd = p$ and $E_s/E_c = n$, Eqs. (3-5a) and (3-8b)

give the following:
$$k = \sqrt{2pn + (pn)^2} - pn$$

From Eqs. (3-8b) and (3-9b) we have
$$f_1 = \frac{6M}{bkd^2(3-k)} \qquad \epsilon_1 = \frac{6M}{bkd^2E_c(3-k)}$$

and from Eq. (3-5a) we have
$$f_s = \frac{6Mn(1-k)}{b(kd)^2(3-k)} \qquad \epsilon_s = \frac{6Mn(1-k)}{b(kd)^2(3-k)E_s}$$

These equations are familiar and, as shown, can be obtained by the general method. It is not suggested that the problem be solved by this approach in practical cases. This example, however, illustrates the generalized solution of the problem.

Prestressed-concrete beam, prestressing force only In this case it is assumed that the only force acting is the prestressing force and the stress-strain relations in concrete and steel are linear.

Let us assume that the overall depth of the beam is h and the prestressing force P is applied at a distance g above the bottom fiber. It is required to calculate the stress in concrete at top and bottom fibers and to determine the position of the neutral axis.

Equation (3-10a) defines the stress-strain diagram in concrete:
$$f = E_c \epsilon \tag{3-10a}$$

Since in this case the prestressing force is assumed known, the stress-strain diagram for steel is not used.

Equations (3-8) and (3-9) will have the following forms:
$$\frac{ba}{\epsilon_1} \int_{\epsilon_2}^{\epsilon_1} E_c \epsilon \, d\epsilon + P = 0$$
and
$$\frac{ba^2}{\epsilon_1^2} \int_{\epsilon_2}^{\epsilon_1} E_c \epsilon^2 \, d\epsilon - P(h - a - g) = 0$$

Substitution of $\epsilon_1 h/(\epsilon_1 - \epsilon_2)$ for a in the above equations and their simplification results in the following:
$$\epsilon_1 = -\frac{2P}{hbE_c} - \epsilon_2$$
and
$$\epsilon_1^2 + \epsilon_1\epsilon_2 + \epsilon_2^2 = -\frac{3P}{hbE_c}\left(\epsilon_2 - \epsilon_2\frac{g}{h} + \epsilon_1\frac{g}{h}\right)$$

A simultaneous solution of the above equations yields the following:

$$\epsilon_1 = \frac{P}{bhE_c}\left(2 - 6\frac{g}{h}\right) \tag{3-14}$$

and

$$\epsilon_2 = \frac{P}{bhE_c}\left(6\frac{g}{h} - 4\right) \tag{3-15}$$

The neutral axis may be obtained from

$$\frac{a}{h} = \frac{\epsilon_1}{\epsilon_1 - \epsilon_2} = \frac{1 - 3g/h}{3 - 6g/h}$$

In the special case when $g/h = 0$, we have

$$\epsilon_1 = \frac{2P}{bhE_c}$$

$$\epsilon_2 = -\frac{4P}{bhE_c}$$

and

$$\frac{a}{h} = \frac{1}{3}$$

Homogeneous elastic rectangular section In this case there is no reinforcement in the beam. A plain concrete beam before failure approximates this condition.

Only Eqs. (3-10), (3-8), and (3-9) are applicable, and the unknowns are ϵ_1, f_1, and a.

Equation (3-10a) defines the stress-strain diagram for concrete,

$$f = E_c\epsilon \tag{3-10a}$$

and from Eq. (3-8) we have

$$\frac{ba}{\epsilon_1}\int_{\epsilon_2}^{\epsilon_1} E_c\epsilon \, d\epsilon = 0$$

or

$$\epsilon_1 = -\epsilon_2$$

and

$$a = \frac{h}{2}$$

Substituting the above expressions in Eq. (3-9) yields

$$M + \frac{bh^2}{4\epsilon_1^2}\int_{-\epsilon_1}^{\epsilon_1} E_c\epsilon^2 \, d\epsilon = 0$$

FLEXURAL BEHAVIOR AND STRENGTH OF PRESTRESSED-CONCRETE BEAMS

from which

$$\epsilon_2 = -\epsilon_1 = \frac{6M}{bh^2 E_c}$$

which is a familiar conclusion.

There are many more cases that can be solved by the above general method. If the stress-strain relationship is linear, the analysis of the problem is simple, as shown in the preceding examples. However, the analysis of a prestressed-concrete beam for any stage of loading between zero and ultimate load is more complicated. Since the stress-strain relations for concrete and prestressed steel are nonlinear, the general solution of the simultaneous equations becomes difficult. The analysis can be made numerically.

In studying the behavior of a beam it is convenient to determine the relationship between moment and curvature as the moment varies from zero to ultimate. The determination of moment-curvature relationship for the entire range of load, in sections with prestressed steel only, can be simplified by assuming a stress and strain in concrete and solving Eqs. (3-5), (3-8), and (3-11) simultaneously in order to obtain ϵ_s, f_s, and a. With these quantities known the moment M can be calculated from Eq. (3-9).

For sections with non-prestressed reinforcement the same general procedure may be used. Stress in non-prestressed steel may be assumed and its validity checked by Eq. (3-12) or (3-13) after the solution.

3-9 APPROXIMATE CONCRETE STRESS–STRAIN DIAGRAM

From the preceding examples it can be seen that Eqs. (3-8) and (3-9) involve the knowledge of the stress-strain diagram for concrete in a functional form. Though it is possible to express the stress-strain diagram for concrete as a combination of a parabola and a straight line, as shown in Fig. 3-10c, its actual integration becomes tedious. It is more convenient to express the actual or idealized stress-strain diagram for concrete by a series of points that are connected by straight lines, as shown in Fig. 3-12a.

The equations of equilibrium of horizontal forces and moments for the stress-strain diagram shown in Fig. 3-12a may be expressed as follows:

$$\frac{ba}{\epsilon_1} \sum_{1}^{n} f_i(\Delta\epsilon)_i + A_s f_s + A_{s1} f_{s1} + A_{s2} f_{s2} = 0 \qquad (3\text{-}16)$$

and

$$M + \frac{ba^2}{\epsilon_1^2} \sum_{1}^{n} f_i(\Delta\epsilon)_i \bar{\epsilon}_i + A_s f_s (d - a) + A_{s1} f_{s1} (a - d_1)$$
$$+ A_{s2} f_{s2} (d_2 - a) = 0 \qquad (3\text{-}17)$$

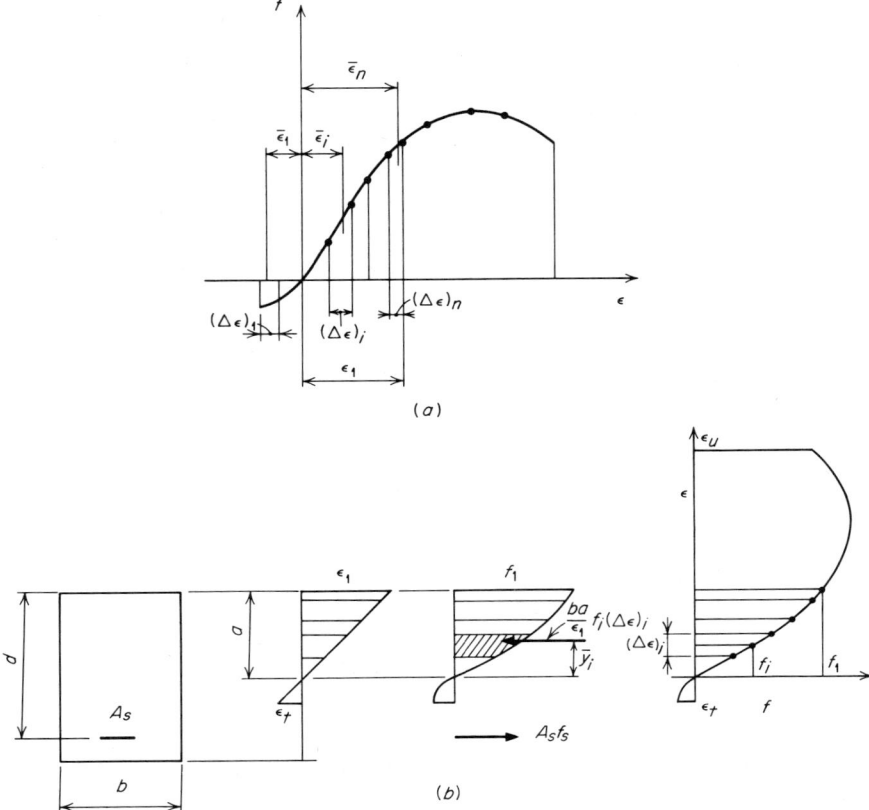

Fig. 3-12 (a) Stress-strain diagram for concrete expressed as a series of straight lines. (b) Relationship between the stress-strain diagram for concrete and the stress distribution in the beam section.

where n = total number of divisions in entire range of strain
f_i = average stress in element i
$(\Delta \epsilon)_i$ = element of strain in ith division
$\bar{\epsilon}_i$ = distance from centroid of ith element to origin

Figure 3-12b shows the relationship between the stress-strain diagram in concrete and the stress distribution in the beam section.

Example Calculate the cracking moment in the plain concrete section shown in Fig. 3-13a. The stress-strain diagram for concrete is shown in Fig. 3-13b.

Solution Cracking occurs when the strain at the bottom fiber reaches 0.00025. From Eq. (3-16) we have

$$\frac{ba}{\epsilon_1}\sum_{1}^{n} f_i(\Delta\epsilon)_i = \frac{6a}{\epsilon_1}[(0.0000913 \times \tfrac{0.4}{2}) + (0.0001587 \times 0.45)$$
$$- \epsilon_1^2 \tfrac{2.19}{0.0005} \times \tfrac{1}{2}] = 0$$

or

$$\epsilon_1 = -0.000202$$

and

$$a = \tfrac{0.000202}{0.000452} \times 12 = 5.36 \text{ in.}$$

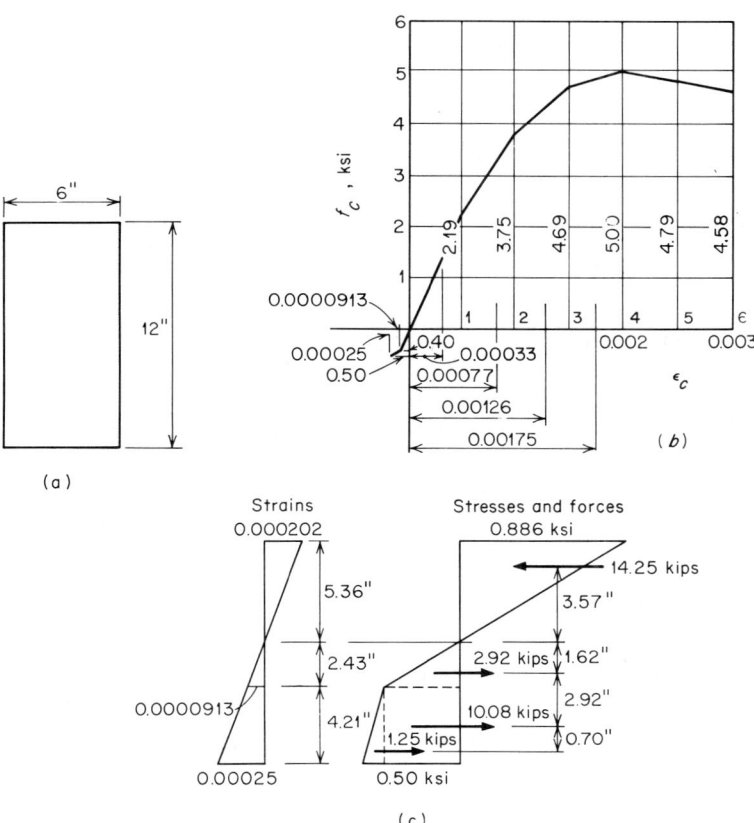

Fig. 3-13 Cracking in a plain concrete section.

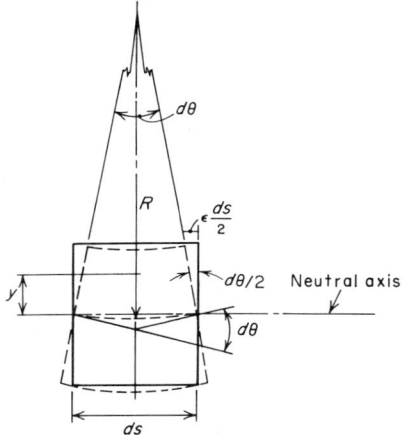

Fig. 3-14 Curvature of a section.

From Eq. (3-17) we have

$$M_c - \frac{ba^2}{\epsilon_1{}^2} \sum f_i(\Delta\epsilon)_i \bar{\epsilon}_i = 0$$

$$M_c - \frac{6 \times (5.36)^2}{(0.000202)^2}[0.2(0.0000913)(0.0000609) + 0.45(0.0001587)(0.0001736) + 0.44(0.000192)(0.000135)] = 0$$

and

$$M_c = 108 \text{ in.-kips}$$

It can be seen that use of Eq. (3-17) is inconvenient. Knowledge of a permits determination of the stress distribution and the forces in the section from which the moment may conveniently be calculated. The strain and stress distributions, as well as individual forces, are shown in Fig. 3-13c.

3-10 CURVATURE

In the study of behavior of beams, curvature provides a convenient measure of deformation in the beam as the acting loads increase.

Figure 3-14 shows an element of a beam in the region of uniform moment. Before the loads are applied, this element is rectangular in shape. After the loads are applied, the uncracked region of the beam becomes trapezoidal, as shown.

The curvature is defined as the angle that the two faces of an element with unit length make with each other after deformation. From

FLEXURAL BEHAVIOR AND STRENGTH OF PRESTRESSED-CONCRETE BEAMS

geometry

$$\varphi = \frac{1}{R} = \frac{d\theta}{ds}$$

where φ = curvature
R = radius of curvature

In Fig. 3-14 the angle $d\theta$ can be obtained from the deformation of the element as follows:

$$\frac{d\theta}{2} = \frac{\epsilon}{y}\left(\frac{ds}{2}\right) = \frac{\epsilon_1}{a}\left(\frac{ds}{2}\right)$$

where ϵ is the strain of a fiber at a distance y from the neutral axis. A transformation of the preceding expression yields the curvature φ as the ratio ϵ/y. This is a general expression that permits the calculation of curvature if the strain at a given fiber and its distance to the neutral axis are known.

In the case of linearly elastic materials, where $\epsilon = f/E$, curvature can be expressed by the familiar ratio M/EI.

3-11 ILLUSTRATIVE PROBLEM 3-1

In order to show the application of the method of analysis developed in the preceding section to a prestressed-concrete beam, the following example will be solved.

Determine the moment-curvature and moment-deflection relationships at midspan for a simply supported beam having a 9-ft span and a 10-ft length. The beam is subjected to two concentrated loads acting at

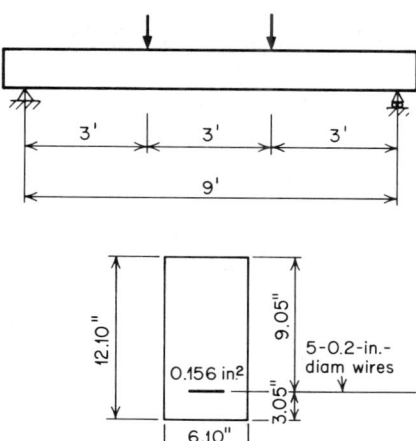

Fig. 3-15 Dimensions of the beam analyzed in Illustrative Problem 3-1.

the third points. The beam is rectangular in section, has an overall depth of 12.10 in., an effective depth of 9.05 in., and a width of 6.10 in. The prestressed wires are straight and have a cross-sectional area of 0.156 in.^{2}. The effective prestress f_{se} is 118.0 ksi, and the beam is a pretensioned bonded construction in which it may be assumed that $F_1 = F_2 = 1.0$. The dimensions of the beam are shown in Fig. 3-15.

The stress-strain diagrams for steel and concrete are shown in Figs. 3-16 and 3-17a, respectively. For this problem the stress-strain diagram for concrete may be idealized as shown in Fig. 3-17b.

Moment-curvature relationship In order to develop the moment-curvature relationship for all stages of loading, it is necessary to analyze the beam for a few cases, from no load on the beam to the load corresponding to failure. From the moment-curvature relationship, the moment-deflection or load-deflection diagram can be obtained easily.

In the following paragraphs the beam will be analyzed for several stages of loading.

a. Prestressing force only In this case it is assumed that the only force acting on the beam is the prestressing force, which is applied by prestressing wires with an effective prestress f_{se} of 118 ksi. The quantity f_{se} is defined as the stress that exists in the prestressing wire at the time of analysis of the beam and corresponds to ϵ_{se} defined previously. It is the amount of prestress after the elastic shortening and the time-dependent losses—due to shrinkage and creep in concrete and relaxation in steel—

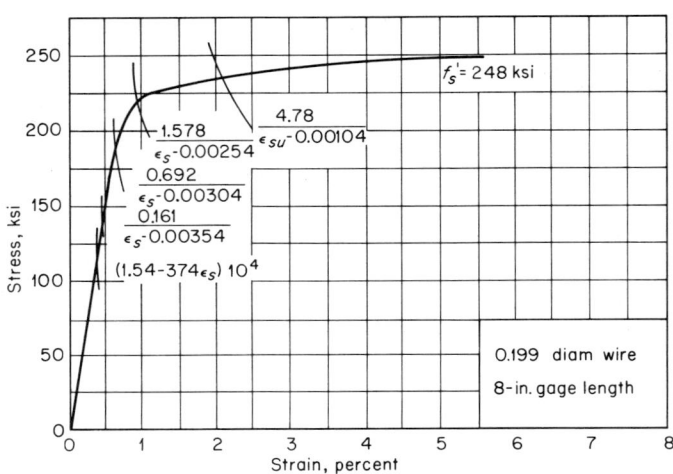

Fig. 3-16 The stress-strain relationship for the prestressing wire in Illustrative Problem 3-1.

FLEXURAL BEHAVIOR AND STRENGTH OF PRESTRESSED-CONCRETE BEAMS 73

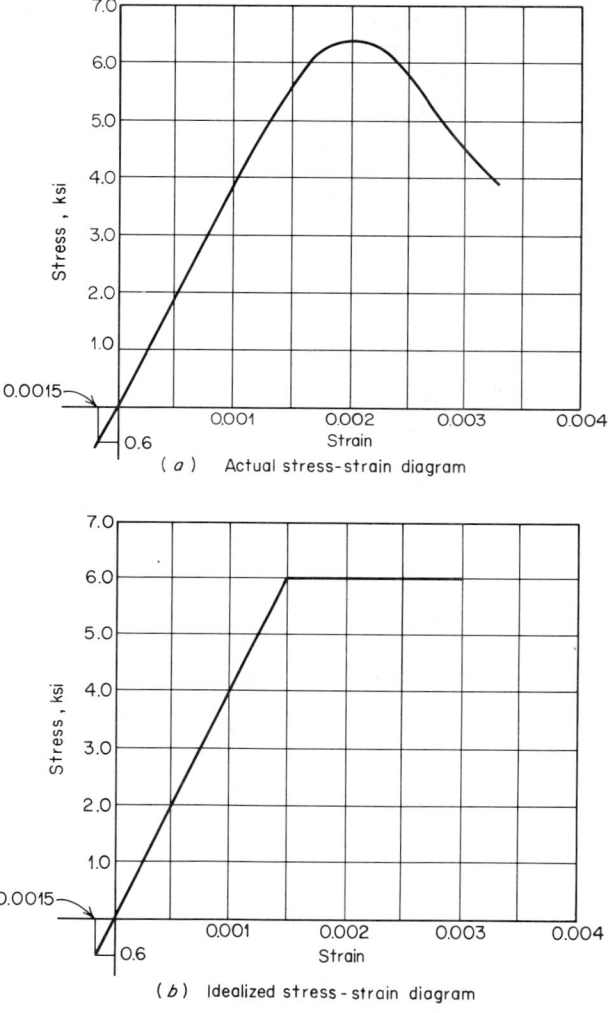

Fig. 3-17 The stress-strain diagram for concrete.

have taken place in concrete. Since the weight of the beam is always present in the type of beam considered here, the prestress will be affected. The weight of the beam causes the concrete fiber at the level of steel to lengthen, thus increasing the prestress slightly. For simplicity it has been assumed here that f_{se} does not include the effect of the weight of the beam.

For the idealized stress-strain diagram shown in Fig. 3-17b, the stress in concrete due to prestressing force will be in the elastic region of the stress-strain diagram. In this case Eqs. (3-14) and (3-15), derived in the preceding section, can be applied in computing the strain in concrete at the top and bottom fibers.

$$\epsilon_1 = \frac{P}{bhE_c}\left(2 - 6\frac{g}{h}\right) \qquad (3\text{-}14)$$

$$\epsilon_2 = \frac{P}{bhE_c}\left(6\frac{g}{h} - 4\right) \qquad (3\text{-}15)$$

In this problem we have

$$P = A_s f_{se} = 0.156 \times 118 = 18.40 \text{ kips}$$
$$b = 6.10 \text{ in.}$$
$$h = 12.10 \text{ in.}$$
$$E_c = 4000 \text{ ksi}$$

and

$$\frac{g}{h} = \frac{3.05}{12.10} = 0.252$$

Substituting these values in Eqs. (3-14) and (3-15), we have

$$\epsilon_1 = 0.0000305$$
$$\epsilon_2 = -0.000155$$
$$a = 1.99 \text{ in.}$$

where the negative sign implies a compressive strain. The stress at the top and bottom fibers may be computed by multiplying the above strains by $E_c = 4000$.

$$f_1 = 0.122 \text{ ksi}$$
$$f_2 = -0.620 \text{ ksi}$$

Figure 3-18 shows the stress and strain distribution in the beam section due to prestressing force only. The quantity ϵ_{ce}, the strain in concrete at the level of steel due to effective prestress, is the following:

$$\epsilon_{ce} = -0.0000305 \times \tfrac{7.06}{1.99} = -0.000108$$

From Fig. 3-18 it can be seen that the moment is zero and curvature is -1.53×10^{-5} in.$^{-1}$. The negative curvature implies that the top fibers of the beam are in tension.

Since in this case the stress distribution through the depth of the beam is linear, the above results may be obtained by calculating the

FLEXURAL BEHAVIOR AND STRENGTH OF PRESTRESSED-CONCRETE BEAMS

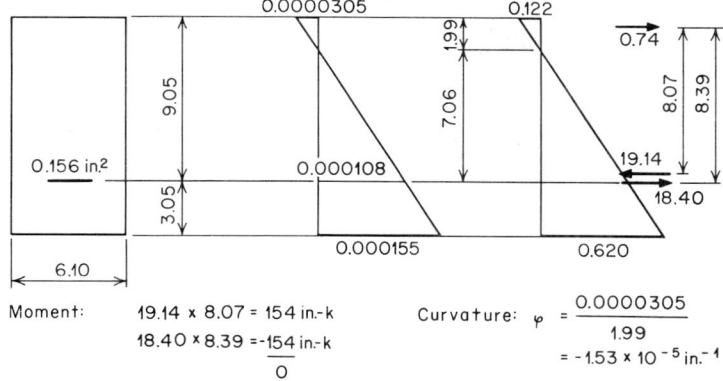

Fig. 3-18 The stress and strain in the beam due to prestressing force only, Illustrative Problem 3-1.

stresses. The section properties of the beam in this problem are the following:

A = gross cross-sectional area = $12.10 \times 6.10 = 73.8$ in.²

I = moment of inertia = $\frac{1}{12}(12.10)^3 6.10 = 900$ in.⁴

$y_t = y_b$ = depth to neutral axis = 6.05 in.

A_s = cross-sectional area of steel = 0.156 in.²

P = effective prestressing force = $118 \times 0.156 = 18.4$ kips

$\dfrac{P}{A}$ = stress due to axial load = $\dfrac{18.4}{73.8} = 0.249$ ksi

$\dfrac{Pey_t}{I}$ = stress due to bending = $\dfrac{18.4 \times 3 \times 6.05}{900} = 0.371$ ksi

Hence

$$f_1 = -\frac{P}{A} + \frac{Pey_t}{I} = -0.249 + 0.371 = 0.122 \text{ ksi}$$

$$f_2 = -\frac{P}{A} - \frac{Pey_b}{I} = -0.249 - 0.371 = -0.620 \text{ ksi}$$

The above solutions do not include the effect of concrete area replaced by steel. This effect may be taken into account by using the section properties of the net section. If, instead of the section properties of the gross cross-sectional area, the properties of the net area were considered, the quantities f_1 and f_2 would become

$f_1 = 0.124$ ksi $f_2 = -0.611$ ksi

b. Load corresponding to zero strain in concrete at the level of steel—stage 2 The unknowns in this case are ϵ_1, f_1, ϵ_s, f_s, and M. Generally, for the determination of these quantities, Eqs. (3-3), (3-10a), (3-11a), (3-8a), and (3-9a) may be solved simultaneously.

Equation (3-3) gives the strain in steel at this stage of loading:

$$\epsilon_s = \epsilon_{se} + F_1 \epsilon_{ce} \tag{3-3}$$

In this problem we have

$$\epsilon_{se} = 0.00393$$
$$\epsilon_{ce} = 0.000108$$
$$F_1 = 1.0$$

Hence

$$\epsilon_s = 0.00404$$

and from the stress-strain diagram in steel

$$f_s = 121.2 \text{ ksi}$$

and

$$P = 18.92 \text{ kips}$$

In addition, we have $b = 6.10$ in., $a = 9.05$ in., $E_c = 4000$ ksi, $\epsilon_2 = -3.05\epsilon_1/9.05 = -0.337\epsilon_1$. From Eq. (3-8b)

$$\frac{ba}{\epsilon_1} \int_{\epsilon_2}^{\epsilon_1} E_c \epsilon \, d\epsilon + P = \frac{6.10 \times 9.05 \times 4000}{2\epsilon_1} [\epsilon_1{}^2 - (-0.337\epsilon_1)^2]$$
$$+ 18.92 = 0$$

Hence

$$\epsilon_1 = -0.000192 \qquad \epsilon_2 = 0.0000648$$

Figure 3-19 shows the stress and strain distribution in the beam section when the strain in concrete at the level of steel is zero.

The bending moment in the section can be calculated from Eq. (3-9) as follows:

$$M + \frac{ba^2}{\epsilon_1{}^2} \int_{\epsilon_2}^{\epsilon_1} E_c \epsilon^2 \, d\epsilon = 0$$

Hence

$$M = -\frac{(6.10)(9.05)^2(4000)}{3}[1-(-0.337)^3](-0.000192)$$
$$= 133.7 \text{ in.-kips}$$

The curvature in this case is the following:

$$\varphi = \frac{0.000192}{9.05} = 2.12 \times 10^{-5} \text{ in.}^{-1}$$

FLEXURAL BEHAVIOR AND STRENGTH OF PRESTRESSED-CONCRETE BEAMS

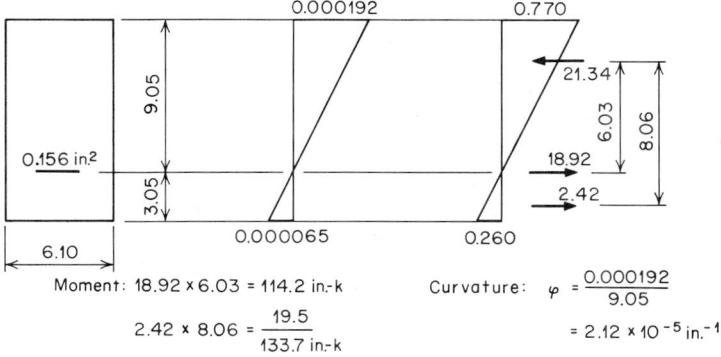

Fig. 3-19 The stress and strain in the beam when strain in concrete at the level of steel is zero, Illustrative Problem 3-1.

Since the stress distribution in concrete is linear in this case, the problem can be solved easier by replacing Eq. (3-8b) by the following expression:

$$\tfrac{1}{2}\epsilon_1 E_c ab + \tfrac{1}{2}\epsilon_2 E_c (h - a)b + P = 0$$

or

$$\tfrac{1}{2}\epsilon_1(4000)(9.05)(6.10) + \tfrac{1}{2}(-0.337\epsilon_1)(4000)(12.10 - 9.05)(6.10) + 18.92 = 0$$

and

$$\epsilon_1 = -0.000192$$

which agrees with the result obtained by the more general Eq. (3-8).

The stress distribution and the forces acting in the section are shown in Fig. 3-19, from which the moment in the section may be calculated directly.

$$\begin{aligned}(18.92)(6.03) &= 114.2 \text{ in.-kips} \\ (2.42)(8.06) &= \underline{19.5} \\ &133.7 \text{ in.-kips}\end{aligned}$$

The above moment also agrees with the one obtained by Eq. (3-9b).

c. Cracking load The cracking load in this case corresponds to a tensile stress of 0.6 ksi or a tensile strain of 0.00015 at the bottom fiber of the beam.

$$\epsilon_2 = \epsilon_t = 0.00015$$

The unknowns in this case are $\epsilon_1, f_1, \epsilon_s, f_s, a$, and M. Sufficient equations are available for the solution.

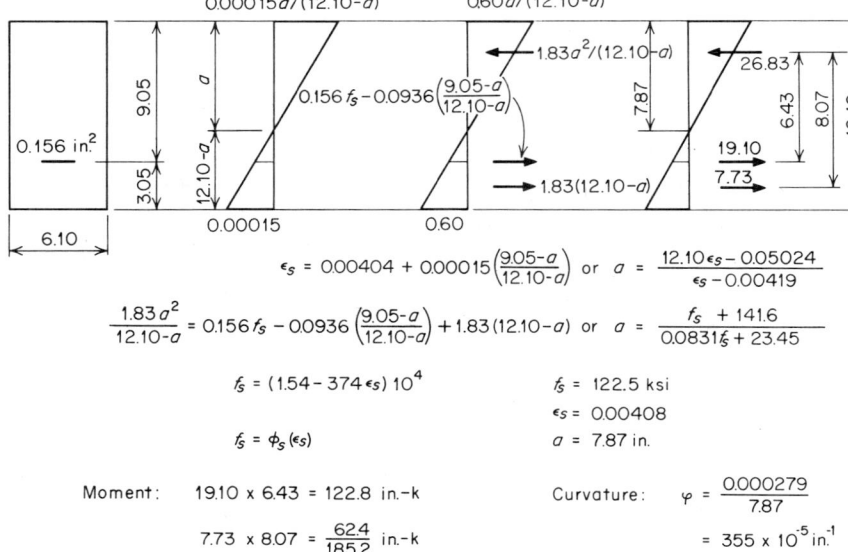

Fig. 3-20 Analysis of the beam at the cracking load, Illustrative Problem 3-1.

Figure 3-20 shows the stress and strain distribution in the section. The strain in steel can be calculated from Eq. (3-5):

$$\epsilon_s = \epsilon_{se} + F_1\epsilon_c + \frac{\epsilon_1}{a}(d-a)F_2 \qquad (3\text{-}5)$$

From a previous calculation we know that the sum of the first two terms on the right-hand side of the above equation is 0.00404; hence the total strain in steel is

$$\epsilon_s = 0.00404 + 0.00015\frac{9.05 - a}{12.10 - a}$$

Since the above expression is in terms of a, the stress in steel f_s at cracking cannot be determined directly from the stress-strain diagram for steel. In this case the above expression can be written in terms of a as follows:

$$a = 12.1 - \frac{0.00046}{0.00419 - \epsilon_s}$$

Since in this case $\epsilon_1 = -0.00015a/(12.10 - a)$ and $\epsilon_2 = 0.00015$, from Fig. 3-20 and Eq. (3-8) we have

$$\frac{ba}{\epsilon_1}\int_{\epsilon_2}^{\epsilon_1} E_c\epsilon\, d\epsilon + P = -\frac{ab}{2}\frac{0.60a}{12.10-a} + \frac{0.60b}{2}(12.10-a)$$

$$-0.00015\frac{9.05-a}{12.10-a}4000 \times 0.156 + 0.156f_s = 0$$

The term $0.00015[(9.05 - a)/(12.10 - a)]4000 \times 0.156$ in the above expression represents the force contributed by the area of concrete replaced by steel.

Eliminating a in the above two equations results in the following:

$$f_s = (1.54 - 374\epsilon_s)10^4$$

A simultaneous solution of the above equation and the equation for the stress-strain diagram in steel, $f_s = \phi_s(\epsilon_s)$, gives the quantities f_s and ϵ_s at cracking.

It is inconvenient to present the stress-strain diagram for steel by an equation. However, f_s and ϵ_s may be obtained by plotting $f_s = (1.54 - 374\epsilon_s)10^4$ with the stress-strain diagram for steel shown in Fig. 3-16. The coordinates of the point of intersection will be f_s and ϵ_s. The following quantities are the coordinates of the point of intersection of this curve with the stress-strain diagram:

$$f_s = 122.5 \text{ ksi}$$
$$\epsilon_s = 0.00408$$

The moment can be calculated from Eq. (3-9a) as

$$M = -\frac{ba^2}{\epsilon_1^2} \int_{\epsilon_t}^{\epsilon_1} E_c \epsilon^2 \, d\epsilon + P(d - a)$$

where $b = 6.10$ in., $a = 7.87$ in., $E_c = 4000$ ksi
$\epsilon_1 = -0.00015 \frac{7.87}{4.23} = 0.000279$, $\epsilon_t = 0.00015$
$P = 0.156 \times 122.5 = 19.12$ kips

Substitution of these quantities in the above equation gives the following value for moment:

$$M = 185.2 \text{ in.-kips}$$

In this case the moment may also be calculated from the forces in the section as shown in Fig. 3-20.

A simple method may be used to obtain the cracking moment from the results of stage 2. To the strain distribution shown in Fig. 3-19 we may add enough strain to produce cracking of the concrete. This amounts to $\Delta\epsilon = 0.000085$ in the bottom fiber. Because tensile stresses in the concrete have been assumed to vary linearly with strains and because compressive stresses are small and therefore also still proportional to strains, the moment ΔM that creates this additional strain distribution can be calculated by using the expression

$$\Delta M = (\Delta\epsilon) E \frac{I}{c}$$

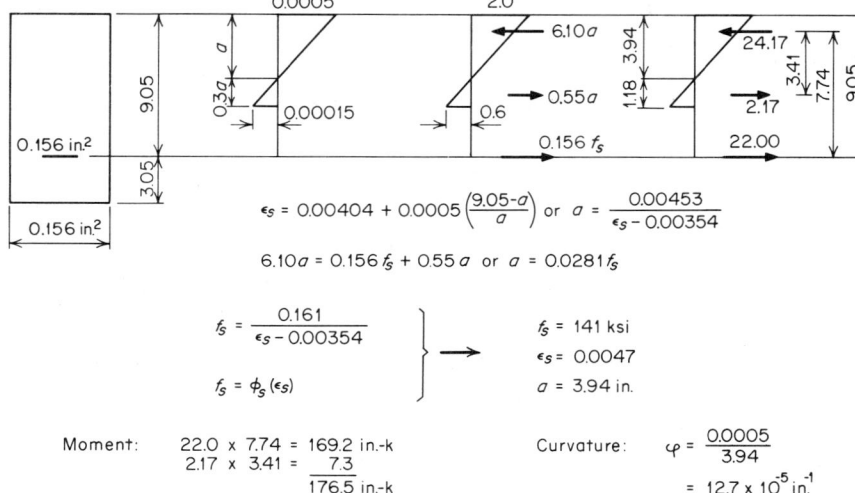

Fig. 3-21 Analysis of the beam at a load corresponding to a compressive strain of 0.0005 at the top fiber, Illustrative Problem 3-1.

which yields
$$\Delta M = (0.000085)(4000)\frac{(6.1)(12.1)^2}{6} = 50.5 \text{ kip-in.}$$

The cracking moment is the sum of the moment at the preceding stage plus ΔM. Therefore $M = 133.7 + 50.5 = 184.2$ kip-in. Because the properties of the transformed section were not used in evaluating the factor I/c, the solution obtained by this method is slightly smaller.

d. Load corresponding to a compressive strain of 0.0005 in concrete at the top fiber In this case the beam is cracked. At any stage of loading between cracking and ultimate that is defined by the stress and strain in concrete at the top fiber, there are four unknowns to be determined: $\epsilon_s, f_s, a,$ and M. These unknowns may be calculated by a simultaneous solution of Eqs. (3-5), (3-11), (3-8a), and (3-9a). Elimination of a between Eqs. (3-5) and (3-8a) results in a relation between stress and strain in steel which can be solved simultaneously with Eq. (3-11), the stress-strain diagram for steel. This solution will yield the stress and strain in steel. From Eqs. (3-8a) and (3-9a) the quantities a and M can be calculated.

Strain in steel in this case is the following:

$$\epsilon_s = 0.00404 + 0.00050\frac{9.05 - a}{a}$$

FLEXURAL BEHAVIOR AND STRENGTH OF PRESTRESSED-CONCRETE BEAMS

or

$$a = \frac{0.00453}{\epsilon_s - 0.00354}$$

Since the stress distribution is still linear, Eq. (3-8a) can be written by equating the summation of all forces acting in the section to zero, as shown in Fig. 3-21.

$$6.10a = 0.156f_s + 0.55a \quad \text{or} \quad a = 0.0281f_s$$

Elimination of a between this equation and

$$a = \frac{0.00453}{\epsilon_s - 0.00354}$$

results in

$$f_s = \frac{0.161}{\epsilon_s - 0.00354}$$

This equation relates the stress in steel with the strain in steel, and if it is solved simultaneously with Eq. (3-11), the stress-strain diagram for steel, the stress and strain in steel for this particular stage of loading can be obtained. A solution can also be obtained by plotting the above equation on the stress-strain diagram for steel. The point of intersection of two curves defines the stress and strain in steel. Figure 3-16 shows the plot of the above equation on the stress-strain diagram for

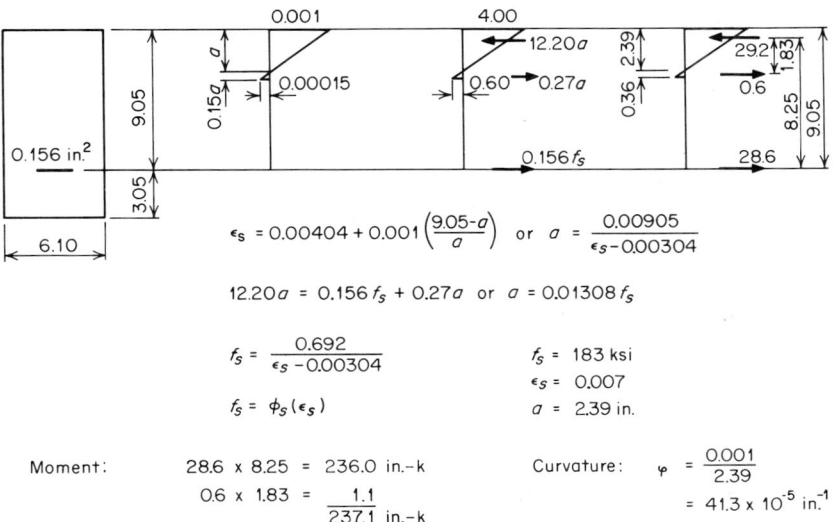

Fig. 3-22 Analysis of the beam at a load corresponding to a compressive strain of 0.001 at the top fiber, Illustrative Problem 3-1.

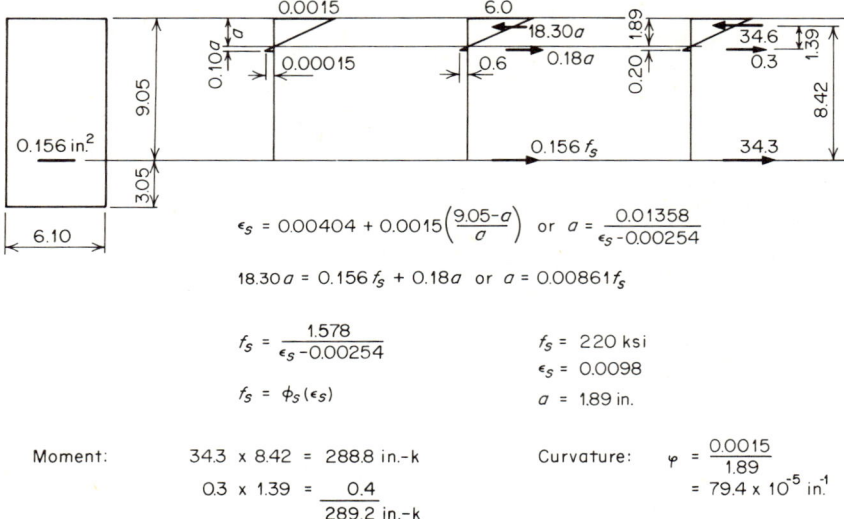

Fig. 3-23 Analysis of the beam at a load corresponding to a compressive strain of 0.0015 at the top fiber, Illustrative Problem 3-1.

steel. The coordinates of the point of intersection of two curves or the stress and strain in steel are

$f_s = 141.0$ ksi

$\epsilon_s = 0.047$

The bending moment can be calculated either from Eq. (3-9b) or directly from the forces in the section, as shown in Fig. 3-21.

From the calculations accompanying Fig. 3-21 it can be seen that in this case the moment is 176.5 in.-kips, and the curvature is 12.7×10^{-5} in.$^{-1}$.

e and f. Loading stages corresponding to $\epsilon_1 = -0.0010$ *and* $\epsilon_1 = -0.0015$ The unknowns and the equations used for their solution are identical with those used in the preceding case. All the calculations corresponding to these stages of loading are shown with Figs. 3-22 and 3-23. It can be seen that in these cases moment and curvature are the following:

$\epsilon_1 = -0.0010:$ $M = 237.1$ in.-kips

$\varphi = 41.30 \times 10^{-5}$ in.$^{-1}$

$\epsilon_1 = -0.0015:$ $M = 289.2$ in.-kips

$\varphi = 79.40 \times 10^{-5}$ in.$^{-1}$

FLEXURAL BEHAVIOR AND STRENGTH OF PRESTRESSED-CONCRETE BEAMS

g. Ultimate load In this case the strain in concrete at the top fiber has reached the limiting compressive strain of 0.003, and the tension carried by concrete is small and may be neglected. The unknowns are the stress and strain in steel, the position of the neutral axis, and the ultimate moment. From Eq. (3-5) we have

$$\epsilon_{su} = 0.00404 + \frac{0.003}{a}(9.05 - a)$$

or

$$a = \frac{0.02715}{\epsilon_{su} - 0.00104}$$

where ϵ_{su} is the strain in steel at ultimate. From Eq. (3-8)

$$\frac{ba}{-0.003}\int_0^{-0.0015} E_c \epsilon \, d\epsilon + \frac{ba}{-0.003}\int_{-0.0015}^{-0.003} -6 d\epsilon + 0.156 f_{su} = 0$$

From Fig. 3-24 the above expression can also be written as follows:

$$18.30a + 9.15a = 0.156 f_{su}$$

or

$$a = 0.00568 f_{su}$$

where f_{su} is the stress in steel at ultimate.

Elimination of a between the above equation and

$$a = \frac{0.02715}{\epsilon_{su} - 0.00104}$$

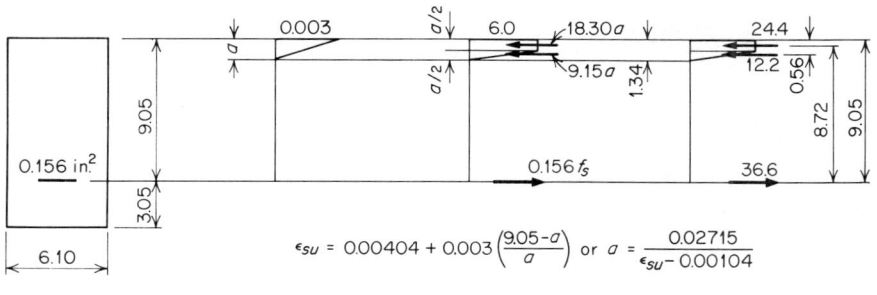

Fig. 3-24 Analysis of the beam at ultimate, Illustrative Problem 3-1.

yields

$$f_{su} = \frac{4.78}{\epsilon_{su} - 0.00104}$$

and from the stress-strain diagram in Fig. 3-9 we have

$$f_{su} = 235 \text{ ksi}$$
$$\epsilon_{su} = 0.0215$$

and

$$a = 1.34 \text{ in.}$$

From Eq. (3-9) the bending moment can be calculated,

$$M_u = 312.2 \text{ in.-kips}$$

and

$$\varphi_u = \frac{0.003}{1.34} = 224 \times 10^{-5} \text{ in.}^{-1}$$

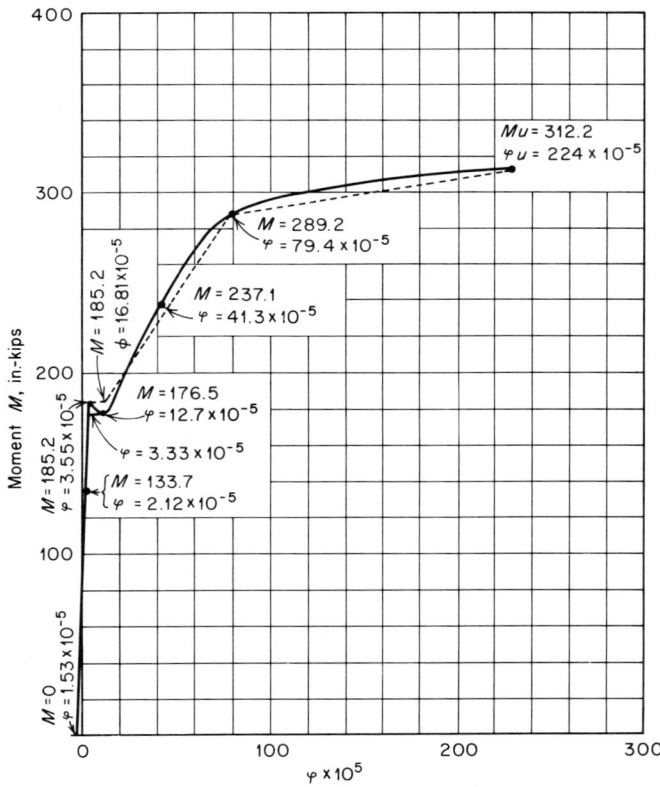

Fig. 3-25 Moment-curvature diagram, Illustrative Problem 3-1.

This stage of loading by definition is the ultimate, which in this particular problem is defined as the load corresponding to a strain of 0.003 in concrete at the top fiber. This definition does not necessarily imply that the beam will completely collapse under this moment. It means that a beam that has a strain of 0.003 at the top fiber has undergone large permanent deformation, and has no practical usefulness.

The method presented here is completely general and as shown is applicable to all stages of loading, whether the materials are in their elastic range or not. Before flexural cracking the beam may be analyzed by assuming linear distribution of stress. For any stage of load beyond the flexural cracking, the beam can be analyzed by the simultaneous solution of Eqs. (3-5), (3-11), (3-8), and (3-9) if there is no non-prestressed reinforcement in the beam—as was the case in this problem. The unknowns are f_s, ϵ_s, a, and M. It should be pointed out that since M does not appear in Eqs. (3-5), (3-11), and (3-8), they may be solved simultaneously to obtain f_s, ϵ_s, and a. From these quantities M can be calculated conveniently.

Table 3-3 shows the quantities f_s, ϵ_s, a, φ, and M for the various stages of load considered. The solid curve in Fig. 3-25 shows moment-curvature relationship as obtained from the above analysis.

A study of moment-curvature relationship indicates that the beam behaves linearly until cracking. At cracking the load drops slightly before it begins to increase with curvature. If stability of a section is defined as the positive rate of change of moment with curvature, the drop in the load indicates that the beam loses stability immediately after cracking and regains the stability with addition of load. Beyond cracking after stability is regained, the load increases with decreasing rate until ultimate is reached. If we assume that the loading for $\epsilon_s = 0.0098$ and $f_s = 220$ ksi corresponds approximately to the yield point of steel, it can be seen that there is no discontinuity in the moment-curvature relationship at this load. However, for loads beyond that corresponding to the yield point of steel any increase in moment is accompanied by an appreciable amount of deformation, which is measured by curvature.

3-12 CALCULATION OF DEFLECTION

The moment-curvature relationship developed in the preceding section is a convenient measure of behavior of a given beam section. However, it does not sufficiently describe the behavior of the beam as a structure.

For the understanding of the behavior of a beam, it is more convenient to study the relationship between midspan moment and midspan deflection. This relationship is particularly useful in the verification

Table 3-3 Summary of the results, Illustrative Problem 3-1*

Loading	$\epsilon_1 \times 10^2$	$f_1,$ ksi	$\epsilon_2 \times 10^2$	$f_2,$ ksi	$\epsilon_s \times 10^2$	$f_s,$ ksi	$a,$ in.	$\varphi \times 10^5$	$M,$ in.-kips
(a) Prestress only	0.003	0.122	−0.015	−0.620	0.393	118.0	1.99	−1.53	0
(b) Strain at level of steel = 0	−0.019	−0.770	0.0065	0.260	0.404	121.2	9.05	2.12	133.7
(c) Cracking	−0.028	−1.116	0.0150	0.600	0.408	122.5	7.87	3.55	185.2
(d) $\epsilon_1 = 0.0005$	−0.050	−2.000			0.470	141.0	3.94	12.7	176.5†
(e) $\epsilon_1 = 0.0010$	−0.100	−4.000			0.700	183.0	2.39	41.3	237.1
(f) $\epsilon_1 = 0.0015$	−0.150	−6.000			0.980	220.0	1.89	79.4	289.2
(g) Ultimate	−0.300	−6.000			2.150	235.0	1.34	224.0	312.2

* Negative quantities indicate compressive strain or stress.
† Instability after cracking.

of computed deflections by tests since it is very convenient to measure deflections at various stages of loading.

In order to calculate the deflection at any point along the span in a simply supported beam, it is necessary to know the moment-curvature relationship at any section along the beam, the span length, and the type of loading. In the discussions that follow, it is assumed that the section of the beam does not change, that is, the moment-curvature relationship is the same for all sections of the beam. Two types of loading are considered here: (1) uniform moment and (2) two concentrated loads acting at the third points of the span. In the following paragraphs, calculation of midspan deflection is discussed briefly for each type of load.

Uniform moment Figure 3-26a shows the type of loading considered. Figure 3-26b, c, and d shows the moment diagram, the curvature diagram, and the deflection along the span, respectively. Figure 3-26e shows the moment-curvature relationship for any section in the beam for the entire range of load. It is assumed that the moment-curvature relationship is the same for all the sections along the span of the beam.

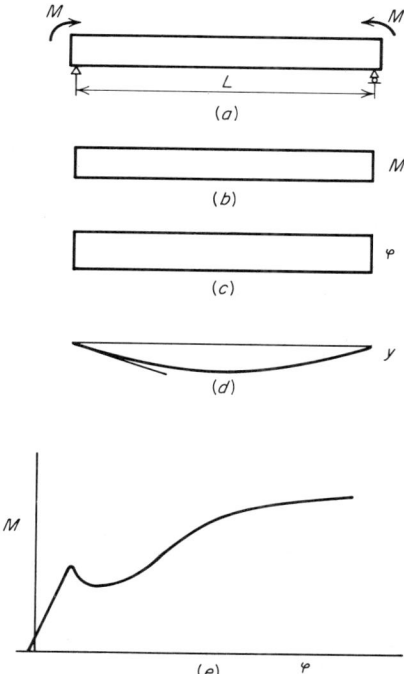

Fig. 3-26 Simply supported beam subject to uniform moment.

From Fig. 3-26c and d the midspan deflection can be written as follows:

$$\Delta = \varphi \frac{L}{2} \frac{L}{4} = \varphi \frac{L^2}{8}$$

where Δ = midspan deflection
L = span length
φ = curvature at midspan

Therefore, in order to obtain the midspan deflection when the beam is subject to a given uniform moment, it is sufficient to multiply the curvature by $L^2/8$. Hence a relationship can be established between the moment and deflection for the entire range of load by multiplying the curvature corresponding to each moment by $L^2/8$. It may be con-

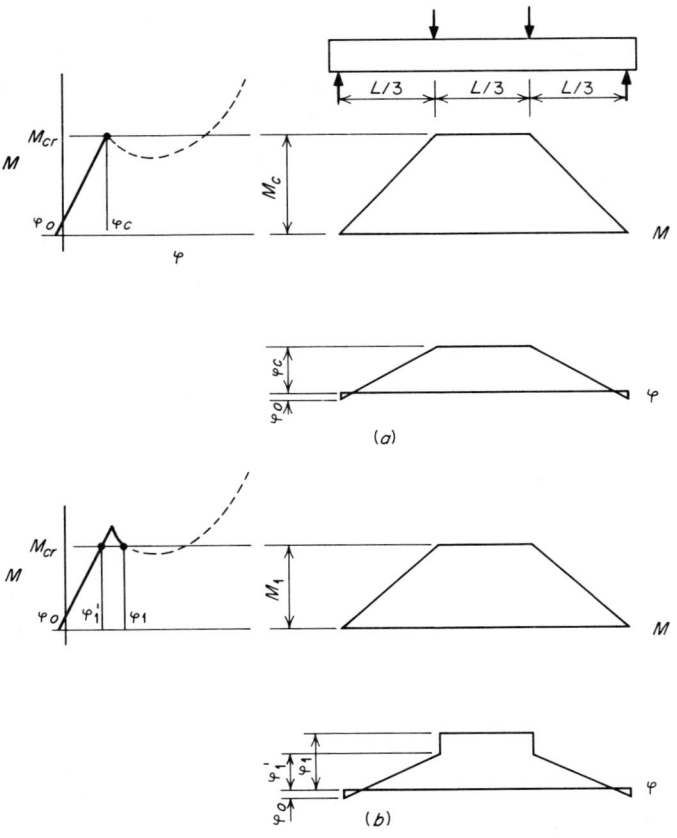

Fig. 3-27 Determination of variation of curvature along the span. Two-point loading. Loading stages (a) and (b).

FLEXURAL BEHAVIOR AND STRENGTH OF PRESTRESSED-CONCRETE BEAMS 89

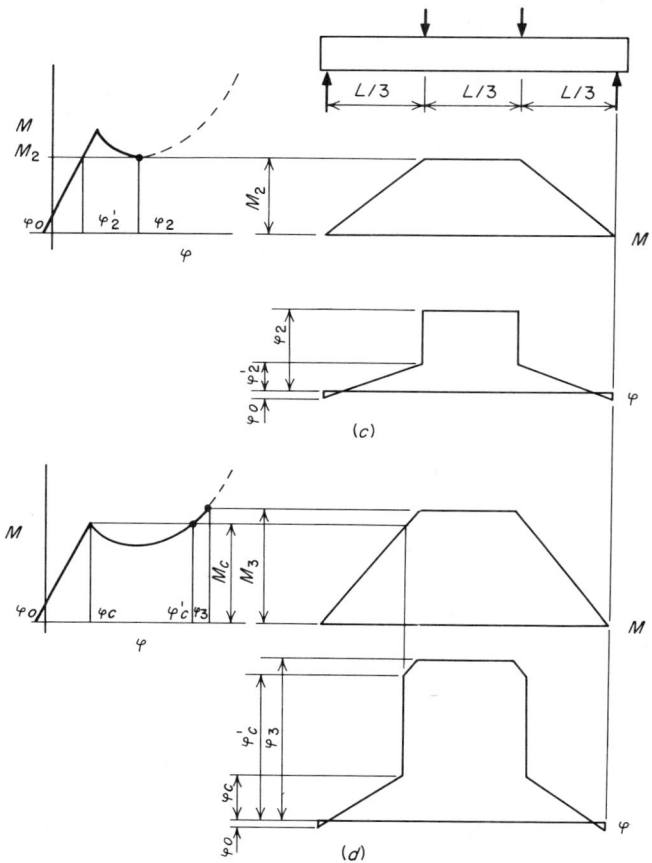

Fig. 3-27 Determination of variation of curvature along the span. Two-point loading. Loading stages (c) and (d).

cluded that when the moment is uniform, the moment-curvature relationship represents qualitatively the moment-deflection diagram.

Two concentrated loads acting at the third points In this case the moment varies along the span, and the moment diagram is trapezoidal. The curvature diagram in this case depends, therefore, upon the bending-moment diagram.

Figure 3-27a shows the moment and curvature diagrams when the moment is equal to the cracking moment M_{cr}. The moment-curvature diagram is the same as that shown in Fig. 3-26e.

Immediately after cracking, the moment in the cracked region (the middle third of the beam) begins to decrease, and the beam becomes momentarily unstable. Figure 3-27b shows the moment and curvature diagrams when the moment is M_{cr}. It can be seen that as the moment in the cracked region decreases, the curvature increases to φ_1, while in the uncracked side of each load as the moment decreases the curvature also decreases to φ_1'.

Figure 3-27c and d shows the moment and curvature diagrams for two further stages of loading. The stage shown in Fig. 3-27c corresponds to the lowest moment after cracking, beyond which the beam regains stability and begins to carry higher loads.

From these curvature diagrams the midspan deflection can be calculated as in the preceding case.

3-13 MOMENT–DEFLECTION RELATIONSHIP FOR ILLUSTRATIVE PROBLEM 3-1

The detailed calculations of deflections are shown in Fig. 3-28a through g. The method is self-explanatory and no comments seem to be necessary.

The moment-deflection diagram from the above calculations is shown by a solid line in Fig. 3-29.

The beam of Illustrative Problem 3-1 was one of many tested to failure at the University of Illinois.[5] The beam was actually designated as 0B.34.043, for which a measured load-deflection curve was obtained.

The measured load-deflection diagram is plotted by dashed lines in Fig. 3-29. A comparison of measured and computed points indicates

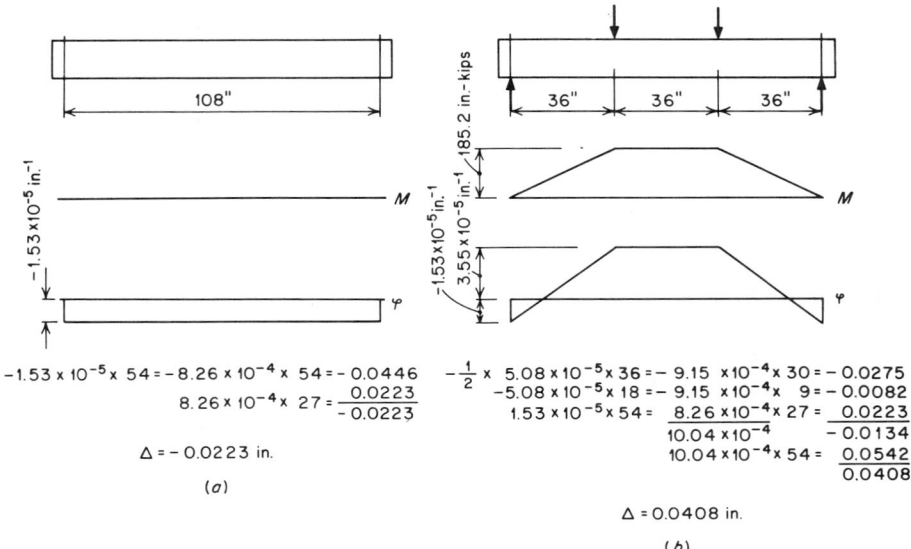

Fig. 3-28 Computation of deflections, Illustrative Problem 3-1. Loading stages (a) and (b).

FLEXURAL BEHAVIOR AND STRENGTH OF PRESTRESSED-CONCRETE BEAMS   91

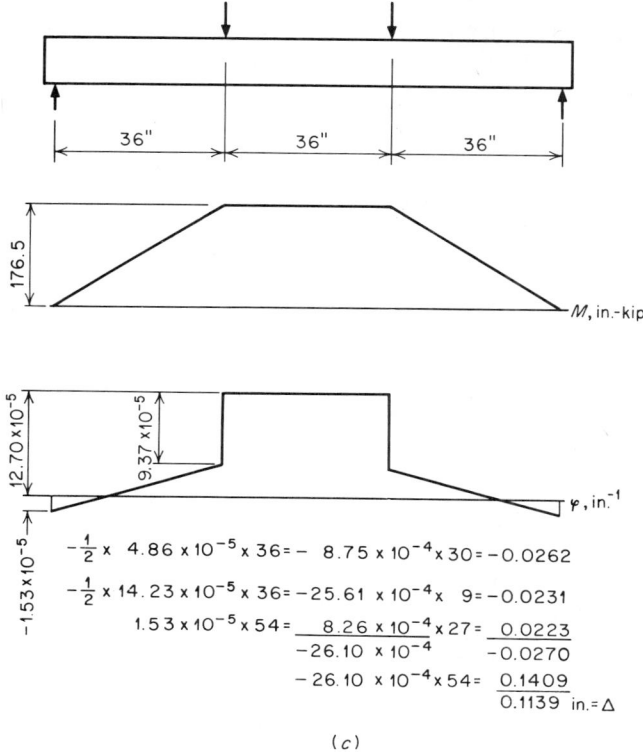

Fig. 3-28 Computation of deflections, Illustrative Problem 3-1. Loading stage (c).

that generally the agreement is close. Minor variations are expected since the analysis is made on the basis of the idealized stress-strain diagram for concrete, $F_2 = 1.0$, $\epsilon_t = 0.00015$, and $\epsilon_u = 0.003$, which are approximations. The abrupt change in load-deflection diagram at cracking often does not appear in the measurement, since the temporary loss of stability at cracking takes place very quickly.

3-14 ILLUSTRATIVE PROBLEM 3-2

Determine the moment-curvature relationship at midspan for the beam in Illustrative Problem 3-1, in which the section now has non-prestressed reinforcement consisting of two No. 3 bars with $A_{s1} = 0.22$ in.^{2} placed 1 in. from the top fiber, as shown in Fig. 3-30. The yield point is given as 50 ksi and the modulus of elasticity as 30,000 ksi. The stress-strain diagrams for concrete and prestressing steel are shown in Figs. 3-17a and

3-16, respectively. For this problem, the idealized stress-strain diagram of Fig. 3-17b may be used.

In the analysis of this problem the same stages of loading are considered as those in Illustrative Problem 3-1.

a. Prestressing force only In this particular case since the stress-strain relations are assumed linear, the above quantities may be calculated by superposition of stresses computed from the equation of bending and direct stress. The properties of the section are:

$A = 75.1$ in.^{2} $y_t = 5.95$ in.

$y_b = 6.15$ in. $I = 934.5$ in.^{4}

$\dfrac{I}{y_t} = 157$ in.^{3} $\dfrac{I}{y_b} = 152$ in.^{3}

$P = 18.4$ kips

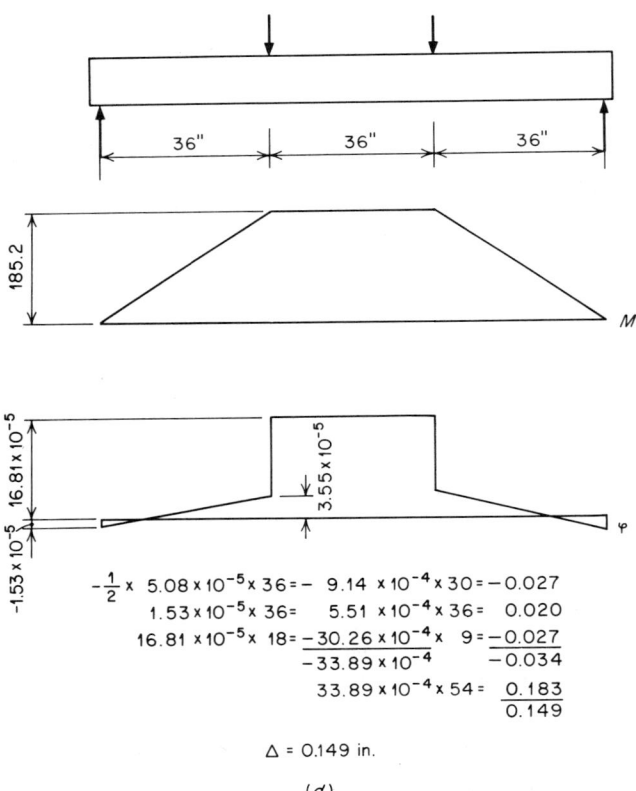

$\Delta = 0.149$ in.

(d)

Fig. 3-28 Computation of deflections, Illustrative Problem 3-1. Loading stage (d).

FLEXURAL BEHAVIOR AND STRENGTH OF PRESTRESSED-CONCRETE BEAMS

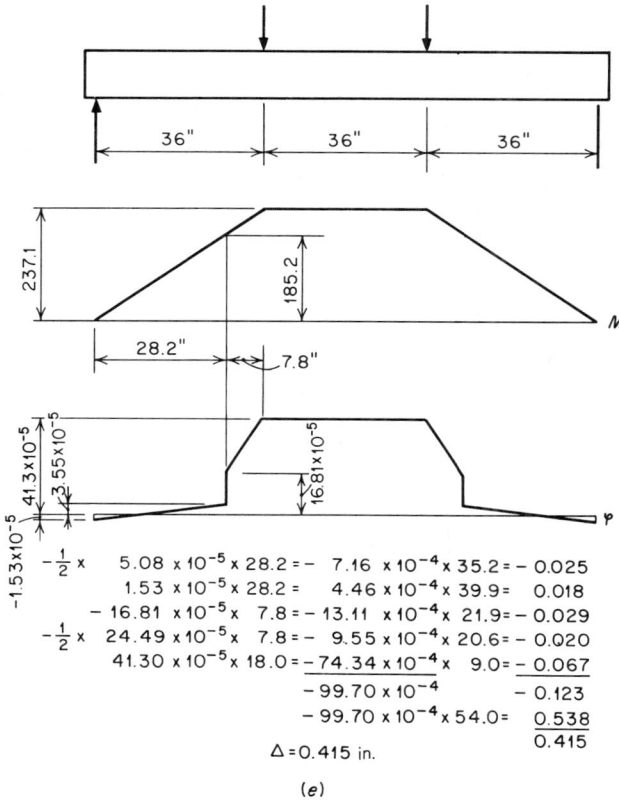

Fig. 3-28 Computation of deflections, Illustrative Problem 3-1. Loading stage (e).

The stresses at the top and bottom fiber are:

$$f_1 = -\frac{18.4}{75.1} + \frac{18.4 \times 3.10}{934.5} \times 5.95 = 0.119 \text{ ksi}$$

$$f_2 = -\frac{18.4}{75.1} - \frac{18.4 \times 3.10}{934.5} \times 6.15 = -0.621 \text{ ksi}$$

Figure 3-31a shows the forces in the section at this stage of loading. The moment in the section is chosen to be zero, and the curvature is calculated.

b. Load corresponding to zero strain in concrete at the level of steel

Figure 3-31b shows the stress and strain distribution in this case. From the equilibrium of horizontal forces the strains at the top and bottom fibers may be computed directly since, in this case, $a = d = 9.05$ in. The calculations accompanying the figure are self-explanatory and show the moment and curvature.

c. Cracking load The calculations accompanying Fig. 3-31c show the solution of the problem is similar to that for the cracking load in Illustrative Problem 3-1. The equation of equilibrium of horizontal forces includes the effect of compressive force in the non-prestressed steel. In this solution it is assumed that stress in non-prestressed steel is at the elastic range.

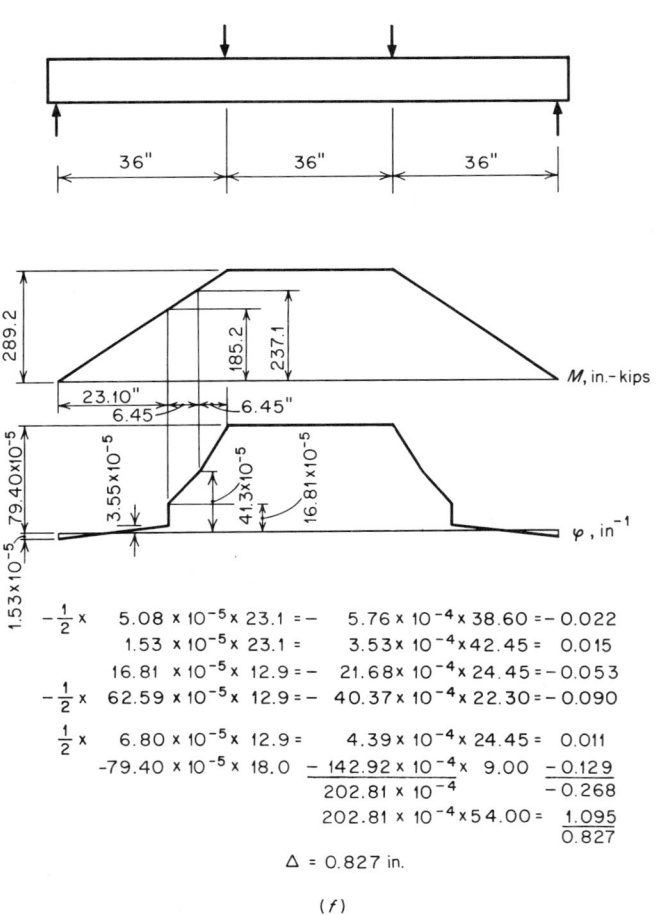

Fig. 3-28 Computation of deflections, Illustrative Problem 3-1. Loading stage (f).

FLEXURAL BEHAVIOR AND STRENGTH OF PRESTRESSED-CONCRETE BEAMS

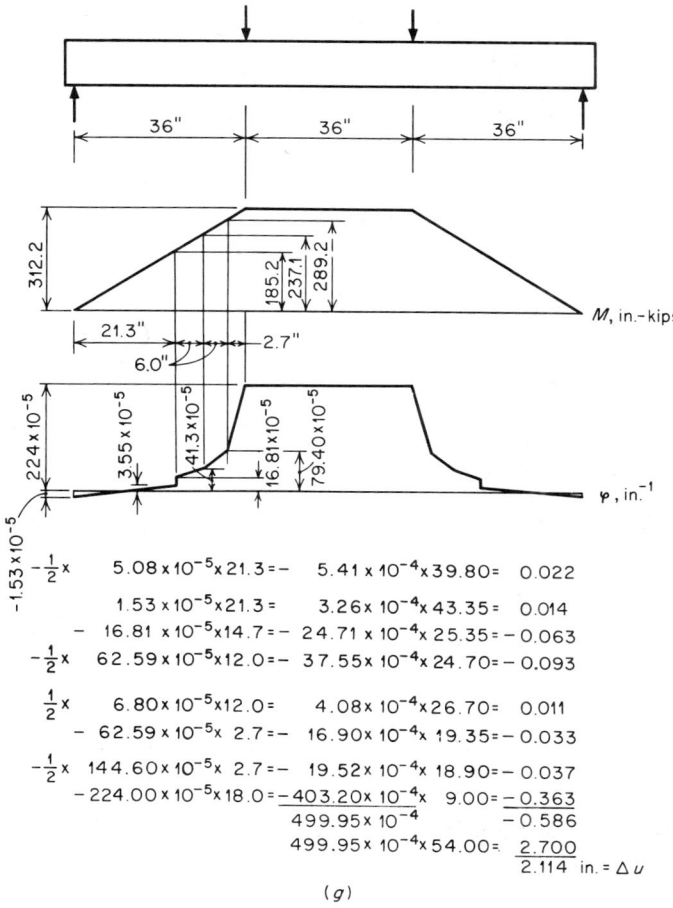

Fig. 3-28 Computation of deflections, Illustrative Problem 3-1. Loading stage (g).

d. Load corresponding to a compressive strain of 0.0005 in concrete at the top fiber
Strain in steel in this case is

$$\epsilon_s = 0.00404 + 0.00050 \frac{9.05 - a}{a}$$

or

$$a = \frac{0.00453}{\epsilon_s - 0.00354}$$

From Fig. 3-31d the equilibrium of horizontal forces yields

$$6.10a + 0.22f_{s1} = 0.156f_s + 0.55a$$

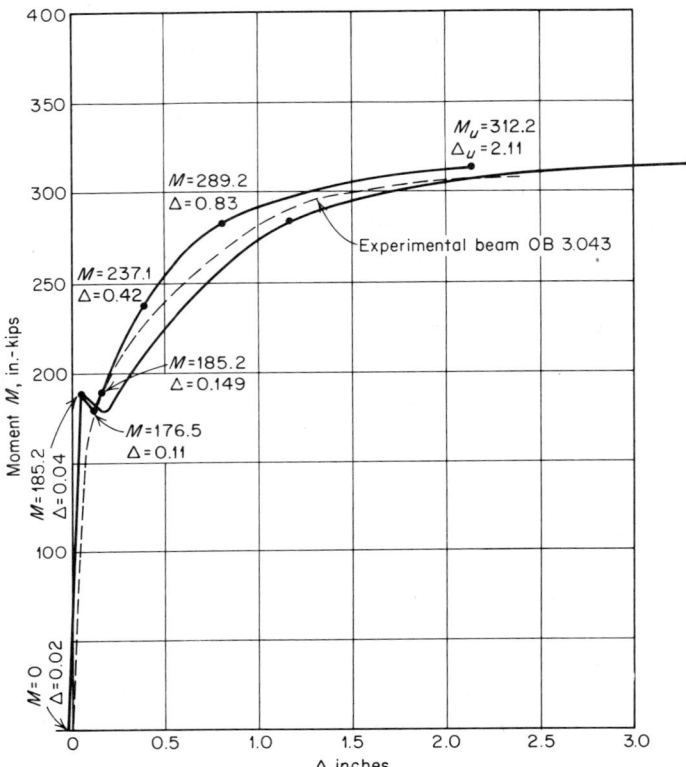

Fig. 3-29 Moment-deflection diagrams, Illustrative Problem 3-1.

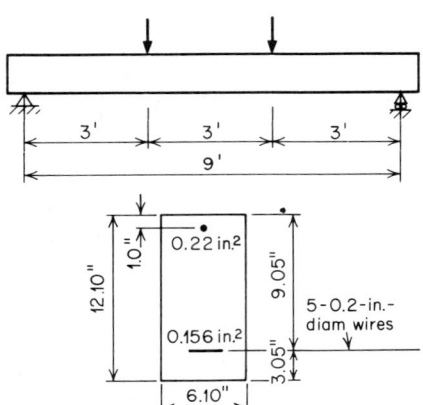

Fig. 3-30 Dimensions of the beam analyzed in Illustrative Problem 3-2.

FLEXURAL BEHAVIOR AND STRENGTH OF PRESTRESSED-CONCRETE BEAMS

or

$$a = 0.0281 f_s - 0.0396 f_{s1}$$

Elimination of a in the above two equations results in

$$f_s = \frac{0.161}{\epsilon_s - 0.00354} + 1.41 f_{s1}$$

For any value of f_{s1} the point of intersection of the above equation with the stress-strain diagram for steel yields the correct values of f_s and ϵ_s. However, the correct value of f_{s1} should satisfy Eqs. (3-6) and (3-12); that is, it should satisfy the following expression if the strain in the non-prestressed steel is in the elastic range:

$$f_{s1} = (0.0005) \frac{a-1}{a} 30{,}000$$

The solution of the problem may be carried out by an iterative process:

First trial Assume $f_{s1} = 10$ ksi. The stress in steel can be expressed as follows:

$$f_s = \frac{0.161}{\epsilon_s - 0.00354} + 14.1$$

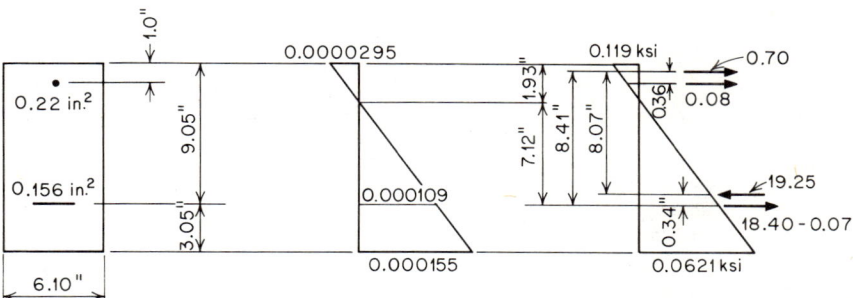

(a) Prestressing force only

Fig. 3-31 Analysis of the beam with non-prestressed reinforcement, Illustrative Problem 3-2. Loading stage (a).

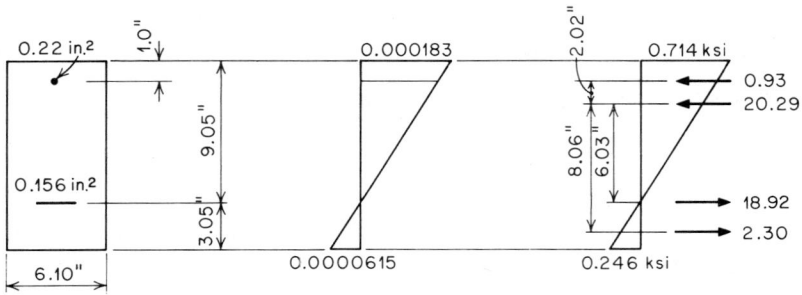

$$\epsilon_s = \epsilon_{se} + \epsilon_{ce} = 0.00393 + 0.00011 = 0.00404$$

Moment: 2.30 × 8.06 = 18.5   Curvature: $\varphi = \dfrac{0.000183}{9.05}$
         18.92 × 6.03 = 114.2
         0.93   2.02 =   1.9   = 2.02×10^{-5} in.$^{-1}$
                        ─────
                        134.6 in.-k

(b) Strain in concrete at the level of steel = 0

Fig. 3-31 Analysis of the beam with non-prestressed reinforcement, Illustrative Problem 3-2. Loading stage (b).

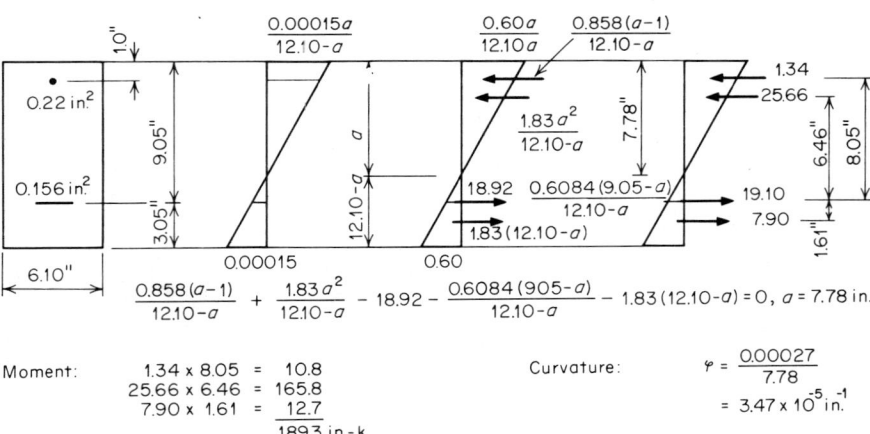

$$\dfrac{0.858(a-1)}{12.10-a} + \dfrac{1.83\,a^2}{12.10-a} - 18.92 - \dfrac{0.6084(905-a)}{12.10-a} - 1.83(12.10-a) = 0, \quad a = 7.78 \text{ in.}$$

Moment: 1.34 × 8.05 = 10.8   Curvature: $\varphi = \dfrac{0.00027}{7.78}$
         25.66 × 6.46 = 165.8
         7.90 × 1.61 = 12.7   = 3.47×10^{-5} in.$^{-1}$
                       ─────
                       189.3 in.-k

(c) Cracking load

Fig. 3-31 Analysis of the beam with non-prestressed reinforcement, Illustrative Problem 3-2. Loading stage (c).

FLEXURAL BEHAVIOR AND STRENGTH OF PRESTRESSED-CONCRETE BEAMS

Solution of the above expression with the stress-strain diagram for steel yields the following:

$$f_s = 143.5 \text{ ksi}$$
$$\epsilon_s = 0.00478$$
$$a = 3.62 \text{ in.}$$

and

$$f_{s1} = 0.0005 \tfrac{2 \cdot 62}{3 \cdot 62} \, 30{,}000 = 10.83 \text{ ksi} > 10.0$$

Hence, the assumption of $f_{s1} = 10$ ksi is not exactly correct but is close to the correct answer.

Second trial Assume $f_{s1} = 10.83$ ksi. The stress in prestressed steel is

$$f_s = \frac{0.161}{\epsilon_s - 0.00354} + 15.3$$

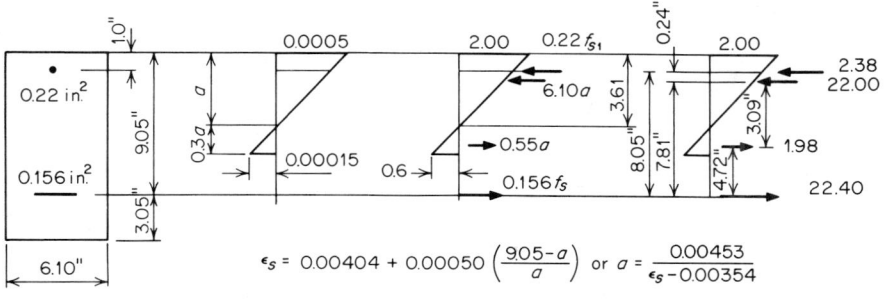

(d) Strain in concrete at the top fiber = 0.0005

Fig. 3-31 Analysis of the beam with non-prestressed reinforcement, Illustrative Problem 3-2. Loading stage (d).

This expression with the stress-strain diagram gives

$$f_s = 143.6 \text{ ksi}$$
$$\epsilon_s = 0.00479$$
$$a = 3.61 \text{ in.}$$

and

$$f_{s1} = 0.0005 \tfrac{2 \cdot 61}{3 \cdot 61} \, 30{,}000 = 10.83 \text{ ksi}$$

which agrees with the original assumption. Figure 3-31d shows the calculation of moment and curvature for this stage of loading.

The problem can also be solved by assuming a value for a and calculating f_s and f_{s1} from Eqs. (3-5), (3-6), (3-11), and (3-12). Substitution of these values of f_s and f_{s1} in Eq. (3-8) yields a new value of a which if correct should be the same as that assumed. If this value of a does not agree with the one assumed, the process should be repeated until convergence is achieved. In the following paragraphs this solution is summarized.

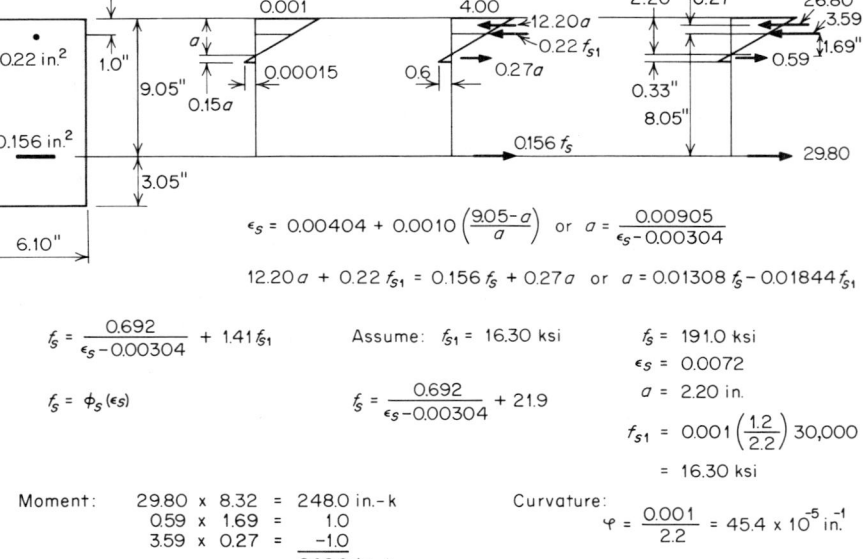

(e) Strain in concrete at the top fiber = 0.001

Fig. 3-31 Analysis of the beam with non-prestressed reinforcement, Illustrative Problem 3-2. Loading stage (e).

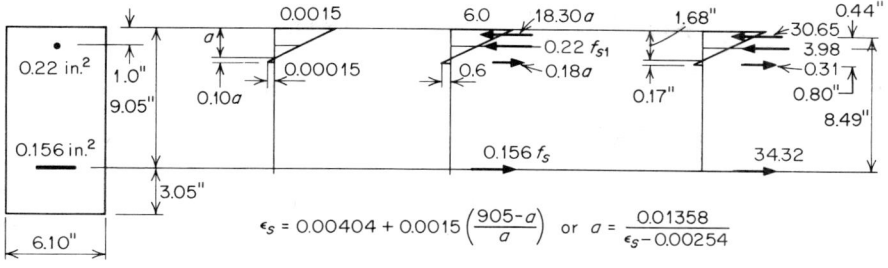

(f) Strain in concrete at the top fiber = 0.0015

Fig. 3-31 Analysis of the beam with non-prestressed reinforcement, Illustrative Problem 3-2. Loading stage (f).

First trial Assume $a = 5$ in. From Eqs. (3-5) and (3-6) we have

$$\epsilon_s = 0.00404 + 0.00050 \frac{9.05 - 5.0}{5.0} = 0.00443$$

$$\epsilon_{s1} = 0.0005 \tfrac{4}{5} = 0.0004$$

and from the stress-strain diagrams for the prestressed and non-prestressed steel we have

$$f_s = 133 \text{ ksi} \qquad f_{s1} = 12 \text{ ksi}$$

Substitution of these quantities in Eq. (3-8) yields the following:

$$a = 0.0281 \times 133 - 0.0396 \times 12 = 3.27 \text{ in.}$$

which does not agree with the assumed value of $a = 5$ in.

Second trial Assume $a = 3.27$ in.; we have

$$\epsilon_s = 0.00404 + 0.005 \tfrac{5.78}{3.27} = 0.004 + 0.00088 \equiv 0.00492$$

$$\epsilon_{s1} = 0.005 \tfrac{2.27}{3.27} = 0.00035$$

and

$$f_s = 145 \text{ ksi} \qquad f_{s1} = 10.5 \text{ ksi}$$
$$a = 0.0281 \times 145 - 0.0396 \times 10.5 = 3.66 \text{ in.}$$

which still does not agree with $a = 3.27$ in.

In the third trial we may assume $a = 3.66$ in. If the process is continued, a will converge to 3.61 in.

e to g. Higher stages of loading Figure 3-31e and f shows the stress and strain distributions in the section for load corresponding to a compressive strain of 0.001 and 0.0015, respectively. The calculation of moment and curvature accompanies the diagrams. Figure 3-31g shows the calculations for the ultimate load, for which the contribution of the tensile strength of concrete has been neglected.

Table 3-4 shows a summary of the quantities ϵ_1, f_1, ϵ_2, f_2, ϵ_s, f_s, ϵ_{s1}, f_{s1}, a, φ, and M for all stages of load considered.

Figure 3-31h shows the moment-curvature relationship for the entire range of load.

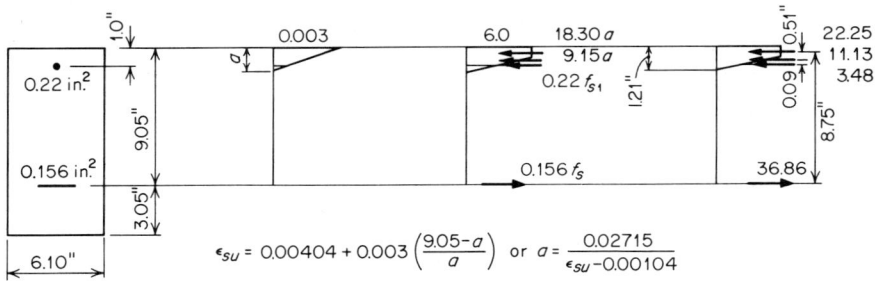

Fig. 3-31 Analysis of the beam with non-prestressed reinforcement, Illustrative Problem 3-2. Loading stage (g).

FLEXURAL BEHAVIOR AND STRENGTH OF PRESTRESSED-CONCRETE BEAMS

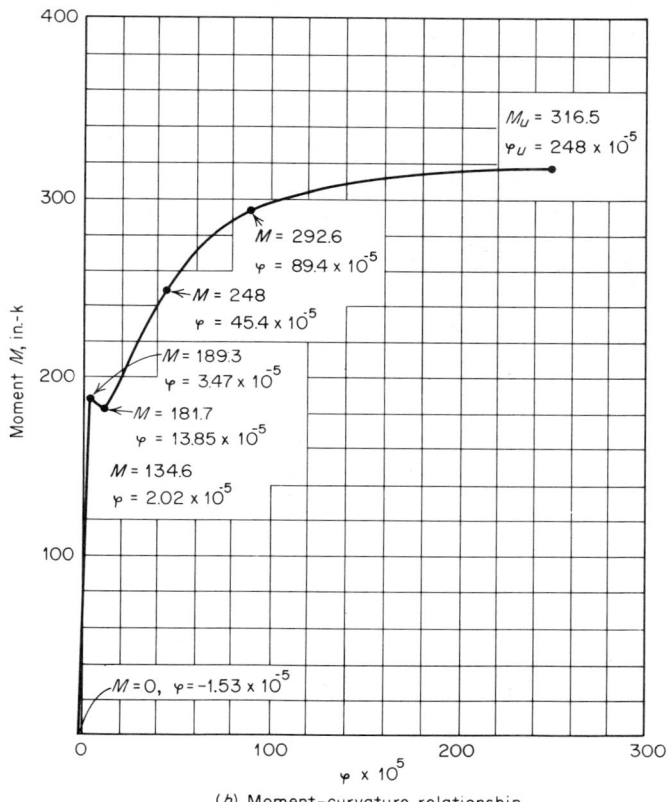

(*h*) Moment–curvature relationship

Fig. 3-31 Analysis of the beam with non-prestressed reinforcement, Illustrative Problem 3-2. Loading stage (*b*).

The effect of non-prestressed steel at the top may be studied by comparing the moment-curvature relationship for Illustrative Problem 3-1 in Fig. 3-25 with that for Illustrative Problem 3-2 in Fig. 3-31*h*. It can be seen that the strength of the section increases only slightly, while the ductility of the section measured by the curvature increases appreciably.

The moment-deflection curve may be obtained as shown for Illustrative Problem 3-1. Evidently the presence of non-prestressed steel at the top will result in larger deflection at ultimate for Illustrative Problem 3-2.

The analysis presented here does not take into account the effect of the area of concrete replaced by steel for the stages of load beyond cracking. This effect is small, but may be included conveniently if accurate results are required.

Table 3-4 Summary of results, Illustrative Problem 3-2*

Loading	$\epsilon_1 \times 10^2$	$f_1,$ ksi	$\epsilon_2 \times 10^2$	$f_2,$ ksi	$\epsilon_s \times 10^2$	$f_s,$ ksi	$\epsilon_{s1} \times 10^2$	$f_{s1},$ ksi	$a,$ in.	$\varphi \times 10^5,$ in.$^{-1}$	$M,$ in.-kips
(a) Prestress only	0.003	0.119	−0.016	−0.621	0.393	118.0	0.0014	0.43	1.93	−1.53	0
(b) Strain at level of steel = 0	−0.018	−0.714	0.0062	0.246	0.404	121.2	−0.0163	−4.88	9.05	2.02	134.6
(c) Cracking	−0.027	−1.080	0.015	0.600	0.408	122.5	−0.0235	−7.05	7.78	3.47	189.3
(d) $\epsilon_1 = 0.0005$	−0.050	−2.000			0.479	143.6	−0.0361	−10.83	3.61	13.85	181.7
(e) $\epsilon_1 = 0.0010$	−0.100	−4.000			0.720	191.0	−0.0542	−16.30	2.20	45.4	248.0
(f) $\epsilon_1 = 0.0015$	−0.150	−6.000			1.06	220.0	−0.0603	−18.10	1.68	89.4	292.6
(g) Ultimate	−0.300	−6.000			2.35	236.0	−0.0526	−15.80	1.21	248.	316.5

* Negative quantities are either compressive strain or stress.

FLEXURAL BEHAVIOR AND STRENGTH OF PRESTRESSED-CONCRETE BEAMS

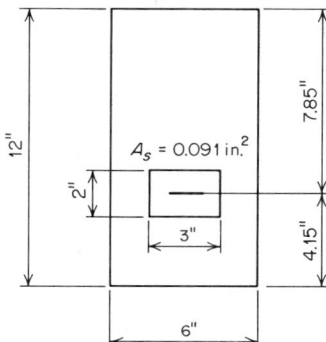

Fig. 3-32 Unbonded beam, Illustrative Problem 3-3.

3-15 ILLUSTRATIVE PROBLEM 3-3

Determine the moment-curvature relationship in the unbonded beam shown in Fig. 3-32. The beam is similar to that in Illustrative Problem 3-1 except that the prestressed reinforcement is unbonded and is placed in a 2- by 3-in. opening. The stress-strain diagrams for steel and concrete are those shown in Figs. 3-5 and 3-17b, respectively. The quantity F_2 decreases with increase in load after cracking, and for this problem the following values of F_2 may be used for each given top fiber strain indicated:

ϵ_1	F_2
0	$F_1 = 0.67$
0.0005	0.60
0.0010	0.53
0.0015	0.47
0.0020	0.40
0.0025	0.33
$\epsilon_u = 0.0030$	0.27

It should be pointed out that the values of F_2 in the above problem are not intended to be real quantities obtained from tests. They are given here to show the effect that reduction of F_2 has on the moment-curvature relationship of the beam.

Since this problem is similar to Illustrative Problem 3-1, the detailed description of its solution has been omitted. However, the results have been summarized in Table 3-5.

A comparison of the results presented in Table 3-5 with those in Table 3-3 indicates that reduction of F_2 results in a reduction in strain in steel and in the moment.

Table 3-5 Summary of results, Illustrative Problem 3-3*

Loading	F_2	$\epsilon_1 \times 10^2$	f_1, ksi	$\epsilon_2 \times 10^2$	f_2, ksi	$\epsilon_s \times 10^2$	f_s, ksi	a, in.	$\varphi \times 10^5$	M, in.-kips
(a) Prestress only	$F_1 = 0.67$	—	−0.013	−0.0079	−0.317	0.393	118.0	—	−0.63	0
(b) Strain at level of steel = 0	$F_1 = 0.67$	−0.016	−0.65	0.0085	0.339	0.396	118.5	7.85	2.04	90.4
(c) Cracking	0.67	−0.022	−0.89	0.0150	0.600	0.400	120.0	7.16	3.09	133.7
(d) $\epsilon_1 = 0.0005$	0.60	−0.050	−2.00			0.439	135.4	4.12	12.14	168.4
(e) $\epsilon_1 = 0.0010$	0.53	−0.100	−4.00			0.563	181.0	2.26	44.20	210.3
(f) $\epsilon_1 = 0.0015$	0.47	−0.150	−6.00			0.699	190.0	1.73	86.78	251.7
(g) Ultimate	0.27	−0.300	−6.00			0.875	216.0	1.31	229.09	288.7

* Negative quantities indicate compressive strain or stress.

3-16 INFLUENCE OF THE VARIABLES ON THE BEHAVIOR OF A PRESTRESSED-CONCRETE BEAM

From a study of Eqs. (3-5) through (3-14) discussed in the preceding sections, several observations and conclusions may be made.

1. An increase in the amount of prestressing steel for a given beam results in an increase in moment that the beam can resist, and a decrease in the ductility of the beam. The instability of the beam at the cracking load becomes more severe for smaller amounts of prestressing steel. It is conceivable to have a beam with such a small amount of prestressed steel that the moment at cracking is never regained after cracking—a condition similar to that in a plain concrete beam. It is also possible to have sufficiently large amounts of steel that the beam remains stable through cracking.
2. A bonded beam shows a more ductile behavior than an equivalent unbonded beam. Since the strain compatibility factors are smaller in unbonded beams, the strain in steel increases slowly with load in the region of maximum moment. It is likely that an unbonded beam may fail in a brittle manner.
3. Non-prestressed reinforcement at the top of the beam does not increase the moment capacity of a ductile section appreciably. However, it causes the beam to become more ductile.
4. Non-prestressed reinforcement at the bottom of the beam increases the moment-carrying capacity of the beam, but causes the beam to become less ductile. It also makes the beam more stable at cracking.
5. An increase in ϵ_{se}, or the level of prestress, increases the cracking moment of the section. However, variation of ϵ_{se} has no influence on the strength or ultimate moment, when the ratio p/f_{cu} is small, where p is the percentage of steel reinforcement and f_{cu} the effective strength of the concrete in the comparison zone of the beam. When the ratio p/f_{cu} is large, any reduction in ϵ_{se} affects the strength of the beam appreciably.
6. The flexural strength of a beam is based upon ϵ_u, the limiting strain in concrete at the top fiber. An increase in ϵ_u increases the strain in steel at failure, and hence increases the ductility of the beam. An increase in ϵ_u usually causes an increase in the flexural strength of the beam, which depends on the amount of prestressed steel.
7. The behavior of the beam depends on the stress-strain diagram for the materials. Idealization of the stress-strain diagram in concrete has a comparatively small influence on the behavior of the beam. The shape of the stress-strain diagram in steel influences the stress in steel, hence the moment and ductility of the beam.

3-17 ILLUSTRATIVE PROBLEM 3-4

Determine the moment-curvature relationship in the beam of Illustrative Problem 3-1, using the actual stress-strain diagram in concrete, which is shown in Fig. 3-17a and is approximated by the diagram shown in Fig. 3-33a.

The abscissa for the compression part of the stress-strain diagram is divided into 14 equal segments, each equal to a strain of 0.00025. The segments are numbered from left to right, and n designates the number of each.

The moment-curvature relationship may be obtained conveniently by arranging the calculations in a tabular form. Tables 3-6a and 3-6b show the entire calculation for the determination of moment-curvature relationship for this problem.

Column 1 of Table 3-6a shows the designation of each segment. Columns 2 and 3 are the stress and strain at the right end of each segment, respectively. Column 4 shows the area of each segment, which is calculated by taking the average of stress in each side of the segment (and multiplying it by 1). For segment 2, for example, we have

$$A_n = (f_n + f_{(n-1)})\tfrac{1}{2} \qquad A_2 = (f_2 + f_1)\frac{1}{2} = \frac{2+1}{2} = 1.50$$

Column 5 is the cumulative sum of column 4 from which the area under the tension part of the stress-strain diagram is subtracted. The maxi-

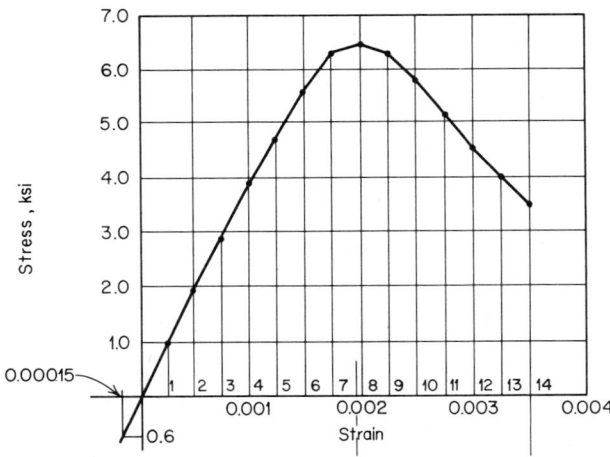

Fig. 3-33 (a) Stress-strain diagram for concrete, Illustrative Problem 3-4.

Table 3-6a Determination of moment-curvature relationship

(1)	(2)	(3)	(4)	(5)	(6)	(7)	(8)	(9)
n	f_n, ksi	$\epsilon_c \times 10^5$	$A_n = (f_n + f_{(n-1)})\frac{1}{2}$	$\sum_1^n A_i - 0.18$	$\bar{f}_n = \frac{1}{n}\sum_1^n (A_i - 0.18)$	$\bar{x}_i = \frac{f_{i-1} + 2f_i}{6A_i}$	$\bar{m}_n = 0.072 + \sum_1^n A_i(\bar{x}_i + i - 1)$	$1 - \frac{\bar{m}_n}{n^2 \bar{f}_n}$
0	0	0	0					
1	1.00	25	0.50	0.32	0.32	0.667	0.4	0.267
2	2.00	50	1.50	1.82	0.91	0.556	2.7	0.247
3	2.95	75	2.48	4.30	1.43	0.532	9.0	0.303
4	3.90	100	3.43	7.73	1.93	0.523	21.1	0.316
5	4.75	125	4.33	12.06	2.41	0.513	40.7	0.324
6	5.60	150	5.18	17.24	2.88	0.513	69.3	0.331
7	6.30	175	5.95	23.19	3.31	0.510	108.1	0.335
8	6.40	200	6.35	29.54	3.69	0.502	155.7	0.342
9	6.30	225	6.35	35.89	3.99	0.499	209.7	0.352
10	5.80	250	6.05	41.94	4.19	0.493	267.1	0.387
11	5.15	275	5.48	47.42	4.31	0.490	314.0	0.398
12	4.50	300	4.83	52.25	4.35	0.489	369.5	0.411
13	4.00	325	4.25	56.50	4.35	0.490	422.6	0.420
14	3.50	350	3.75	60.25	4.30	0.489	473.2	0.439

Table 3-6b Determination of moment-curvature relationship

(1)	(2)	(3)	(4)	(5)	(6)	(7)	(8)	(9)	(10)	(11)
n	f_n, ksi	$\epsilon_c \times 10^5$	$f_s = \dfrac{353.5\bar{f}_n \epsilon_c}{\epsilon_s - 0.00404 + \epsilon_c}$	f_s, ksi	ϵ_s	$A_s f_s$, kips	a, in.	Moment arm, in.	M, in.-kips	$\varphi \times 10^5$ in.$^{-1}$
1	1.00	25		141	0.0047	27.0	3.96	8.08	178.0	12.6
2	2.00	50	$0.161/(\epsilon_s - 0.00354)$	166	0.0055	25.9	2.96	8.15	211.0	25.3
3	2.95	75	$0.382/(\epsilon_s - 0.00329)$	178	0.0069	27.8	2.36	8.30	231.0	42.4
4	3.90	100	$0.683/(\epsilon_s - 0.00304)$	200	0.0081	31.2	2.12	8.36	261.0	59.0
5	4.75	125	$1.067/(\epsilon_s - 0.00279)$	214	0.0097	33.4	1.90	8.42	281.0	79.1
6	5.60	150	$1.529/(\epsilon_s - 0.00254)$	222	0.0115	34.6	1.71	8.48	293.0	102.4
7	6.30	175	$2.040/(\epsilon_s - 0.00229)$	226	0.0136	35.2	1.57	8.51	300.0	127.3
8	6.40	200	$2.615/(\epsilon_s - 0.00204)$	228	0.0157	35.5	1.46	8.53	302.0	154.0
9	6.30	225	$3.170/(\epsilon_s - 0.00179)$	230	0.0176	35.9	1.41	8.50	305.0	177.5
10	5.80	250	$3.700/(\epsilon_s - 0.00154)$	231	0.0195	36.0	1.38	8.50	306.0	199.0
11	5.15	275	$4.175/(\epsilon_s - 0.00129)$	233	0.0208	36.4	1.37	8.49	308.0	219.0
12	4.50	300	$4.610/(\epsilon_s - 0.00104)$	234	0.0222	36.5	1.38	8.47	309.0	235.5
13	4.00	325	$5.000/(\epsilon_s - 0.00079)$	235	0.0225	36.3	1.40	8.43	308.0	250.0
14	3.50	350	$5.310/(\epsilon_s - 0.00054)$							

FLEXURAL BEHAVIOR AND STRENGTH OF PRESTRESSED-CONCRETE BEAMS 111

mum abscissa in the tension part is 0.00015, which is $\frac{0.00015}{0.00025} = 0.6$ of one unit of the division. The area under the tension part will be the following:

$$\tfrac{1}{2} \times 0.6 \times 0.6 = 0.18$$

Each line in column 5 is the area of all segments from the left end of the stress-strain diagram to the segment corresponding to the line. For segment 2 we have

$$0.50 + 1.50 - 0.18 = 1.82$$

Column 6 is the average stress for the part of the stress-strain diagram to the left of each segment. Column 7 is the centroid of each segment, measured from the left end of each segment.

Each line in column 8 is the moment of the area of all segments to the left of the corresponding segment with respect to the origin, and includes the moment of the area under the tension part of the stress-strain diagram, which is $0.18 \times \tfrac{2}{3} \times 0.6 = 0.072$. For segment 2 we have

$$0.50 \times 0.667 = 0.333$$
$$1.50 \times 1.556 = 2.334$$
$$\underline{0.072}$$
$$2.739$$

Column 9 gives the position of the centroid of the area of segments to the left of the particular segment in a convenient dimensionless form.

The calculations are continued in Table 3-6b. The first three columns are the same as those in Table 3-6a, and are repeated for convenience. The first two lines are not recorded since they correspond to the uncracked condition of the beam where the materials are elastic. The moment and curvature for these stages of loading may be calculated by simple formulas.

Column 4 represents the relationship between stress and strain in steel. In this case strain in steel is

$$\epsilon_s = 0.00404 + \epsilon_c \frac{9.05 - a}{a} \quad \text{or} \quad a = \frac{9.05\epsilon_c}{\epsilon_s - 0.00404 + \epsilon_c}$$

From equilibrium of horizontal forces we have

$$a \times 6.10 \bar{f}_n = 0.156 f_s \quad \text{or} \quad a = \frac{0.0256 f_s}{\bar{f}_n}$$

where $\bar{f}_n$ is the average stress given in column 6 of Table 3-6a. Elimination of a between the above two equations yields

$$f_s = \frac{353.5 \bar{f}_n \epsilon_c}{\epsilon_s - 0.00404 + \epsilon_c}$$

Columns 5 and 6 give the stress and strain in steel and were obtained by plotting the above equation on the stress-strain diagram of Fig. 3-16. Columns 7 through 11 give the tension force, position of neutral axis, moment arm, moment, and curvature, respectively. Figure 3-33b gives the moment-curvature diagram, which does not differ greatly from the diagram in Fig. 3-25, which was based on an idealized stress-strain diagram for concrete.

By a similar tabulation it is possible to calculate moment-deflection relationship.

3-18 FLANGED SECTIONS

The preceding discussion for rectangular sections can be generalized to apply to a flanged section. Flanged sections are used more commonly in practice since they are more efficient than the rectangular sections.

Figure 3-34 shows a typical flanged section, which is an idealized I section in which the thickness of each flange is equal to t and is uniform. The width of bottom flange is assumed to be k times the width of top flange. The quantity b' is the web thickness.

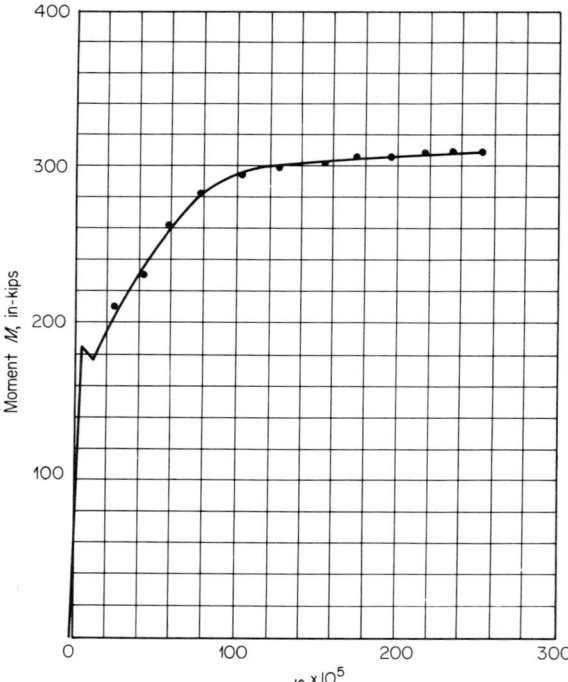

Fig. 3-33 (b) Moment-curvature relationship, Illustrative Problem 3-4.

FLEXURAL BEHAVIOR AND STRENGTH OF PRESTRESSED-CONCRETE BEAMS

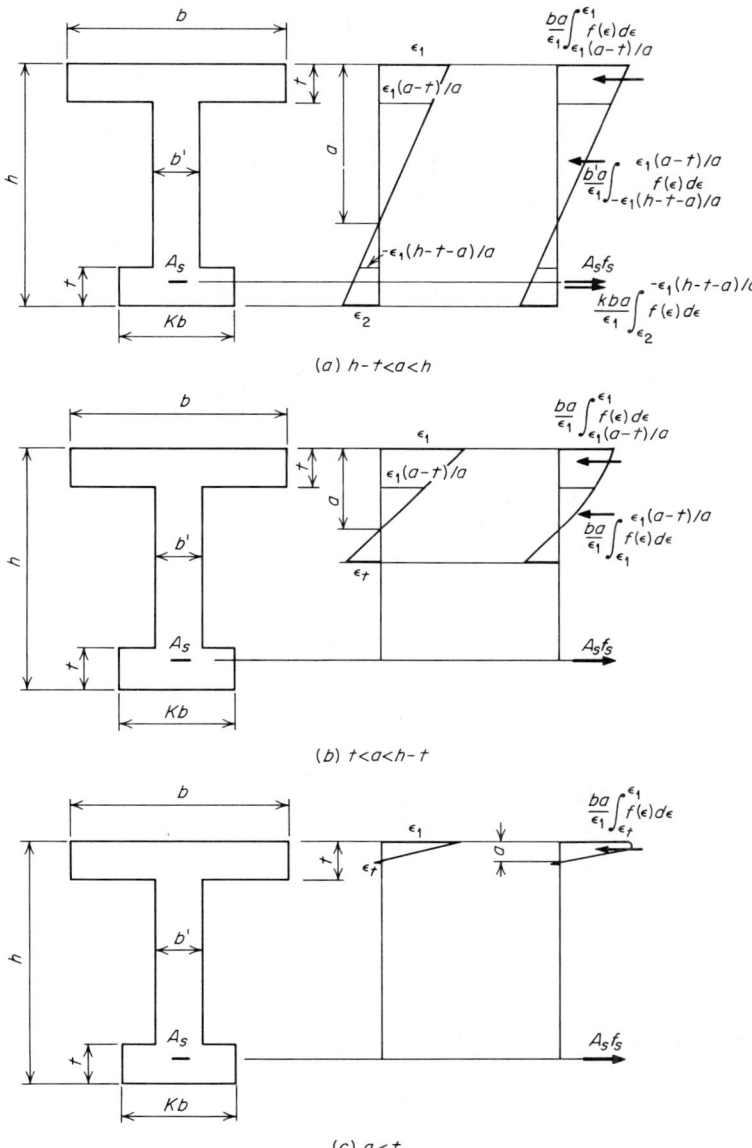

Fig. 3-34 Flanged sections.

In the analysis of flanged sections all assumptions of Sec. 3-7 are assumed to apply. The assumption that the strain is distributed linearly with depth in the uncracked region of the beam is taken as valid even if there are disturbances at the point of connection of web and flange. Equations (3-5) through (3-7) are applicable to flanged sections. However, Eqs. (3-8) and (3-9), which are the equations of equilibrium of horizontal forces and moments, respectively, will be modified since the beam has variable width.

Before flexural cracking, the equilibrium of horizontal forces can be written in the following form, if we assume that there is only non-prestressed steel in the section:

$$\frac{kba}{\epsilon_1} \int_{\epsilon_2}^{-\epsilon_1(h-t-a)/a} f(\epsilon)\, d\epsilon + \frac{b'a}{\epsilon_1} \int_{-\epsilon_1(h-t-a)/a}^{\epsilon_1(a-t)/a} f(\epsilon)\, d\epsilon$$
$$+ \frac{ba}{\epsilon_1} \int_{\epsilon_1(a-t)/a}^{\epsilon_1} f(\epsilon)\, d\epsilon + A_s f_s = 0 \quad (3\text{-}18)$$

In the above equation, ϵ_1 and ϵ_2 are positive when they are tensile; tensile forces are taken as positive and compressive forces as negative. Figure 3-34a shows a typical idealized I section before flexural cracking.

In this case the equilibrium of moments results in

$$M + \frac{kba^2}{\epsilon_1^2} \int_{\epsilon_2}^{-\epsilon_1(h-t-a)/a} f(\epsilon)\epsilon\, d\epsilon + \frac{b'a^2}{\epsilon_1^2} \int_{-\epsilon_1(h-t-a)/a}^{\epsilon_1(a-t)/a} f(\epsilon)\epsilon\, d\epsilon$$
$$+ \frac{ba^2}{\epsilon_1^2} \int_{\epsilon_1(a-t)/a}^{\epsilon_1} f(\epsilon)\epsilon\, d\epsilon + A_s f_s(d-a) = 0 \quad (3\text{-}19)$$

In analysis of an I section before cracking, Eqs. (3-8) and (3-9) should be replaced by Eqs. (3-18) and (3-19).

After flexural cracking, the exact form of the equations of equilibrium will depend upon the position of the fiber at the root of the crack. The following three cases are possible.

  a. $h - t < a(1 - \epsilon_1/\epsilon_2)$ In this case Eqs. (3-18) and (3-19) apply, except that the quantity ϵ_2 should be replaced by ϵ_t.

  b. $t < a(1 - \epsilon_1/\epsilon_t) < h - t$ Figure 3-34b shows this condition. Equilibrium of horizontal forces results in

$$\frac{b'a}{\epsilon_1} \int_{-\epsilon_t}^{\epsilon_1(a-t)/a} f(\epsilon)\, d\epsilon + \frac{ba}{\epsilon_1} \int_{\epsilon_1(a-t)/a}^{\epsilon_1} f(\epsilon)\, d\epsilon + A_s f_s = 0 \quad (3\text{-}18a)$$

A similar equation can be written for the equilibrium of moments.

  c. $a < t$ This condition is shown in Fig. 3-34c. The equation of equilibrium of horizontal forces is the following:

$$\frac{ba}{\epsilon_1} \int_{\epsilon_t}^{\epsilon_1} f(\epsilon)\, d\epsilon + A_s f_s = 0 \quad (3\text{-}18b)$$

FLEXURAL BEHAVIOR AND STRENGTH OF PRESTRESSED-CONCRETE BEAMS

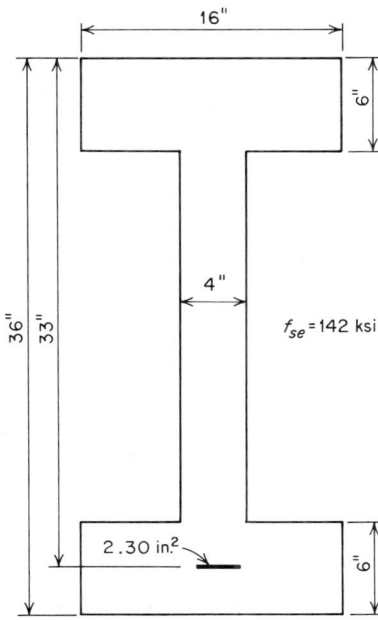

Fig. 3-35 (*a*) Analysis of flanged section, Illustrative Problem 3-5.

(*a*) Flanged section considered

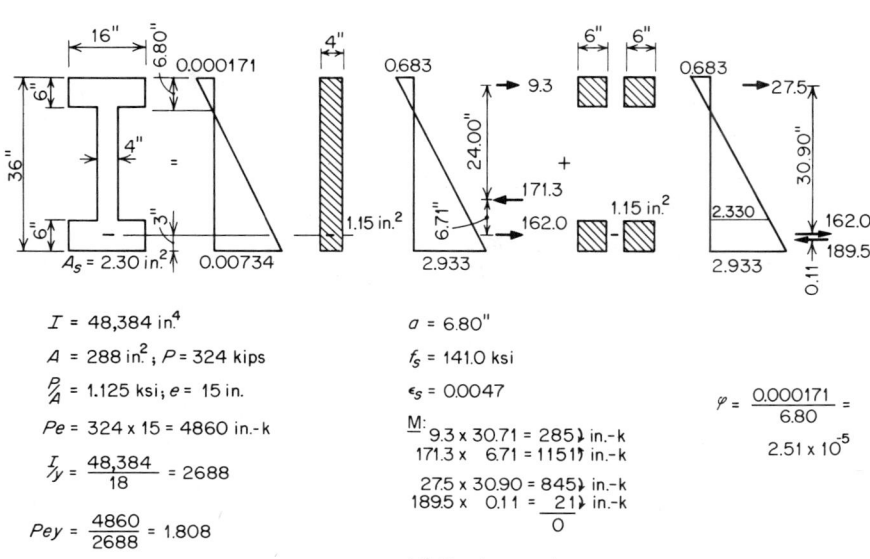

$I = 48{,}384$ in.^{4}

$A = 288$ in.^{2}; $P = 324$ kips

$\dfrac{P}{A} = 1.125$ ksi; $e = 15$ in.

$Pe = 324 \times 15 = 4860$ in.-k

$\dfrac{I}{y} = \dfrac{48{,}384}{18} = 2688$

$Pey = \dfrac{4860}{2688} = 1.808$

$a = 6.80''$

$f_s = 141.0$ ksi

$\epsilon_s = 0.0047$

$\underline{M:}$
$9.3 \times 30.71 = 285\}$ in.-k
$171.3 \times 6.71 = 1151\}$ in.-k

$27.5 \times 30.90 = 845\}$ in.-k
$189.5 \times 0.11 = \underline{\ \ 21}\}$ in.-k
0

$\varphi = \dfrac{0.000171}{6.80} =$
2.51×10^{-5}

(*b*) Prestress only

Fig. 3-35 (*b*) Analysis of flanged section, Illustrative Problem 3-5. Action of prestress only.

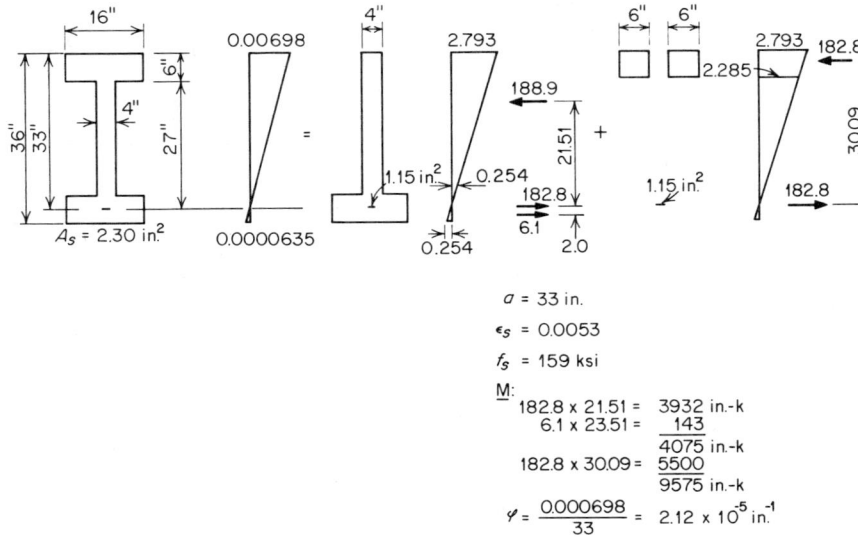

(c) Strain in concrete at the level of steel = 0

Fig. 3-35 (c) Analysis of flanged section, Illustrative Problem 3-5. Strain in concrete at the level of steel = 0.

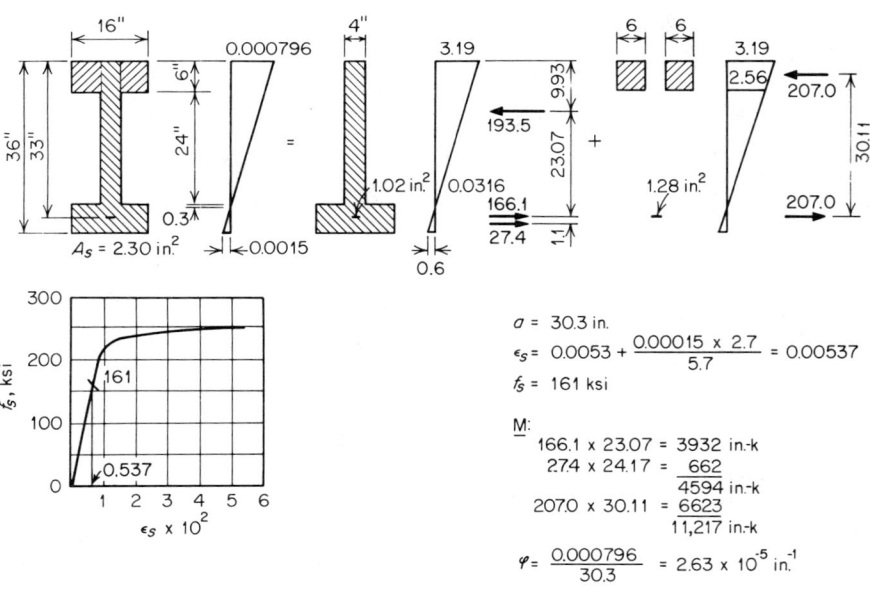

(d) Cracking load

Fig. 3-35 (d) Analysis of flanged section, Illustrative Problem 3-5. Cracking load.

FLEXURAL BEHAVIOR AND STRENGTH OF PRESTRESSED-CONCRETE BEAMS

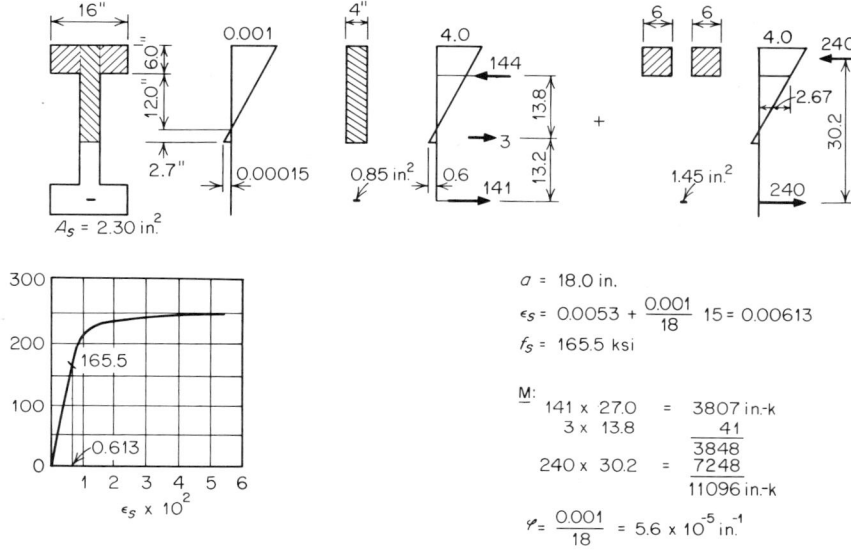

(e) Strain in concrete at the top fiber = 0.001

Fig. 3-35 (e) Analysis of flanged section, Illustrative Problem 3-5. Strain in concrete at the top fiber = 0.001.

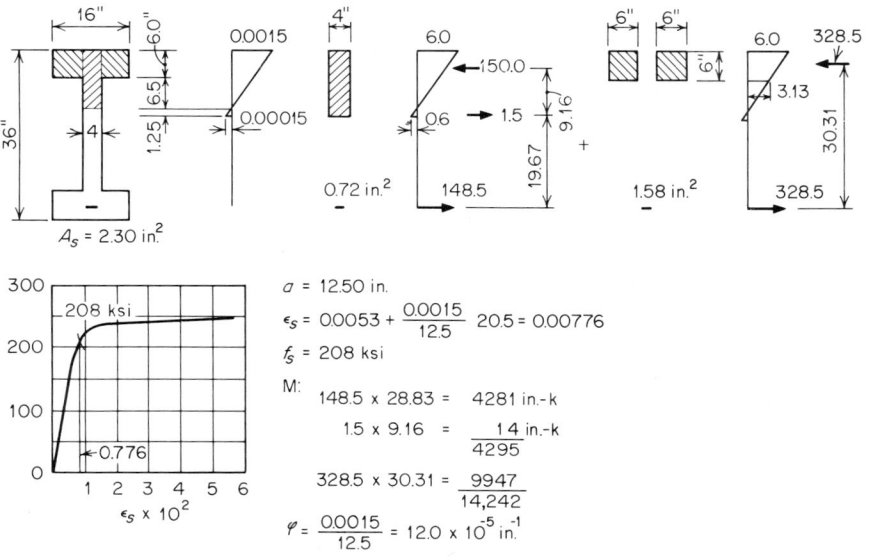

(f) Strain in concrete at the top fiber = 0.0015

Fig. 3-35 (f) Analysis of flanged section, Illustrative Problem 3-5. Strain in concrete at the top fiber = 0.0015.

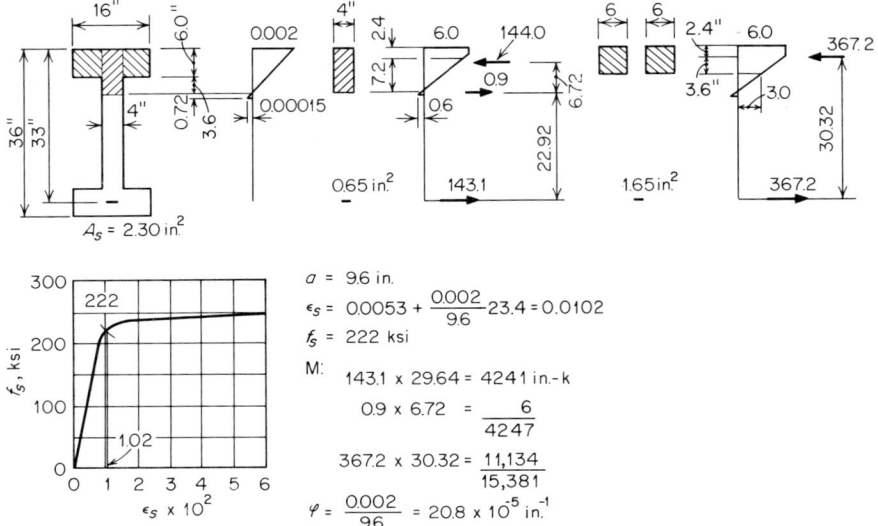

(g) Strain in concrete at the top fiber = 0.002

Fig. 3-35 (g) Analysis of flanged section, Illustrative Problem 3-5. Strain in concrete at the top fiber = 0.002.

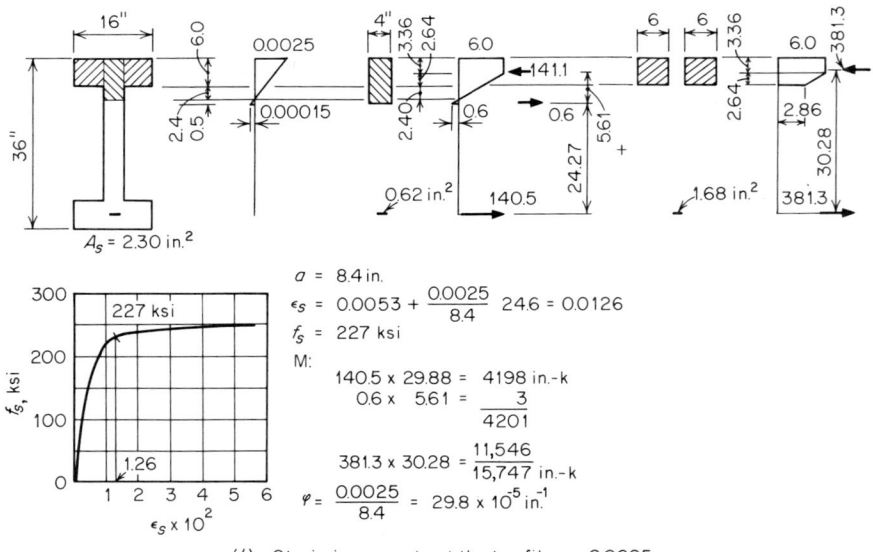

(h) Strain in concrete at the top fiber = 0.0025

Fig. 3-35 (h) Analysis of flanged section, Illustrative Problem 3-5. Strain in concrete at the top fiber = 0.0025.

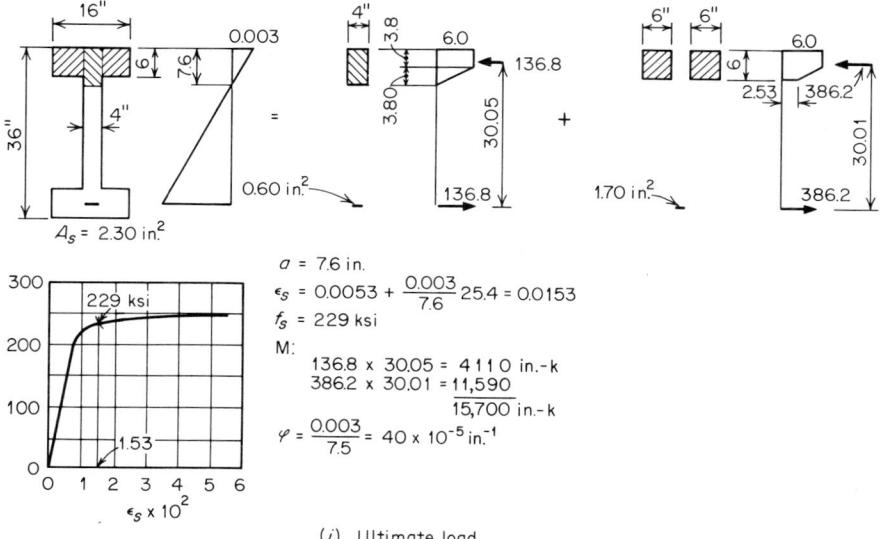

(i) Ultimate load

Fig. 3-35 (*i*) Analysis of flanged section, Illustrative Problem 3-5. Ultimate load.

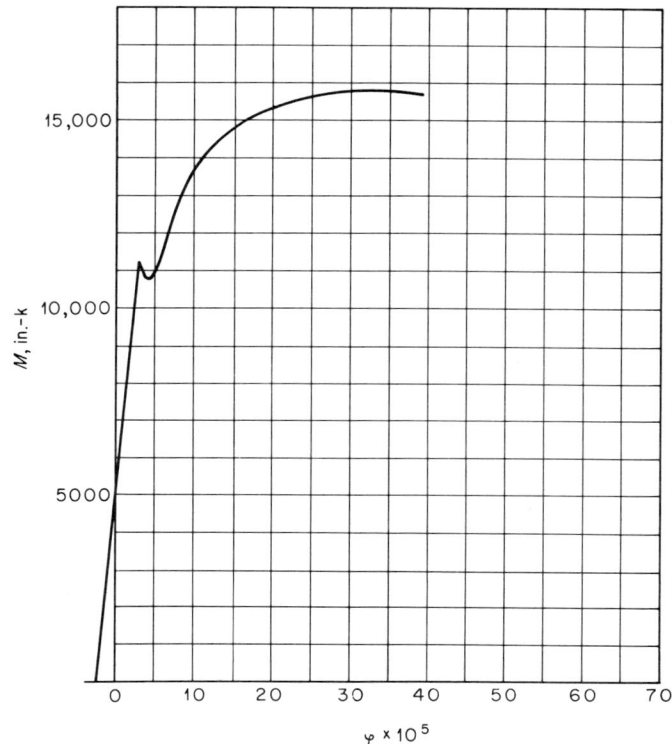

Fig. 3-35 (*j*) Analysis of flanged section, Illustrative Problem 3-5. Moment-curvature diagram.

Table 3-7 Summary of results, Illustrative Problem 3-5*

Loading	$\epsilon_1 \times 10^2$	f_1, ksi	$\epsilon_2 \times 10^2$	f_2, ksi	$\epsilon_s \times 10^2$	f_s, ksi	a, in.	M, in.-kips	$\varphi \times 10^5$ in.$^{-1}$
Prestress only	0.0171	0.683	−0.0734	2.933	0.470	141.0	6.8	0	−2.51
Strain at level of steel = 0	−0.0698	−2.793	−0.0064	0.254	0.530	159.0	33.0	9,580	2.12
Cracking	−0.0976	−3.190	−0.0150	0.600	0.537	161.0	30.3	11,220	2.63
$\epsilon_1 = 0.0010$	−0.1000	−4.000			0.613	165.5	18.0	11,090	5.60
$\epsilon_1 = 0.0015$	−0.1500	−6.000			0.776	208.0	12.5	14,240	12.00
$\epsilon_1 = 0.0020$	−0.2000	−6.000			1.020	222.0	9.6	15,380	20.80
$\epsilon_1 = 0.0025$	−0.2500	−6.000			1.260	227.0	8.4	15,750	29.80
Ultimate	−0.3000	−6.000			1.530	229.0	7.6	15,700	40.00

* Negative quantities indicate compressive strain or stress.

FLEXURAL BEHAVIOR AND STRENGTH OF PRESTRESSED-CONCRETE BEAMS

The equation of equilibrium of moments can be written in a similar way.

It can be seen that the analysis of flanged sections is not significantly different from the analysis of rectangular sections; however, it is more tedious.

Actually, it is not necessary to use Eqs. (3-18) and (3-19) for the equilibrium of horizontal forces and moments. The equilibrium may be established directly by calculations of tensile and compressive forces in the section. Equations (3-18) and (3-19) are given to show that the general method of analysis discussed in Sec. 3-7 can be applied to flanged sections.

3-19 ILLUSTRATIVE PROBLEM 3-5

Determine the moment-curvature relationship for the flanged section shown in Fig. 3-35a. The stress-strain diagrams for steel and concrete are shown in Figs. 3-16 and 3-17b, respectively.

The entire solution of this problem is shown in Fig. 3-35b through j, and the results are summarized in Table 3-7. The solution is self-explanatory.

PROBLEMS

3-1. A simply supported plain concrete beam is subjected to two concentrated loads, as shown in Fig. 3-36a. The two sections shown in Fig. 3-36b and c are to be considered. The stress-strain diagram for concrete is shown in Fig. 3-37.

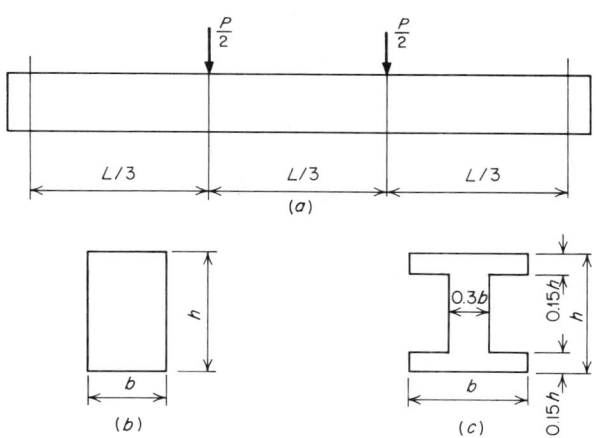

Fig. 3-36 Simply supported beam in Probs. 3-1 through 3-3.

122   PRESTRESSED CONCRETE

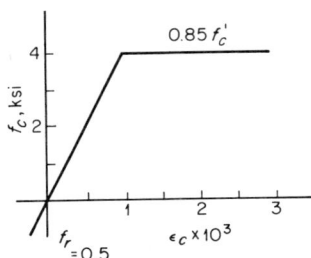

Fig. 3-37 Stress-strain diagram for concrete in Probs. 3-1 and 3-3.

(a) Draw the load-deflection diagram for each section, taking $M/f'_c bh^2$ as ordinate and $\Delta h/L^2$ as abscissa.

(b) Plot the load-crack depth relationship for each section, taking $M/f_r bh^2$ as ordinate and c/h as abscissa.

3-2. Solve Prob. 3-1 using a stress-strain diagram for concrete as shown in Fig. 3-38.

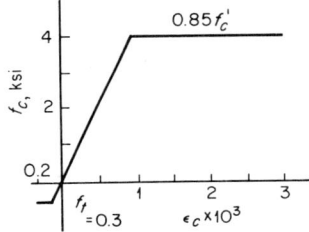

Fig. 3-38 Stress-strain diagram for concrete in Prob. 3-2.

3-3. The beam of Prob. 3-1 is subjected to prestressing force only (in addition to its weight).

(a) Plot the load-crack depth relationship for the top fiber at midspan and at the ends, for the sections shown in Fig. 3-36b and c, assuming that the prestressing force is acting at the bottom fiber. Take $P/f_r bh$ as ordinate and c/h as abscissa.

(b) Plot the load-crack depth relationship for the top fiber at midspan and at the ends as in (a), but assume that the prestressing force is acting $0.10h$ above the bottom fiber.

The stress-strain diagram for concrete is shown in Fig. 3-37.

3-4. Draw the moment-deflection diagram for the reinforced-concrete beam shown in Fig. 3-39, taking M as ordinate and deflection Δ as abscissa. Consider the instability

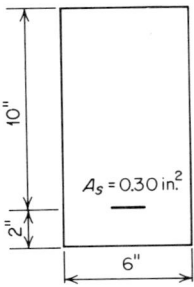

Fig. 3-39 The reinforced-concrete section in Prob. 3-4.

FLEXURAL BEHAVIOR AND STRENGTH OF PRESTRESSED-CONCRETE BEAMS 123

at cracking. The reinforcement is intermediate-grade steel with a yield-point stress of 50,000 psi and a modulus of elasticity of 30 million psi. The stress-strain diagram for concrete is shown in Fig. 3-17b.

3-5. Calculate the moments and curvatures in each of the prestressed-concrete sections A, B, and C shown in Fig. 3-40, for the following stages of loading.
    (a) Prestressing force only.
    (b) Sufficient load to make the strain in concrete at the level of steel equal zero.
    (c) Flexural cracking load.
    (d) Sufficient load to produce a concrete fiber strain of 0.0015.
    (e) Sufficient load to produce the yield stress in the prestressing steel.
    (f) Ultimate.

Plot moments with the corresponding curvatures for each section; take M as ordinate and φ as abscissa.

Plot moments with the corresponding deflections at the midspan, for beams A, B, and C; take moment M as ordinate and deflection Δ as abscissa.

The effective prestress $f_{se} = 120$ ksi. The stress-strain diagrams for the prestressing steel and concrete are shown in Figs. 3-16 and 3-17a, respectively. The non-prestressed compression reinforcement is intermediate grade with a yield-point stress of 50,000 psi and a modulus of elasticity of 30 million psi.

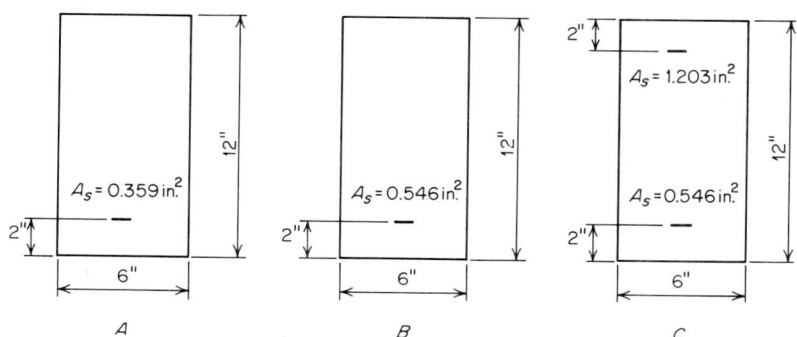

Fig. 3-40 The prestressed-concrete sections in Prob. 3-5.

3-6. (a) Calculate curvature φ_u for each of the sections shown in Fig. 3-41.
    (b) Is it possible to make φ_u for section B equal to that of section A by addition of intermediate-grade compression steel in the compression zone of section B? If so, how much compression steel is needed if it is to be placed 2 in. from the top fiber? The yield-point stress of the intermediate-grade steel may be assumed as 50,000 psi.

The stress-strain diagrams for prestressing steel and concrete are shown in Figs. 3-16 and 3-17a, respectively. $f_{se} = 120$ ksi. For this problem, assume $\epsilon_{ce} = 0$.

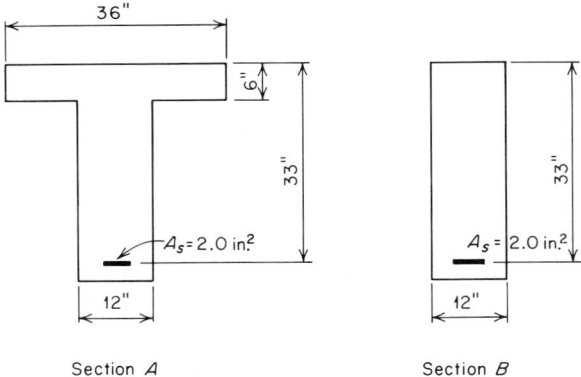

Section *A*   Section *B*

Fig. 3-41 The sections in Prob. 3-6.

3-7. Figure 3-42 shows a section of prestressed-concrete beam.

(a) Determine the area of prestressed steel required so that the stress in steel at failure, f_{su}, will be 225 ksi.

(b) Determine the area of prestressed steel required so that the stress in steel at failure, f_{su}, will be 248 ksi.

(c) Determine the area of non-prestressed tensile reinforcement required in (b) to make its ductility equal to that of (a).

$f_{se} = 120.0$ ksi. The stress-strain diagrams for steel and concrete are shown in Figs. 3-16 and 3-17b, respectively.

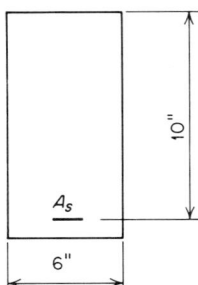

Fig. 3-42 The section in Prob. 3-7.

REFERENCES

1. Hognestad, E., N. W. Hanson, and Douglas McHenry: Concrete Stress Distribution in Ultimate Strength Design, *J. Am. Concrete Inst.*, December, 1955; *Proc.*, vol. 52, pp. 455–479.
2. Billet, D. F., and J. H. Appleton: Flexural Strength of Prestressed Concrete Beams, *J. Am. Concrete Inst.*, June, 1954.
3. Hognestad, E.: A Study of Combined Bending and Axial Load in Reinforced Concrete Members, *Univ. Illinois Eng. Expt. Sta. Bull.* 399, 1951.

4. Warwaruk, J., M. A. Sozen, and C. P. Siess: Strength and Behavior in Flexure of Prestressed Concrete Beams, *Univ. Illinois Eng. Expt. Sta. Bull.* 464, 1962.
5. Standards, American Society for Testing Materials, vol. 4, 1968.
6. Whitney, C. S.: Plastic Theory in Reinforced Concrete Design, *Trans. Am. Soc. Civil Engrs.*, vol. 107, pp. 251–282, 1942.
7. Ramaley, D., and D. McHenry: Stress-Strain Curves for Concrete Strained beyond the Ultimate Load, *Bur. Reclamation Lab. Rept.* SP-12, March, 1947.
8. Jensen, Vernon P.: Ultimate Strength of Reinforced Concrete Beams as Related to the Plasticity Ratio of Concrete, *Univ. Illinois Eng. Expt. Sta. Bull.* 345, 1943.

4
Behavior of Prestressed-concrete Beams in the Region of Combined Stresses

4-1 INTRODUCTION

The behavior of prestressed-concrete beams discussed in the preceding chapter is for sections in which only flexural stresses exist and no shearing stress is present.

In this chapter our attention will be directed to the region of the beam in which flexural and shearing stresses occur simultaneously.

Figure 4-1 shows a beam subjected to two concentrated forces acting at the third points. The part of the beam between the support and the point of action of concentrated load is the region where both moment and shear are present. This region is often called the *shear span*. We shall call it the region of combined stresses. The part of the beam between the two concentrated forces is the region of flexure where the amount of shear due to the weight of the beam is small. At midspan, shear is zero.

The analysis based on pure flexure presented in the preceding chapter assumed no shear in a section. This assumption simplified the

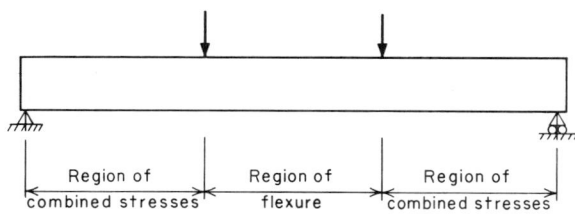

Fig. 4-1 Regions of flexure and combined stresses.

analysis considerably since the flexural or normal stresses were also the principal stresses, and acted in a horizontal direction.

Presence of shear in addition to moment introduces one other unknown, namely, the distribution of shearing stress in a section. Presence of shearing stresses in addition to flexural stresses means that at each point there are principal tensile and compressive stresses. These principal stresses vary in direction and magnitude along the span, and through the depth of the beam. A complete analysis of a beam involves determination of magnitude and direction of principal stresses at every point in the region of combined stresses. Lines which describe the direction of principal tensile and compressive stresses along the span are called stress trajectories. Differential equations describing the stress trajectories for the principal tensile and compressive stresses can be derived if the flexural and shearing stresses can be calculated at any point.

Determination of the magnitude and direction of the principal stresses when the region of combined stresses is uncracked is comparatively simple. The shearing stresses for this stage of loading may be calculated by assuming that the flexural stresses are distributed linearly with depth. Shearing and flexural stresses thus calculated are satisfactory when the beam is uncracked, and yield principal stresses that are reasonably accurate.

The first cracking in the beam occurs in the region of pure flexure since the principal tensile stress has its highest value at the bottom fiber at midspan. However, as the loads are further increased, cracks develop in the region of combined stresses in different ways, which will be discussed later. After cracking, determination of shearing and normal stresses based on the assumption that the flexural stresses are distributed linearly with depth is no longer satisfactory. Since the magnitude of the principal tensile stress is greatly dependent upon the shearing stress, slight inaccuracies in estimating the shearing stress result in comparatively large variations in the magnitude and direction of the principal stresses.

The region of combined stresses has the peculiarity of having a different state of stress in each section. Since in each section along the span the ratio of moment to shear changes, there are many combinations of shearing and flexural stresses to be considered.

Presence of shear in the region of combined stresses causes development of inclined cracks that form suddenly. These cracks progress a long distance with no increase in load in an inclined direction before they stop. They lead to the failure of the beam in the region of combined stresses usually before the flexural capacity of the beam is fully developed. Failures in the region of combined stresses are brittle and often violent. Furthermore, because of the sensitivity of the behavior of the beam in the region of combined stresses to the various quantities involved, it is not possible to predict the failure load with the same degree of precision that is possible for the region of flexure.

Since the section subjected to pure flexure provides higher strength and can be controlled more accurately, according to our present practice it is made the weakest link of the beam. This means that failure is not permitted to occur in the region of combined stresses before the full flexural capacity of the beam is developed. Such a failure is prevented by provision of vertical reinforcement which tends to contain inclined cracks. The vertical reinforcement thus provided is called web reinforcement or stirrups.

The problem, therefore, reduces to the determination of the amount of web reinforcement such that the beam will fail in flexure before any section in the region of combined stresses fails.

In the following paragraphs the behavior of beams in the region of combined stresses is discussed in detail, first for beams without web reinforcement, second for sections with web reinforcement. Subsequently, the criteria for the design of web reinforcement are presented.

4-2 BEHAVIOR OF BEAMS BEFORE CRACKING

The discussions in this section are for the low stages of loading at which the beam has not yet cracked. The study of the condition of the beam in the region of combined stresses when the beam is uncracked is not practically useful, except that it permits determination of points of maximum stress where potential cracks may develop.

Figure 4-2a shows part of a rectangular beam before cracking. Beam width is denoted by b, and overall depth is denoted by h. The section at the left end of the beam is the section of zero moment. A positive shearing force acts at this section, and distributed or concentrated loads may be acting along the beam to the right of this section. The origin of the x and y axes is taken at the section of zero moment and at middepth of the beam, as shown in Fig. 4-2a.

BEHAVIOR OF PRESTRESSED-CONCRETE BEAMS

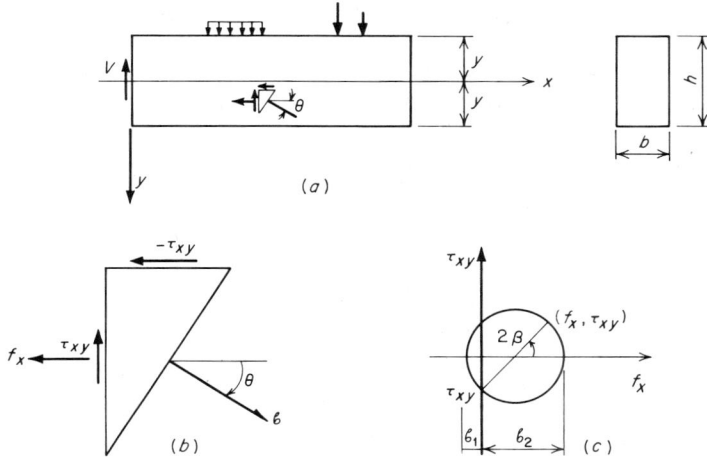

Fig. 4-2 State of stress at a point in the region of combined stresses.

Before cracking occurs, the maximum tensile stress σ_1 makes an angle θ with the beam axis, as shown in Fig. 4-2b.

From Fig. 4-2c, which shows the Mohr's circle of stress for any uncracked point, we have

$$\tan 2\beta = \frac{2\tau_{xy}}{f_x}$$

where τ_{xy} = shearing stress at point considered
f_x = flexural stress at same point

and

$$\tan 2\beta = \tan 2\theta = \frac{2 \tan \theta}{\tan^2 \theta - 1}$$

Since

$$\tan \theta = \frac{dy}{dx}$$

we have

$$\frac{2(dy/dx)}{(dy/dx)^2 - 1} = \frac{2\tau_{xy}}{f_x} \tag{4-1}$$

or

$$\left(\frac{dy}{dx}\right)^2 - \frac{f_x}{\tau_{xy}}\left(\frac{dy}{dx}\right) - 1 = 0 \tag{4-2}$$

The above differential equation describes the magnitude and direction of the principal stresses. The quantities f_x and τ_{xy} are the normal and shearing stresses, respectively, at any point, and are therefore functions of x and y.

Equation (4-2) is in quadratic form and may be solved in terms of dy/dx, which is the slope of the stress trajectories. Solution of Eq. (4-2) for dy/dx yields the following two expressions:

$$\frac{dy}{dx} = \frac{f_x}{2\tau_{xy}} + \left[\left(\frac{f_x}{2\tau_{xy}}\right)^2 + 1\right]^{\frac{1}{2}} \quad (4\text{-}3)$$

and

$$\frac{dy}{dx} = \frac{f_x}{2\tau_{xy}} - \left[\left(\frac{f_x}{2\tau_{xy}}\right)^2 + 1\right]^{\frac{1}{2}} \quad (4\text{-}4)$$

The above expressions give the slope of the principal tensile and compressive stresses along the span, and describe the paths of principal stresses, called stress trajectories.

It can be shown that Eq. (4-3) describes the stress trajectories for the tensile stresses and Eq. (4-4) the stress trajectories for the compressive stresses. The product of both expressions is -1, indicating that tension and compression stress trajectories are normal to each other.

Equations (4-2) to (4-4) represent a generalized formulation of the direction and magnitude of the principal tensile and compressive stresses at every point. Of course, the same information may be obtained by construction of Mohr's circle for every point of the beam.

Equation (4-2) represents a large class of differential equations in which the quantities f_x and τ_{xy} are functions of both x and y. The exact form of these functions depends upon the type of loading as well as the shape of the section.

The beam shown in Fig. 4-3 is subject to two concentrated loads acting at the third points. The centroid of the prestressing force is parabolic, and P is the prestressing force. For this beam the quantities f_x and τ_{xy} for any point between the left support and the left concentrated load can be expressed as follows:

$$f_x = \frac{12}{bh^3}\left[-W\left(x + \frac{L}{2}\right) + Pe_2 - 4P(e_1 - e_2)\frac{x^2}{L^2}\right]y - \frac{P}{bh} \quad (4\text{-}5)$$

$$\tau_{xy} = \frac{48P(e_1 - e_2)}{L^2bh^3}y^2x - \frac{6W}{bh^3}y^2 - \frac{12P(e_1 - e_2)}{L^2bh}x + \frac{3W}{2bh} \quad (4\text{-}6)$$

where W = magnitude of one of concentrated loads
e_1 = end eccentricity
e_2 = eccentricity at midspan
L = length of simple span

Substitution of Eqs. (4-5) and (4-6) in Eq. (4-2) for f_x and τ_{xy}, respectively, will yield the differential equation for the stress trajectories

BEHAVIOR OF PRESTRESSED-CONCRETE BEAMS

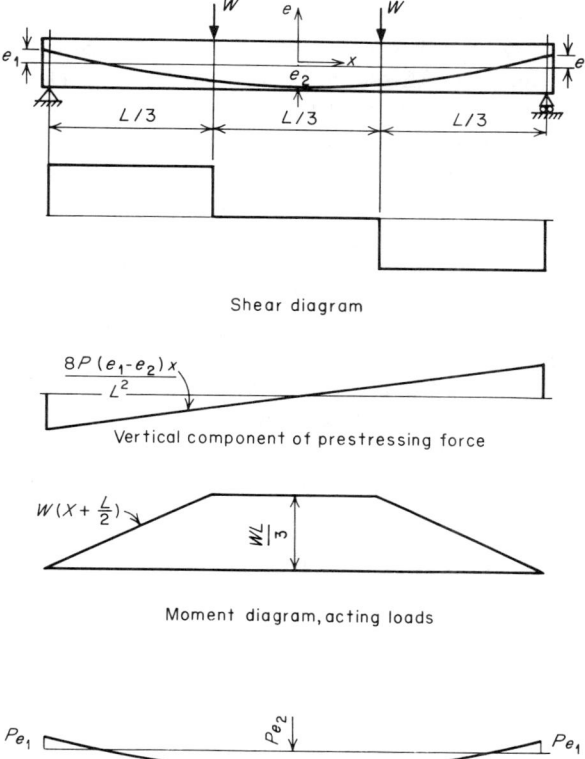

Fig. 4-3 Shear and moment diagrams due to acting loads and prestressing force.

at the shear span for the particular loading and beam section. Similar equations can be developed for different types of loading and beam sections.

The closed-form solution of the class of differential equations described by Eq. (4-2) is difficult; however, a convenient numerical solution is possible.[1]

From Eqs. (4-5) and (4-6) it can be seen that the determination of stress trajectories, as well as the actual values of principal tensile stresses, depends upon many variables. The variables are the prestressing force, the shape of the profile of the prestressing elements, the type of loading, and the shape of the cross-sectional area of the beam.

A numerical study of these equations for all the points of the beam indicates that the highest principal tensile stress usually occurs at the bottom fiber of the beam in the region of flexure. If we assume that the tensile strength of the beam is approximately the same throughout the beam, this means that the cracking of the beam first occurs at the bottom fiber somewhere in the region of flexure. Therefore, in the early stages of loading, the behavior of the beam is the same as that for a beam which fails in flexure, and usually there are no cracks in the region of combined stresses. However, as the load is increased, the principal tensile stress begins to increase in various points in the region of combined stresses. The maximum principal tensile stress may occur either at the bottom fiber or in the web, somewhere in the region of combined stresses, depending upon the relative magnitude of the variables involved.

Usually if the prestress is small, the section rectangular, the web thick, and the prestressing elements draped, the maximum principal tensile stress will occur at the bottom fiber somewhere in the shear span. Reinforced-concrete beams fall in this category.

On the other hand, if prestress is large and there is no draping and the section of the beam has a thin web, the maximum principal tensile stress will occur somewhere in the web.

In summary, we may state that the flexural cracks which form at the bottom fiber in the region of flexure are always first to occur. The cracks in the region of combined stresses may begin either at the bottom fiber or in the web.

4-3 CRACKING OF BEAMS

The cracks develop when the stress or strain corresponding to the principal tensile stress or strain reaches a limiting value. The cracks are perpendicular to the direction of principal tensile stress, and at their inception they are unstable.

The stability of cracks may be studied as a measure of behavior of the beam. If the crack propagates only with increase in load, it is stable. If the crack propagates a certain distance while load decreases in value, the crack is unstable.

In the region of flexure the cracks develop in a vertical direction and are unstable until a certain depth is reached, beyond which the stability is regained and the crack progresses only with additional load on the beam. The distance which the crack progresses initially in the region of flexure before stability is regained depends upon the amount of reinforcement and the proximity of reinforcement to the bottom fiber of the beam. In plain concrete beams, once the crack begins it remains unstable until it reaches the top fiber of the beam. On the other hand, it is possible to put a large amount of reinforcement sufficiently close to the bottom fiber of the beam that the crack is stable as it forms and progresses only with the addition of load. For practical percentages and positions of steel, however, the crack is always unstable and should reach a certain depth before the beam has regained stability.

After the flexural cracks are developed, further increase in load causes cracks to appear in the region of combined stresses. If the crack in the region of combined stresses starts at the bottom fiber, the progress of the crack is somewhat similar to that of cracks formed in the region of flexure.

As the crack forms at the bottom fiber somewhere in the region of combined stresses, it is unstable until it reaches a certain depth before stability is regained. The cracks are vertical at initiation but become inclined as they move up. The presence of shear causes the principal tensile stress to be inclined, and the crack is perpendicular to the direction of principal stress. After the crack is stabilized, it moves up only with addition of load. Often the crack may become unstable again and propagate a large distance in an inclined direction before stability is regained. This type of crack usually leads the beam to shear-compression failure, which means that the crack has reached so high in the beam that the compression zone is reduced in size, and the concrete fails in compression. This type of failure is similar to flexural failure, except that the cracks are inclined and the failure takes place at a load lower than that corresponding to flexural failure.

Reinforced-concrete beams and prestressed-concrete beams with thick unreinforced webs and low levels of prestress usually fail in this fashion. Occasionally, this failure is violent: after the crack has developed at the bottom fiber, it becomes stable and progresses only with addition of load, but it becomes unstable again and progresses suddenly all the way to the top fiber of the beam, causing a violent failure of the beam. This type of failure is more common in reinforced-concrete beams.

For beams in which the crack in the region of combined stresses develops at the bottom fiber and progresses with load, Eq. (4-2) is still applicable if a reasonable estimate is made of the normal and shearing stresses at the root of the crack. By use of Eq. (4-2) it is possible to study the relationship between the crack depth and load and estimate the failure load at each section. One such study has been made for reinforced-concrete beams.[1]

If the crack in the region of combined stresses originates in the web, usually it is unstable and causes a violent formation of inclined crack of large size before stability is regained. In this case, the magnitude of crack is such that after its formation the entire mechanism of behavior of the beam changes. Analysis of this type of beam is difficult.

This type of behavior is peculiar to beams with thin web and high prestress. This type of failure is called web-distress failure.

4-4 BEHAVIOR OF BEAMS AFTER CRACKING

Precise theoretical analysis of beams in the region of combined stresses is nearly impossible after cracking. In a shear-compression failure, it is difficult to estimate the shearing stress at the root of a crack. Furthermore, the available experimental results indicate that the strain in the uncracked zone does not vary linearly with depth after the inclined cracks are developed. Hence, it is not possible to estimate the magnitude of normal stress with a reasonable degree of precision. In web-distress type of failure it is difficult to simulate the precise load-carrying mechanism of the beam.

These difficulties associated with the analysis of the beam in the region of combined stresses have led to a more empirical approach of the analysis of concrete beams.

A large number of beams without web reinforcement were tested at the University of Illinois.[2,3] These beam specimens were designed to simulate the range of variables encountered in practice. A large class of these beams failed in the region of combined stresses under the combined influence of shear and moment.

The tests were carried out, in general, on simply supported beams with one or two concentrated loads. The variables considered were the strength of concrete, the percentage of reinforcement, the amount of prestress, the shape of cross-sectional area of the beam, and the shape of the profile of prestressing elements.

Observation of the results of these tests shows fundamental differences between the behavior of the beam in the region of flexure and that in the region of combined stresses after cracking has developed.

In the region of flexure the crack is caused by the principal tensile stress, which is horizontal and is initiated at the bottom fiber. The crack propagates in a vertical direction, and as it progresses upward vertically, the reinforcement carries an increasingly larger portion of tension and the root of crack is near the neutral axis of the beam. The linearity of strain distribution in the uncracked zone is not greatly disturbed as the crack moves up.

On the other hand, in the region of combined stresses the cracks may not necessarily originate at the bottom fiber. They may start at a point in the region of combined stresses where the principal tensile stress (or strain) exceeds the tensile strength (or limiting strain) of concrete. An inclined crack may be initiated either before or after a vertical crack.

As long as a beam action is maintained, the propagation of inclined

Fig. 4-4 Inclined tension crack originating from flexure crack.

tension crack is accompanied by an increase in the inclined tensile stresses, which is not effectively taken over by the longitudinal reinforcement. Longitudinal reinforcement is not usually placed in a way that will be perpendicular to the inclined cracks, as in the case of pure flexure. The inclined cracks propagate to a level above the neutral axis of the beam, and they distort the strain distribution over the depth of the cross section severely. After the full development of inclined crack, the compatibility factor decreases greatly and the beam action is changed to arch action.

Shear-compression failure Shear-compression failure is similar to flexural failure. The inclined cracks penetrate very high into the compression zone, and failure occurs by crushing of the concrete. Figure 4-4 shows a typical shear-compression failure. This type of failure originates from a flexure crack which develops somewhere in the shear span.

The measurements of strains indicate severe strain concentration at the top of inclined cracks. The failure is more violent for beams with a higher percentage of steel and lower concrete strength. In practical sections this type of failure is more likely to occur.

Web-distress failure This type of failure usually occurs in the region of combined stresses in beams with very thin webs. In this category may be included three mechanisms of failure: (1) crushing of the web under high compressive stresses due to arch action developed after the bond is destroyed; (2) separation of the tension flange from the web, in which inclined cracks near loading points extend horizontally toward the supports, destroying the bond; (3) secondary inclined tension cracking which forms near the supports and separates the top flange from the web.

Web-distress failures are more violent than the shear-compression failures.

Figure 4-5 shows a web crushing failure.

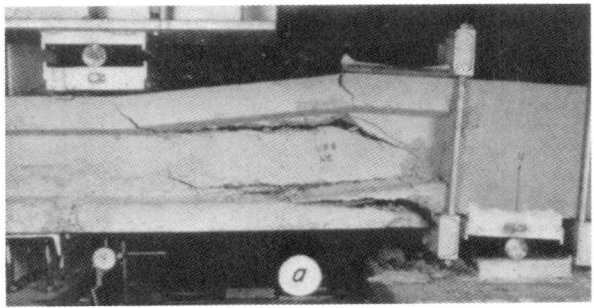

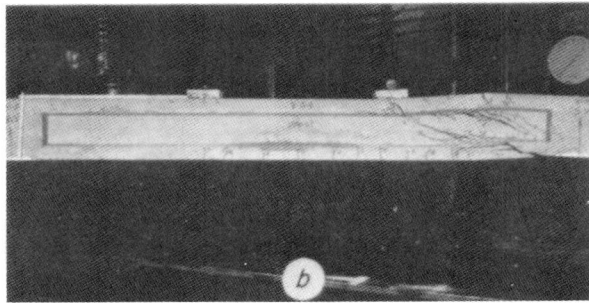

Fig. 4-5 Failure as a result of secondary inclined tension cracking.

Inclined cracks The formation of inclined cracks in the region of combined stresses marks a very important stage in the behavior of a concrete beam. Although there are different mechanisms of failure which lead to inclined cracks, once the inclined cracks are fully developed the behavior of the beam changes significantly.

Immediately after formation of inclined cracks the compatibility factor of the prestressed reinforcement is reduced considerably, causing the strain in steel to increase with load at a considerably reduced rate. In addition, the linearity of strain distribution at the uncracked zone is severely disturbed.

Though ultimate load of the beam is usually greater than the cracking load, the behavior of the beam between the cracking load and ultimate is changed considerably.

The behavior of the beam after inclined cracking is modified to such an extent that the inclined cracking load, rather than the ultimate load, is considered as the useful capacity of the beam.

In the following paragraphs, expressions will be developed for the total shear in a section corresponding to inclined cracking. It should

BEHAVIOR OF PRESTRESSED-CONCRETE BEAMS

be pointed out that this information is empirical in nature and is based almost entirely upon observation of test specimens.[4]

4-5 SHEAR AT INCLINED CRACKING—SHEAR-COMPRESSION FAILURE

Noncomposite sections In this type of beam the beam section carries both dead load and live load. Even if the dead load includes a cast-in-place slab, it does not interact with the beam in supporting the live load.

Observation of test results indicates that a flexure crack originating at the bottom fiber eventually may develop into an inclined crack in the web of the beam. Test results show that if an inclined crack is to change the behavior of the beam significantly, it is necessary for the horizontal projection of the crack to be longer than d, the effective depth of the beam. A flexure crack less than a distance $d/2$ from the point of action of a concentrated load usually does not increase tensile stresses sufficiently to develop inclined crack. For a section considered, a flexure crack at a distance $d/2$ in the direction of decreasing moment may be assumed as being critical.

Figure 4-6 shows an idealized inclined crack which affects the behavior of the beam. The beam is subjected to concentrated load, and the inclined crack occurs in the region of combined stresses. The crack

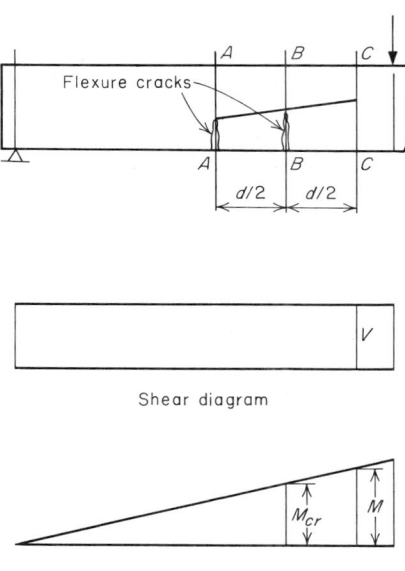

Fig. 4-6 Inclined crack initiated by a flexural crack.

begins at section A-A and ends at section C-C, and its horizontal projection is equal to d, the effective depth of the beam. A flexural crack at section B-B is considered to be critical.

Let us consider section C-C, in which the live-load moment is M and shear is V. The live load may be considered as any load acting on the beam in addition to prestressing force and the total dead load. Section B-B is a distance $d/2$ in the direction of decreasing moment. At B-B the flexural cracking moment is M_{cr} and the shear is V_{cr}. The quantity M_{cr} is the moment in addition to dead-load moment that should be applied at section B-B in order that the flexural crack will occur. From Fig. 4-6 we have

$$M - M_{cr} = \tfrac{1}{2}(V + V_{cr})\frac{d}{2}$$

When the acting loads are concentrated as shown in Fig. 4-6, the above expression can be written as follows:

$$M - M_{cr} = \frac{d}{2} V$$

or

$$\frac{M}{V} - \frac{M_{cr}}{V} = \frac{d}{2}$$

and

$$V = \frac{M_{cr}}{M/V - d/2}$$

where V is the shear due to the live load at section C-C. It should be pointed out that for a particular fixed position of loads considered, the ratio M/V is constant for any proportional increase in loads.

The total shear in section C-C when flexural crack develops at B-B is

$$V = \frac{M_{cr}}{M/V - d/2} + V_d$$

where V_d is shear due to the total dead load in section C-C.

Test results also indicate that the load required to develop inclined crack is greater than the load required to develop the flexural crack at section B-B.

The shear V_{cf} in section C-C which causes the inclined crack to occur has been found to equal the shear V plus a shear which is a function of the dimensions of the cross section and the strength of the concrete. This increment is given by $b'd \sqrt{f'_c}$.

Hence, the shear corresponding to inclined cracking originated

BEHAVIOR OF PRESTRESSED-CONCRETE BEAMS

from flexural cracks may be expressed as follows:

$$V_{cf} = \frac{M_{cr}}{M/V - d/2} + V_d + b'd\sqrt{f'_c} \qquad (4\text{-}7)$$

The above equation applies when the profile of prestressed steel is straight. When the profile of steel is draped, the vertical component of the prestressing force at the section under consideration should be included in the expression for V_{cf}. Hence, for a beam in which the profile of the prestressing steel is draped, the following equation should be used:

$$V_{cf} = \frac{M_{cr}}{M/V - d/2} + V_d + b'd\sqrt{f'_c} + V_p \qquad (4\text{-}7a)$$

In Eqs. (4-7) and (4-7a) the cracking moment M_{cr} may be calculated from the following expression:

$$f_d + \frac{M_{cr}y_b}{I} - f_p = f_r = 6\sqrt{f'_c} \qquad (4\text{-}8)$$

where f_d = stress at bottom fiber due to dead load at section where flexural crack takes place
M_{cr} = cracking moment of section at distance $d/2$ from section under consideration
y_b = distance from center of gravity of beam to bottom fiber
I = moment of inertia of beam section
f_p = stress at bottom fiber due to effective prestress at section where flexural crack occurs
$f_r = 6\sqrt{f'_c}$ = modulus of rupture of concrete

Equation (4-8) may be presented conveniently in terms of M_{cr}, as follows:

$$M_{cr} = \frac{I}{y}(6\sqrt{f'_c} + f_p - f_d) \qquad (4\text{-}9)$$

Composite sections Equations (4-7) and (4-7a) can be generalized and applied to composite sections. In composite sections the cast-in-place slab is placed on the top of precast beams. The total dead load is carried by the beam section, and it is assumed that there is sufficient shear connection between the top fiber of the beam and the bottom of the slab that the live load is carried by the composite section consisting of the beam and the slab.

In applying Eqs. (4-7) and (4-7a) to composite sections, M_{cr} should be calculated by the following expression:

$$M_{cr} = \frac{I_c}{y_{bc}}(6\sqrt{f'_c} + f_p - f_d) \qquad (4\text{-}10)$$

where I_c = moment of inertia of composite section
y_{bc} = distance from centroid of composite section to bottom fiber
All other terms are the same as those used in noncomposite construction.

4-6 SHEAR AT INCLINED CRACKING—WEB-DISTRESS FAILURE

Noncomposite sections In some prestressed-concrete beams it is possible to have the inclined crack occurring in the web before the flexural crack. Observations indicate that this type of crack occurs when the principal tensile stress (or strain) in the web reaches the tensile strength (or limiting tensile strain) in concrete. The principal tensile stress at a distance equal to $\frac{3}{4}h$ away from the reactions or point of application of concentrated load may be expressed as follows:

$$\sigma = \frac{f_x}{2} + \sqrt{\tau_{xy}^2 + \left(\frac{f_x}{2}\right)^2} \tag{4-11}$$

where σ = principal tensile stress
f_x = normal stress, positive if tensile
τ_{xy} = shearing stress

The term f_x, the normal stress, may be expressed as follows:

$$f_x = -\frac{P}{A} - \frac{Pey}{I} - \frac{V_{cs}y}{I}\left(\frac{M}{V}\right) \tag{4-12}$$

where P = effective prestressing force
A = area of beam
e = eccentricity of prestressing force at section considered

The shearing stress τ_{xy} may be expressed as follows:

$$\tau_{xy} = \frac{V_{cs}Q}{Ib'} \tag{4-13}$$

where Q = static moment of area of section of beam above point considered about centroidal axis of beam section
b' = web thickness
V_{cs} = shear corresponding to inclined cracking, web-distress failure
y = the distance of the point under consideration from the centroid of the section

With the aid of Eqs. (4-12) and (4-13) the principal tensile stress at any point in the beam may be determined. If it is assumed that initiation of crack is dependent upon a limiting stress, it follows that the shear crack will form when the largest principal tensile stress exceeds the tensile strength of concrete.

Calculation of principal stress in a large number of beams indicated that the inclined crack initiating in the web occurs at the centroid or below the centroid. The beams in which the maximum principal stress occurred below the centroid were the ones in which there was a large likelihood that the flexural crack would also occur.

Hence, the inclined cracking load may be estimated with sufficient accuracy by using the principal tensile stress at the centroid of the section. In this case, Eqs. (4-12) and (4-13) may be expressed as follows:

$$f_x = -\frac{P}{A}$$

$$\tau_{xy} = \frac{V_{cs}Q}{Ib'}$$

The principal tensile stress may now be calculated from Eq. (4-11).

$$f_t = -\frac{P}{2A} + \sqrt{\left(\frac{V_{cs}Q}{Ib'}\right)^2 + \left(\frac{P}{2A}\right)^2}$$

We can solve for V_{cs} from the above equation and, adding V_d, we shall have

$$V_{cs} = \frac{Ib'f_t}{Q}\sqrt{1 + \frac{P}{Af_t}} + V_d \tag{4-14}$$

The tensile strength of concrete, f_t, may be taken as $5\sqrt{f'_c}$.

If the prestressing steel is draped, the vertical component of the prestressing force, V_p, should be included in the expression for V_{cs}:

$$V_{cs} = \frac{Ib'f_t}{Q}\sqrt{1 + \frac{P}{Af_t}} + V_d + V_p \tag{4-14a}$$

Composite sections As before, Eqs. (4-14) and (4-14a) can be generalized to apply to composite sections.

The normal stress for the composite section can be expressed as follows:

$$f_x = -\frac{P}{A} - \frac{Pey}{I} - \frac{M_d y}{I} - \frac{(V_{cs} - V_d)(M/V)y_c}{I_c} \tag{4-15}$$

where M_d is the dead-load moment at the section considered and y_c is the distance of the point under consideration from the centroid of composite section.

The shearing stress τ_{xy} at the inclined cracking at any point in the web can be expressed as follows:

$$\tau_{xy} = \frac{V_d Q}{Ib'} + \frac{(V_{cs} - V_d)Q_c}{I_c b'} \tag{4-16}$$

where Q_c is the static moment of the area of the beam above the section under consideration about the centroidal axis of the composite section.

In the composite beams tested, the centroid of composite section was found to be in the flange of the precast I beam. If inclined crack occurred, it was observed at the connection of web and flange in the beam. If the centroid is in the flange, the shear may be calculated at the connection of web and flange.

Substituting Eqs. (4-15) and (4-16) in Eq. (4-11) and solving for V_{cs}, we have

$$V_{cs} = \frac{I_c b'}{Q_c}\left(f_t\sqrt{1 - \frac{f_x}{f_t}} - V_d\frac{Q}{Ib'}\right) + V_d \quad (4\text{-}17)$$

In all the equations where f_t occurs, its value may be taken as $5\sqrt{f'_c}$.

If the prestressing steel is draped, the vertical component of prestressing force, V_p, should be added to V_{cs}, as follows:

$$V_{cs} = \frac{I_c b'}{Q_c}\left(f_t\sqrt{1 - \frac{f_x}{f_t}} - V_d\frac{Q}{Ib'}\right) + V_d + V_p \quad (4\text{-}17a)$$

Figures 4-7 and 4-8 show that the above equations agree very closely with the test results.

Equations (4-7), (4-7a), (4-17), and (4-17a) describe the load at which inclined cracking occurs. Actually, the ultimate load may be as high as twice the inclined cracking load. However, the condition of the beam

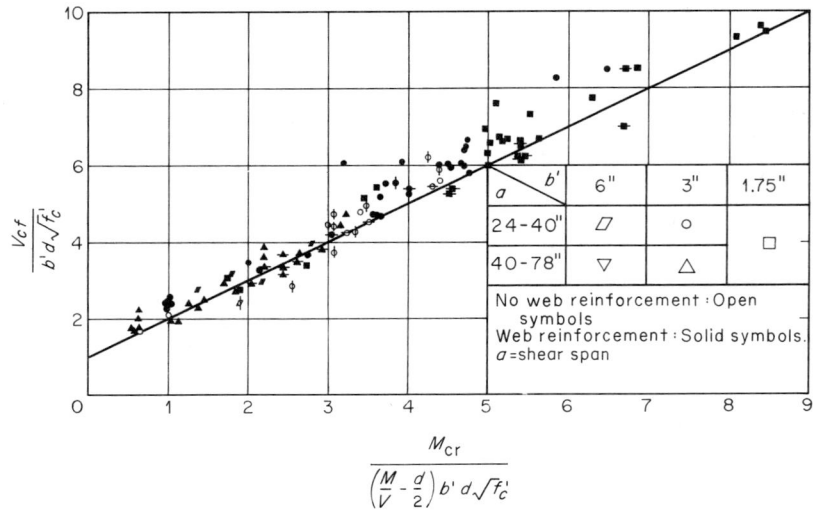

Fig. 4-7 Shear at flexure—shear cracking.

BEHAVIOR OF PRESTRESSED-CONCRETE BEAMS

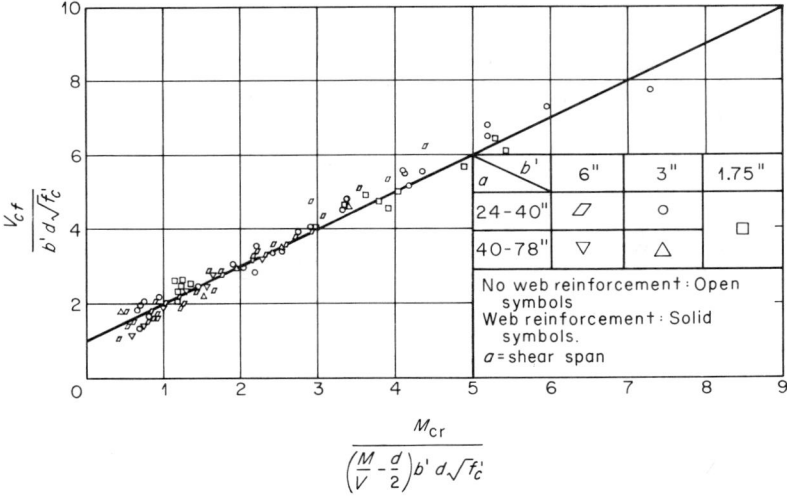

Fig. 4-8 Shear at flexure—shear cracking.

is so radically changed after the formation of inclined cracks that the load corresponding to inclined cracking may be considered the useful-limit load. As a rule, it is very difficult to calculate the ultimate load after the inclined crack has taken place. By simplifying assumptions, it is possible to predict the ultimate load-carrying capacity of the beam.[2] However, from a practical standpoint it is not useful.

Since the flexural capacity of the section usually provides the highest strength and ductility in the section and, as shown in Chap. 3, it can be predicted and controlled accurately, a beam should be designed to fail in flexure. If the inclined cracking load is smaller than the flexural capacity, web reinforcement must be provided to develop both flexural strength and ductility.

4-7 BEAMS WITH WEB REINFORCEMENT

Available data indicate that in beams in which the inclined cracking load is smaller than the flexural capacity, sufficient web reinforcement can be provided that both flexural capacity and ductility can be developed.

Presence of web reinforcement complicates the problem of analysis considerably. It is difficult to formulate an analytical method by which the ultimate load of a beam with web reinforcement can be predicted if such failure is to occur in the region of combined stresses. From a practical point of view, we are not greatly interested in this problem since web reinforcement is provided specifically to develop the flexural strength and ductility of the beam. Therefore, the problem in which we are primarily interested is to determine the amount of web reinforcement that should be provided in order that failure in the region of combined stresses can be prevented until the beam has failed in flexure.

The experimental work available is directed specifically toward solving this problem.

Available experimental results indicate that if the inclined cracking load of the unreinforced web is less than the flexural capacity, web reinforcement increases the load-carrying capacity to a value greater than the inclined cracking load.

The web reinforcement is conveniently measured by the following ratio:

$$r = \frac{A_v}{sb}$$

where A_v = total area of web reinforcement in a section
s = longitudinal spacing of web reinforcement
b = width of top flange

As the amount of web reinforcement is increased, the load-carrying capacity of the beam is increased until r assumes a sufficiently large value to permit development of the flexural strength of the beam.

$$V_u = V_c + rf_y bd \tag{4-18}$$

where V_u = shear at a section due to ultimate load corresponding to flexural failure
V_c = shear corresponding to inclined cracking load, V_{cf} or V_{cs}, whichever is smaller
f_y = yield-point stress of web reinforcement

The above expression assumes that the useful capacity of the beam is reached when the web reinforcement yields. However, this expression underestimates the amount of shear, since the beam can carry additional load even after the web reinforcement has yielded. On the other hand, in some exceptional cases it is possible to develop the flexural capacity without developing the maximum ductility.

Equation (4-18) represents a reasonably simple approach for design of web reinforcement in prestressed-concrete beams.

It should be emphasized that distribution of web reinforcement is an important consideration. If an inclined crack can develop without crossing at least one stirrup, the beam can behave as if no web reinforcement at all was provided. It would be best to have spacing of web reinforcement equal to a very small fraction of the beam depth. However,

BEHAVIOR OF PRESTRESSED-CONCRETE BEAMS

if the spacing is as much as one-half of depth, no appreciable difference can be noticed.

Equation (4-18) can be expressed as follows:

$$rf_y bd = V_u - V_c = \frac{A_v}{s} f_y d$$

or

$$A_v = \frac{V_u - V_c}{f_y d} s \qquad (4\text{-}19)$$

In cases where $V_u \leq V_c$, some web reinforcement should be provided.

4-8 ILLUSTRATIVE EXAMPLE 4-1

Let us consider the beam in Illustrative Problem 3-5. The section shown is used on a span of 60 ft. The effective prestress is 142 ksi for a total $P = 324$ kips, the center of gravity of the reinforcement is shown, and the beam weighs 0.3 klf and is subject to a superimposed dead load of 0.8 klf, hence $W_d = 1.1$ klf, and the live load W is 0.4 klf.

Calculate the inclined cracking loads and the required web reinforcement such that the flexural capacity will be developed.

Determination of V_{cf} and V_{cs}   The section properties of the beam are

$$A = 288 \text{ in.}^2 \qquad I = 48{,}400 \text{ in.}^4$$

$$\frac{I}{A} = \frac{48{,}400}{288} = 168 \text{ in.}^2 \qquad \frac{I}{y} = \frac{48{,}400}{18} = 2690 \text{ in.}^3$$

and at the centroidal section

$$Q = 1728 \text{ in.}^3$$

Figure 4-9 shows the left half of the beam with the shear and moment diagrams and variation of eccentricity along the span.

Inclined cracking load initiated by flexural cracking may be calculated by Eq. (4-7a):

$$V_{cf} = \frac{M_{cr}}{M/V - d/2} + V_d + V_p + b'd \sqrt{f'_c} \qquad (4\text{-}7a)$$

Let us consider sections B and C, which have distances of X and Z from the left support, respectively. Section B is located at a distance $d/2$ from section C in the direction of decreasing moment; therefore, $X = Z - d/2$. The quantity d is the effective depth of the beam at point C.

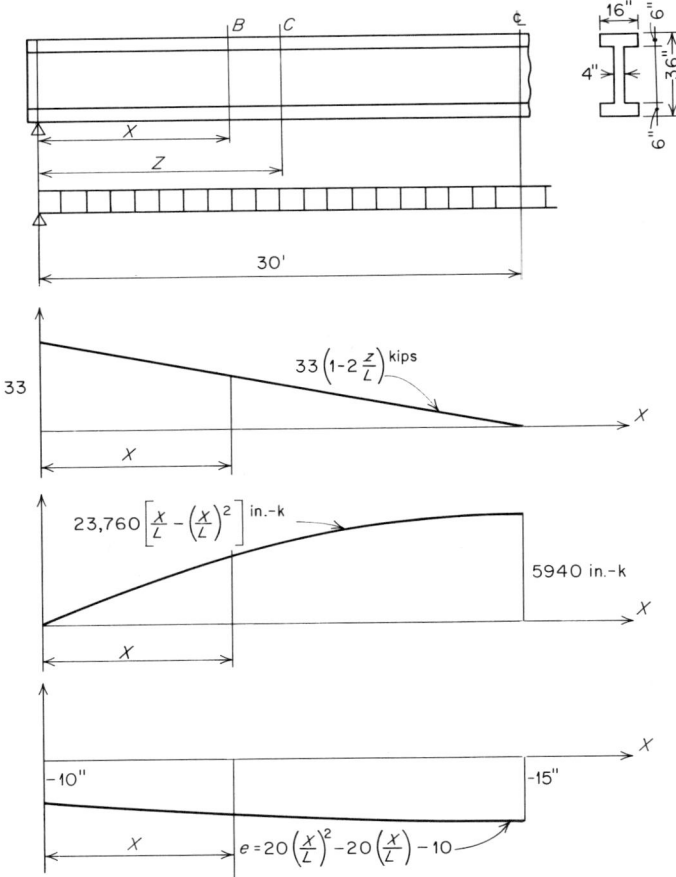

Fig. 4-9 Illustrative Problem 4-1.

For point B we have

$$M_d = \frac{W_d L}{2} X - \frac{W_d X^2}{2} = \frac{W_d L^2}{2}\left[\frac{X}{L} - \left(\frac{X}{L}\right)^2\right]$$

$$= \frac{1.10 \times (60)^2 \times 12}{2}\left[\frac{X}{L} - \left(\frac{X}{L}\right)^2\right]$$

$$= 23{,}760\left[\frac{X}{L} - \left(\frac{X}{L}\right)^2\right] \quad \text{in.-kips}$$

$$f_d = \frac{23{,}760}{2690}\left[\frac{X}{L} - \left(\frac{X}{L}\right)^2\right] = 8.84\left[\frac{X}{L} - \left(\frac{X}{L}\right)^2\right] \quad \text{ksi}$$

BEHAVIOR OF PRESTRESSED-CONCRETE BEAMS

$$e = 20\left(\frac{X}{L}\right)^2 - 20\frac{X}{L} - 10 \text{ in.}$$

$$\frac{Pey}{I} = \frac{324}{2690}\left[-20\left(\frac{X}{L}\right)^2 + 20\frac{X}{L} + 10\right] \quad \text{ksi}$$

$$= 2.41\left[-\left(\frac{X}{L}\right)^2 + \frac{X}{L} + \frac{1}{2}\right] \quad \text{ksi}$$

$$M_{cr} = \frac{I}{y}\left(6\sqrt{f'_c} + \frac{P}{A} + \frac{Pey}{I} - M_d\frac{y}{I}\right)$$

$$= 2690\left[\frac{6\sqrt{5000}}{1000} + 1.125 - 2.41\left(\frac{X}{L}\right)^2 + 2.41\frac{X}{L} + 1.20\right.$$

$$\left. - 8.84\frac{X}{L} + 8.84\left(\frac{X}{L}\right)^2\right] \quad \text{in.-kips}$$

$$M_{cr} = 17{,}300\left[0.427 + \left(\frac{X}{L}\right)^2 - \frac{X}{L}\right] \quad \text{in.-kips}$$

For point C we have

$$d = \left[-20\left(\frac{Z}{L}\right)^2 + 20\frac{Z}{L} + 28\right] \quad \text{in.}$$

$$\frac{M}{V} = \frac{(WL/2)Z - WZ^2/2}{WL/2 - WZ} = L\frac{(Z/L)^2 - Z/L}{2Z/L - 1}$$

$$V_d = \frac{W_d L}{2} - W_d Z = \frac{W_d L}{2}\left(1 - \frac{2Z}{L}\right)$$

$$= 33\left(1 - 2\frac{Z}{L}\right) \quad \text{kips}$$

$$b'd\sqrt{f'_c} = \frac{4\sqrt{5000}}{1000}\left[-20\left(\frac{Z}{L}\right)^2 + 20\frac{Z}{L} + 28\right] \quad \text{kips}$$

$$b'd\sqrt{f'_c} = -5.66\left(\frac{Z}{L}\right)^2 + 5.66\frac{Z}{L} + 7.92$$

$$V_p = \frac{P}{L}\left(-40\frac{Z}{L} + 20\right) = -18\frac{Z}{L} + 9$$

The above quantities, as well as V_{cf}, are listed in Table 4-1 for 3 ft intervals along the beam.

The inclined cracking load for web-distress type of failure may be obtained from Eq. (4-14a) as follows:

$$V_{cs} = \frac{Ib'f_t}{Q}\sqrt{1 + \frac{P}{Af_t}} + V_d + V_p \quad (4\text{-}14a)$$

Table 4-1 Calculation of V_{cf}, V_{cs}, and V_u for Illustrative Problem 4-1

Z, in.	$\dfrac{Z}{L}$	d, in.	X, in.	$\dfrac{X}{L}$	M_{cr}, in.-kips	$\dfrac{M}{V} - \dfrac{d}{2}$, in.	$\dfrac{M_{cr}}{\frac{M}{V} - \frac{d}{2}}$, kips	V_d, kips	V_p, kips	$b'd\sqrt{f_c'}$, kips	V_{cf}, kips	V_{cs}, kips	V_u, kips
0	0	28.00											
36	0.05	28.95	21.5	0.030	6870	23.5	280.0	29.7	8.1	8.2	321.0	118.8	78.6
72	0.10	29.80	57.1	0.079	6130	66.1	92.8	26.4	7.2	8.4	134.8	114.6	69.8
108	0.15	30.55	92.2	0.128	5450	115.8	47.1	23.1	6.3	8.6	85.1	110.4	61.1
144	0.20	31.20	128.4	0.178	4840	176.4	27.4	19.8	5.4	8.8	61.4	106.2	52.4
180	0.25	31.75	164.1	0.228	4340	253.1	17.1	16.5	4.5	9.0	47.1	102.0	43.6
216	0.30	32.20	199.9	0.278	3910	361.9	10.8	13.2	3.6	9.1	36.7	97.8	34.9
252	0.35	32.55	235.7	0.327	3580	529.7	6.8	9.9	2.7	9.2	28.6	93.6	26.2
288	0.40	32.80	271.6	0.377	3320	847.6	3.9	6.6	1.8	9.3	21.6	89.4	17.5
324	0.45	32.95	307.5	0.427	3160	1765.5	1.8	3.3	0.9	9.3	15.3	85.2	8.7
360	0.50	33.00	343.5	0.476	3090	∞	0	0	0	9.3	9.3	81.0	0

BEHAVIOR OF PRESTRESSED-CONCRETE BEAMS

where

$$f_t = 5\sqrt{f'_c} = \frac{5\sqrt{5000}}{1000} = 0.354 \text{ ksi}$$

and at midspan:

$$V_{cs} = \frac{48,400 \times 4 \times 0.354}{1728}\sqrt{1 + \frac{1.125}{0.354}} = 81.0 \text{ kips}$$

Other values of V_{cs} are also listed in Table 4-1.

Selection of web reinforcement In order to determine V_u, we shall first calculate the load corresponding to the flexural strength of the section. At each section V_u may be calculated as the shear at that section due to this load.

From Illustrative Problem 3-5 the ultimate moment of this section is 15,700 in.-kips. The magnitude of the uniformly distributed load corresponding to this moment is the following:

$$w_u = \frac{8M}{L^2} = 8 \times \frac{15,700}{720 \times 720} \times 12 = 2.91 \text{ klf}$$

and

$$V_u = \tfrac{2.91}{2}L\left(1 - 2\frac{Z}{L}\right) = 87.3\left(1 - 2\frac{Z}{L}\right)$$

The values of V_u are listed in Table 4-1. It can be seen that in this case V_u is always smaller than either V_{cf} or V_{cs}. Calculations therefore indicate that web reinforcement is not needed.

Since there is a considerable uncertainty associated with failure of the beam in the region of combined stresses, it is desirable to have a minimum amount of web reinforcement. In this case No. 3 stirrups at 12-in. centers may be used.

4-9 ILLUSTRATIVE PROBLEM 4-2

The composite beam shown in Fig. 4-10a is a section of bridge deck having a span of 60 ft. The weight of the beam as well as the slab is carried by the beam section. The live load is carried by the composite section. There is sufficient shear connection between the beam and the slab to develop the flexural strength of the composite section. The profile of the center of gravity of steel is shown in Fig. 4-10b. The moving live load is shown in Fig. 4-10c.

Calculate the inclined cracking loads V_{cf} and V_{cs}, and determine the amount of web reinforcement required for the flexural capacity of the composite section to be developed.

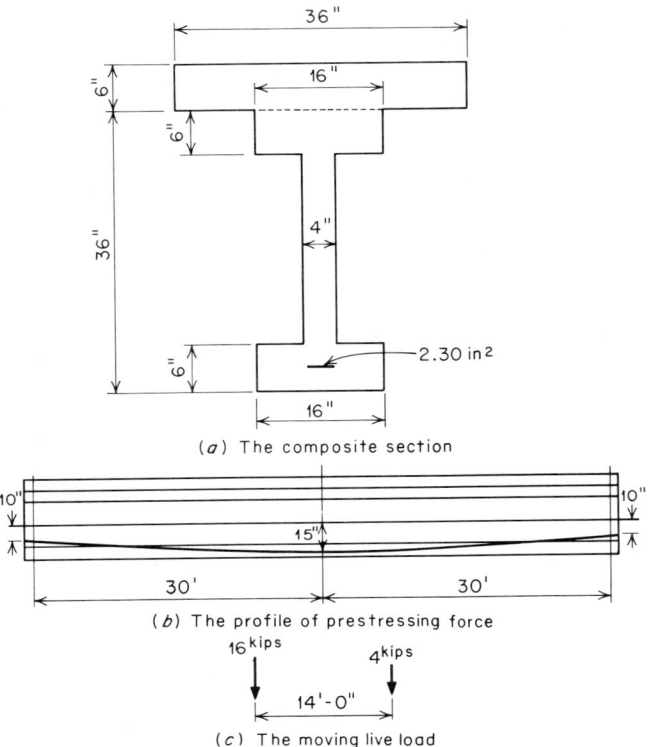

Fig. 4-10 Illustrative Problem 4-2.

Solution Figure 4-10 shows the cross sections of beam and composite sections. The section properties of the beam section are

$$A = 288 \text{ in.}^2 \qquad I = 48{,}400 \text{ in.}^4$$

$$\frac{I}{A} = \frac{48{,}400}{288} = 168 \text{ in.}^2 \qquad \frac{I}{y} = \frac{48{,}400}{18} = 2690 \text{ in.}^3$$

and at the centroidal section

$$\frac{Q}{Ib'} = \frac{1728}{48{,}400 \times 4} = 0.00893 \text{ in.}^{-2}$$

The section properties of the composite section are

$$A_c = 504 \text{ in.}^2 \qquad y_{bc} = 27.0 \text{ in.} \qquad I_c = 103{,}500 \text{ in.}^4$$

$$W_d = 0.525 \text{ klf} \qquad \frac{I_c}{y_{bc}} = \frac{103{,}500}{27.0} = 3830 \text{ in.}^3$$

BEHAVIOR OF PRESTRESSED-CONCRETE BEAMS

and at the centroid of the composite section

$$Q_c = 3186 \text{ in.}^3 \qquad \frac{I_c b'}{Q_c} = \frac{103{,}500 \times 4}{3186} = 130 \text{ in.}^2$$

$$M_{cr} = 19{,}150 \left[0.427 + \left(\frac{X}{L}\right)^2 - \frac{X}{L} \right]$$

The quantity M/V at each section may be calculated by considering the position of wheel loads, which give the maximum live-load shear on the section.

Figure 4-11 shows the position of wheels which produce the maximum shear in section C. It can be seen that the quantity M/V for the section at a distance Z from the left support is equal to Z for this loading. Hence,

$$\frac{M}{V} - \frac{d}{2} = 10\left(\frac{Z}{L}\right)^2 + 710\frac{Z}{L} - 14 \text{ in.}$$

The quantities M_{cr}, $(M/V - d/2)$, and V_{cf} are listed in Table 4-2 for various sections along the span. The quantity V_{cs} to be computed subsequently is also listed.

The quantity V_{cs} may be calculated from Eq. (4-17a):

$$V_{cs} = \frac{I_c b'}{Q_c}\left(f_t\sqrt{1 - \frac{f_x}{f_t}} - V_d \frac{Q}{Ib'}\right) + V_d + V_q \qquad (4\text{-}17a)$$

This quantity may be calculated at the center of gravity of the composite section. In this case

$$f_p = -\frac{P}{A} - \frac{Pe(y_{bc} - y)}{I} = -1.125 - \frac{324 \times (27.0 - 18.0)}{48{,}400} e$$

and substituting $e = 20(Z/L)^2 - 20Z/L - 10$ in the above expression we obtain

$$f_p = -1.21\left(\frac{Z}{L}\right)^2 + 1.21\frac{Z}{L} - 0.52 \text{ ksi}$$

The quantity f_d may be computed from $-M_d(y_b - y_c)/I$, where

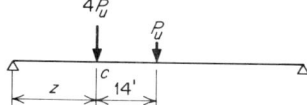

Fig. 4-11 Position of wheels for maximum shear at section C.

Table 4-2 Calculation of V_{cf} for Illustrative Problem 4-2

$\dfrac{Z}{L}$	$\dfrac{X}{L}$	M_{cr}, in.-kips	$\dfrac{M}{V} - \dfrac{d}{2}$, in.	$\dfrac{M_{cr}}{\dfrac{M}{V} - \dfrac{d}{2}}$, kips	V_d, kips	V_{cf}, kips	V_{cs}, kips	V_u, kips
0.025							80.0	103.23
0.05	0.026	7600	18.52	410	14.2	440.5	80.3	100.45
0.10	0.075	6790	54.10	125	12.6	153.2	80.6	94.90
0.15	0.125	6040	89.73	67.2	11.0	93.1	81.5	89.35
0.20	0.174	5350	125.40	42.7	9.4	66.3	81.5	83.80
0.25	0.224	4800	161.13	29.8	7.9	51.2		78.30
0.30	0.274	4330	196.90	22.0	6.3	41.0		72.70
0.35	0.323	3960	232.73	17.0	4.7	33.6		67.20
0.40	0.373	3680	268.60	13.7	3.2	28.0		61.60
0.45	0.423	3500	304.53	11.5	1.6	23.3		56.10
0.50	0.473	3420	340.50	10.1	0	19.4		50.50

$M_d = [(0.525 \times 60^2 \times 12)/12][Z/L - (Z/L)^2]$. Then,

$$f_d = 2.11\left(\dfrac{Z}{L}\right)^2 - 2.11\dfrac{Z}{L} \quad \text{ksi}$$

$$f_x = f_p + f_d = 0.9\left(\dfrac{Z}{L}\right)^2 - 0.9\dfrac{Z}{L} - 0.52 \text{ ksi}$$

The values of V_p and V_d are listed in Tables 4-1 and 4-2 respectively.

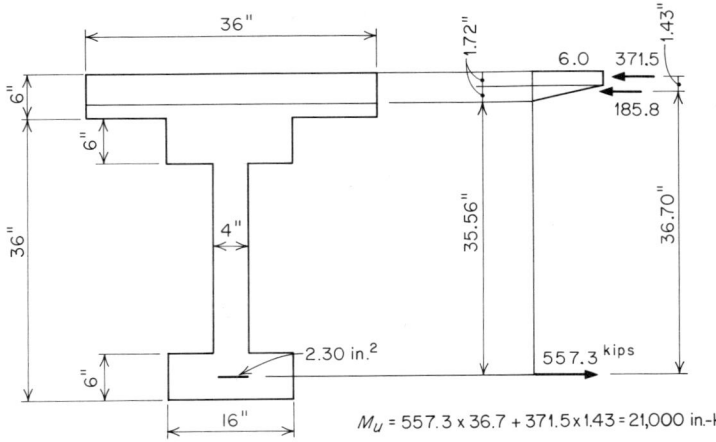

Fig. 4-12 Ultimate moment of the composite section, Illustrative Problem 4-2.

BEHAVIOR OF PRESTRESSED-CONCRETE BEAMS

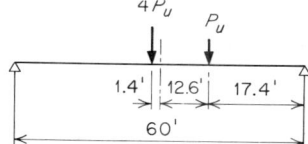

Fig. 4-13 Position of wheels corresponding to the maximum moment.

The various quantities necessary for the calculation of V_{cs} as well as V_{cs} are listed in Table 4-3 for a few sections near the end of the beam. The values of V_{cs} are also listed in Table 4-2 so that they can be compared with the corresponding values of V_{cf}.

From Fig. 4-12 it can be seen that the flexural strength of the composite section is 21,000 in.-kips.

In order to calculate the magnitude of the ultimate live load which would develop the above moment, we shall place the loads in the position shown in Fig. 4-13, and calculate the bending moment under the larger wheel load.

From Fig. 4-13 we have

$$2.39 P_u \times 28.6 \times 12 = 21,000 - \frac{0.525(60)^2}{8} 12$$

or

$$P_u = 22.2 \text{ kips}$$

Now we shall place the wheels in such positions as to obtain the maximum shear. Figure 4-11 shows the position of wheels which gives the maximum shear in a section at a distance Z from the left support.

Hence, from Fig. 4-11 we have

$$V_u = 106 - 1.85Z$$

Values of V_u for various Z values are listed in Table 4-2.

From Table 4-2 it can be seen that the maximum difference $V_u - V_{cf}$ or $V_u - V_{cs}$ is 34 kips.

Table 4-3 Calculation of V_{cs} for Illustrative Problem 4-2

$\frac{Z}{L}$	f_x, ksi	$f_t\sqrt{1-\frac{f_x}{f_t}}$	$V_d \frac{Q}{Ib'}$	V_{cs}
0.025	−0.543	0.561	0.127	80.0
0.05	−0.563	0.568	0.127	80.3
0.10	−0.601	0.578	0.113	80.6
0.15	−0.635	0.590	0.098	81.5
0.20	−0.664	0.600	0.084	81.5

Hence, the spacing of web reinforcement if No. 3 bars are used will be, from Eq. (4-19),

$$s = \frac{0.11 \times 40 \times 39}{34} \sim 5 \text{ in.}$$

PROBLEMS

4-1. Derive general expressions for the stress trajectories for the principal tensile and compressive stresses before cracking for the reinforced-concrete beam shown in Fig. 4-14. Assume a linear stress-strain diagram for concrete in tension and in compression. Assume the origin of the x axis at the left support, and y axis at middepth. For the special case $L = 36$ in., $b = 6$ in., and $h = 12$ in., calculate the principal tensile stress at 2 in. intervals through the depth at sections 18 and 54 in. from the left support.

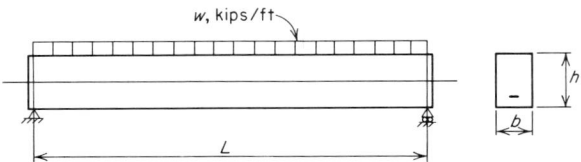

Fig. 4-14 The beam of Prob. 4-1.

4-2. The section shown in Fig. 4-15 is subject to a prestressing force of 324 kips. Calculate the principal tensile stresses at several points through the depth at sections A-A and B-B. Where does the maximum principal tensile stress occur in these sections?

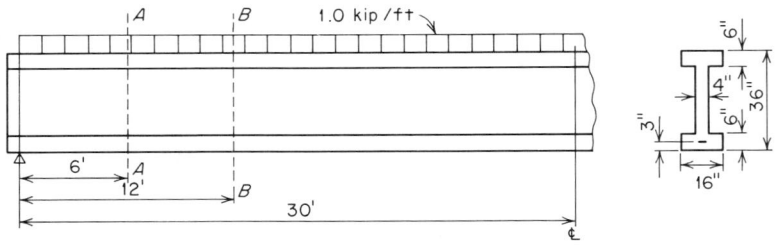

Fig. 4-15 The beam of Prob. 4-2.

4-3. Solve Prob. 4-2 for a prestressing force of 450 kips and the section shown in Fig. 4-15.

4-4. The beam shown in Fig. 4-15 is subject to three moving loads, shown in Fig. 4-16.
(a) Calculate P corresponding to M_u.
(b) Calculate V_u at 6 ft intervals along the span.
The stress-strain diagrams for steel and concrete are shown in Figs. 3-16 and 3-17b.

Fig. 4-16 The moving loads of Prob. 4-4.

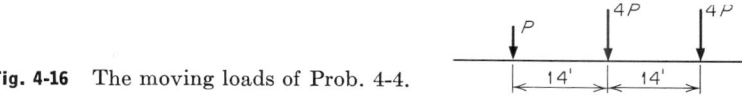

4-5. Calculate P corresponding to the cracking moment M_{cr} along the span for each of the two sections shown in Fig. 4-17. Prestressing force may be taken as 200 kips. For this problem assume $f_t = 6\sqrt{f'_c}$.

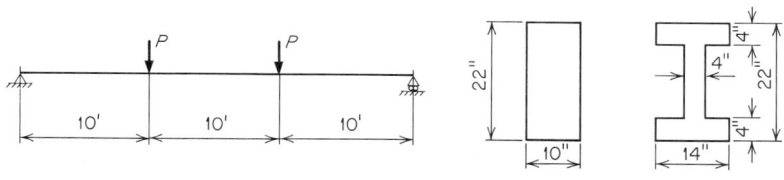

4-6. Calculate V_{cs} and V_{cf} for the sections shown in Fig. 4-17.

REFERENCES

1. Krahl, N. W., N. Khachaturian, and C. P. Siess: Stability of Tensile Cracks in Concrete Beams, *J. Structural Div., Am. Soc. Civil Engrs.*, vol. 93, no. ST1, pp. 235-254, February, 1967.
2. Sozen, M. A., E. M. Zwoyer, and C. P. Siess: Strength in Shear of Beams without Web Reinforcement, *Univ. Illinois Eng. Expt. Sta. Bull.* 452, April, 1959.
3. MacGregor, J. G., M. A. Sozen, and C. P. Siess: Effect of Draped Reinforcement on Behavior of Prestressed Concrete Beams, *Proc. Am. Concrete Inst.*, vol. 57, pp. 649-677, December, 1960.
4. Olsen, S. E., M. A. Sozen, and C. P. Siess: Strength in Shear of Beams with Web Reinforcement, *Univ. Illinois Eng. Expt. Sta. Bull.* 493, July, 1967.

5
Bond and Anchorage

5-1 INTRODUCTION

In the preceding chapters, we discussed the behavior of the beam in the regions of pure flexure and combined stresses. The discussions in these chapters depended upon the assumption that the prestressing force was satisfactorily transferred to the concrete and that sufficient bond existed between concrete and steel to develop the ultimate strain in the steel. In this chapter we shall first examine the stress-transfer and flexural bonds in pretensioned beams and then look into the local stresses developed at the anchorage zones of pretensioned and post-tensioned beams.

5-2 STRESS-TRANSFER BOND IN PRETENSIONED CONCRETE BEAMS

In pretensioned beams, the transfer of prestressing force from steel to concrete is carried out entirely by means of the bond between steel and concrete. Bond in this case has a very important function and is referred to as stress-transfer bond.

BOND AND ANCHORAGE

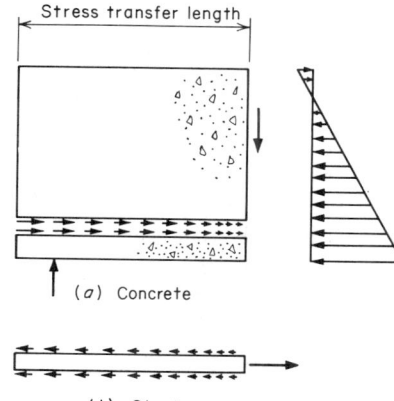

Fig. 5-1 Free-body diagram of forces acting on concrete and steel.

Stress-transfer bond exists at each end of a pretensioned beam in the region where steel is anchored to concrete. The stress in the steel increases from zero at each end to the prescribed amount of prestress at a certain distance from each of the ends. Bond stress, therefore, is present only when the prestress in steel changes; and beyond the section for which prestress reaches a constant value, there is no further stress-transfer bond. The distance from the end of the beam necessary to develop the prestress is the transfer length. Figure 5-1a and b shows free-body diagrams of forces acting on concrete and steel.

There are three factors that contribute to develop bond between steel and concrete: (1) adhesion, (2) friction, and (3) mechanical shear similar to that provided by the deformations in reinforcement bars. Though steel used in prestressing has no deformations, the surface shape of the strands creates similar effect.

Test data are available for the determination of transfer length of both wires and strands. In the following paragraphs, the variables associated with the transfer length will be discussed for each type of steel.

Wires Available test data show that immediately after the wires are cut the steel sinks in the beam; that is, a certain amount of slip takes place between the concrete and steel. This indicates that the adhesion between concrete and steel is destroyed after cutting of wires.[1] Furthermore, since the contact surface of wires used in prestressing is smooth, the mechanical shear connection does not completely exist. Hence, it is reasonable to assume that the stress transfer from steel to concrete is provided to a large extent by the friction between concrete and steel.

An analysis of the stress-transfer bond which leads to the determination of the required transfer length has been developed, assuming that the stress transfer is provided by friction only.[1] In the following paragraphs this method is presented as a guide to understanding the stress-transfer mechanism. The method applies primarily to wires.

As the wires are cut and tension is transferred from steel to concrete, the steel begins to slip and, because of loss in prestress, the diameter of the wire is increased slightly, resulting in a radial pressure exerted on the concrete by the steel. Hence, the frictional force between concrete and steel is increased.

Change of tension at any point along the pretensioned wire or strand in the end zone results in a change in the radius of wire. If the wire were free to expand, the change in the radius could be expressed as follows:

$$\Delta r_1 = r(f_{se} - f_s) \frac{\nu_s}{E_s} \qquad (5\text{-}1)$$

where Δr_1 = change in radius of wire
r = radius of wire
f_{se} = effective prestress
f_s = stress in steel at any point within transfer length
ν_s = Poisson ratio of wire
E_s = modulus of elasticity of wire

However, the wire is not free to expand. The surrounding concrete restrains this expansion by applying radial stress on the wire. If we assume that the concrete section is large enough in comparison with the diameter of the wire, we can calculate the increase in the radius of wire by using the elastic theory of a thick-wall cylinder, as follows:

$$\Delta r_2 = r\sigma_r \frac{1 + \nu_c}{E_c} \qquad (5\text{-}2)$$

where Δr_2 = change in radius due to constraining radial stress
σ_r = radial stress at contact surface between concrete and steel
ν_c = Poisson ratio of concrete
E_c = modulus of elasticity of concrete

Hence, the radial stress can be expressed as follows:

$$\sigma_r = \frac{\Delta r_1 - \Delta r_2}{r} E_s = \left[r(f_{se} - f_s) \frac{\nu_s}{E_s} - r\sigma_r \left(\frac{1 + \nu_c}{E_c} \right) \right] \frac{E_s}{r}$$

or

$$\sigma_r = \frac{(f_{se} - f_s)\nu_s}{1 + (1 + \nu_c)E_s/E_c} \qquad (5\text{-}3)$$

If bond is assumed to be due entirely to friction, then equilibrium of an infinitesimal portion of wire requires that $u(dl)(2\pi r) = (df_s)(\pi r^2)$,

BOND AND ANCHORAGE

from which bond stress at any point in the end zone may be expressed as follows:

$$u = \frac{df_s}{dl}\frac{r}{2} = \mu\sigma_r \tag{5-4}$$

where u = unit bond stress
l = distance from end of pretensioned member
μ = coefficient of friction

From Eqs. (5-3) and (5-4) we have

$$dl = \frac{r}{2\mu\sigma_r}\,df_s = \frac{[1 + (1 + \nu_c)E_s/E_c]r}{2(f_{se} - f_s)\mu\nu_s}\,df_s$$

or

$$l = \frac{-r}{2\mu\nu_s}\left[1 + (1 + \nu_c)\frac{E_s}{E_c}\right]\log\frac{f_{se} - f_s}{f_{se}} \tag{5-5}$$

For $E_s/E_c = 8.5$, $\nu_c = 0.15$, and $\nu_s = 0.3$ we have

$$\frac{l}{r} = \frac{17.96}{\mu}\log\left(1 - \frac{f_s}{f_{se}}\right) \tag{5-5a}$$

or

$$\frac{f_s}{f_{se}} = 1 - \exp\left(-\frac{l}{r}\frac{\mu}{17.96}\right) \tag{5-5b}$$

Table 5-1 shows the values of L/r for various values of f_s/f_{se} calculated from Eq. (5-5a). Figure 5-2 shows the relationship between l/r and f_s/f_{se} for $\mu = 0.2$, 0.4, and 0.6.

Table 5-1 Values of $\dfrac{l}{r}$ for various values of $\dfrac{f_s}{f_{se}}$

$\dfrac{f_s}{f_{se}}$	$\mu = 0.2$	$\mu = 0.4$	$\mu = 0.6$
0	0	0	0
0.1	9.4	4.7	3.1
0.2	20.0	10.0	6.7
0.3	32.0	16.0	10.7
0.4	46.0	23.0	15.4
0.5	62.2	31.1	20.8
0.6	82.1	41.1	27.5
0.7	108.2	54.1	36.1
0.8	142.5	71.3	48.2
0.9	207.0	103.5	69.0
1.0	∞	∞	∞

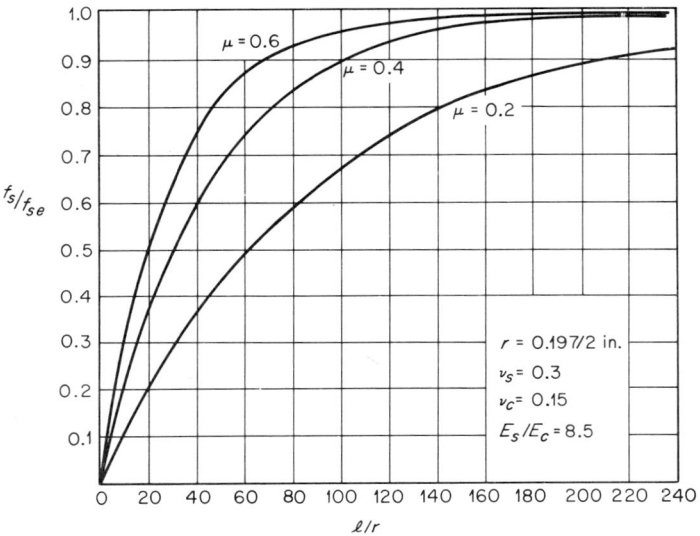

Fig. 5-2 Theoretical variation of stress at the end zone.

Figure 5-2 shows that stress is transferred over a shorter distance when the coefficient of friction μ is large and the diameter of wire is small.

It should be pointed out that the above elastic analysis is based on the assumption that only the friction between concrete and steel is responsible for transfer of stress from steel to concrete. In view of its simplicity, this analysis should be considered only as a rough qualitative guide and not as a formula for accurate determination of stress at any point along the stress-transfer zone.

However, available tests show that there is a reasonable agreement between test results and the theoretical work presented in Fig. 5-2.

The test results further indicate that the transfer length increases with the diameter of wire. The strength of concrete appears to have small influence upon the length of transfer required, though higher strength in concrete tends to decrease the required transfer length slightly. Wires with surface rust help to improve bond properties and require a smaller transfer length. Rusted mill scale, however, does not improve bond properties.

Strands Tests have also been carried out on various sizes of clean strands.[2] These tests indicate that the transfer length varies with the strand size; however, strength of concrete does not have an appreciable influence

BOND AND ANCHORAGE 161

on the transfer length. Sudden transfer of stress tends to lengthen the required transfer length, but gradual releasing results in a decrease in the required transfer length.

Equations (5-5a) and (5-5b), developed for wires, yield transfer lengths that are approximately correct. In comparison with wires, the strands have smaller contact area in relation with their cross-sectional area, but their surface irregularities provide, to a certain extent, mechanical shear connection.

The following expression seems to give a reasonable value for the required transfer length of strand:

$$l_t = f_{se} \frac{D}{3} \tag{5-6}$$

where l_t = transfer length, in.
f_{se} = effective prestress, ksi
D = diameter of strand, in.

5-3 FLEXURAL BOND IN PRETENSIONED CONCRETE BEAMS

Flexural bond exists in pretensioned beams immediately after the beam is subjected to load when prestressing elements are released. However, flexural bond of significant magnitude exists only after flexural cracking has taken place. A concentration of bond stress exists in the vicinity of cracks, which is transferred toward the ends of the beam as the load is increased. If the bond stresses become very large in the vicinity of the region of transfer length, failure may be initiated at the anchorage, leading to general slip. Limitation of flexural bond, therefore, is necessary in order to prevent this type of failure.

Bond stress cannot be calculated in any degree of accuracy at any given point of the beam. However, the average bond stress along the length of strand from the free end of the beam to the section of maximum stress may be used as a convenient measure for evaluating bond in design.

The average bond stress may be expressed as follows:

$$u_a = \frac{f_s A_s}{l \Sigma 0} \tag{5-7}$$

where u_a = average bond stress
f_s = stress in steel
l = embedment length, distance from end of beam to point under consideration
$\Sigma 0$ = sum of nominal circumferences of strands; may be taken as $\frac{4}{3}\pi D$

Equation (5-7) indicates that a short transfer length introduces a higher average bond stress. Also very high average bond stresses are necessary to develop the full tensile strength of the steel in a reasonable transfer length. Figure 5-3 shows the relation between average bond strength and necessary length of embedment to develop the tensile strength of $\frac{3}{8}$- and $\frac{1}{2}$-in. strands. The nominal perimeters of the strands were taken as 1.58 and 2.10 in., respectively, and 250 ksi was assumed as the tensile strength of the steel.

For design purposes it is necessary to develop the flexural strength of the beam to prevent any possible failure in bond. Analysis indicates that if the actual average bond for a given transfer length is less than that shown in Fig. 5-3, the full strength of the steel may be developed and consequently the beam will fail in flexure before it fails in bond. However, the average bond stresses required in Fig. 5-3 are conservative since in practical sections the stress in the steel when flexural failure of the beam occurs is usually less than 250 ksi.

Tests on prestressed-concrete beams give the relationship between stress in steel, at general bond slip, and strand embedment length.[3]

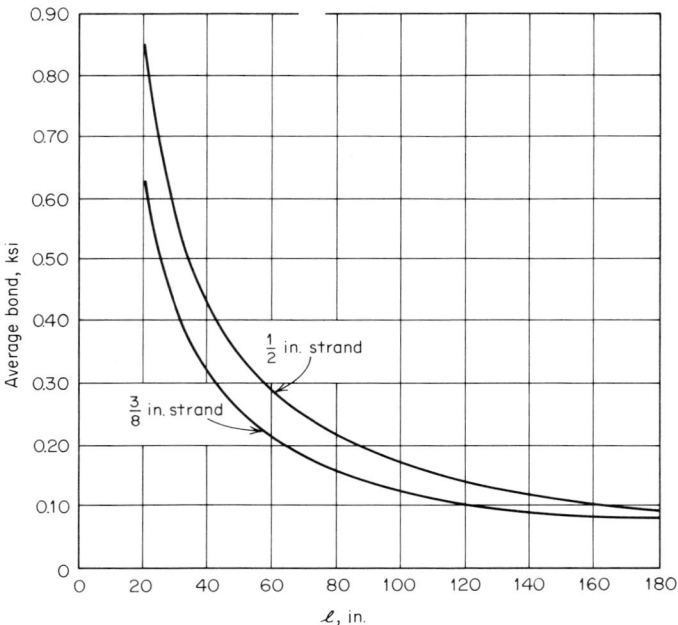

Fig. 5-3 Relationship between average bond and length of embedment.

BOND AND ANCHORAGE

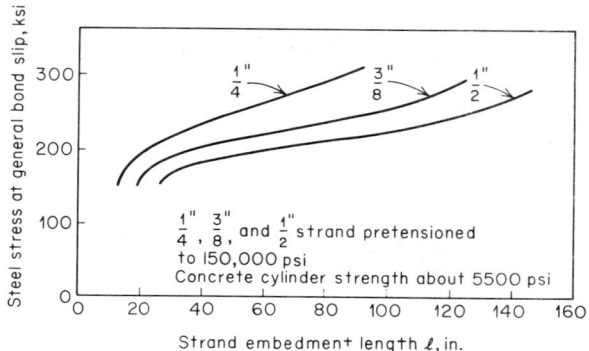

Fig. 5-4 Relation of steel stress at general bond slip to strand embedment length l_u.

Figure 5-4 is a reproduction of the results of these tests, specifically for an effective prestress of 150 ksi.

In a given pretensioned beam, if f_{su}, the stress in steel at ultimate, is known, Fig. 5-4 can be used to determine if a general slip can occur.

The 1963 ACI code specifies that the strand shall be bonded to the concrete from the cross section under consideration for a distance l in inches so that

$$l \geq (f_{su} - \tfrac{2}{3}f_{se})D \tag{5-8}$$

where D, the nominal strand diameter, is in inches and f_{su} and f_{se} are expressed in kips per square inch.

5-4 ANCHORAGE-ZONE STRESSES

Considerable attention has been given during the past few years to the stresses caused by the large forces acting at the anchorage zone. The interest for this problem is due to a large extent to the designers' concern about the formation of visible longitudinal cracks at the ends of the beam. Figure 5-5 shows such a crack in a pretensioned beam without vertical reinforcement.[8]

The determination of the magnitude of stresses is a difficult problem which is somewhat different in pretensioned as compared with posttensioned construction. Figure 5-6a shows a free-body diagram of all the forces acting on concrete at one of the ends of a pretensioned prestressed-concrete beam with rectangular cross section. Figure 5-6b shows a similar diagram for a post-tensioned beam.

From Fig. 5-6a and b, it can be seen that the forces acting on concrete at the ends consist of the transfer forces (as applied by the steel

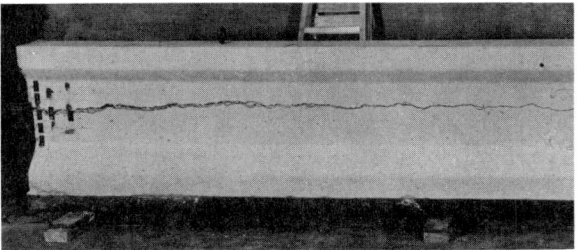

Fig. 5-5 Severe horizontal cracking in a girder without vertical stirrup reinforcement.

to the concrete) and the vertical reaction. In addition, there are shearing (distribution not shown) and normal stresses at the section considered. The largest and the most critical force, however, is the transfer force, which is several times larger than the vertical reaction.

The problem consists in determining the stresses created in the anchorage zones by the forces shown in Fig. 5-6. Since there is a concentration of large forces, high stresses are developed in the concrete at

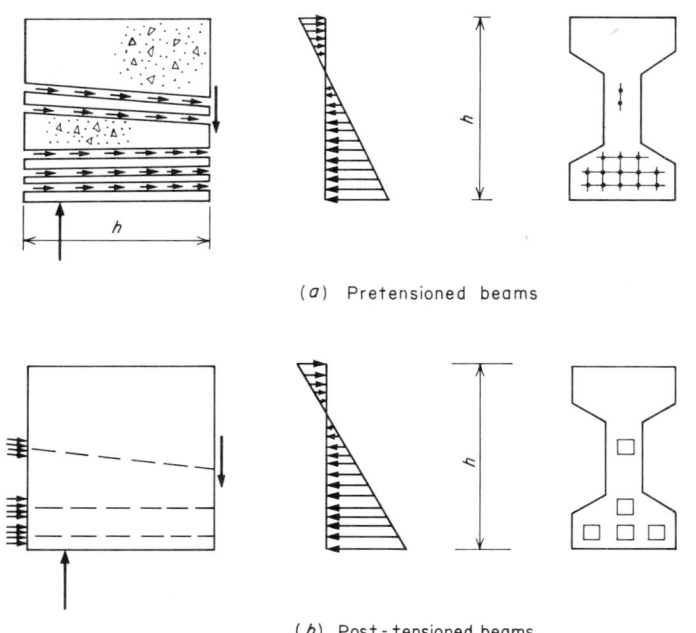

(a) Pretensioned beams

(b) Post-tensioned beams

Fig. 5-6 Anchorage forces.

BOND AND ANCHORAGE

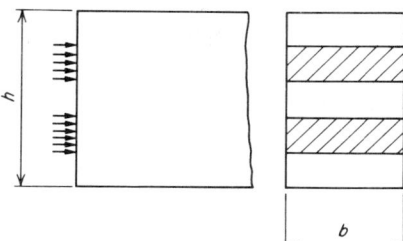

Fig. 5-7 Idealization of the anchorage zone for two-dimensional analysis.

the beam ends. At a distance from the end equal to about the overall depth of the section, beam action is restored. This region is called the transfer zone.

The stresses vary appreciably in value and nature along the span, depth, and width of the beam in the transfer zone. There are many theoretical and practical solutions of this problem based on various simplifying assumptions and idealizations.

Among the several solutions available, the two-dimensional solutions based on the theory of elasticity are noteworthy.[4] In this approach, the influence of the vertical reaction is ignored, and the prestressing force is assumed to be distributed as a line load acting on the entire width of the beam. Figure 5-7 shows the idealized section that has been analyzed by using this approach.

Results of these investigations clearly indicate that loads of the type shown in Fig. 5-7 cause transverse tensile stresses at various points in the transfer zone.

Figure 5-8 shows a typical idealized end block subjected to two concentrated symmetrical forces. Transverse tensile stresses appear in the following two regions:

1. Immediately next to the point of application of prestressing force inside the transfer zone, the exact position depending upon the relative area of the bearing plate. These tensile stresses are called bursting stresses.
2. At the end sections of the beam, between the bearing plates and near the top and bottom. These tensile stresses are called spalling stresses.

Figure 5-8 shows qualitatively the regions where these stresses occur. The shaded area represents the region of compressive stresses.

Photoelastic studies on specimens such as the one shown in Fig. 5-7 verify these results.[5]

Approximate methods to be used in the determination of stresses

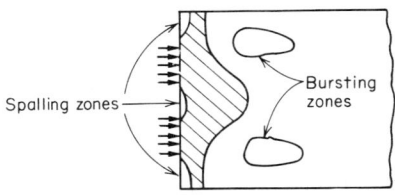

Fig. 5-8 Tensile stresses at the spalling and bursting zones.

in the anchorage zone were developed by Magnel[6] and Guyon.[7] Magnel considered the anchorage zone as a beam in which the stress distribution followed a cubic curve. Guyon introduced a simplified two-dimensional theory.

The results obtained from even the accurate methods are correct only to a certain extent. For small tensile strains, lower than the cracking strain of concrete, these analyses give useful qualitative results. However, since concrete cannot carry large tensile stresses, horizontal cracks are developed, and the analysis becomes incorrect for the cracked end section.

Furthermore, the purpose of the analysis is to develop a method to determine the amount of vertical reinforcement that should be placed in the end portion of the beam. Since the transverse tensile stresses are very high and concrete cracks at low stresses, the purpose of reinforcement can only be to contain such cracks rather than to stop them. The reinforcement cannot be effective before the horizontal cracks have occurred.

Experimental work on rectangular and I-section end blocks indicates the nonlinear nature of the stresses and the formation of longitudinal cracks in most of the beams tested.[8,9] Tests also indicate that there is no clear advantage in the rectangular section over the I section at the anchorage zone.[9]

To take into account the effect of cracking, a physical analog was developed based on beams on elastic foundation.[10]

5-5 DETERMINATION OF TRANSVERSE REINFORCEMENT

The method presented here is a simple, practical approach developed recently.[9]

Let us consider the end block shown in Fig. 5-9a. The distance B represents the lead-in distance, such that the section A-A is far enough from the end that it will not be influenced by the concentrated forces at the anchorage. Let us ignore the vertical reaction and assume that the prestressing force is acting at a distance g from the bottom fiber of the beam.

Any longitudinal section through the anchorage zone at a distance

BOND AND ANCHORAGE

y from the bottom fiber is subjected to a bending moment, which can be determined from the forces that are acting on the ends of the block.

The bending moment may be determined for any longitudinal section as follows. For the end block shown in Fig. 5-9a, two cases can be considered:

1. $y < g$ Figure 5-9b shows a free-body diagram of a longitudinal section in which $y < g$. The bending moment in this section, from Fig. 5-9b, is

$$M = \left[2\left(\frac{y}{h}\right)^3 - \left(3 + \frac{h}{2e}\right)\left(\frac{y}{h}\right)^2 \right] Pe$$

where the convention adopted considers clockwise moments as positive. In dimensionless form the above equation can be written as follows:

$$\frac{M}{Pe} = \left(\frac{y}{h}\right)^2 \left[2\frac{y}{h} - \left(3 + \frac{h}{2e}\right) \right] \tag{5-9}$$

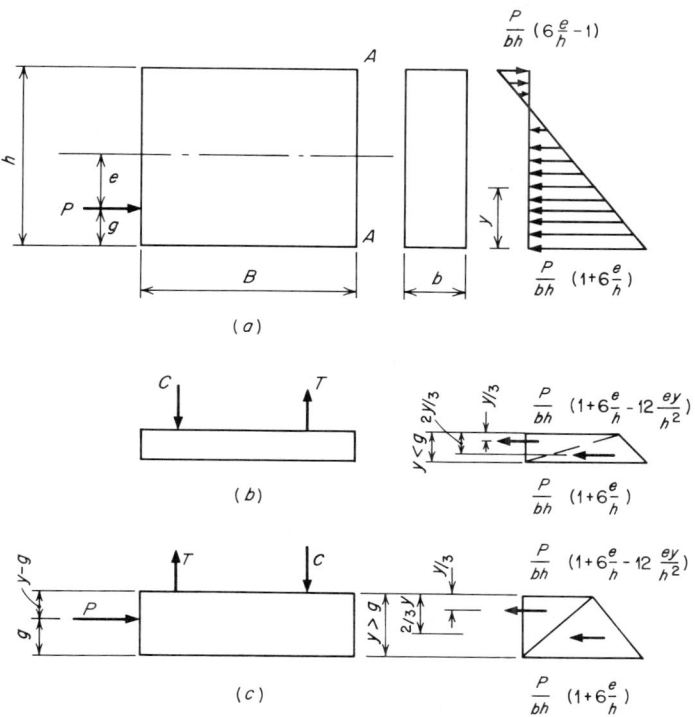

Fig. 5-9 Moment at a longitudinal section.

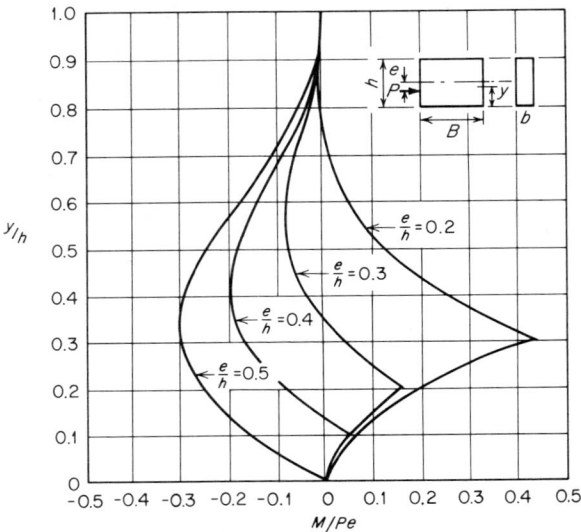

Fig. 5-10 Bending moment in longitudinal sections.

2. $y > g$ Figure 5-9c shows the free-body diagram of the longitudinal forces. In this case, the bending moment in dimensionless form is

$$\frac{M}{Pe} = 2\left(\frac{y}{h}\right)^3 - \left(3 + \frac{h}{2e}\right)\left(\frac{y}{h}\right)^2 + \frac{h}{e}\frac{y}{h} + \left(1 - \frac{h}{2e}\right) \quad (5\text{-}10)$$

In the general case of post-tensioning forces applied at different levels, it may prove helpful to draw free-body diagrams to obtain the necessary expressions for the variation of moment with depth.

Figure 5-10 shows the plots of Eqs. (5-8) and (5-9) for e/h values of 0.5, 0.4, 0.3, 0.2, and 0.167. The negative moment, or M/Pe ratio, corresponds to the moment of the sign shown in Fig. 5-9b. The positive moment corresponds to the moment shown in Fig. 5-9c.

Equations (5-9) and (5-10) give the moments in the longitudinal sections, but are not sufficient to indicate the exact shape of stress distribution in these sections. However, the total tensile or compressive force created by the moment may be determined, if a logical estimate can be made of the internal moment arm. Stirrups then may be designed to carry the total tensile force. The allowable stress in the stirrups can be limited to provide a control on the opening of the longitudinal cracks.

For design purposes, the following has been recommended, which gives a reasonably simple approach to design of the transverse reinforcement at the anchorage zone.[9]

BOND AND ANCHORAGE

1. Transverse reinforcement shall be provided within a distance $h/2$ from the end of the beam in the form of closed stirrups to carry the total force T given by the following expression:

$$F_T = \frac{M_m}{h-z} \qquad (5\text{-}11)$$

where F_T = total tensile force
M_m = maximum longitudinal moment
z = distance between end of beam and centroid of stirrups that are within $h/2$ from end

2. The allowable stress in the stirrups should not exceed the following:

$$f_s = \left(\frac{4E_s \sqrt{f'_c}\, w}{A_s}\right)^{\frac{1}{2}} \qquad (5\text{-}12)$$

where f_s = allowable stress in stirrup, psi
w = crack width, in.
A_s = area of stirrup, in.^{2}

As previously defined f'_c and E_s are the cylinder strength of concrete and the modulus of elasticity of steel, respectively, both measured in pounds per square inch.

Equation (5-12) is an empirical relationship between crack width and the stress in stirrup.[9]

5-6 ILLUSTRATIVE PROBLEM 5-1

Figure 5-11 shows the end portion of a post-tensioned prestressed-concrete beam. The beam has an interior I-shaped section and solid rectangular end blocks. The arrangement of the post-tensioning steel is as shown in the figure. There are six cables anchored at the ends, exerting 60 kips each on the beam. It is required to design the vertical reinforcement for the end block.

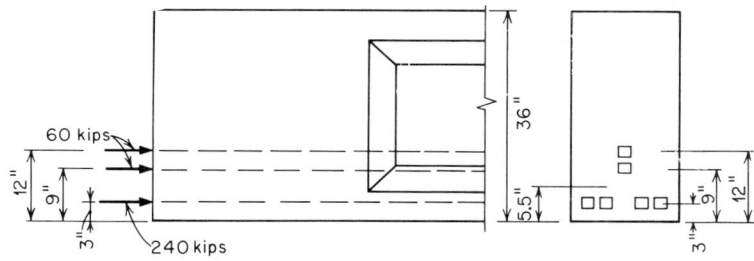

Fig. 5-11 Illustrative Problem 5-1.

In this solution, the effects of the vertical reaction and that of any vertical component of the prestressing force will be ignored. The end forces are considered to be distributed through the width of the end block at their respective levels of application.

Moment at various horizontal planes can be calculated on the basis of the discussion in the preceding section as follows:

For $0 < y \leq 3$ in. the bending moment may be calculated from Eq. (5-9), taking $P = 360$ kips, $e = 12.5$ in., $h = 36$ in., as follows:

$$M = 360 \times 12.50 \left[2\left(\frac{y}{36}\right)^3 - \left(3 + \frac{36}{2 \times 12.5}\right)\left(\frac{y}{36}\right)^2 \right]$$

from which

$$M = -0.1929y^3 + 15.417y^2$$

When $3 < y < 9$,

$$M = -0.1929y^3 + 15.417y^2 - 240y + 720$$

and when $9 < y < 12$,

$$M = -0.1929y^3 + 15.417y^2 - 300y + 1260$$

and when $y > 12$,

$$M = -0.1929y^3 + 15\ 417y^2 - 360y + 1980$$

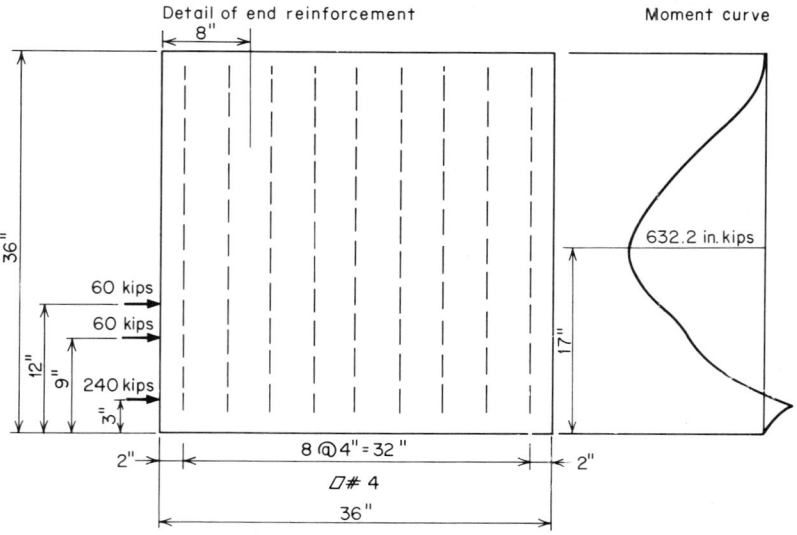

Fig. 5-12 Anchorage zone.

BOND AND ANCHORAGE

Table 5-2 Moment at the various longitudinal sections—Illustrative Problem 5-1

y, in.	M, in.-kips	y, in.	M, in.-kips
1.0	+15.2	19.0	−617.7
2.0	+60.1	20.0	−596.5
3.0	+133.5	21.0	−567.7
4.0	−56.8	22.0	−532.3
5.0	−118.7	23.0	−491.6
6.0	−206.7	24.0	−446.6
7.0	−270.7	25.0	−398.6
8.0	−312.1	26.0	−348.7
9.0	−333.1	27.0	−298.1
10.0	−391.2	28.0	−247.9
11.0	−431.3	29.0	−199.2
12.0	−453.3	30.0	−153.3
13.0	−518.4	31.0	−111.3
14.0	−567.7	32.0	−74.3
15.0	−602.3	33.0	−43.5
16.0	−623.5	34.0	−20.1
17.0	−632.2	35.0	−5.2
18.0	−630.0	36.0	0

Figure 5-12 shows a plot of the above moments throughout the depth and the beam, and Table 5-2 shows these moments for 1 in. increments of y.

From Fig. 5-12 and Table 5-2 it can be seen that the maximum moment is 632.2 in.-kips and occurs 17.0 in. from the bottom of the beam.

We shall use No. 4 bars with $A_s = 0.2$ in.^{2} and assume $f'_c = 5000$ psi, $E_s = 30$ million psi, and a crack width of 0.005 in.

From Eq. (5-12) we have

$$f_s = \left(\frac{4E_s \sqrt{f'_c}\, w}{A_s}\right)^{\frac{1}{2}} = \left[\frac{(4)(30{,}000{,}000)(\sqrt{5000})(0.005)}{0.2}\right]^{\frac{1}{2}}$$
$$= 14{,}600 \text{ psi}$$

We can assume $z = 8$ in. and calculate the total tensile force in the stirrups:

$$F_T = \frac{632.2}{36 - 8} = 22.6 \text{ kips}$$

The total number of No. 4 closed stirrups is

$$\frac{22.6}{14.6 \times 0.2 \times 2} \approx 4$$

These four stirrups should be placed in the exterior end portion of the block. If a 2-in. necessary cover is left for the first stirrup, a 4-in. spacing for the other three required stirrups would still comply with the assumption made for z, thereby requiring no recalculations.

Figure 5-12 shows the detailed arrangement of the stirrups. It can be seen that the stirrups have been extended to cover the whole block. This was done because of the presence of moments of opposite sign at the bottom fibers which create tensile stresses in the interior end of the block. For practical reasons, even when these moments require a smaller amount of transverse reinforcement, the same size, type, and spacing of stirrups should be specified.

PROBLEMS

5-1. Solve Illustrative Problem 5-1 by including the effect of end reaction. For this problem assume that the beam has a span of 60 ft and weighs 0.35 klf, and the applied load 1.5 klf. The center of the reaction may be assumed 12 in. from the end of the beam.

5-2. Design the end stirrups for the pretensioned beam shown in Fig. 6-13.

REFERENCES

1. Janney, Jack R.: Nature of Bond in Pre-tensioned Prestressed Concrete, *J. Am. Concrete Inst.*, May, 1954; *Proc.*, vol. 50, pp. 717–736.
2. Kaar, P. H., R. W. LaFraugh, and M. A. Mass: Influence of Concrete Strength on Strand Transfer Length, *J. Prestressed Concrete Inst.*, vol. 8, no. 5, pp. 47–67, October, 1963.
3. Hanson, N. W., and P. H. Kaar: Flexural Bond Tests of Pretensioned Prestressed Beams, *J. Am. Concrete Inst.*, January, 1959; *Proc.*, vol. 55, pp. 783–802.
4. Iyengar, K. T. S. R.: Two-dimensional Theories of Anchorage Zone Stresses in Post-tensioned Prestressed Beams, *J. Am. Concrete Inst.*, October, 1962; *Proc.*, vol. 59, pp. 1443–1446.
5. Christodoulides, S. P.: The Distribution of Stresses around the End Anchorages of Prestressed Concrete Beams: Comparison of Results Obtained Photoelastically, with Strain Gauge Measurements and Theoretical Solutions, *Intern. Assoc. Bridge Structural Eng.*, vol. 16, pp. 55–70, 1956.
6. Magnel, Gustave: "Prestressed Concrete," 3d ed., pp. 69–72, McGraw-Hill Book Company, New York, 1954.
7. Guyon, Y.: "Prestressed Concrete," pp. 127–212, John Wiley & Sons, Inc., New York, 1953.
8. Marshall, W. T., and A. H. Mattock: Control of Horizontal Cracking in the Ends of Pretensioned Prestressed Concrete Girders, *J. Prestressed Concrete Inst.*, vol. 7, no. 5, pp. 56–74, October, 1962.
9. Gergeley, P., M. A. Sozen, and C. P. Siess: The Effect of Reinforcement on Anchorage Zone Cracks in Prestressed Concrete Members, *Univ. Illinois Dept. Civil Eng. Structural Res. Ser.* 271, July, 1963.
10. Welsh, W. A., and M. A. Sozen: Investigation of Prestressed Reinforced Concrete for Highway Bridges, Part V: Analysis and Control of Anchorage Zone Cracking in Prestressed Concrete, *Univ. Illinois Civil Eng. Studies, Structural Res. Ser.* 309, June, 1966.

6
Losses and Long-time Deflections

6-1 INTRODUCTION

This chapter consists of two parts, namely, discussion of loss of prestress and long-time deflections.

The object of the first part is to study in detail the sources of loss of prestress in steel, and to present numerical examples to show how these losses can be determined. Losses may take place instantaneously or in a long time. Examples of losses in prestress that occur instantaneously are those due to frictional forces between the prestressing steel and the concrete, elastic shortening of the concrete, and anchorage set.

After the initial losses occur, the force in the steel is gradually reduced by additional losses due to shrinkage and creep effects in the concrete and relaxation in the steel. These losses take place in a long time (up to 3 years) and not only cause a reduction in the prestressing force but also affect the transverse deflections of the beam.

The object of the second part is to discuss the long-time deflections of prestressed-concrete beams, and present simple methods for their calculation. Two cases are considered: prestressed beams without transverse loads and prestressed beams with the transverse loads.

6-2 LOSS OF PRESTRESS DUE TO FRICTION IN POST-TENSIONED BEAMS

Curved cables are subject to normal forces that are created by contact with their enclosure or the surrounding concrete. These normal stresses are directly proportional to the curvature of the cable and the tensile force in the cable. Section 10-10 shows the derivation of the relationship between curvature and normal stresses. In post-tensioned beams, in addition to the primary curvature of the cable due to the variation of eccentricity, secondary curvatures due to small vertical and horizontal deviations from the theoretical path may occur because of faulty construction. This latter effect creates additional normal stresses and is called the wobble effect.

The presence of normal stresses induces friction between the sliding cables and the surrounding material. The tensile force at the ends of the cables is reduced by friction accumulating as one moves from the jacking end toward the center of the beam. The variation of the tensile force in the cable and its minimum value along the span are of interest to the designer for various reasons. First, it is necessary to know the extent of required overstressing of the cable to compensate for this loss; and second, an accurate estimate of the final value of the tension in the cable is necessary to determine its effects on the stresses under service loads.

Figure 6-1 presents an infinitesimal portion of a cable of length ds subjected to a tensile force P at the left end (the closest to the pulling end) and a force $P + dP$ at the other end. The inflection angle in the length ds is $d\theta$, the principal curvature of the cable is $d\theta/ds$, and normal

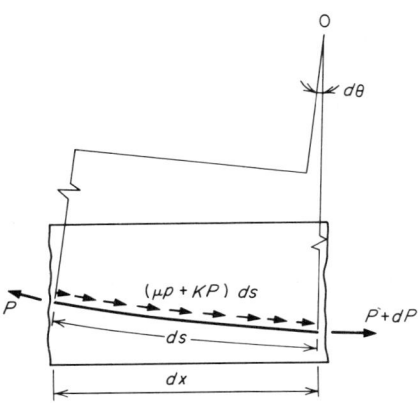

Fig. 6-1 Loss of prestress due to friction.

Table 6-1 Values of wobble and friction coefficients

Type of steel	Usual range of observed values		Suggested design values	
	K	μ	K	μ
Wire cables	0.0005–0.0030	0.15–0.35	0.0015	0.25
High-strength bars	0.0001–0.0005	0.08–0.30	0.0003	0.20
Galvanized strands	0.0005–0.0020	0.15–0.30	0.0015	0.25

stresses of intensity $p = P(d\theta/ds)$ are created. If μ is the coefficient of friction between the cable and its surrounding material, the frictional stress created per unit length is μp. Additional frictional stresses are induced by the wobble effect. If K, the wobble coefficient, is defined as the friction stress created per unit tensile force in the cable per unit length, then KP is the friction stress per unit length induced by the wobble effect. For metal sheathing, values of μ and K which are generally employed are shown in Table 6-1.[1]

The equilibrium of forces along the cable results in the following equation:

$$P + dP + (\mu p + KP)\, ds - P = 0 \qquad (6\text{-}1)$$

Substituting $p\, ds = P\, d\theta$ and dividing by P, we have

$$dP + \mu P\, d\theta + KP\, ds = 0 \qquad (6\text{-}2)$$

Dividing by P and integrating, we have

$$\int_{P_0}^{P_x} \frac{dP}{P} + \int_0^{\alpha} \mu\, d\theta + \int_0^s K\, ds = 0 \qquad (6\text{-}3)$$

where P_0 = tensile force in cable at jacking end of beam
$\quad P_x$ = tensile force in cable in a section at a distance x from the jacking end of beam
$\quad \alpha$ = total angle of inflection of cable between jacking-end section and section considered
$\quad s$ = length of cable between end of beam and section considered

The solution is obtained after integration of Eq. (6-3) as follows:

$$\log(P_x - P_0) + \mu\alpha + Ks = 0 \qquad (6\text{-}4)$$

from which

$$P_x = P_0 e^{-(\mu\alpha + Ks)} \qquad (6\text{-}5)$$

Usually s is not very much different from x, the distance between the end of the beam and the section under consideration, and the preceding expression can be transformed into

$$P_x = P_0 e^{-(\mu\alpha + Kx)} \tag{6-6}$$

Equation (6-6) can be expressed in the following simple approximate form:

$$P_x = P_0(1 - \mu\alpha - Kx) \tag{6-7}$$

6-3 ILLUSTRATIVE EXAMPLE 6-1

Figure 6-2a shows a longitudinal cross section of a two-span continuous beam with the profile of one of its cables. Post-tensioning takes place only at end A, where a force of 100 kips is applied. Determine the variation of the tension force along the cable if the coefficient of friction $\mu = 0.3$ and the wobble coefficient $K = 0.0002$ ft^{-1}.

The application of Eq. (6-7) gives the following losses:

Between A and B, uniformly distributed:

$(100)(0.0002)(50) = 1$ kip

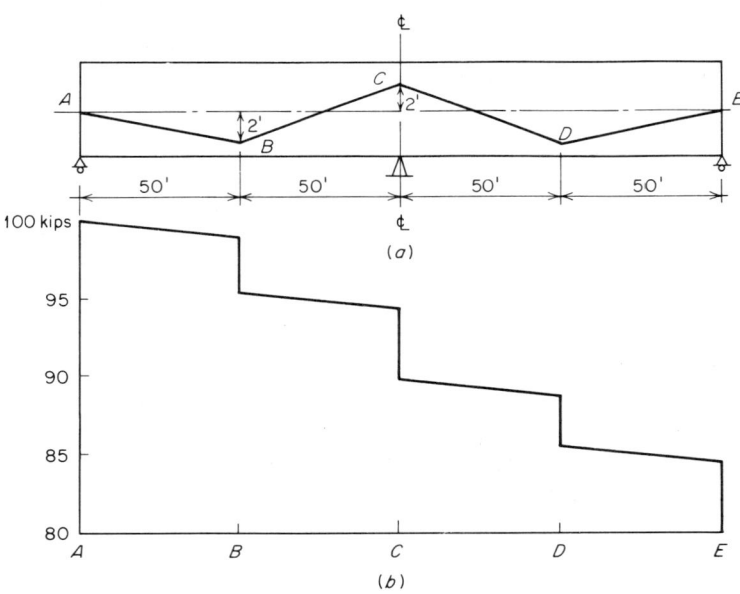

Fig. 6-2 The beam of Illustrative Example 6-1 showing loss of prestress due to friction.

LOSSES AND LONG-TIME DEFLECTIONS

At B, concentrated:
$$\alpha = \frac{2}{50} + \frac{2+2}{50} = \frac{6}{50}$$
and
$$\frac{(99)(0.3)(6)}{50} = 3.56 \text{ kips}$$

Between B and C, uniformly distributed:
$$(95.44)(0.0002)(50) = 0.95 \text{ kip}$$

At C, concentrated:
$$\frac{(94.49)(0.3)(4+4)}{50} = 4.54 \text{ kips}$$

Between C and D, uniformly distributed:
$$(89.95)(0.0002)(50) = 0.90 \text{ kip}$$

At D, concentrated:
$$\frac{(89.05)(0.3)(4+2)}{50} = 3.21 \text{ kips}$$

Between D and E, uniformly distributed:
$$(85.84)(0.0002)(50) = 0.86 \text{ kip}$$

Tension at end E is, therefore, approximately 85 kips, and the variation of tensile force in the cable is shown in Fig. 6-2b.

6-4 LOSS OF PRESTRESS DUE TO ELASTIC SHORTENING

In post-tensioned beams, not all cables are tensioned at the same time, for practical reasons. As tensioning of cables proceeds, the cables that were initially tensioned suffer a loss in prestress due to the elastic shortening of the beam. A loss occurs, therefore, for each cable due to the stressing of subsequent cables.

Let N be the number of cables in a given beam. To find the loss of prestress that occurs in the ith cable after cables numbered $i+1$ to N are tensioned, the loss in stress in the concrete, Δf_{ci}, at the level of the ith cable must be evaluated:

$$\Delta f_{ci} = \sum_{j=i+1}^{N} \frac{f_{sj} A_{sj}}{A} \left(1 + \frac{e_j y_i}{r^2}\right) \tag{6-8}$$

where f_{sj} is the stress in the jth cable after the last cable has been tensioned. A_{sj} and e_j are the area and eccentricity, respectively, of the

jth cable, y_i is the distance of the centroid of the ith cable to the centroid of the section, and A and r are the area and radius of gyration of the section of the beam, respectively.

The loss in prestress Δf_{si} in the ith cable can be estimated as $n(\Delta f_{ci})$, where n, the modular ratio E_s/E_c, is determined with not more than one significant digit.

The determination of f_{sj} requires an estimate of loss in each cable caused by the post-tensioning of the successive cables. Actually, it is better to start the computation with the last cable which itself is not subject to loss due to elastic shortening, but causes elastic shortening in all other cables. By proceeding from the last cable to the first one in inverse order of prestressing, the determination of the final prestressing force after elastic shortening can be obtained. However, in view of the approximate nature of n, an accurate determination of the final value of prestress in a particular cable may be simplified, and a conservative value obtained, if the initial values of prestress in the successive cables are used to determine the losses induced.

The same general method applies to pretensioned beams, for which, instead of post-tensioning cables, strands are used and prestressing is applied by successive cutting of the strands. The average loss, however, may be estimated by the following expression:

$$\Delta f_s = n \frac{\Sigma f_s A_s}{A} \left(1 + \frac{e^2}{r^2}\right) \tag{6-9}$$

where e in this case is the distance of the centroid of all cables to the neutral axis of the section.

6-5 ILLUSTRATIVE PROBLEM 6-2

The section shown in Fig. 6-3 is post-tensioned by three 1 in.² cables each with an initial prestress of 150 ksi. Find the losses that occur in each cable caused by the elastic shortening, assuming that the prestressing is applied by successive post-tensioning. Cables are numbered in the order in which they are post-tensioned.

Losses due to post-tensioning are as follows:

In the second and first cables due to the third cable:

$$\Delta f_{s1,3} = \Delta f_{s2,3} = 6 \frac{150 \times 1}{36 \times 16} = 1.56 \text{ ksi}$$

In the first cable due to the second cable:

$$\Delta f_{s1,2} = 6 \frac{(150 - 1.56)1}{36 \times 16} \left(1 + \frac{6 \times 12}{108}\right) = 2.58 \text{ ksi}$$

LOSSES AND LONG-TIME DEFLECTIONS 179

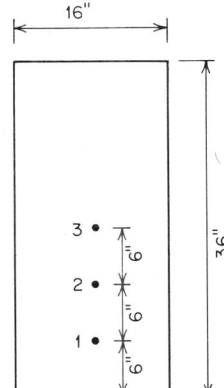

Fig. 6-3 The section in Illustrative Problem 6-2.

The total loss of prestress in the first cable is then $1.56 + 2.58 = 4.14$ ksi.

The final tension in each cable after post-tensioning of the last cable is as follows:

$P_3 = 150$ kips

$P_2 = 148.4$ kips

$P_1 = 145.9$ kips

Initial overstressing of all cables but the last one is necessary to overcome losses due to elastic shortening and to maintain the desired level of prestressing.

6-6 LOSS OF PRESTRESS DUE TO ANCHORAGE SET

This loss is present only in post-tensioned construction. The anchorage devices are used to hold the cable at the ends after tensioning occurs and to transmit this force to the concrete. When the jack is released, the pulling force of the cable is transmitted to the anchorage device in some fashion depending on the particular procedure used. A small deformation or slip, called set, occurs in the anchorage device because of the action of the force, which induces a certain loss of prestress.

A typical example of this type of loss occurs for the Freyssinet system with the anchorage cones. The interior cone is pushed inward by the jack before releasing the cable. However, the interlocking effect whereby the wires of the cable are gripped by the interior cone is developed only after a certain amount of slip has occurred. The average value of slip is approximately $\frac{1}{4}$ in. For any system of post-tensioning for which anchorage devices set at transfer, the designer should have the value of slip to estimate the loss in prestress.

Though this loss is not very large for long spans, it may be appreciable for short spans.

6-7 LONG-TIME EFFECTS

The following discussions concern the losses of prestress and transverse deflections of prestressed-concrete beams, caused by the prestressing force and the superimposed dead and live loads, under the effect of time. The instantaneous losses in prestress were studied in the preceding sections; and the deflections under instantaneous load, known as short-time deflections, have been considered in Chap. 3, where the behavior of prestressed beams under short-time loads was studied. The term *long-time effects* refers to the losses in prestress and the deflections of the beam some time after applications of prestressing with or without transverse loads.[2]

In reinforced-concrete structures subjected to a constant load, deformations increase with time as a result of shrinkage and creep in the concrete. In prestressed-concrete beams, this is also true, with the additional effect of the relaxation of the prestressing reinforcement, which causes increased deformations.

The individual effects of shrinkage, creep, and relaxation on the deformations of prestressed-concrete beams will be studied first, followed by a study of the combined effects as they occur in practical situations. An approximate method of predicting long-time deflections will be presented, along with an illustrative example.

6-8 SHRINKAGE

Shrinkage strain is the deformation of the concrete which occurs in unstressed-concrete members.

Shrinkage of concrete is mainly due to the evaporation of the mixing water. Therefore, the amount of water placed in the mix and the humidity of the surrounding air affect the magnitude of the resulting shrinkage. Many factors influence shrinkage, the most important being the humidity of the surrounding air, the water-cement ratio, the proportions of the mix, the curing conditions, and the size and shape of the concrete member.

If H is the relative humidity in percent and ϵ_{sh} is the shrinkage strain, the following formula, which is due to Schorer,[3] gives reasonable results for design purposes:

$$\epsilon_{sh} = \frac{(12.5)(90 - H)}{10^6} \tag{6-10}$$

LOSSES AND LONG-TIME DEFLECTIONS

In the case of $H = 50$, Eq. (6-10) yields $\epsilon_{sh} = 0.0005$, which represents a reasonable value for the shrinkage strain.

6-9 CREEP

The nature of creep is not yet fully understood. Results of observation show that strain in concrete resulting from sustained loads increases with time.[4] Strain continues to increase for several years; however, a large percentage of deformations takes place in the first 2 years. Table 6-2 shows the approximate values of C_u, defined as the ratio of ultimate creep strain to initial strain, for normal-weight concrete under different average relative humidities for ordinary concrete and high-strength concrete. The larger values shown correspond to an earlier loading stage.

The following two parameters are defined:

The unit creep strain δ_t is defined as the creep per unit stress, and the creep coefficient C_t is defined as the ratio of creep strain to initial strain. According to these definitions

$$\delta_t = \frac{\epsilon_t}{\sigma} \tag{6-11}$$

and

$$C_t = \frac{\epsilon_t}{\epsilon_i}$$

where ϵ_t = creep strain
ϵ_i = initial strain
σ = constant stress

The creep coefficient C_t may be expressed in the following convenient form:

$$C_t = \delta_t E_c \tag{6-12}$$

where $E_c = \sigma/\epsilon_i$.

Table 6-2 C_u: ratio of ultimate creep strain to initial strain

Concrete strength	Average relative humidity		
	100%	70%	50%
Ordinary	1.0–2	1.5–3	2.0–4
High	0.7–1.5	1.0–2.5	1.5–3.5

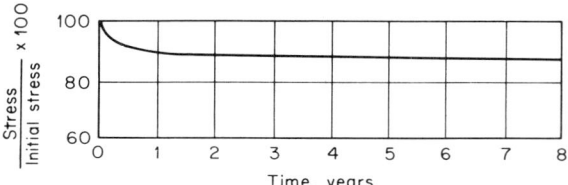

Fig. 6-4 Stress variation with time in a wire due to relaxation.

6-10 RELAXATION

This phenomenon is the loss in stress in steel that occurs at constant strain. Relaxation continues for many years, although the rate is reduced greatly during the first year. In Fig. 6-4 the stress variation in a wire prestressed to 79 percent of its 1 percent strain is shown as a fraction of the initial stress. After a large initial loss during the first 6 months, the loss rate decreases with time.

The amount of relaxation loss to be expected in a given steel depends on many factors, of which some important ones occur during manufacture. In spite of the large variation that may take place in the amount of relaxation loss, it is possible to estimate this loss reasonably for practical purposes. The following expression is given for the percentage of relaxation loss for high-strength wire and strand reinforcement:[5]

$$\frac{f_s}{f_{si}} = 1 - \frac{\log t}{10}\left(\frac{f_{si}}{f_{sy}} - 0.55\right) \tag{6-13}$$

where f_s = steel stress at time t
t = hours after initial stressing to f_{si}
f_{si} = initial prestress
f_{sy} = yield-point stress of steel

Figure 6-4 shows relaxation loss for wires. Equation (6-13) may be used to determine the relaxation loss for a wire 1 year after initial prestress. Taking $f_{si} = 0.79 f_{sy}$ and $t = 8760$ hr, we have

$$\frac{f_s}{f_{si}} = 1 - \frac{\log 8760}{10}(0.79 - 0.55) = 0.92 \tag{6-14}$$

The critical parameter appears to be the ratio of initial prestress to the yield-point stress of the steel.

6-11 DETERMINATION OF EFFECTIVENESS

The effectiveness is defined as the ratio of the final prestressing force to the prestressing force at transfer and is designated as η. In the determination of η it is the prestressing force at the midspan or the section of maximum moment that should be considered. Determination of η requires knowledge of losses discussed in the preceding sections. In the following paragraphs the problems associated with the determination of η will be presented.

In the manufacture of a prestressed-concrete beam, the required amount of prestressing force is usually applied by means of a jack at one of the anchorages. The only measurement of the actual magnitude of the prestressing force takes place at the time of tensioning at the jacking anchorage. Since this is the highest prestressing force, the allowable stress in steel is given at the jacking anchorage immediately after the tensioning is completed.

On the other hand, in design of a beam it is necessary to know the prestressing force at the section of maximum moment at transfer, which may or may not be equal to the prestressing force at the jacking anchorage at the time of tensioning. The difference between the prestressing force at jacking anchorage and at the section of maximum moment is caused by friction. In post-tensioned construction the time when tensioning is complete is the same as the time of transfer; however, in pretensioned construction the time of tensioning may precede the time of transfer by a few days.

In most specifications the total losses are measured from the time when tensioning is complete at anchorages. In the Commentary to the 1963 ACI code the loss in steel stress, not including losses due to friction, is given as 35,000 psi in pretensioning and 25,000 psi in post-tensioning.[6] The difference is due to the fact that in pretensioning the time of tensioning is not the same as the time of transfer when the wires are cut. These values are not exact, but it can be shown that they are reasonable.

In post-tensioned construction, the time of tensioning and transfer being the same, the losses do not include the effect of elastic shortening of the beam and include only part of the shrinkage effect. In pretensioned construction the losses are measured from the time the steel wires are pulled and anchored—at which time no beam exists yet—to the time when all losses have taken place. In this case, in calculating losses the elastic shortening as well as shortening due to full shrinkage is taken into account. In pretensioned construction, therefore, it is not clear how much of the 35,000-psi loss in prestress takes place between the time of pulling and the time of cutting the prestressing steel.

It may be reasoned, however, that the condition of a pretensioned beam immediately after the steel is cut is roughly similar to that of a post-tensioned beam at the time of prestressing. The quality and age of concrete at the time of cutting in a pretensioned beam is the same as that at the time of prestressing in a post-tensioned beam. In both cases, a certain amount of shrinkage has taken place. Furthermore, in both cases the elastic shortening takes place simultaneously during the stressing of concrete, and it should not be taken into account if losses are measured from transfer.

It may be said, therefore, that in calculating η both in pretensioned and in post-tensioned construction, losses measured from the transfer may be taken as 25,000 psi.

It should be pointed out that there is some inaccuracy in this assumption. In a pretensioned beam at transfer some relaxation has perhaps taken place, while in a post-tensioned beam at transfer there has been no time for the relaxation to take place. It is difficult to determine the amount of relaxation that has taken place in a pretensioned beam at transfer. If the losses measured from transfer are 25,000 psi in a post-tensioned beam, it is true that in a pretensioned beam they will be smaller.

In the determination of losses for calculation of η, frictional losses are not involved. Since the ratio of stress at anchorage to that at the section of maximum moment may be assumed to be the same at all times, $P/P_t = \eta$ may be taken at any point along the beam. Evidently, if they are taken at the jacking end the frictional effect vanishes.

Hence, the following values of η may be used in design:

Pretensioned construction: $\quad \eta = 1 - \dfrac{25{,}000}{f_{si} - 10{,}000}$

Post-tensioned construction: $\quad \eta = 1 - \dfrac{25{,}000}{f_{si}}$

where f_{si} is the allowable stress in steel at the time of pulling at the jacking anchorage.

6-12 LONG-TIME DEFLECTIONS

The effects of creep and shrinkage in the concrete and relaxation in the steel on the long-time deflections of a prestressed-concrete beam are easier to study when the beam is subject to one load. When the beam is subjected to the prestressing force and the transverse load, the problem of determining the long-time deflections becomes difficult. For convenience, two separate cases will be considered. In the first case it will be assumed that the beam is subjected to the prestressing force only; in the second, that the beam is subjected to the prestressing force and transverse loads.

Effects of shrinkage, creep, and relaxation with prestressing alone The section shown in Fig. 6-5a is subjected to an initial prestressing force P_i with an eccentricity e which causes a linear distribution of strain with depth, as shown in Fig. 6-5b. The stress distribution may be assumed

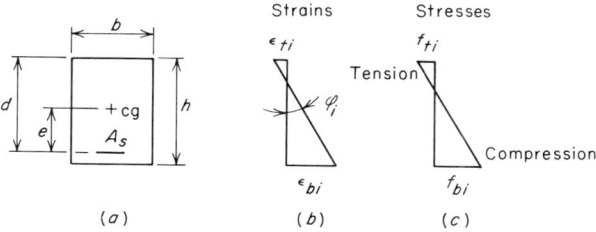

Fig. 6-5 A section of a concrete beam immediately after application of prestressing force.

to be linear, as concrete stresses caused by prestressing force alone are usually low. If ϵ_{bi} and ϵ_{ti} are the strains at top and bottom fibers, the instantaneous curvature φ_i caused by the prestressing force can be expressed as follows:

$$\varphi_i = \frac{\epsilon_{bi} - \epsilon_{ti}}{h} \equiv \frac{P_i e}{E_c I} \tag{6-15}$$

Creep and shrinkage in the concrete and relaxation in the reinforcement take place, which changes the strain and stress distributions with time, as shown in Fig. 6-6. Even though these effects are continuous and are interdependent, they will be considered separately, so that their individual influences can be understood.

The shrinkage strain may be assumed to be uniform along the span. The net result of shrinkage strain at the level of the steel is a loss in the amount of prestress. The equivalent effect on the concrete cross section is that of applying a tension force equal to the loss in prestress at the level of the steel. The strain and stress distributions resulting from shrinkage are shown in Fig. 6-7.

Relaxation losses in steel result in a reduction of prestressing force, with an effect similar to that of shrinkage. If ΔP is the loss in prestress

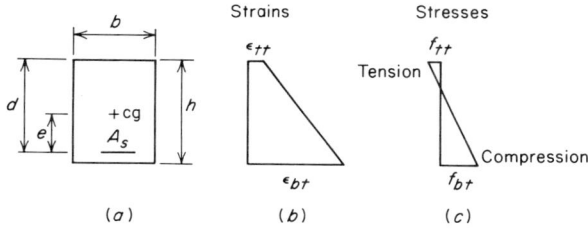

Fig. 6-6 A section of a concrete beam at a time t after application of prestressing force.

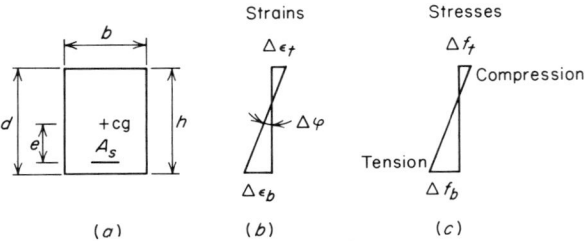

Fig. 6-7 Strain and stress distributions due to shrinkage and relaxation.

due to shrinkage of the concrete and relaxation in steel, and $\Delta\epsilon_b$ and $\Delta\epsilon_t$ are the resulting changes in strain at the bottom and top fibers, respectively, at a time t, the corresponding change in curvature is

$$\Delta\varphi = \frac{\Delta\epsilon_b - \Delta\epsilon_t}{h} = \frac{(\Delta P)e}{E_c I} \tag{6-16}$$

Both shrinkage of the concrete and relaxation of the steel reduce the curvature indirectly, as a consequence of the reduction of prestressing force.

Creep is not as simple to study because the change in stresses caused by a reduction of prestress results in a continuously different rate of creep strain. At the beginning the change in strain caused by creep is directly proportional to the instantaneous strain distribution. An initial creep effect is to increase the compressive strain in the concrete at the level of steel, thereby effecting a reduction in the level of prestress. This causes the first decrease in the normal stresses, which in turn reduces the rate of creep.

This process continues until the increase in curvature caused by creep is a multiple of the instantaneous curvature. However, because the increment of curvature is obtained under a decreasing stress, the ratio of creep curvature to instantaneous curvature is smaller than the ratio of creep strain to instantaneous strain in a cylinder of comparable concrete subjected to constant stress.

As shown in Fig. 6-7, the stress in the top fiber may be tensile after a certain time t, as its value depends only on the prestressing force. However, the total strain in the fiber may be a shortening due to the combined effects of shrinkage and creep in the concrete. It can be seen that creep tends to increase curvature directly, though there is a small reduction in curvature through loss of prestress. The net effect, however, is that of an increase in curvature.

The combined time effects of shrinkage, creep, and relaxation depend,

LOSSES AND LONG-TIME DEFLECTIONS

as in the case of short-time deflections, upon the magnitude of the stress gradient over the depth of the section after release of prestress. If the stress gradient is small, as in the case of small eccentricity, creep effects will be small, in which case the effects due to shrinkage and relaxation will dominate and the beam may deflect downward. Usually, however, the stress gradient is large and creep dominates, causing a simply supported beam to deflect upward.

Effects of shrinkage, creep, and relaxation with a transverse load on the beam Consider the section shown in Fig. 6-8a. The instantaneous strain distribution shown in Fig. 6-8b is caused by the stress distribution in an uncracked section under the action of transverse loads only. If we assume that creep characteristics are identical in tension and compression, the strain distribution after a given time t will be as shown in Fig. 6-8c. By ignoring the tendency of the neutral axis to shift toward the steel and the corresponding small reduction in stresses, it may be safely assumed that the stress distribution is constant with time. Therefore, in this case, increase of curvature is obtained under constant stress and the ratio is the same as the ratio of creep strain to instantaneous strain in a comparable specimen of concrete under constant stress.

The deflection of the beam caused by the prestress and the transverse loads may be determined by superposition. In the case of a uniformly loaded, simply supported beam shown in Fig. 6-9a, the distribution of curvature due to transverse load is a parabola, as shown in Fig. 6-9b. If the profile of steel in the beam is such that the eccentricity is constant in the middle third and variable in the rest of the beam (because the steel is draped at the third points), the curvature diagram due to the prestressing force would be as shown in Fig. 6-9c. The combined curvature diagram, shown in Fig. 6-9d, will change with time as the component diagrams change. The deflections corresponding to the two individual

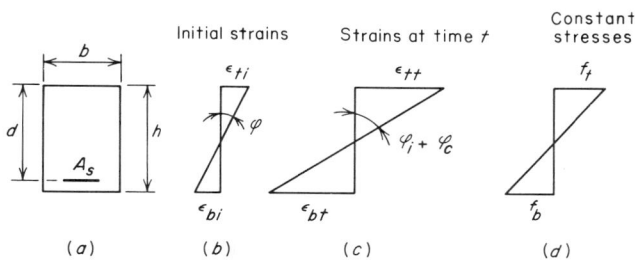

Fig. 6-8 Strain and stress distributions due to the transverse load only.

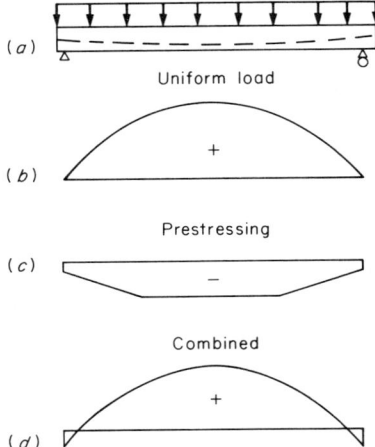

Fig. 6-9 Variation in curvature with span.

systems as they vary with time are shown in Fig. 6-10. Curves A and B show the variation of deflection with time due to the prestress and the transverse load, respectively. If at $t = 0$ the transverse load is applied to the beam, the resulting variation of the deflection with time is the algebraic sum of the ordinates of curves A and B. If the transverse load is applied at $t = t_0$, the resulting variation in the deflection will be given by curve A from $t = 0$ to $t = t_0$ and from there on by curve $A + B$, of which each ordinate is obtained as the algebraic sum of the ordinate of curve A at time t and the ordinate of curve B for a time $t - t_0$.

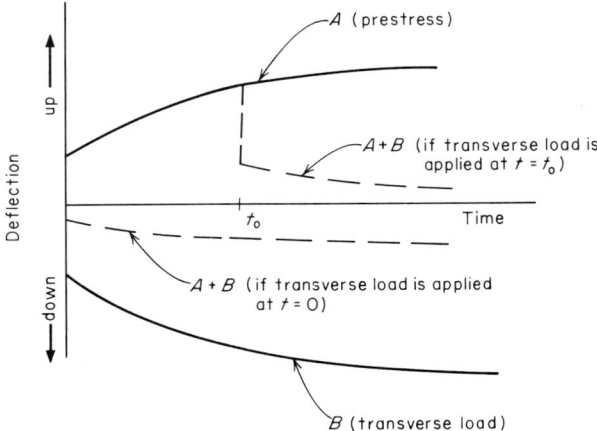

Fig. 6-10 Combined action of prestress and transverse load.

6-13 CALCULATION OF LONG-TIME DEFLECTIONS

It has been shown that the time-dependent deflections are the sum of two opposing effects: that of the prestress and that of the transverse load. Thus, the net curvature at a section at a time t can be represented as follows:

$$\varphi_t = \varphi_{mt} + \varphi_{pt} \tag{6-17}$$

where φ_{mt} is the curvature caused by the transverse load and φ_{pt} is the curvature caused by the prestress.

It is necessary to determine ways of predicting the variation of φ_{mt} and φ_{pt} with time. One important difference exists between the two. It has been shown already that the curvatures caused by the transverse load occur under constant stress, if the minor variations that take place due to a change in the prestress and a small shift in the neutral axis are ignored. While the curvature caused by the prestressing force occurs under variable stress, the change in compressive stress is usually very small and the assumption of constant stress is justified. Therefore, because of the action of transverse loads only, concrete creeps under constant stress. It is possible then to obtain the creep strain, at any time t, by multiplying the instantaneous strain by the creep coefficient C_t for that particular time. Therefore

$$\varphi_{mt} = (1 + C_t)\varphi_i = (1 + C_t)\frac{M_x}{EI} = \frac{M_x}{[E/(1 + C_t)]I} \tag{6-18}$$

The above equation can be considered to use a reduced modulus of elasticity, $E/(1 + C_t)$. This is approximately right if the load is applied all at the same time, for which E and C_t have unique values. If the load is applied in increments, the variation in E and C_t at the various times may be large.

Unlike the preceding case, the determination of φ_{pt}, the curvature caused by the prestress, is not simple because it involves the effects of creep, shrinkage, and relaxation with the resulting variation in the prestressing force. The value of φ_{pt} can be considered as the sum of three components: (1) instantaneous curvature at transfer, (2) change in curvature occurring as a direct result of loss of prestress, and (3) change in curvature resulting from creep under prestress. Parts 2 and 3 are interrelated, as creep affects the loss in prestress while the amount of prestress affects the magnitudes of stresses in the section, which in turn affect creep.

There are two different methods to calculate creep strain in concrete subjected to varying stress: the rate-of-creep method and the superposition method. The rate-of-creep method[2] is illustrated in Fig. 6-11. The first portion of the diagram assumes a concrete cylinder loaded with a stress σ, while the creep strain increases from zero to a value $\sigma\delta_{t1}$ at time

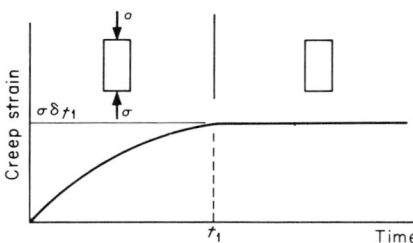

Fig. 6-11 Rate-of-creep method for determination of long-time deflections.

$t = t_1$. If at this instant the stress is removed, the creep strain is assumed to remain constant thereafter at the value $\sigma \delta_{t1}$.

The superposition method[7] is illustrated in Fig. 6-12. The first portion of the diagram is identical to that of the rate-of-creep method; all similarity ends, however, at time $t = t_1$ when the stress is removed. It is assumed that removal of stress does not take place actually but instead a tensile stress of equal value is applied at $t = t_1$ and that the specimen creeps under two opposing fictitious stresses, as shown in Fig. 6-12. If the creep properties of concrete in tension and compression are assumed to be the same, curves A and B would represent the individual variations in creep strain beyond $t = t_1$ due to the compressive and tensile values of σ, respectively, if they were acting alone. Curve C, obtained by subtracting the ordinates of curve B from the ordinates of curve A, is the actual variation of creep strain according to the superposition method. This method appears to be more accurate than the rate-of-creep method; however, its use is laborious and is worthwhile only if the basic data for creep, shrinkage, and relaxation are reliable.[8]

Because of the continuous variations in the prestressing force and the creep, the accurate determination of φ_{pt}, the time curvature due to prestress, requires a summation procedure that recognizes such changes. This is extremely laborious even for the rate-of-creep method and also is not justified for practical purposes. Instead of the rigorous method, a simple approximate method to estimate φ_{pt} can be obtained if the total

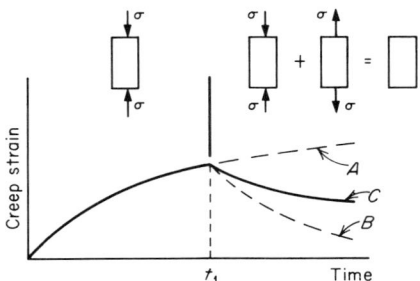

Fig. 6-12 Superposition method for determination of long-time deflections.

LOSSES AND LONG-TIME DEFLECTIONS

loss in prestress is calculated. Curvature φ_{pt} at any section may be expressed as follows:

$$\varphi_{pt} = -\frac{P_i e}{EI} + \frac{P'_t e}{EI} - \left(P_i - \frac{P'_t}{2}\right) C_t \frac{e}{EI} \qquad (6\text{-}19)$$

where P'_t = total loss in prestress
e = eccentricity at any section

The first term in Eq. (6-19) represents the instantaneous curvature, the second term is the loss of initial curvature due to the loss in prestress caused by creep shrinkage and relaxation, and the third term is the additional curvature resulting from the direct effect of creep. The last term has been obtained by assuming that creep occurs under a constant prestressing force equal to the mean of the initial and final prestressing forces. Hence,

$$\Delta\varphi_{pt} = \frac{P_i + (P_i - P'_t)}{2} C_t \frac{e}{EI}$$

$$= \left(P_i - \frac{P'_t}{2}\right) C_t \frac{e}{EI} \qquad (6\text{-}20)$$

The loss in prestressing force, P'_t, can be estimated by the following expression:

$$P'_t = r_s f_{se} + \epsilon_{sh} E_s + C_t n f_{ci} \left(1 - C_t n \frac{f_{ci}}{2 f_{se}}\right) A_s \qquad (6\text{-}21)$$

where $r_s = 1 - f_s/f_{si}$.

From Eq. (6-21) it can be seen that P'_t is the sum of three quantities. The first term is the relaxation loss $r_s f_{se}$; the second term is due to shrinkage, $\epsilon_{sh} E_s$; and the third term is an estimate of the loss due to the direct effect of creep.

6-14 ILLUSTRATIVE EXAMPLE 6-3

The beam shown in Fig. 6-13 is used to illustrate the computation of long-time deflections. The beam has an I section 36 in. deep, spans 54 ft, and is simply supported at the ends. The dead load w_g is 376 lb per ft, and the applied live load w_a is 1600 lb per ft. The effective prestress force after release, P_i, is 371 kips, and the effective prestress in the steel, f_{se}, is 157 ksi. The profile of the steel is that of a horizontal line in the middle third with an effective depth e of 14.10 in. Between the third point section and the end section, because of draping, the steel shows an inclined profile with a minimum eccentricity at the end of section, e_2, of 10.33 in.

The geometrical properties of the gross cross section are the following: area = 361 in.^{2}, moment of inertia = 57,100 in.^{4}, distance from cen-

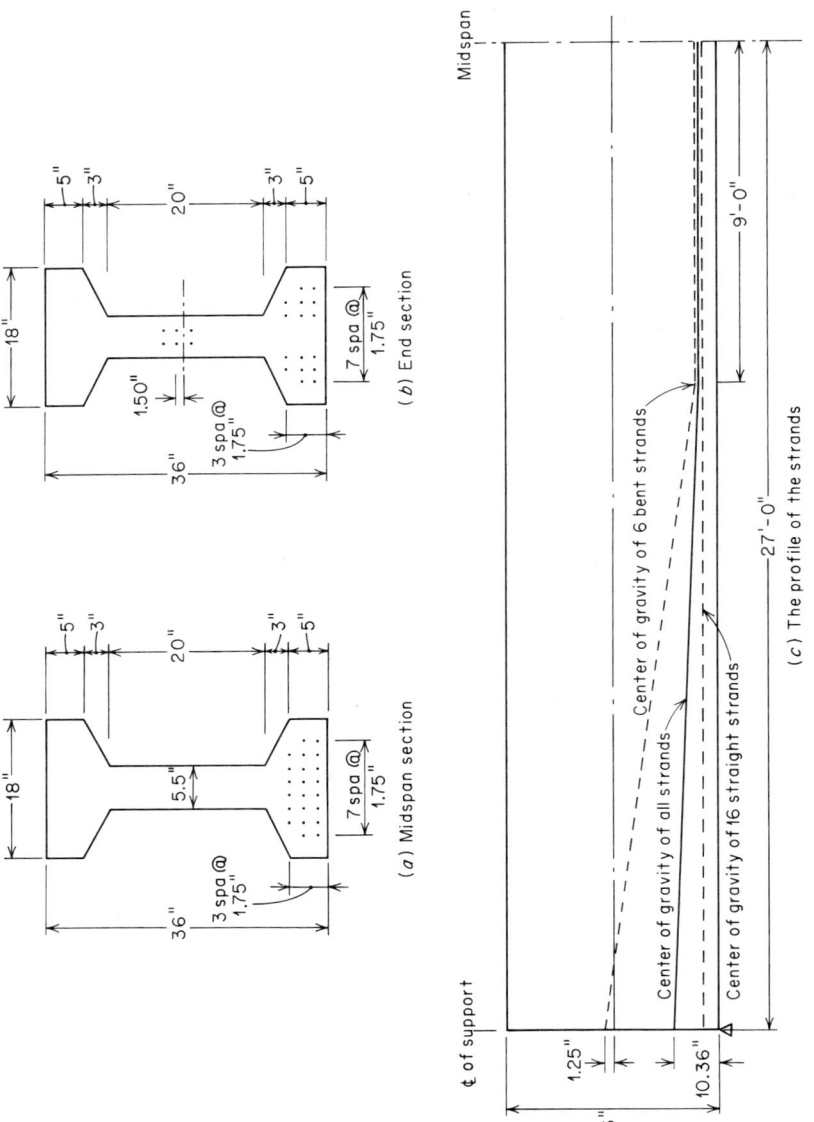

Fig. 6-13 Details of the arrangement of the prestressing steel, Illustrative Problem 6-3.

LOSSES AND LONG-TIME DEFLECTIONS

troidal axis to top or bottom fiber = 18 in. The material properties are taken as follows: concrete strength = 4000 psi at release of prestressing steel and 5000 psi when live load is applied to the beam, modulus of elasticity of concrete E_c = 3800 ksi at release and E_{ci} = 4300 ksi under service load conditions, modulus of elasticity of embedded strand reinforcement E_s = 30,000 ksi, concrete creep coefficient C_t = 2.0, concrete shrinkage strain ϵ_{sh} = 0.0005 (after release of prestress), and steel relaxation loss r_s = 5 percent of effective prestress after release.

It is required to estimate (1) the long-time deflections of the beam under prestress and dead load only; (2) the long-time deflections of the beam under live load, applied 2 weeks after release of prestress; and (3) an upper bound for the deflection under total load.

The total midspan deflection Δ_t will be the sum of Δ_{mt}, the midspan deflection caused by the transverse load, and Δ_{pt}, the midspan deflection caused by the prestress. The deflection due to transverse load can be calculated as follows:

$$\Delta_{mt} = \Delta_{mi}(1 + C_t)$$

where Δ_{mi}, the instantaneous deflection caused by the transverse load, is given by the well-known elastic expression

$$\Delta_{mi} = \frac{5}{384}\frac{wL^4}{EI}$$

Substituting in the above expressions yields, for the dead load,

$$\Delta_{mt} = \frac{5}{384}\frac{(376)(54)^4(12)^3}{(3800)(1000)(57,100)}(1+2) = 1.0 \text{ in. downward}$$

and, for the live load,

$$\Delta_{mt} = \frac{5}{384}\frac{(1600)(54)^4(12)^3}{(4300)(1000)(57,100)}(1+2) = 3.6 \text{ in. downward}$$

The deflection caused by the prestressing force requires a more lengthy determination. Because of the linear variation in the outer thirds of the span the deflection at midspan can be expressed as follows:

$$\Delta = \frac{L^2}{216}(23\varphi_1 + 4\varphi_2)$$

where ϕ_1 and ϕ_2 are the curvatures at midspan and at the end section, respectively. The instantaneous curvatures caused by the prestressing are

$$\varphi_1 = \frac{P_i e_1}{E_{ci}I} \quad \text{and} \quad \varphi_2 = \frac{P_i e_2}{E_{ci}I}$$

Substituting $P_i = 371$ kips, $E_{ci} = 3800$ ksi, $e_1 = 14.1$ in., $e_2 = 10.33$ in., and $I = 57,100$ in.^{4} in the preceding expressions yields

$$\varphi_1 = 2.41 \times 10^{-5} \text{ in.}^{-1}$$

and $\varphi_2 = 1.77 \times 10^{-5}$ in.$^{-1}$, respectively. The instantaneous deflection due to the prestressing force is

$$\Delta_{pi} = \frac{(54)^2(12)^2}{216}(23 \times 2.41 + 4 \times 1.77)(10^{-5}) = 1.2 \text{ in. upward}$$

The determination of the long-time deflection due to the prestressing force requires calculation of the loss of prestress. This loss is not constant along the span, as the center of gravity of steel varies in the outer thirds of the span. Hence, loss of prestress caused by creep has to be evaluated at the middle third and at the end section. The initial compressive stress at the center of gravity of steel in the middle third of the span is

$$f_{ci} = P_i\left(\frac{1}{A} + \frac{e_1{}^2}{I}\right) = 371\left(\frac{1}{361} + \frac{14.33^2}{57,100}\right) = 2.36 \text{ ksi}$$

At the support

$$f_{ci} = 371\left(\frac{1}{361} + \frac{10.33^2}{57,100}\right) = 1.72 \text{ ksi}$$

The unit loss in prestress is determined as a summation of the individual losses due to relaxation, shrinkage, and creep. In the middle third of the span the prestress loss is obtained as follows:

$$\frac{P'_t}{A} = r_s f_{se} + \epsilon_{sh} E_s + C_t n f_{ci}\left(1 - \frac{C_t n f_{ci}}{2 f_{se}}\right)$$

Substituting in the preceding equation yields

$$\frac{P'_t}{A} = (0.05)(157) + (0.0005)(30,000)$$
$$+ (2)(7)(2.36)\left(1 - \frac{2 \times 7 \times 2.36}{2 \times 157}\right)$$
$$= 7.9 + 15.0 + 29.6 = 52.5 \text{ ksi}$$

$$\frac{P'_t}{P_i} = \frac{52.5}{157} = 0.33$$

At the support

$$\frac{P'_t}{A} = (0.05)(157) + (0.0005)(30,000)$$
$$+ (2)(7)(1.72)\left(1 - \frac{2 \times 7 \times 1.72}{2 \times 157}\right)$$
$$= 7.9 + 15.0 + 22.2 = 45.1 \text{ ksi}$$

$$\frac{P'_t}{P_i} = \frac{45.1}{157} = 0.30$$

The total curvature at the two sections considered is evaluated by using

$$\varphi_{pt} = \varphi\left[1 - \frac{P'_t}{P_i} + \left(1 - \frac{P'_t}{2P_i}\right)C_t\right]$$

In the middle third of the span

$$(\varphi_{pt})_1 = (2.41)(10^{-5})[1 - 0.33 + (1 - \tfrac{0.33}{2})2] = (5.64)(10^{-5}) \text{ in.}^{-1}$$

At the end section

$$(\varphi_{pt})_2 = (1.77)(10^{-5})[1 - 0.30 + (1 - \tfrac{0.30}{2})2] = (4.25)(10^{-5}) \text{ in.}^{-1}$$

Total long-time deflection at midspan caused by the prestress is then

$$\Delta_{pt} = \frac{L^2}{216}[23(\varphi_{pt})_1 + 4(\phi_{pt})_2]$$
$$= \frac{(54)^2(12)^2}{216}[(23)(5.64) + (4)(4.25)](10^{-5})$$
$$= 2.8 \text{ in. upward}$$

If the live load is not always present, as in a bridge girder, it is not included in the computation of the long-time midspan deflection. Therefore,

$$\Delta_t = 1.0 - 2.8 = -1.8 \text{ in. upward}$$

If the live load is applied some time after the beam is prestressed, as is usually the case, part of the time-dependent deflection under prestress and dead load will have already occurred. Unless better data are available, it may be assumed that time-dependent deflections occur as given by the following table:

Time	Deflection
2 weeks	$\tfrac{1}{4}\Delta$
3 months	$\tfrac{1}{2}\Delta$
1 year	$\tfrac{3}{4}\Delta$

If we assume that the live load is placed 2 weeks after release of prestress and left there indefinitely, the total deflection under the live load referred to the position of the beam just before the application of the live load is as follows:

$$\Delta_t = \tfrac{3}{4}(\text{time-dependent deflection under prestress and dead load})$$
$$+ \text{ instantaneous and time-dependent deflection under live load}$$

Hence,

$$\Delta_t = \frac{3}{4}\left[1.0\,\frac{2}{1+2}\text{ downward } - (2.8 - 1.2)\text{ upward}\right] + 3.6\text{ in. downward}$$

$$= 2.9 \text{ in. downward}$$

An upper bound for the deflection under total load, measured from the original horizontal line at the casting yard, can be conservatively estimated as $3.6 - 1.8 = 1.8$ in. downward. This corresponds to approximately $\frac{1}{360}$ of the span, a ratio which is considered acceptable under any design standards.

PROBLEMS

6-1. Study the effects on long-time deflections in the beam of Illustrative Example 6-3 for the following cases:
  (a) Straight horizontal strands, $e = 9.2$ in.
  (b) Parabolic strands, $e = 9.2$ in. at midspan and $e = 0$ at the ends.
  (c) Straight sloping strands, $e = 9.2$ in. at midspan and $e = 0$ at the ends.

6-2. Find the ideal profile of strands for a given beam such that under dead and live loads, the short-time transverse deflections are zero anywhere in the beam. Discuss the effects of long-time deflections on such a beam.

6-3. For the beam of Illustrative Example 6-3, determine the loss of prestress due to friction and wobble effects if $K = 0.0015$ and $\mu = 0.25$.

REFERENCES

1. Building Code Requirements for Reinforced Concrete, American Concrete Institute (ACI 318-63), Detroit, 1963.
2. Deflections of Prestressed Concrete Members, *J. Am. Concrete Inst.*, December, 1963; *Proc.*, vol. 60, no. 12, pp. 1697–1728.
3. Schorer, Herman: Prestressed Concrete, Design Principles and Reinforcing Units, *J. Am. Concrete Inst.*, July, 1943; *Proc.*, vol. 39, no. 4, pp. 493–528.
4. Troxell, G. D., J. M. Raphael, and R. E. Davis: Long Time Creep and Shrinkage Tests of Plain and Reinforced Concrete, *Proc. ASTM*, vol. 58, pp. 1–20, 1958.
5. Magura, D. D., M. A. Sozen, and C. P. Siess: A Study of Stress Relaxation in Prestressing Reinforcement, *Univ. Illinois Civil Eng. Studies, Structural Res. Ser.* 237, Urbana, September, 1962.
6. Commentary on Building Code Requirements for Reinforced Concrete (ACI 318-63), p. 77, American Concrete Institute, 1965.
7. McHenry, D.: A New Aspect of Creep in Concrete and Its Application to Design, *Proc. ASTM*, vol. 43, pp. 1969–1984, 1943.
8. Corley, W. G., M. A. Sozen, and C. P. Siess: Time-dependent Deflections of Prestressed Concrete Beams, *Highway Res. Board Bull.* 307, pp. 1–25, 1961.

7
Working-stress Design of Simply Supported Prestressed-concrete Beams

7-1 INTRODUCTION

Prestressed-concrete beams are usually designed and proportioned on the basis of criteria which ensure certain low stresses in the beam at service loads. Design on the basis of these criteria is known as *working-load design*, and sometimes it is called *service-load design*. In the subsequent discussions the term *working-stress design* will be used.

In working-stress design the stresses in the beam due to all possible combinations of service loads are calculated on the basis of the assumption that concrete is an elastic material. These calculated stresses are required to be less than or equal to certain limiting stresses known as allowable stresses.

There is no clear-cut procedure for the determination of the allowable stresses. They are generally chosen by experience so that the structure will be usable and safe under service conditions while a reasonably economical use is made of the material. Limitation of stresses, however, controls short-time as well as long-time deflections, and assumes a certain safety factor in an indirect way.

Prestressed-concrete beams usually are not cracked under service conditions. The calculation of stresses on the basis of the assumption that concrete is an elastic material is perhaps more nearly correct for prestressed concrete than for reinforced concrete. However, it should be emphasized that, as the term indicates, the working-stress design criteria are conveniently used for the express purpose of design and proportioning. They are not intended as means of analyzing the beam or studying its behavior.

The working-stress design of prestressed-concrete beams is based upon the section of maximum moment where there are only normal stresses caused by axial load and bending. The shear stresses and inclined principal stresses caused by normal and shear stresses are not considered in service-load design.

The design procedure presented in this chapter is for simply supported, prestressed-concrete beams most commonly used in practice and is applicable to important load-carrying members. Generally the most common prestressed-concrete beams have the following characteristics.

1. In addition to the prestressing force the beam is subjected to its own weight, superimposed dead load, and live load—including impact if any—which act in the same direction. The superimposed dead load may or may not be present.
2. The prestressing operation is carried out at one time, and during prestressing the only load acting is the weight of the beam. In common practice, during the prestressing of a beam the only load present is the weight of the beam; this is almost always the case in pretensioned construction. However, in post-tensioned construction it is possible to cast the superimposed dead load on the beam before the application of the prestressing force. The advantage gained in this practice in increasing the prestressing force is offset by the requirement of heavy falsework.
3. The centroid of the prestressing force is considerably below the lower kern point of the section. This feature signifies a well-designed beam. Placing the centroid of the prestressing steel below the lower kern point of a section not only results in an efficient design but also improves the behavior of the beam under load.

Since the above features are the ones most commonly encountered in practice, they will be adopted as bases for the relations which are to be developed. It should be pointed out that the inclusion of these features is not essential for the development of a design procedure. They are brought up here for the clarification of the problem rather than its simplification. Furthermore, once the fundamental relations based upon the most common limitations are understood, design procedures may be developed for any specialized case with comparatively small effort.

7-2 CONDITIONS OF LOADING

From the time it comes into being through its service life a prestressed-concrete beam is actually subjected to a large variety of conditions of loading. However, there are two ideal and limiting conditions in the life of a prestressed-concrete beam which most interest the designer: the transfer condition and the final condition.

At transfer, which is the time when the fabrication of a prestressed-concrete beam is complete, the prestressing force has its highest magnitude, for the losses have not begun to take place. The concrete strength, on the other hand, is the lowest and is below the 28-day strength. In manufacturing a prestressed-concrete beam it is customary and desirable to prestress the beam soon after the concrete is cast, so that the beam can be removed and the prestressing bed reused. The prestressing operation is seldom carried out after the concrete has gained its 28-day strength. The strength of concrete at transfer, which is designated as f'_{ci}, is usually taken in the neighborhood of 4000 psi. The allowable stresses in concrete at transfer are given as percentages or functions of f'_{ci}. The condition of the beam at transfer is often referred to as the condition of the beam before losses.

Immediately after transfer the prestressing force begins to decrease at a decreasing rate until all the losses have taken place. The total decrease in the prestressing force, which is primarily due to shortening of the beam—caused by shrinkage and creep in concrete—and relaxation in steel, takes place mostly during the first few years after the manufacture of the beam. It can be assumed that for all practical purposes, all the losses have taken place 3 years after the beam is manufactured, at which time the prestressing force is at its minimum value. Evidently, at the time when all the losses have taken place the concrete strength can be conservatively estimated as f'_c, its 28-day strength. The condition of a prestressed-concrete beam corresponding to the minimum prestressing force and maximum concrete strength is referred to as the final condition. In specifications the final condition of the beam is referred to as the condition of the beam after losses. After the final condition is reached, no changes take place in the prestressing force. The allowable stresses in concrete corresponding to the final condition are given as percentages or functions of f'_c.

From the above discussion one may conclude that there are two ideal and limiting sets of conditions in the life of a prestressed-concrete beam: the transfer condition and the final condition. Since a prestressed-

Table 7-1 Idealized conditions of loading

Loading condition	Loads acting*	Prestressing force	Strength of concrete
1	$P + G$		
2	$P + G + S$	Maximum	Minimum
3	$P + G + S + L + I$		
4	$P + G$		
5	$P + G + S$	Minimum	Maximum
6	$P + G + S + L + I$		

*P = prestressing force; G = weight of the beam; S = superimposed dead load; L = live load; I = impact.

concrete beam in addition to the prestressing force is subject to its own weight, superimposed dead load, and live load (including impact), the limiting and idealized conditions of loading presented in Table 7-1 may be defined.

Conditions 1, 2, and 3 are temporary and correspond to transfer.

Loading condition 1 corresponds to transfer when the beam is subjected to the prestressing force and its own weight. All prestressed-concrete beams go through this condition. In pretensioned beams this condition results immediately after the wires or strands are cut. In post-tensioned construction this condition corresponds to the time when prestressing has just been completed.

Loading condition 2 corresponds to a beam in which at transfer both the weight of the beam and the superimposed dead load are acting. Although this case is rare, it is possible in beams that are to be post-tensioned in place after the application of the superimposed dead load—if the falsework is sufficiently strong to carry the weight of the beam and the superimposed dead load. This condition may also apply to beams in which the superimposed dead load is applied at transfer or immediately thereafter.

Loading condition 3 is possible only if at transfer or immediately thereafter the beam is subjected to the live load and to impact, if any. Although this condition may be approximated in practice, it is only of academic value.

Conditions 4 to 6 are the final loading conditions.

Loading condition 4 is possible in a precast, prestressed-concrete beam if it is stored for a long period of time prior to its use or in a cast-in-place construction, when the work is discontinued for an indefinite period, which allows all the losses to take place. This condition also applies to structures which have no superimposed dead load and which have been in place for several years, at a time when they are not subjected to live load.

Loading condition 5 corresponds to all beams which have been in place for several years, carrying superimposed dead load, at a time when they are not subject to live load.

Loading condition 6 represents the fully loaded structure sometime after the construction has been completed.

It can be shown that conditions 1 and 6 are the most important conditions of loading. Condition 1 always governs among the first three conditions, while condition 6 governs among the second three.

7-3 ALLOWABLE STRESSES AND STRESS COEFFICIENTS

As discussed in the preceding section, there are two idealized conditions in the life of a prestressed-concrete beam: the transfer and the final. Since the characteristics of the beam are different in each of these conditions, it is necessary to have two sets of allowable stresses for concrete.

The present specifications give allowable stresses in concrete at transfer, or before losses, as functions of f'_{ci}, the concrete strength at transfer. However, for the sake of convenience in the subsequent discussions the allowable stresses in concrete at transfer are taken as percentages of f'_c, the 28-day concrete strength. This modification can be made conveniently since the strength of concrete at transfer and at 28 days is known. Throughout this text the allowable compressive stress in concrete at transfer will be designated as $c'_t f'_c$, and the allowable tensile stress at transfer as $a'_t f'_c$. The dimensionless quantities c'_t and a'_t are defined as the stress coefficients at transfer that correspond to the allowable compressive and tensile stresses in concrete, respectively. Generally, the range of c'_t may be from 0.40 to 0.50, while a'_t may vary between 0 and 0.10.

Current specifications give allowable stresses in concrete in final conditions, or after losses, directly as functions of f'_c. In the following discussions the allowable compressive stress in concrete in final conditions will be designated as $c'f'_c$, and the allowable tensile stress in final conditions as $a'f'_c$. The dimensionless quantities c' and a' are defined as the final stress coefficients that correspond to the allowable compressive and tensile stresses, respectively, in concrete. The range of c' is between 0.40 and 0.45, and a' may vary between 0 and 0.10. A negative sign for a' indicates a compressive stress.

For convenient reference the computed stresses in concrete are also presented as coefficients of f'_c. The computed tensile and compressive stresses in concrete at transfer are designated as $a_t f'_c$ and $c_t f'_c$, respectively. The computed final tensile and compressive stresses are denoted as $a f'_c$ and $c f'_c$, respectively. The dimensionless quantities a_t, c_t, a and c are the stress

Table 7-2 A summary of allowable stresses in concrete in the prestressed-concrete specifications in the United States

		Allowable stress at transfer		Allowable stress after losses	
Type of structure	Method of prestressing	Compressive	Tensile	Compressive	Tensile
AASHO Specifications for Highway Bridges 1961[1]					
Single-element beams without non-prestressed reinforcement in tension zone	Pretensioned	$0.60 f'_{ci}$	$3\sqrt{f'_{ci}}$	$0.40 f'_c$	0
	Post-tensioned	$0.55 f'_{ci}$	$3\sqrt{f'_{ci}}$	$0.40 f'_c$	0
Single-element beams with non-prestressed reinforcement in tension zone	Pretensioned	$0.60 f'_{ci}$	$6\sqrt{f'_{ci}}$	$0.40 f'_c$	0
	Post-tensioned	$0.55 f'_{ci}$	$6\sqrt{f'_{ci}}$	$0.40 f'_c$	0
Segmental-element beams without non-prestressed reinforcement in tension zone	Pretensioned	$0.60 f'_{ci}$	0	$0.40 f'_c$	0
	Post-tensioned	$0.55 f'_{ci}$	0	$0.40 f'_c$	0
Segmental-element beams with non-prestressed reinforcement in tension zone	Pretensioned	$0.60 f'_{ci}$	$3\sqrt{f'_{ci}}$	$0.40 f'_c$	0
	Post-tensioned	$0.55 f'_{ci}$	$3\sqrt{f'_{ci}}$	$0.40 f'_c$	0
ACI Building Code 1963[2]					
Single-element beams without non-prestressed reinforcement in tension zone, protected*		$0.60 f'_{ci}$	$3\sqrt{f'_{ci}}$†	$0.45 f'_c$	$6\sqrt{f'_c}$
Single-element beams without non-prestressed reinforcement in tension zone, unprotected*		$0.60 f'_{ci}$	$3\sqrt{f'_{ci}}$†	$0.45 f'_c$	0

* Protected members are those not exposed to freezing temperatures nor to a corrosive environment, which contain bonded prestressed or unprestressed reinforcement located so as to control cracking. All other members are designated as unprotected.
† Where calculated tension stress exceeds this value, reinforcement shall be provided to resist the total tension force in concrete computed on the assumption of an uncracked section.

Table 7-3 A summary of stress coefficients prepared from the prestressed-concrete specifications in the United States as shown in Table 7-2

$f'_c = 5000$ psi, $f'_{ci} = 4000$ psi

Type of structure	Method of prestressing	c'_t	a'_t	c'	a'
AASHO Specifications for Highway Bridges 1961					
Single-element beams without non-prestressed reinforcement in tension zone	Pretensioned	0.48	0.038	0.40	0
	Post-tensioned	0.44	0.038	0.40	0
Single-element beams with non-prestressed reinforcement in tension zone	Pretensioned	0.48	0.076	0.40	0
	Post-tensioned	0.44	0.076	0.40	0
Segmental-element beams without non-prestressed reinforcement in tension zone	Pretensioned	0.48	0	0.40	0
	Post-tensioned	0.44	0	0.40	0
Segmental-element beams with non-prestressed reinforcement in tension zone	Pretensioned	0.48	0.038	0.40	0
	Post-tensioned	0.44	0.038	0.40	0
ACI Building Code 1963					
Single-element beams without non-prestressed reinforcement in tension zone, protected		0.48	0.038	0.45	0.085
Single-element beams without non-prestressed reinforcement in tension zone, unprotected		0.48	0.076	0.45	0

coefficients corresponding to the computed stresses. It is the purpose of working-stress design to proportion the section in such a way as to make each computed stress equal to or less than the corresponding allowable stress. This can be expressed as follows:

$$a_t \leq a'_t \quad c_t \leq c'_t$$
$$c \leq c' \quad a \leq a'$$

Table 7-2 contains a summary of allowable stresses used for various types of construction in current specifications in the United States. In Table 7-3 the stress coefficients are listed for the allowable stresses shown in Table 7-2. In Table 7-3 it is assumed that $f'_c = 5000$ psi and $f'_{ci} = 4000$ psi.

7-4 THE FOUR BASIC REQUIREMENTS

As mentioned before, loading condition 1 always governs among the three loading conditions at transfer; that is, if the stresses in the beam are satisfied in loading condition 1, they will automatically be satisfied in loading conditions 2 and 3. This conclusion can be drawn easily since the allowable concrete stresses at transfer are primarily for the limitation of the stresses in the beam caused by the prestressing force. In loading condition 1 the effect of prestressing is dominant.

Similarly, as stated before, loading condition 6 governs among the three final loading conditions. In other words, if the stresses in the beam are satisfied in loading condition 6, they will automatically be satisfied in loading conditions 4 and 5. This conclusion is justified since the allowable final concrete stresses tend to limit the acting loads, and in loading condition 6 the effect of acting loads is dominant.

From a designer's point of view, therefore, there are only two loading conditions to be taken into account, namely, 1 and 6. In each loading condition, the top and bottom fiber stresses (extreme fiber stresses) must be equal to or less than the corresponding allowable stress in concrete. Since there are two governing loading conditions, there will be four requirements to be met. These requirements are as follows:

Requirement 1 For loading condition 1, the tensile stress at the top fiber $(a_t f'_c)$ must be less than or equal to the allowable tensile stress at transfer $(a'_t f'_c)$.

Requirement 2 For loading condition 1, the compressive stress at the bottom fiber $(c_t f'_c)$ must be less than or equal to the allowable compressive stress at transfer $(c'_t f'_c)$.

Requirement 3 For loading condition 6, the compressive stress at the top fiber $(c f'_c)$ must be less than or equal to the allowable final compressive stress $(c' f'_c)$.

Requirement 4 For loading condition 6, the tensile stress at the bottom fiber $(a f'_c)$ must be less than or equal to the allowable final tensile stress $(a' f'_c)$.

Although the above requirements must be satisfied in all points in the beam, in the discussions that follow the stresses at the center of a simply supported beam will be studied. By draping of reinforcement and other means the stresses at the ends can be kept within the allowable stresses.

The above four requirements can be stated algebraically as follows:

$$\frac{P_t}{A}\left(\frac{ey_t}{r^2} - 1\right) - M_g \frac{y_t}{I} = a_t f'_c \leq a'_t f'_c \tag{7-1}$$

$$\frac{P_t}{A}\left(\frac{ey_b}{r^2} + 1\right) - M_g \frac{y_b}{I} = c_t f'_c \leq c'_t f'_c \tag{7-2}$$

$$-\eta \frac{P_t}{A}\left(\frac{ey_t}{r^2} + 1\right) + M_t \frac{y_t}{I} = cf'_c \leq c'f'_c \tag{7-3}$$

$$-\eta \frac{P_t}{A}\left(\frac{ey_b}{r^2} + 1\right) + M_t \frac{y_b}{I} = af'_c \leq a'f'_c \tag{7-4}$$

where P_t = prestressing force at transfer
A = gross cross-sectional area of beam
e = eccentricity of prestressing force
r = radius of gyration of section
y_t = distance of top fiber from centroidal axis of gross section
y_b = distance of bottom fiber from centroidal axis of gross section
I = moment of inertia of gross section
M_g = moment due to weight of beam, taken as $A\gamma L^2/8$, where γ is the unit weight of concrete and L is length of simple span
M_t = moment due to all vertical loads acting on beam, taken as $M_a + M_g$, where M_a is sum of moments due to superimposed dead load and live load
a_t, c_t, c, a = dimensionless coefficients of f'_c that represent computed stresses in beam
η = effectiveness, taken as P/P_t, where P is final or effective prestressing force, that is, prestressing force after losses

It should be remembered that in developing the above four requirements it has been assumed that the center of gravity of steel is considerably below the lower kern point of the section. If the center of gravity of steel were in the core of the section, the first requirement would be satisfied automatically.

The quantities A, y_t, y_b, and I as used in the above requirements are defined on the basis of the gross cross-sectional area of the beam. Although from a practical point of view this is a convenient approximation, it is not exactly correct.

The correct values for quantities A, y_t, y_b, and I in a pretensioned bonded construction should be based on the fully transformed cross-sectional area of the beam. In post-tensioned grouted construction the correct values of these quantities in expressions (7-1) and (7-2) should be

Table 7-4 Correct sections in common types of construction

Requirement	Pretensioned bonded	Post-tensioned grouted		Post-tensioned unbonded
	P, G, S, L	P, G	S, L	P, G, S, L
1	Transformed	Net	Transformed	Net
2	Transformed	Net	Transformed	Net
3	Transformed	Net	Transformed	Net
4	Transformed	Net	Transformed	Net

computed by using the net cross-sectional area of the beam since at transfer or immediately thereafter the grout is not present or effective in the preformed hole. In expressions (7-3) and (7-4), however, the correct values of these quantities are computed on the basis of the net area for the prestressing force and dead load, and the fully transformed section for superimposed dead load and live load. In unbonded construction the net cross-sectional area is used to determine the section properties in all expressions. The above statements are summarized in Table 7-4; the sections to be used are given specifically for the three common types of construction.

Calculation of the section properties on the basis of the gross cross-sectional area of the beam is a convenient and close approximation. In the subsequent discussions A, y_b, y_t, and I may be taken as those corresponding to the gross cross-sectional area of the beam.

If in any of the four basic requirements the computed stress is equal to the allowable stress, it is said that *that requirement is satisfied exactly*. On the other hand, if a computed stress is less than the allowable stress, it is said that *the particular requirement is satisfied by a margin*.

For example, when $a_t f'_c = a'_t f'_c$ or $a_t = a'_t$, requirement 1 is satisfied exactly; and when $a_t f'_c < a'_t f'_c$ or $a_t < a'_t$, this requirement is satisfied by a margin. In a prestressed-concrete beam some of the four requirements are satisfied exactly and the others are satisfied by a margin. It is possible to proportion a beam so that all requirements are satisfied exactly. Usually in an actual beam all requirements are satisfied by a margin.

7-5 Problems in Design

There are usually two types of problem in the design of prestressed-concrete beams. In the first type the beam is a standard precast section with known section properties, and the problem is to determine the required prestressing force and the eccentricity. The standard sections available can be used for a wide range of loads and span lengths by varying the prestressing force. Information concerning the properties of standard sections is available in manuals prepared by the manufacturers.

The second type of problem arises when standard precast sections are not available or are not to be used and it is required to determine the shape of the beam as well as the magnitude and position of the prestressing force.

Although the use of standard sections does not assume a least-weight design, it results in some economy since there is no necessity of manufacturing special forms. Standard sections are particularly suitable when a small number of short-span beams are required. However, for long spans it may become necessary to proportion a least-weight section, which requires manufacturing of special forms. When a particular section is repeated many times in a structure—as is the case in large buildings that have equal simple spans—proportioning for the least weight becomes economical.

In the following paragraphs both of these problems are discussed in detail.

DESIGN USING STANDARD SECTIONS

7-6 DETERMINATION OF PRESTRESSING FORCE AND ECCENTRICITY

Design by use of standard sections may be carried out conveniently by requirements 1 through 4 since in these requirements P_t and e are the only unknowns. In problems of this type the largest eccentricity available may be adopted and P_t computed from the four requirements.

Since P_t decreases with eccentricity, the smallest amount of prestressing force is obtained, with the largest possible eccentricity, provided that the distance from the center of gravity of steel to the bottom fiber is sufficiently large to permit the arrangement of the prestressing steel. In practical sections the distance between the center of gravity of steel and the bottom fiber is in the neighborhood of one-tenth of the overall depth of the beam.

For an assumed eccentricity each of the four expressions (7-1) through (7-4) will give a range of possible answers for P_t. Expressions (7-1) and (7-2) limit the upper bound, and expressions (7-3) and (7-4) limit the lower bound of the prestressing force. Hence if the larger of the two P_t values given by expressions (7-3) and (7-4) is less than P_t given by expressions (7-1) and (7-2), it is the answer.

The following illustrative example shows the application of this method.

7-7 ILLUSTRATIVE PROBLEM 7-1

The standard section shown in Fig. 7-1a is to be used on a simply supported beam of 62 ft span and is to be subjected to a superimposed dead load and live load of 0.984 klf. Determine the eccentricity and prestressing force.

For this problem assume the following:

$$f'_c = 5000 \text{ psi} \qquad \gamma = 150 \text{ pcf} \qquad a'_l = 0.038 \qquad c'_t = 0.48$$
$$c' = 0.45 \qquad a' = 0.085 \qquad \eta = 0.85$$

Solution The properties of the section shown in Fig. 7-1a are as follows:

$$A = 472 \text{ in.}^2 \qquad r^2 = 74.0 \text{ in.}^2$$
$$y_t = 10.88 \text{ in.} \qquad \frac{I}{y_t} = 3205 \text{ in.}^3$$
$$y_b = 13.10 \text{ in.} \qquad \frac{I}{y_b} = 2667 \text{ in.}^3$$
$$I = 34{,}936 \text{ in.}^4 \qquad w_g = 0.492 \text{ klf}$$

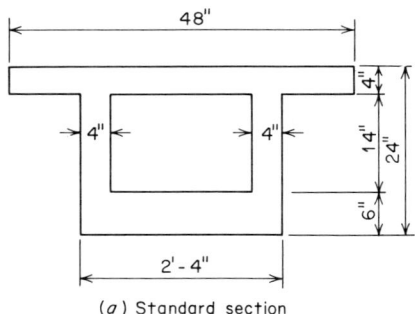

(a) Standard section

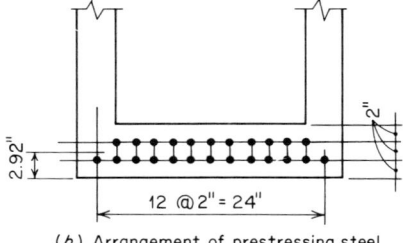

(b) Arrangement of prestressing steel

Fig. 7-1 Standard section used in Illustrative Problem 7-1.

WORKING-STRESS DESIGN OF SUPPORTED PRESTRESSED-CONCRETE BEAMS

The moments and stresses are:

$$M_g = 0.492 \frac{(62)^2}{8} 12 = 2840 \text{ in.-kips}$$

$$M_a = 0.984 \frac{(62)^2}{8} 12 = 5670 \text{ in.-kips}$$

$$\frac{M_g y_t}{I} = 0.886 \text{ ksi}$$

$$\frac{M_g y_b}{I} = 2.655 \text{ ksi}$$

$$\frac{(M_a + M_g)y_t}{I} = 1.065 \text{ ksi}$$

$$\frac{(M_a + M_g)y_b}{I} = 3.191 \text{ ksi}$$

Substituting the above quantities in requirements (7-1) through (7-4) and taking $e = 10.18$ in., we shall obtain the following ranges for P_t:

$$\frac{P_t}{472}\left(\frac{10.18 \times 10.19}{74.02} - 1\right) - 0.886 \leq 0.190$$

$$P_t \leq 1272 \text{ kips} \quad (7\text{-}1)$$

$$\frac{P_t}{472}\left(\frac{10.18 \times 13.10}{74.02} + 1\right) - 1.065 \leq 2.400$$

$$P_t \leq 585 \text{ kips} \quad (7\text{-}2)$$

$$-0.85\frac{P_t}{472}\left(\frac{10.18 \times 10.19}{74.02} - 1\right) + 2.655 \leq 2.250$$

$$P_t \geq 562 \text{ kips} \quad (7\text{-}3)$$

$$-0.85\frac{P_t}{472}\left(\frac{10.18 \times 13.10}{74.02} + 1\right) + 3.191 \leq 0.425$$

$$P_t \geq 549 \text{ kips} \quad (7\text{-}4)$$

Hence $P_t = 562$ kips is the answer. The number of strands needed, if $\frac{1}{2}$-in. strands are used, can be calculated as follows:

$$\frac{562}{163 \times 0.1438} = 24 \text{ half-in. strands}$$

where the allowable stress in steel is taken as 163 ksi and the area of $\frac{1}{2}$-in. strand as 0.1438 in.^{2}.

Figure 7-1b shows the arrangement of steel in the section. It can be seen that the distance of the center of gravity of steel from the bottom fiber is 2.92 in., which corresponds to $e = 10.18$ in., the value used in calculating P_t.

7-8 VARIATION OF ECCENTRICITY WITH THE PRESTRESSING FORCE

For a given section by studying the variation of eccentricity with the prestressing force one may obtain the range of eccentricity and the corresponding force that can be used for the given loads and span lengths.[3] It is more convenient to study the relationship between $1/P_t$ and e, since they are related linearly.

When $e \geq r^2/y_t$, the four requirements can be expressed in the following form:

$$\frac{1}{P_t} \geq \frac{ey_t/r^2 - 1}{(M_g y_t/I + a'_t f'_c)A} \tag{7-1}$$

$$\frac{1}{P_t} \geq \frac{ey_b/r^2 + 1}{(M_g y_b/I + c'_t f'_c)A} \tag{7-2}$$

$$\frac{1}{P_t} \leq \frac{(ey_t/r^2 - 1)\eta}{(M_t y_t/I - c' f'_c)A} \tag{7-3}$$

$$\frac{1}{P_t} \leq \frac{(ey_b/r^2 + 1)\eta}{(M_t y_b/I - a' f'_c)A} \tag{7-4}$$

If these inequalities are presented as equations, they will represent lines which define the variation of $1/P_t$ with e at the limit of the require-

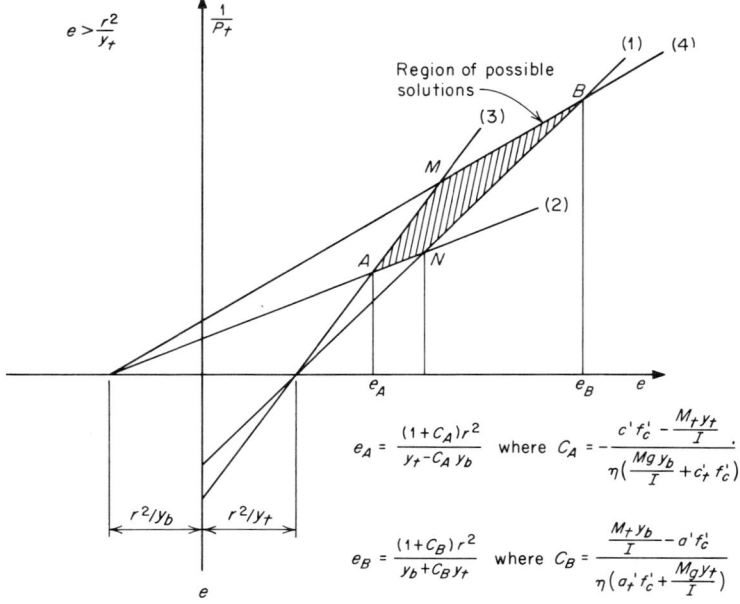

Fig. 7-2 The region of possible solutions.

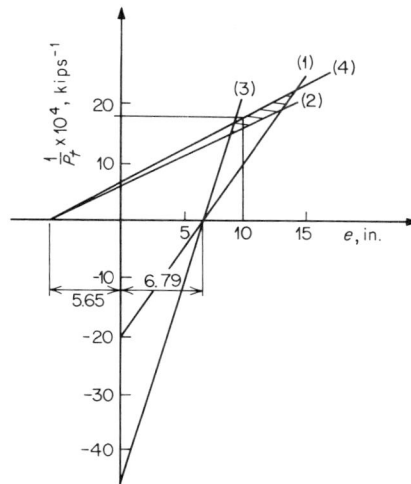

Fig. 7-3 The region of possible solutions, Illustrative Problem 7-1.

ments. Figure 7-2 shows these lines as well as the region within which an answer is possible. The maximum eccentricity is at point B, which corresponds to the minimum P_t. The eccentricity at point A is the minimum and corresponds to the maximum P_t.

For a given section the largest eccentricity should be used, provided that sufficient space is available for placing the steel. From Fig. 7-2 it can be seen that the minimum P_t results when the design falls on line 4, which means that requirement 4 should be satisfied exactly. If this is not possible, the design should fall on line 3.

Variation of eccentricity with prestressing force may be studied in a similar way for Illustrative Problem 7-1.

The four requirements for Illustrative Problem 7-1 may be written as follows:

$$\frac{1}{P_t} \geq 0.000271e - 0.001975 \tag{7-1}$$

$$\frac{1}{P_t} \geq 0.000108e + 0.000611 \tag{7-2}$$

$$\frac{1}{P_t} \leq 0.000612e - 0.004480 \tag{7-3}$$

$$\frac{1}{P_t} \leq 0.000115e + 0.000651 \tag{7-4}$$

The region of possible solutions based upon the above inequalities is shown in Fig. 7-3. An eccentricity of 10.18 in. corresponding to $P_t = 562$ kips is a possible solution.

From Fig. 7-2 it can be seen that requirement 3 provides the lower bound for the eccentricity. However, if the live load is too small for the standard section, the top fiber stress after losses when the beam is fully loaded may still be in tension. In this case requirement 3 will vanish, and the lower bound of eccentricity will fall in the range $e < r^2/y_b$. The study may be extended to the range $e < r^2/y_t$ in order to establish the lower bound for the eccentricity, though practically we are seldom interested in this range.

LEAST-WEIGHT DESIGN[4]

7-9 THE DIMENSIONLESS VARIABLES

In design of a prestressed-concrete beam for least weight, it is convenient to consider the variables in dimensionless form.

Introducing h as the overall depth of the beam and noting that

$$M_g = \gamma \frac{AL^2}{8}$$

and

$$M_t = M_a + M_g$$

we can write expressions (7-1) through (7-4) in the form of equations as follows:

$$\frac{P_t}{Af_c'}\left(\frac{e}{h}\frac{h^2}{r^2}\frac{1}{y_b/y_t + 1} - 1\right) - \frac{\gamma L^2}{8hf_c'}\frac{1}{y_b/y_t + 1}\frac{h^2}{r^2} = a_t$$

$$\frac{P_t}{Af_c'}\left(\frac{e}{h}\frac{h^2}{r^2}\frac{y_b/y_t}{y_b/y_t + 1} + 1\right) - \frac{\gamma L^2}{8hf_c'}\frac{y_b/y_t}{y_b/y_t + 1}\frac{h^2}{r^2} = c_t$$

$$-\eta \frac{P_t}{Af_c'}\left(\frac{e}{h}\frac{h^2}{r^2}\frac{1}{y_b/y_t + 1} - 1\right) + \frac{\gamma L^2}{8hf_c'}\frac{1 + M_a/M_g}{y_b/y_t + 1}\frac{h^2}{r^2} = c$$

$$-\eta \frac{P_t}{Af_c'}\left(\frac{e}{h}\frac{h^2}{r^2}\frac{y_b/y_t}{y_b/y_t + 1} + 1\right) + \frac{\gamma L^2}{8hf_c'}\frac{y_b/y_t(1 + M_a/M_g)}{y_b/y_t + 1}\frac{h^2}{r^2} = a$$

The following dimensionless quantities are introduced:

$$\frac{P_t}{Af_c'} = m \qquad\qquad \frac{e}{h} = \epsilon$$

$$\frac{y_b}{y_t} = \Delta \qquad\qquad \frac{r^2}{h^2} = \rho$$

$$\frac{M_a}{M_g} = R \qquad\qquad \frac{hf_c'}{\gamma L^2} = \omega$$

By substituting these quantities in the above equations and rearranging, the four basic requirements can be written as follows:

$$m\left[\frac{\epsilon}{\rho(1+\Delta)} - 1\right] - \frac{1}{8\rho\omega(1+\Delta)} = a_t \qquad (7\text{-}1a)$$

$$m\left[\frac{\epsilon\Delta}{\rho(1+\Delta)} + 1\right] - \frac{\Delta}{8\rho\omega(1+\Delta)} = c_t \qquad (7\text{-}2a)$$

$$-\eta m\left[\frac{\epsilon}{\rho(1+\Delta)} - 1\right] + \frac{1+R}{8\rho\omega(1+\Delta)} = c \qquad (7\text{-}3a)$$

$$-\eta m\left[\frac{\epsilon\Delta}{\rho(1+\Delta)} + 1\right] + \frac{\Delta(1+R)}{8\rho\omega(1+\Delta)} = a \qquad (7\text{-}4a)$$

If the values of a_t, c_t, c, a, and η are known or assumed, there will be six dimensionless variables in the above four equations, namely, Δ, m, ϵ, R, ρ, and ω.

In order to understand the physical significance of the six dimensionless unknowns, they are discussed briefly in the following paragraphs.

The depth factor ω Although ω has been considered an unknown, a reasonable value for it can be established. The expression for $\omega = hf'_c/\gamma L^2$ contains the unit weight of concrete γ, the span length L, the 28-day concrete strength f'_c, and the depth of the beam, h. Of these four quantities, only h is an unknown. Therefore, the quantity ω is actually an expression for the depth of the beam. The depth of the beam, on the other hand, is frequently controlled by clearance or architectural requirements, and sometimes it is dictated by stiffness requirements. By making a reasonable estimate for the depth of the beam, h, one can determine the value of ω.

In practice ω varies between unity for very long spans and about 10 for short spans.

The efficiency ρ The variable ρ is a measure of the efficient distribution of cross-sectional area. From the expression $\rho = r^2/h^2$, one can conclude that for a given depth the greater r becomes the greater is the efficiency of the section. Theoretically ρ varies between zero and 0.25. In a hypothetical section in which all the area is concentrated at the centroidal axis, ρ is equal to zero. On the other hand, a maximum ρ of 0.25 would result if all the area were concentrated at the extreme fibers.

The practical range of ρ in prestressed-concrete beams is from 0.08 to 0.14. For a rectangular section ρ is a constant value and is equal to 0.0833. For I sections ρ varies between 0.10 and 0.14, while for T sections and inverted T sections its range is between 0.08 and 0.10.

It should be pointed out that although in rolled or built-up steel sections ρ can get as high as 0.20, in prestressed concrete a section with a ρ of

0.14 is considered highly efficient; steel beams have considerably more slender webs than prestressed-concrete beams.

The shape factor Δ The shape factor Δ is a measure of the position of the centroidal axis of a section. Although theoretically Δ may vary between zero and infinity, its practical range is limited. For rectangular sections, symmetrical I sections, and all sections in which the centroidal axis is at middepth, Δ is equal to unity. For practical T sections and unsymmetrical I sections in which the top flange is heavier than the bottom flange, its range is from 1.2 to about 1.6. For inverted T sections and sections with heavy bottom flange, it may vary between 0.6 and 0.9.

In some cases where T sections are used as beams and slabs (in roofs), Δ may be as high as 4.

The moment ratio R The quantity R is the ratio of the moment caused by the applied loads (M_a) to the moment due to the weight of the beam (M_g). Theoretically it can vary between zero and infinity; however, its practical range of variation is from zero to about 10.

It can also be stated that R is an expression for the cross-sectional area of the beam. If the beam is assumed to be prismatic, from the definition of R the following will result:

$$R = \frac{M_a}{M_g} = \frac{8M_a}{A\gamma L^2}$$

All the terms have been defined previously.

Since in a given problem M_a, γ, and L are known, R becomes proportional with the reciprocal of A, the cross-sectional area.

The reinforcement ratio m The reinforcement ratio m is the ratio of stress at the centroid to the 28-day concrete strength, P_t/Af'_c. Its theoretical range is from zero to infinity; however, practically it is from 0.15 to 0.40.

For a given A and f'_c, m is proportional with P_t, the prestressing force at transfer.

The eccentricity ratio ϵ The eccentricity ratio ϵ is a measure of effective utilization of the prestressing force. For a given section the greater ϵ becomes, the less is the required prestressing force.

From the expression $\epsilon = e/h$, it is seen that for a given depth, ϵ increases with eccentricity. The quantity ϵ varies between $r^2/y_t h$ and y_b/h. The lower limit, $r^2/y_t h$, corresponds to the case in which the center of gravity of steel coincides with the lower kern point of the section. This limit has been set in developing the four basic requirements. However, in practical sections the lower limit is never reached. The upper limit

WORKING-STRESS DESIGN OF SUPPORTED PRESTRESSED-CONCRETE BEAMS

corresponds to the hypothetical condition in which the center of gravity of steel coincides with the bottom fiber of the section. Practically it is impossible to reach the upper limit. Generally ϵ varies between 0.25 and 0.55.

It is possible to express ϵ in terms of the ratio of the distance of center of gravity of steel from the bottom fiber to the depth of the beam, as follows:

$$\epsilon = \frac{y_b}{h} - \frac{g}{h}$$

where g is the distance of the center of gravity of steel to the bottom fiber. Other terms have been defined.

The above expression can be written as follows:

$$\epsilon = \frac{y_b}{y_b + y_t} - \frac{g}{h} = \frac{\Delta}{\Delta + 1} - \frac{g}{h}$$

For most pretensioned beams the value of g/h is in the neighborhood of 0.10. In post-tensioned beams it may be somewhat lower. From the above expression one may conclude, therefore, that for a given value of Δ it is possible to estimate a reasonable value for ϵ. In symmetrical sections in which $\Delta = 1$, the value of ϵ is about 0.4.

7-10 RELATIONS AMONG THE DIMENSIONLESS UNKNOWNS

Of the six dimensionless unknowns discussed above, the first four define the section properties and the last two define the magnitude and location of the prestressing steel once the section properties of the beam are known.

From the discussions in the preceding section it is seen that reasonable values for ω and ρ can be established. The quantity ω is an expression for the depth of the beam, and a reasonable value for the depth can be established. The quantity ρ varies in a comparatively narrow range, and a reasonable estimate may be made of its magnitude. If we adopt reasonable values for ω and ρ, they may be taken as known quantities, and Eqs. (7-1a) through (7-4a) will contain only four dimensionless unknowns, namely, Δ, R, m, and ϵ.

A simultaneous solution of Eqs. (7-1a) through (7-4a) yields the following expressions for these four dimensionless unknowns:

$$\Delta = \frac{\eta c_t + a}{\eta a_t + c} \tag{7-5}$$

$$R = 8\rho\omega[(a + c) + \eta(a_t + c_t)] - (1 - \eta) \tag{7-6}$$

$$m = \frac{c_t c - a_t a}{(a + c) + \eta(a_t + c_t)} \tag{7-7}$$

$$\epsilon = \left[\rho(a_t + c_t) + \frac{1}{8\omega}\right]\frac{(a + c) + \eta(a_t + c_t)}{c_t c - a_t a} \tag{7-8}$$

Appendix A-1 shows the derivation of the above expressions.

Equations (7-5) and (7-6) define the section properties of a prestressed-concrete beam as far as the concrete section is concerned. Equations (7-7) and (7-8) define the magnitude and position of the prestressing force in terms of the section properties.

It is interesting to note that Eqs. (7-5) and (7-7) for Δ and m, respectively, are only functions of the stress coefficients and the effectiveness and do not contain any of the dimensionless quantities. Equations (7-6) and (7-8), however, are functions of ρ and ω.

Equations (7-5) through (7-8) are used in designing a prestressed-concrete beam. In the following paragraphs alternative forms of the above equations are given to facilitate the discussion of design criteria that is to follow.

Elimination of the terms $\eta a_t + c$ and $a + \eta c_t$ between Eqs. (7-5) and (7-6), respectively, results in the following expressions for R:

$$R = 8\rho\omega(a + \eta c_t)\left(\frac{1}{\Delta} + 1\right) - (1 - \eta) \tag{7-6a}$$

$$R = 8\rho\omega(\eta a_t + c)(\Delta + 1) - (1 - \eta) \tag{7-6b}$$

By eliminating ρ between Eqs. (7-6) and (7-8) the following is obtained:

$$R = \frac{8\omega\epsilon(c_t c - a_t a) - [(a + c) + \eta(a_t + c_t)]}{a_t + c_t} - (1 - \eta) \tag{7-9}$$

Elimination of a and c_t between Eqs. (7-5) and (7-7), respectively, results in the following expressions for m:

$$m = \frac{c_t - \Delta a_t}{1 + \Delta} \tag{7-7a}$$

$$m = \frac{\Delta c - a}{\eta(1 + \Delta)} \tag{7-7b}$$

Eliminating c_t between Eqs. (7-5) and (7-8), we have

$$\epsilon = \frac{1}{m}\left[\rho(1 + \Delta)a_t + \frac{\rho}{\eta}(\Delta c - a) + \frac{1}{8\omega}\right] \tag{7-8a}$$

Eliminating a_t between Eqs. (7-5) and (7-8a), we have

$$\epsilon = \frac{1}{m}\left[2\rho c_t - \frac{\rho}{\eta}(\Delta c - a) + \frac{1}{8\omega}\right] \tag{7-8b}$$

7-11 DESIGN FOR ECONOMY

It may be assumed that the design which requires the least quantity of material, both concrete and steel, is the most economical. Although in practice a least-weight design may not necessarily be the most economical, the quantity of materials needed is, nevertheless, always an important economic consideration. Labor costs, however, play an important part.

The least area of concrete The parameter R implicitly defines the area of the concrete; R is the ratio of the moment caused by the applied loads (superimposed dead load and live load) to the moment produced by the weight of the beam. For a simply supported beam

$$R = \frac{M_a}{M_g} = \frac{8M_a}{A\gamma L^2}$$

For a given applied load, unit weight of concrete, and span length, an increase in R will result in a decrease in the cross-sectional area of the beam.

From Eq. (7-6) it may be observed that for given h, ρ, and η, in order to reduce the concrete area, stress coefficients as near the allowable as possible would be desirable, and in order to obtain the minimum concrete area, the following must be satisfied:

$$a_t = a'_t \qquad c_t = c'_t$$
$$c = c' \qquad a = a'$$

This means that for a given h, ρ, and η, the exact satisfaction of the four requirements will result in the minimum concrete area. However, from Eq. (7-5) it may be seen that the four requirements can be satisfied exactly only when Δ assumes the following value:

$$\Delta_e = \frac{\eta c'_t + a'}{\eta a'_t + c'}$$

The quantity Δ_e, in which the subscript e indicates "exact"—implying exact satisfaction of all requirements—is a function of stress coefficients and the effectiveness, which are known. For a given set of allowable stresses and a reasonable value for the effectiveness the value of Δ_e can be determined. In many cases it is possible to make $\Delta = \Delta_e$. However, for long spans it is necessary to make $\Delta > \Delta_e$ to provide sufficient cover for the prestressing force.

It can be seen that for values of Δ other than Δ_e one or more of the requirements will have to be satisfied by a margin.

For a given value of Δ that is greater than Δ_e, from Eqs. (7-5) and (7-6a) it can be seen that in order to obtain the maximum value of R or the minimum area of concrete section the following must be correct:

$$c_t = c'_t \qquad a = a'$$

Similarly, for a given value of Δ that is smaller than Δ_e, from Eqs. (7-5) and (7-6b) it can be seen that in order to obtain the maximum value of R corresponding to the minimum area of concrete section the following must hold:

$$a_t = a'_t \quad c = c'$$

One may conclude, therefore, that in order to obtain the smallest concrete area for given h and ρ the following must be satisfied:

When $\Delta > \Delta_e$: Either or both of requirements 1 and 3 may be satisfied by a margin, while 2 and 4 must be satisfied exactly.
When $\Delta < \Delta_e$: Either or both of requirements 2 and 4 may be satisfied by a margin, while 1 and 3 must be satisfied exactly.
When $\Delta = \Delta_e$: All four requirements must be satisfied exactly.

The preceding discussion has the following graphical interpretation. Equations (7-6a) and (7-6b), representing the relations between R and Δ for the allowable stresses, are plotted in Fig. 7-4. These two curves are the upper bounds of two families of curves which can be obtained for values of the stress coefficients smaller than the allowable. The intersection of any two curves yields the value of R and Δ that defines a certain section. The region of possible solutions lies below the lowest of the two curves of Fig. 7-4, for any given Δ.

The intersection of the two curves of Fig. 7-4 occurs when the four requirements are met exactly, i.e., when $\Delta = \Delta_e$. At this point R is maximum, and a minimum area for the section of the beam A is obtained. If Δ has to be larger than Δ_e, it can be seen from Fig. 7-4 that the smaller values of A are obtained when requirements 2 and 4 are met exactly. For the case where Δ must be smaller than Δ_e, the smaller values of A are obtained when requirements 1 and 3 are met exactly.

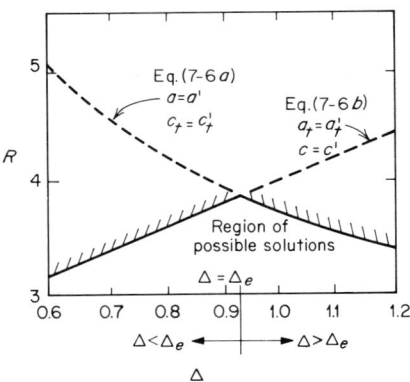

Fig. 7-4 Graphical interpretation of least-weight design.

The amount of steel The reinforcement ratio $m = P_t/Af'_c$ is a measure of the magnitude of the prestressing force and the area of steel required. For given values of A and f'_c the smaller m becomes, the smaller is the required steel area.

Since $P_t = mAf'_c$, for a given f'_c the least prestressing force or the least amount of prestressing steel is obtained when A and m have their least values. In the preceding discussion the criteria resulting in the least value of A were presented for different ranges of Δ. For convenience the criteria resulting in the minimum value of m will be studied in connection with the same ranges of Δ.

$\Delta > \Delta_e$ In the preceding discussion it was shown that for given $\Delta > \Delta_e$, ω, and ρ, in order to obtain the least area of concrete it is necessary that $c_t = c'_t$ and $a = a'$. The magnitude of m as presented in the following paragraphs is based upon the minimum area of concrete thus obtained.

From Eq. (7-7a) it is seen that since $c_t = c'_t$, for a given value of Δ, m decreases with an increase in a_t; that is, the higher a_t becomes, the smaller is m or the amount of the prestressing steel required. Evidently, the least area of prestressing steel will result when $a_t = a'_t$.

On the other hand, since $c_t = c'_t$ for a given value of Δ, the maximum m or amount of prestressing steel will result when a_t is the smallest. From Eq. (7-5) it can be shown that for a given Δ, when $c_t = c'_t$ and $a = a'$, the least value of a_t and the maximum amount of prestressing steel will result when $c = c'$.

It may be stated, therefore, that for given $\Delta > \Delta_e$, ω, and ρ, in order to have the minimum area of concrete and the lowest corresponding steel area, requirements 1, 2, and 4 must be satisfied exactly, and 3 by a margin; for the highest corresponding steel area, requirements 2 to 4 must be satisfied exactly, and 1 by a margin.

$\Delta < \Delta_e$ As shown previously in this case for given $\Delta < \Delta_e$, ω, and ρ, in order to obtain the least concrete area it is necessary that $a_t = a'_t$ and $c = c'$.

From Eq. (7-7b) it can be seen that since $c = c'$ for a given value of Δ, m or the area of the prestressing steel is a minimum when a is as large as possible, that is, when $a = a'$. Similarly, the maximum prestressing force will result when a is minimum. From Eq. (7-5) it can be seen that for a given Δ, when $a_t = a'_t$ and $c = c'$, the least value of a and the maximum amount of prestressing steel will result when $c_t = c'_t$.

It may be stated that for given $\Delta < \Delta_e$, ω, and ρ, in order to have the minimum area of concrete and the lowest corresponding steel area, requirements 1, 3, and 4 must be satisfied exactly, and 2 by a margin; for the highest corresponding steel area requirements 1 to 3 must be satisfied exactly, and 4 by a margin.

The least-weight design criteria For convenient reference the least-weight design criteria are summarized in the following paragraphs.

When Δ is not equal to Δ_e for given Δ, ρ, and ω values, the following four criteria will produce the least-weight design.

Criterion 1 When $\Delta > \Delta_e$, criterion 1 corresponds to a design which produces the least area of concrete and the lowest corresponding steel area. In this case it is necessary to satisfy requirements 1, 2, and 4 exactly and 3 by a margin.

Criterion 2 When $\Delta > \Delta_e$, criterion 2 corresponds to a design which produces the least area of concrete and the highest corresponding steel area. In this case it is necessary to satisfy requirements 2 to 4 exactly and 1 by a margin.

Criterion 3 When $\Delta < \Delta_e$, criterion 3 corresponds to a design which produces the least area of concrete and the lowest corresponding steel area. In this case it is necessary to satisfy requirements 1, 3, and 4 exactly and 2 by a margin.

Table 7-5 Summary of the least-weight design criteria

Δ	Criterion	Stress coefficients	Area of concrete for given ω and ρ	Corresponding area of steel	Corresponding eccentricity
$\Delta = \Delta_e$		$a_t = a'_t$ $c_t = c'_t$ $c = c'$ $a = a'$	Minimum		
$\Delta > \Delta_e$	Criterion 1	$a_t = a'_t$ $c_t = c'_t$ $c < c'$ $a = a'$	Minimum	Minimum	Maximum
	Criterion 2	$a_t < a'_t$ $c_t = c'_t$ $c = c'$ $a = a'$	Minimum	Maximum	Minimum
$\Delta < \Delta_e$	Criterion 3	$a_t = a'_t$ $c_t < c'_t$ $c = c'$ $a = a'$	Minimum	Minimum	Maximum
	Criterion 4	$a_t = a'_t$ $c_t = c'_t$ $c = c'$ $a < a'$	Minimum	Maximum	Minimum

Criterion 4 When $\Delta < \Delta_e$, criterion 4 corresponds to a design which produces the minimum area of concrete and the highest corresponding steel area. In this case, it is necessary to satisfy requirements 1 to 3 exactly and 4 by a margin.

The above criteria have also been summarized in Table 7-5.

It should be pointed out that for a given $\Delta > \Delta_e$, ρ, and ω, criteria 1 and 2 give the same concrete area since in both criteria requirements 2 and 4 are satisfied exactly. Similarly, criteria 3 and 4 yield the same concrete area.

The relationship between the eccentricity and prestressing force for least-weight design criteria is shown in Fig. 7-5a through c. Figure 7-5c shows graphically the condition in which all requirements are satisfied exactly.

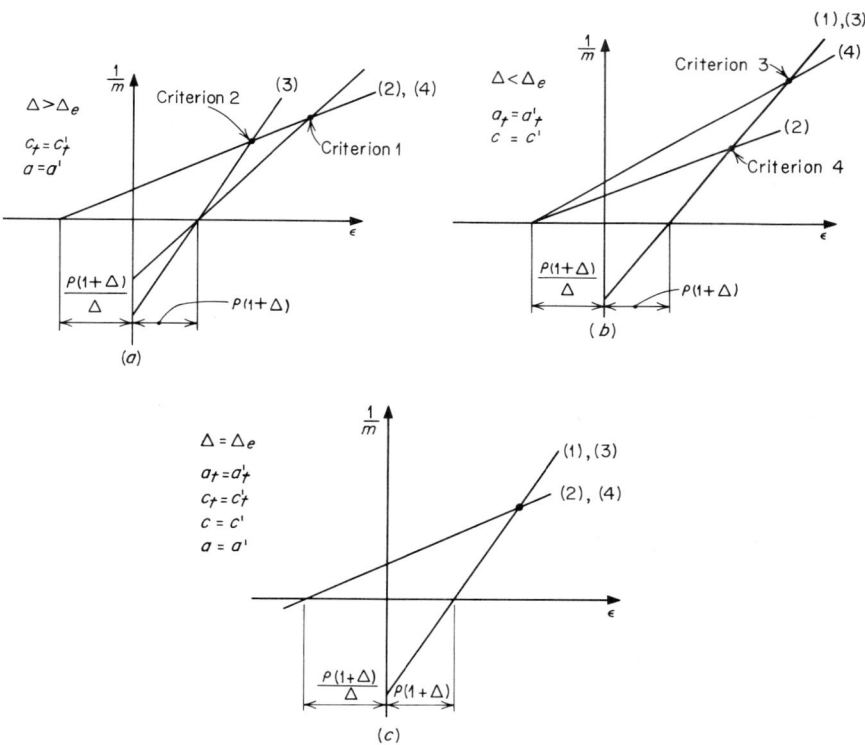

Fig. 7-5 Variation of eccentricity with the prestressing force, least-weight sections.

7-12 APPLICABILITY OF THE LEAST-WEIGHT DESIGN CRITERIA

As shown in the preceding discussion, not all four least-weight design criteria are applicable at one time. Criteria 1 and 2 are applicable only when $\Delta > \Delta_e$, while criteria 3 and 4 apply only when $\Delta < \Delta_e$. The value of Δ_e, therefore, is all-important in determining the criteria which apply.

In order to show the practical range of Δ_e, Table 7-6 is presented.

Table 7-6 Values of Δ_e and m for the stress coefficients shown in Table 7-3
$f'_c = 5000$ psi, $f'_{ci} = 4000$ psi

Type of structure	Method of prestressing	$\eta = 0.80$		$\eta = 0.85$	
		Δ_e	m	Δ_e	m
AASHO Specifications for Highway Bridges 1961					
Single-element beams without non-prestressed reinforcement in tension zone	Pretensioned	0.90	0.236	0.94	0.228
	Post-tensioned	0.83	0.225	0.87	0.218
Single-element beams with non-prestressed reinforcement in tension zone	Pretensioned	0.86	0.227	0.88	0.220
	Post-tensioned	0.78	0.217	0.81	0.210
Segmental-element beams without non-prestressed reinforcement in tension zone	Pretensioned	0.96	0.245	1.02	0.238
	Post-tensioned	0.88	0.235	0.94	0.227
Segmental-element beams with non-prestressed reinforcement in tension zone	Pretensioned	0.90	0.236	0.94	0.228
	Post-tensioned	0.83	0.225	0.87	0.218
ACI Building Code 1963					
Single-element beams without non-prestressed reinforcement in tension zone, protected		0.98	0.245	1.02	0.237
Single-element beams without non-prestressed reinforcement in tension zone, unprotected		0.76	0.241	0.80	0.234

Table 7-6 shows the values of Δ_e and m for the allowable stresses shown in Table 7-2. The values of Δ_e are computed for two values of η, namely, 0.80 and 0.85.

A study of Table 7-6 indicates that the range of Δ_e is between 0.71 and 1.02, and in most cases $\Delta_e < 1.0$.

When $\Delta_e < 1.0$, criteria 1 and 2 are applicable to rectangular sections, symmetrical I sections, T sections, and all sections in which the top flange is heavier than the bottom flange—sections in which $\Delta \geq 1.0$. In this case criteria 3 and 4 are applicable to inverted T sections and all sections in which the bottom flange is heavier than the top flange—sections in which $\Delta < 1.0$.

When $\Delta_e > 1.0$, criteria 3 and 4 are applicable to rectangular sections and symmetrical I sections as well as to sections in which $\Delta > 1.0$. However, they have a limited application to sections in which the top flange is heavier than the bottom flange since, from Table 7-6, the highest value of Δ_e is 1.02.

From a practical point of view, in most sections used (in noncomposite construction) $\Delta > 1.0$. In sections in which Δ is appreciably less than 1.0, the top flange, which is the compression zone, tends to be weak. Beams of this type have a tendency to fail in compression and have a limited ductility, which is an undesirable feature.

Furthermore, in sections in which the bottom flange is heavier than the top flange for beams of moderate span length, it becomes difficult to provide sufficiently high value for g, the distance from the center of gravity of prestressing force to the bottom fiber. In many cases in order to provide an adequate value for g it may become necessary to reduce the efficiency of the section, ρ, or increase the depth of the beam. Either solution has practical limitations.

One may conclude that from a practical point of view sections in which $\Delta < 1.0$ are not desirable. Consequently, criteria 3 and 4 have only a limited application to sections in which Δ is in the neighborhood of unity and when Δ_e is close to 1.02.

When $\Delta = \Delta_e$, all requirements can be satisfied exactly, resulting in the minimum area of concrete for given ρ and ω values. When Δ_e is in the neighborhood of 1.0 or greater than 1.0, this solution is ideal.

7-13 RELATIONSHIP AMONG THE DIMENSIONLESS VARIABLES

Design of prestressed-concrete beams can be expedited by plotting Eqs. (7-6) and (7-9) for a given set of stress coefficients. Such a presentation is convenient in studying the relationship among the dimensionless variables which affect the design of a beam.

$$R = 8\rho\omega[(a + c) + \eta(a_t + c_t)] - (1 - \eta) \quad (7\text{-}6)$$

$$R = \frac{8\omega\epsilon(c_t c - a_t a) - [(a + c) + \eta(a_t + c_t)]}{a_t + c_t} - (1 - \eta) \quad (7\text{-}9)$$

Let us use the AASHO specifications for highway bridges and consider a pretensioned single-element beam without non-prestressed reinforcement. From Table 7-3, if all requirements are satisfied exactly, we have

$$a_t = a'_t = 0.038 \qquad c_t = c'_t = 0.48$$
$$c = c' = 0.40 \qquad a = a' = 0$$

Assuming $\eta = 0.85$, from Table 7-6 we have $\Delta_e = 0.94$, and when $\Delta = \Delta_e = 0.94$, from Eq. (7-7a)

$$m = \frac{c_t - \Delta a_t}{1 + \Delta} = \frac{0.48 - 0.94 \times 0.038}{1.94} = 0.229$$

Substitution of the above stress coefficients and effectiveness in Eqs. (7-6) and (7-9) results in the following expressions:

$$R = 6.74\rho\omega - 0.15$$
$$R = 2.97\epsilon\omega - 1.77$$

Figure 7-6 shows the plots of these equations using R as ordinate and ω as abscissa. The first of the above equations—which is Eq. (7-6) for specific values of stress coefficients, effectiveness, and Δ—relates R to

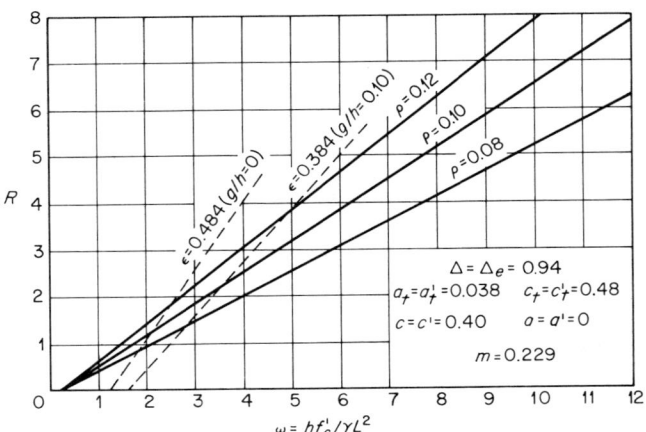

Fig. 7-6 Variation of R with ω when $\Delta = 0.94$ and $\Delta_e = 0.94$.

ω and ρ, and is plotted in Fig. 7-6 by solid lines for three values of ρ: 0.08, 0.10, and 0.12. The second equation—which is Eq. (7-9) for the same values of stress coefficients, effectiveness, and Δ—relates R to ω and ϵ and is plotted in the same figure by dashed lines for two values of ϵ: 0.485 and 0.385. These values of ϵ are adopted in order to make g/h equal to zero and 0.10, respectively. As shown in Sec. 7-9, g/h and ϵ are related by the following expression:

$$\frac{g}{h} = \frac{\Delta}{1+\Delta} - \epsilon$$

In Fig. 7-6 the line corresponding to $\epsilon = 0.485$ or $g/h = 0$ designates a hypothetical condition in which the center of gravity of prestressing steel is coincident with the bottom fiber of the beam.

Equations (7-6) and (7-9) can also be plotted for values of Δ other than Δ_e for any one of the criteria defined. In the following paragraphs three values of Δ—1.00, 1.25, and 0.75—are considered.

$\Delta = 1.00$ This case corresponds to rectangular sections, symmetrical I sections, and all sections in which the neutral axis is at middepth. Since in this case $\Delta > \Delta_e$, criteria 1 and 2 apply.

Criterion 1 In this criterion requirements 1, 2, and 4 are satisfied exactly, or

$$a_t = a'_t = 0.038$$
$$c_t = c'_t = 0.48$$
$$a = a' = 0$$

From Eq. (7-5)

$$c = \frac{\eta c_t - a}{\Delta} - \eta a_t = \frac{0.85 \times 48}{1.0} - 0.85 \times 0.038 = 0.376$$

and from Eq. (7-7a)

$$m = \frac{c_t - \Delta a_t}{1 + \Delta} = \frac{0.48 - 0.038}{2} = 0.222$$

Substituting the above stress coefficients and effectiveness in Eqs. (7-6) and (7-9), we have

$$R = 6.54\rho\omega - 0.15$$
$$R = 2.77\epsilon\omega - 1.73$$

Figure 7-7a shows the plot of the above equations for various values of ρ and ϵ (or g/h) similar to Fig. 7-6.

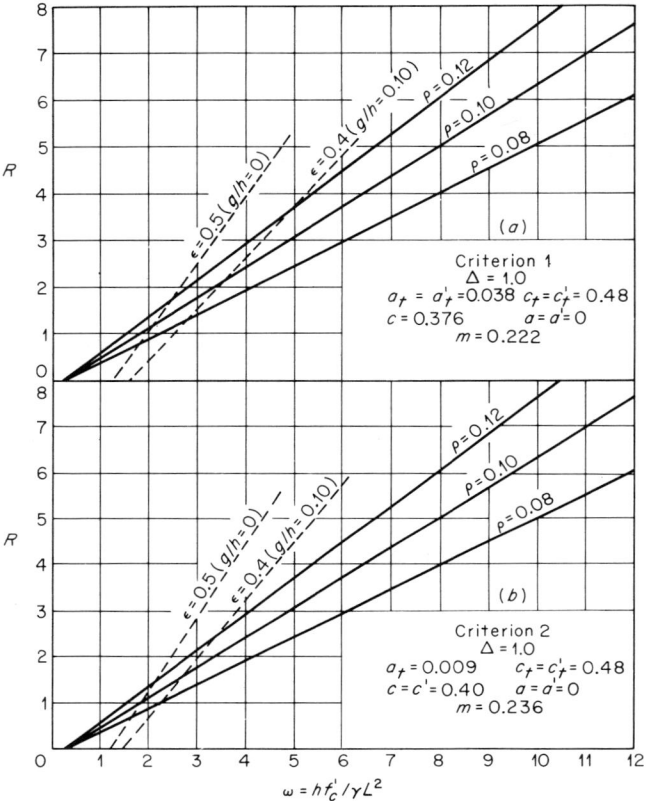

Fig. 7-7 Variation of R with ω for criteria 1 and 2 when $\Delta = 1.0$ and $\Delta_e = 0.94$.

Criterion 2 In this criterion requirements 2 to 4 are satisfied exactly, and we have

$$c_t = c'_t = 0.48$$
$$c = c' = 0.40$$
$$a = c' = 0$$

From Eq. (7-5)

$$a_t = \frac{1}{\eta}\left(\frac{\eta c'_t + a'}{\Delta} - c'\right) = \frac{1}{0.85}\left(\frac{0.85 \times 0.48}{1.0} - 0.40\right) = 0.009$$

WORKING-STRESS DESIGN OF SUPPORTED PRESTRESSED-CONCRETE BEAMS

and from Eq. (7-7a)

$$m = \frac{c_t - \Delta a_t}{1 + \Delta} = \frac{0.48 - 0.009}{2} = 0.236$$

Substituting the above stress coefficients and $\eta = 0.85$ in Eqs. (7-6) and (7-9), we have

$$R = 6.54\rho\omega - 0.15$$
$$R = 3.14\epsilon\omega - 1.82$$

Plots of these equations (similar to those in the preceding cases) are given in Fig. 7-7b. It should be pointed out that Eq. (7-6) is the same in criteria 1 and 2; therefore, in Fig. 7-7a and b the solid lines are identical. However, Eq. (7-9) is not the same in the two criteria, indicating that the difference in criteria 1 and 2 is in eccentricity and prestressing force only. As seen in Fig. 7-7, the dashed lines in criterion 1 have smaller slope than those in criterion 2. From this figure it can be seen that for given values of ρ and ω, criterion 2 results in smaller ϵ and larger g/h values.

Equations (7-6) and (7-9), as well as the stress coefficients and reinforcement factors for each criterion, are given in Table 7-7 for convenient reference.

$\Delta = \mathbf{1.25}$ Derivation of Eqs. (7-6) and (7-9) in this case for criteria 1 and 2 is similar to the preceding case. These equations and the stress coefficients for each criteria are given in Table 7-7. Figure 7-8a and b shows plots of these equations for criteria 1 and 2, respectively.

$\Delta = \mathbf{0.75}$ Since in this case $\Delta < \Delta_e$, criteria 3 and 4 apply.

Table 7-7 A summary of least-weight design relations for various values of Δ

$a_t' = 0.038$ $c' = 0.40$
$c_t' = 0.48$ $a' = 0$
$\Delta_e = 0.94$ $\eta = 0.85$

Δ	Criterion	a_t	c_t	c	a	m	Eq. (7-6)	Eq. (7-9)
0.94	$\Delta = \Delta_e$	0.038	0.48	0.40	0	0.229	$R = 6.74\rho\omega - 0.15$	$R = 2.97\epsilon\omega - 1.77$
1.00	1	0.038	0.48	0.376*	0	0.222	$R = 6.54\rho\omega - 0.15$	$R = 2.77\epsilon\omega - 1.73$
	2	0.009	0.48	0.40	0	0.236		$R = 3.14\epsilon\omega - 1.82$
1.25	1	0.038	0.48	0.294	0	0.198	$R = 5.87\rho\omega - 0.15$	$R = 2.18\epsilon\omega - 1.57$
	2	-0.087	0.48	0.40	0	0.262		$R = 3.91\epsilon\omega - 1.92$
0.75	3	0.038	0.382	0.40	0	0.202	$R = 6.05\rho\omega - 0.15$	$R = 2.91\epsilon\omega - 1.95$
	4	0.038	0.48	0.40	-0.84	0.258		$R = 3.01\epsilon\omega - 1.61$

* Stress coefficients underlined correspond to those that are satisfied by a margin.

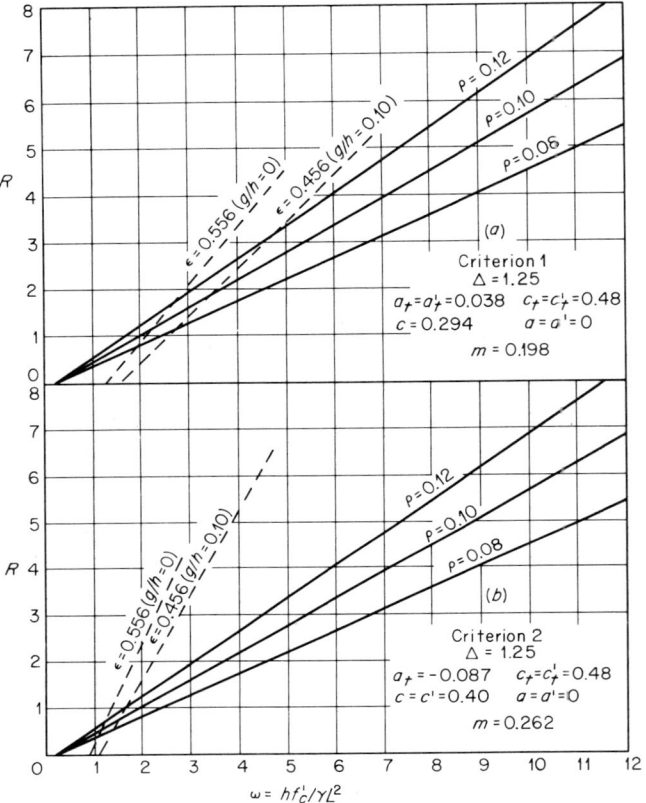

Fig. 7-8 Variation of R with ω for criteria 1 and 2 when $\Delta = 1.25$ and $\Delta_e = 0.94$.

Criterion 3 In this criterion requirements 1, 3, and 4 are satisfied exactly and

$$a_t = a'_t = 0.038$$
$$c = c' = 0.40$$
$$a = a' = 0$$

From Eq. (7-5)

$$c_t = a_t \Delta + \frac{c\Delta - a}{\eta} = 0.038 \times 0.75 + \frac{0.40 \times 0.75}{0.85} = 0.382$$

WORKING-STRESS DESIGN OF SUPPORTED PRESTRESSED-CONCRETE BEAMS

and from Eq. (7-7a)

$$m = \frac{c_t - \Delta a_t}{1 + \Delta} = \frac{0.382 - 0.75 \times 0.038}{1.75} = 0.202$$

Substituting the above values in Eqs. (7-6) and (7-9), we have

$$R = 6.05\rho\omega - 0.15$$
$$R = 2.91\epsilon\omega - 1.95$$

Figure 7-9a shows the plot of the above equations for various values of ρ and ϵ.

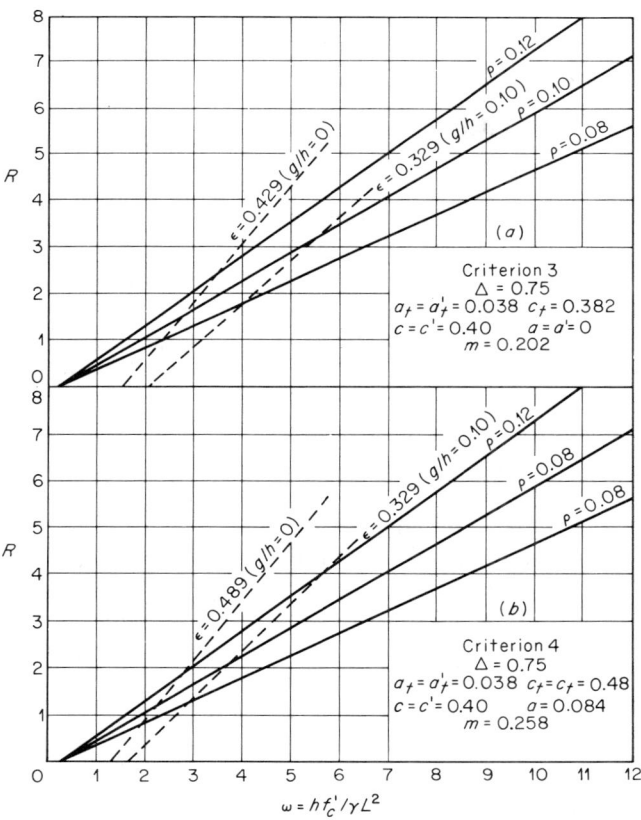

Fig. 7-9 Variation of R with ω for criteria 3 and 4 when $\Delta = 0.75$ and $\Delta_e = 0.94$.

Criterion 4 In this criterion requirements 1 to 3 are satisfied exactly, and we have

$$a_t = a'_t = 0.038$$
$$c_t = c'_t = 0.48$$
$$c = c' = 0.40$$

From Eq. (7-5)

$$a = \Delta(\eta a'_t + c') - \eta c'_t = 0.75(0.85 \times 0.038 + 0.40) - 0.85 \times 0.48$$
$$= 0.084$$

and from Eq. (7-7a)

$$m = \frac{c_t - \Delta a_t}{1 + \Delta} = \frac{0.48 - 0.75 \times 0.038}{1.75} = 0.258$$

Hence Eqs. (7-6) and (7-9) will be as follows:

$$R = 6.05\rho\omega - 0.15$$
$$R = 3.01\epsilon\omega - 1.61$$

These equations are plotted in Fig. 7-9b.

Equation (7-6) is the same for criteria 3 and 4 since these criteria give the same concrete area, and the solid lines in Fig. 7-9a and b are identical. Equations (7-6) and (7-9), as well as the stress coefficients and reinforcement factors for criteria 3 and 4, are also listed in Table 7-7.

Figures 7-6 through 7-9 show the relationship between the cross-sectional area of the beam and the eccentricity or the prestressing force. This relationship is an important guide in design of beams.

Figure 7-6 corresponds to a section in which $\Delta = \Delta_e = 0.94$ and the four requirements are satisfied exactly. In this figure the solid lines represent the relationship between the maximum R, or the minimum area of concrete, and ω. The dashed lines show the eccentricities available. The part of the solid lines to the left of the line designated $\epsilon = 0.484$, or $g/h = 0$, has no practical significance since it corresponds to sections in which the center of gravity of steel is below the bottom fiber of the section. From this figure it can be seen that for values of ω less than about 3.0, the four requirements cannot be satisfied exactly since it is not possible to provide a sufficiently small eccentricity for the prestressing steel unless the efficiency is reduced.

Figure 7-7a and b corresponds to criteria 1 and 2, respectively, for sections in which $\Delta = 1.0$. The solid lines representing the least concrete area are identical in these figures. By comparing the solid lines in Fig. 7-7 with those in Fig. 7-6, it is seen that by increasing Δ from 0.94 to 1.00

the slope of solid lines is reduced. Therefore, for particular values of ρ and ω, as Δ deviates from Δ_e the area of concrete increases.

The dashed lines in Fig. 7-7a and b show the limiting extremes of the eccentricity and the prestressing force. The dashed lines in Fig. 7-7a correspond to criterion 1 and give the values of ϵ and g/h for the maximum possible eccentricity or minimum prestressing force. The dashed lines in Fig. 7-7b are for criterion 2 and indicate ϵ and g/h values for the maximum eccentricity. From Fig. 7-7a it can be seen that if ω is small, it may not be possible to use criterion 1 for lack of sufficient cover or g/h value, in which case criterion 2 may be used.

Let us assume $\Delta = 1.0$, $\omega = 3.0$, and in order to place the prestressing steel satisfactorily, g/h should be at least 0.10. With criterion 1 from Fig. 7-7a, if g/h is to be as high as 0.10, ρ cannot exceed about 0.09. By using criterion 2 from Fig. 7-7b, for $\omega = 3.0$ it is possible to use $\rho = 0.11$ and still provide $g/h = 0.10$. Although in this case criterion 2 gives a reasonable solution, it may not be necessary to use it since it corresponds to the extreme case in which the prestressing force is maximum. By using the least concrete area the eccentricity and prestressing force may be chosen between the limits designated by criteria 1 and 2.

Figure 7-8 is similar to Fig. 7-7 except that in this case, since $\Delta = 1.25$, the difference between the two criteria is more pronounced. The relationship between the area of concrete and ω is the same in Fig. 7-8a and b for criteria 1 and 2, respectively. Since in this case $\Delta = 1.25$ and has deviated even farther from Δ_e, the slope of solid lines in Fig. 7-8 is less than that in Fig. 7-7. For particular values of ρ and ω, the least area of concrete increases with Δ.

The slope of the dashed lines in Fig. 7-8a is considerably smaller than that in Fig. 7-8b. Consequently, as Δ increases, the difference between criteria 1 and 2 becomes significant. When criterion 1 is used, there is no significant difference in the required eccentricity as Δ is increased from 1.00 to 1.25. However, when criterion 2 is used, there is a marked decrease in the required eccentricity when Δ is increased from 1.00 to 1.25. It can be concluded, therefore, that by increasing Δ and using criterion 2 it is possible to reduce the eccentricity and thus provide a sufficient space for the prestressing steel.

It should be emphasized that although it may not be possible to apply criterion 1 because of lack of sufficient cover, it is not necessary to use criterion 2 since it may result in a high value for the prestressing force.

Figure 7-9 is similar to the plots discussed above, and it corresponds to $\Delta = 0.75$. As stated before, this type of section is not practical in noncomposite construction because of its possible brittle behavior. However, Fig. 7-9 is of academic interest, and it shows that in this case also since Δ is deviated from Δ_e the solid lines have smaller slope in comparison with those in Fig. 7-6. The dashed lines in this case are similar to those discussed above.

Figures 7-6 through 7-8 are useful for guidance in design of beams. Although these are prepared for a particular combination of stress coefficients, the conclusions drawn here are of a general nature and can be applied to all specifications.

The conclusions in this section can be summarized as follows:

1. Increasing Δ from Δ_e results in an increase in the required concrete area, but provides means to obtain sufficient cover for placing the prestressing steel.
2. Increasing ρ results in a decrease in the concrete area, but tends to limit the cover.
3. Decreasing Δ from Δ_e results in an increase in the required area, but does not help to increase the cover.

These conclusions can also be drawn from Eqs. (7-6) to (7-8); however, plots of Eqs. (7-6) and (7-9) provide a more convenient means to determine the design variables for a particular problem.

7-14 THE IDEALIZED SECTIONS

The actual shapes of sections used in noncomposite prestressed-concrete construction are roughly equivalent to unsymmetrical I sections, T sections, symmetrical I sections, and rectangular sections. Usually the flanges in the actual sections are tapered, and sometimes there are fillets at the points of connection of flange to the web. These practical sections, however, can be idealized as sections having web and flanges of

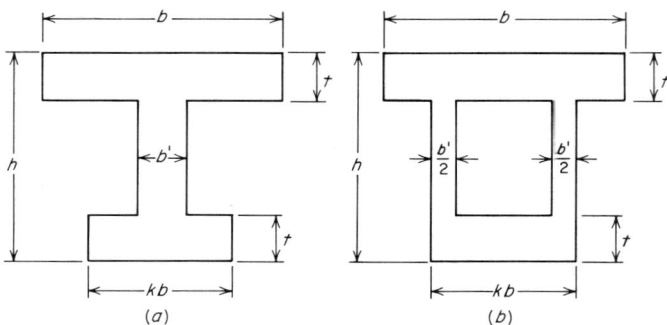

Fig. 7-10 Idealized I sections, $\Delta > 1.0$.

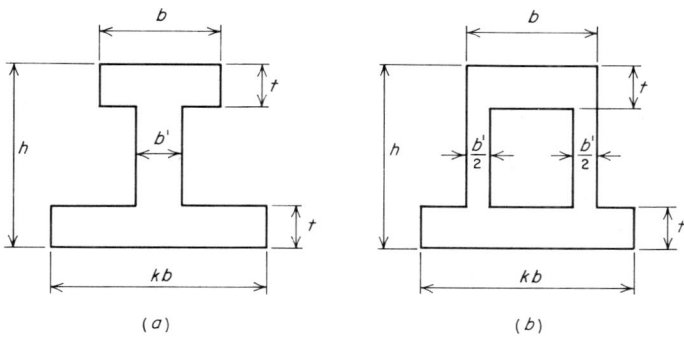

Fig. 7-11 Idealized I sections, $\Delta < 1.0$.

uniform thickness with no fillets at the points where the flange and web meet.

In the following paragraphs the idealized sections used in prestressed concrete are discussed.

The typical prestressed-concrete section can be idealized in the form of an unsymmetrical I section. Figure 7-10a and b shows unsymmetrical I sections in which $\Delta > 1.0$ since the top flange is heavier than the bottom flange. A brief study of these diagrams will indicate that although they are different, they represent the same shape as far as flexure about the horizontal axis is concerned. In these sections the thickness of flange t is the same for the top and bottom flanges. The width of top flange is b and that of bottom flange is kb, and $k < 1.0$. The width of web is b'.

Figure 7-11a and b shows unsymmetrical I sections in which $\Delta < 1.0$ since the bottom flange is heavier than the top flange. These sections can be obtained by inverting the sections in Fig. 7-10a and b. In this case all the notations are the same and $k > 1.0$. Unsymmetrical I sections

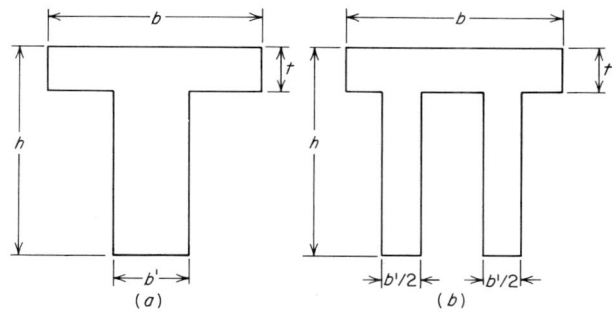

Fig. 7-12 Idealized T sections.

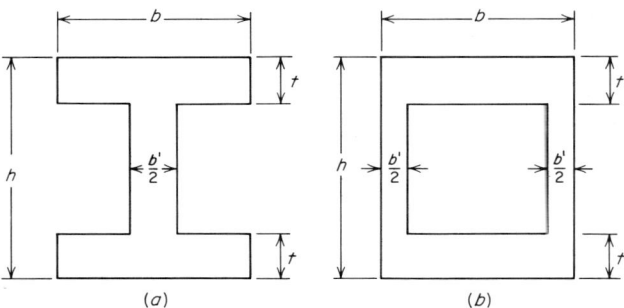

Fig. 7-13 Idealized symmetrical I sections.

in which the bottom flange is heavier than the top flange are seldom used in noncomposite construction, especially if Δ is very much less than 1.0.

The T sections represented by Fig. 7-12a and b can be considered a special case of unsymmetrical I sections of Fig. 7-10a and b in which $kb = b'$. The section shown in Fig. 7-12b is actually an idealized double T section; for our purpose, however, it is the same as the idealized T section shown in Fig. 7-12a.

The symmetrical I section is another special case of an unsymmetrical I section in which $b = kb$ or $k = 1.0$. Figure 7-13a and b shows idealized I sections; however, the section in Fig. 7-13b is called an idealized symmetrical box section. Rectangular sections may be considered a special case of symmetrical I sections when $b = b'$ or $t = 0.5h$.

7-15 RELATIONSHIP AMONG ρ, Δ, AND THE SECTION PROPERTIES OF IDEALIZED SECTIONS

Since both Δ and ρ are section properties, they are related and their relationship is important in the design of prestressed-concrete beams. In the following paragraphs the relationships between Δ and ρ are presented and discussed for a few commonly used idealized sections.

Idealized unsymmetrical I sections It can be shown that for the sections shown in Fig. 7-10a and b the following expressions are correct:

$$\rho = \frac{\left(1 + k - 2\frac{b'}{b}\right)\left(\frac{t}{h}\right)^3 + \frac{b'}{b} + 3\frac{t}{h}\left(1 - \frac{t}{h}\right)\left(1 - \frac{b'}{b}\right)}{3\left[\left(1 + k - 2\frac{b'}{b}\right)\frac{t}{h} + \frac{b'}{b}\right]} - \left(\frac{\Delta}{1 + \Delta}\right)^2$$

(7-10)

$$\Delta = \frac{\dfrac{b'}{b}\left(1 - 2\dfrac{t}{h}\right) + 2\dfrac{t}{h} - \dfrac{t^2}{h^2}(1-k)}{\dfrac{b'}{b}\left(1 - 2\dfrac{t}{h}\right) + 2k\dfrac{t}{h} + \dfrac{t^2}{h^2}(1-k)} \tag{7-11}$$

The derivation of the above expressions is shown in Appendix A-2.

Equation (7-10) indicates that ρ is a function of t/h, b'/b, k, and Δ. Since from Eq. (7-11) Δ is a function of t/h, b'/b, and k, it may be concluded that both ρ and Δ are functions of t/h, b'/b, and k.

Equations (7-10) and (7-11) are plotted in Figs. 7-14 through 7-16 for values of k equal to 0.667, 0.50, and 0.4, corresponding to unsym-

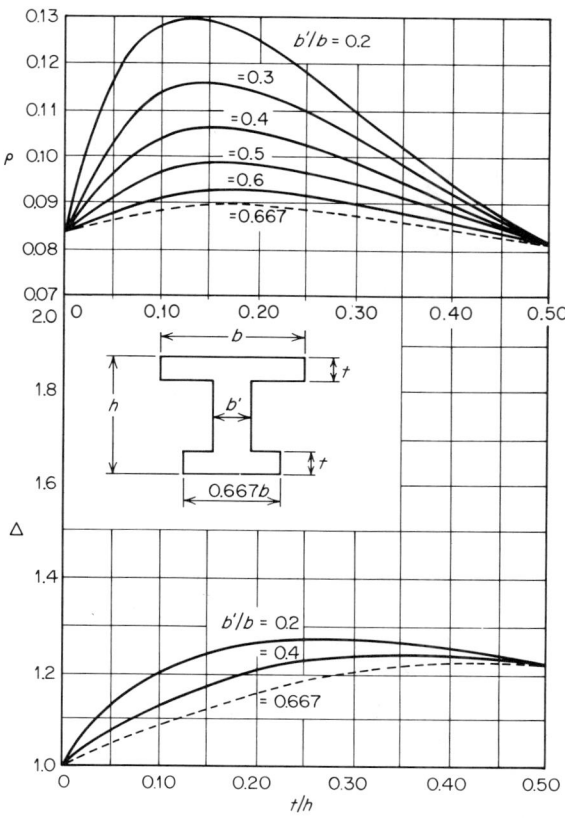

Fig. 7-14 The relations among the geometric properties of an unsymmetrical I section in which the flanges are of equal thickness and the width of the top flange is $1\frac{1}{2}$ times greater than that of the bottom flange.

metrical I sections in which the top flange is 50, 100, and 150 percent wider than the bottom flange, respectively.

Figure 7-14 is a graphical presentation of Eqs. (7-10) and (7-11) for the specific value of $k = 0.667$. This case corresponds to an unsymmetrical I section in which the top flange is $1\frac{1}{2}$ times wider than the bottom flange. The curves presented at the top of Fig. 7-14 represent Eq. (7-10). The quantity ρ is taken as ordinate and t/h is taken as abscissa. The curves are drawn for six values of b'/b, namely, 0.2, 0.3, 0.4, 0.5, 0.6, and 0.667. The curve corresponding to $b'/b = 0.667$ is shown by dashed lines since it corresponds to a T section. This curve represents the practical higher limit of b'/b, since curves drawn for values of $b'/b > 0.667$ are of no practical significance.

The curves shown in the bottom of Fig. 7-14 represent Eq. (7-11). In this case Δ is taken as ordinate while the abscissa is t/h, as in the

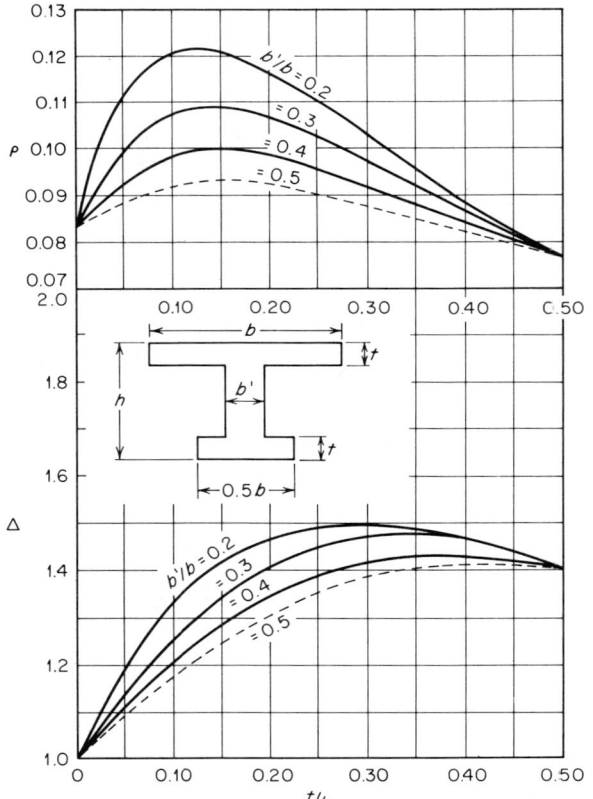

Fig. 7-15 The relations among the geometric properties of an unsymmetrical I section in which the flanges are of equal thickness and the width of the top flange is twice that of the bottom flange.

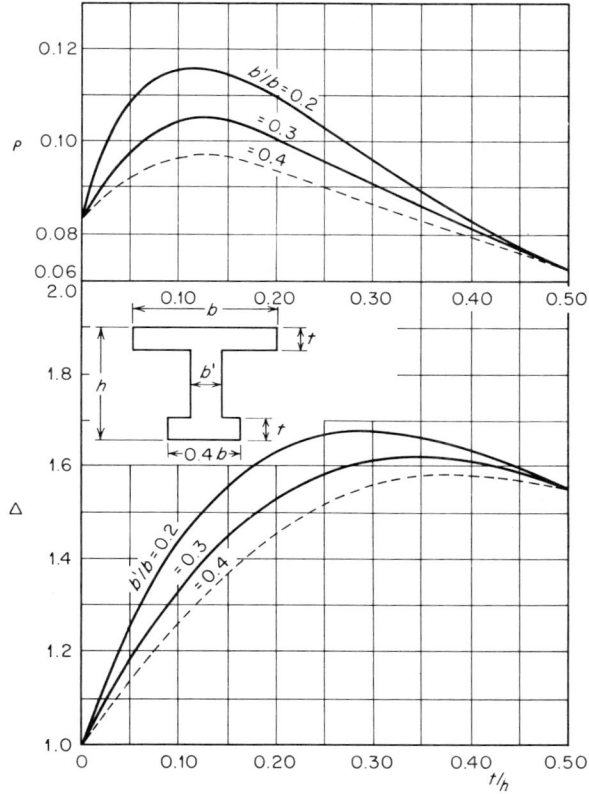

Fig. 7-16 The relations among the geometric properties of an unsymmetrical I section in which the flanges are of equal thickness and the width of the top flange is $2\frac{1}{2}$ times greater than that of the bottom flange.

curves above. This equation is plotted for the same values of b'/b, that is, 0.2, 0.3, 0.4, 0.5, 0.6, and 0.667. As before, the curve corresponding to $b'/b = 0.667$ represents a T section and is shown by dashed lines.

The two sets of curves in Fig. 7-14 representing Eqs. (7-10) and (7-11) completely define the relations among the geometric properties of an idealized unsymmetrical I section in which $k = 0.667$. By using these curves, for a given Δ value and a reasonably assumed t/h value, b'/b and ρ can be determined. Similarly, if ρ and t/h are known, consistent values of Δ and b'/b can be obtained.

Figures 7-15 and 7-16 show plots of Eqs. (7-10) and (7-11) for $k = 0.50$ and $k = 0.40$, respectively.

In plotting Eqs. (7-10) and (7-11) in Figs. 7-14 through 7-16, for convenience, the lower limit of b'/b is always taken as 0.2. The higher limit of b'/b in each case is the k value for that case.

Idealized T sections It can be shown that for the T sections of Fig. 7-12a and b the following expressions are correct:

$$\rho = \frac{\left(1-\frac{b'}{b}\right)\left[3\frac{t}{h}\left(1-\frac{t}{h}\right)+\left(\frac{t}{h}\right)^3\right]+\frac{b'}{b}}{3\left[\left(1-\frac{b'}{b}\right)\frac{t}{h}+\frac{b'}{b}\right]} - \left(\frac{\Delta}{1+\Delta}\right)^2 \quad (7\text{-}12)$$

and

$$\Delta = \frac{\frac{b'}{b}\left(1-\frac{t}{h}\right)^2 + \frac{2t}{h}\left(1-\frac{t}{2h}\right)}{\frac{b'}{b}\left(1-\frac{t^2}{h^2}\right)+\frac{t^2}{h^2}} \quad (7\text{-}13)$$

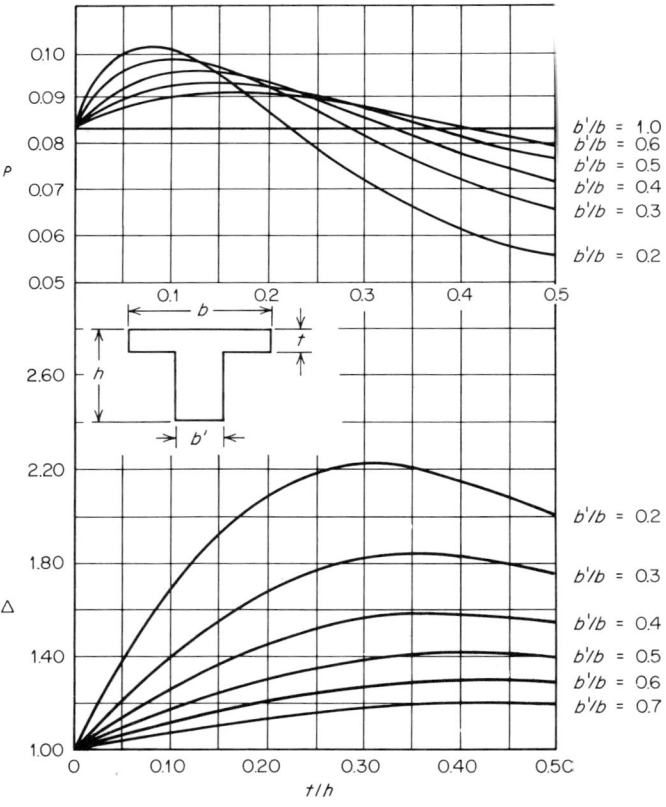

Fig. 7-17 The relations among the geometric properties of a T section.

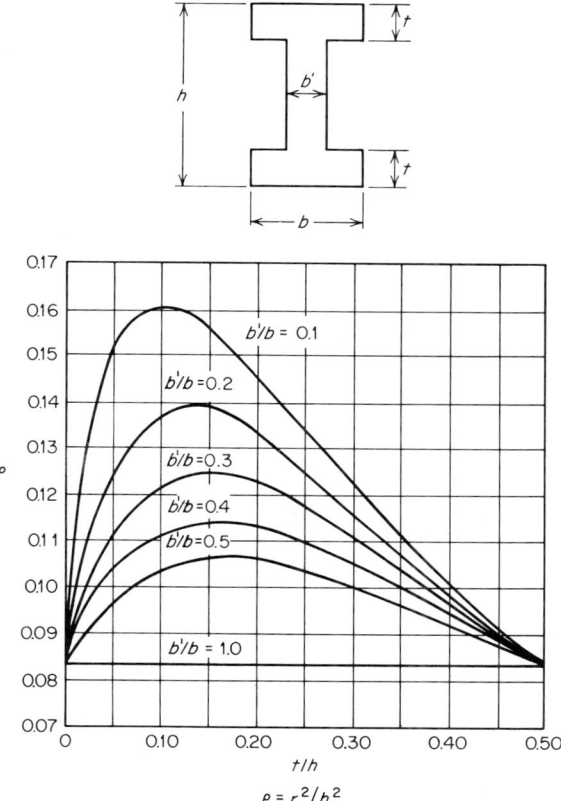

Fig. 7-18 The relations among the geometric properties of a symmetrical I section.

The above equations can be obtained by substituting $b'/b = k$ in Eqs. (7-10) and (7-11).

Equation (7-12) is plotted in the top of Fig. 7-17 for a wide range of b'/b values, taking ρ as ordinate and t/h as abscissa. Equation (7-13) is plotted in the bottom of the same figure for the same values of b'/b, varying Δ with t/h.

It should be noted that the lines which are marked $b'/b = 1.0$ correspond to rectangular sections.

Idealized symmetrical I sections By substituting $\Delta = 1.0$ and $k = 1.0$ in Eq. (7-10), the following expression will be obtained:

$$\rho = \frac{1 - (1 - b'/b)(1 - 2t/h)^3}{12[1 - (1 - b'/b)(1 - 2t/h)]} \qquad (7\text{-}14)$$

Equation (7-14) has been plotted in Fig. 7-18 for different values of b'/b, taking ρ as ordinate and t/h as abscissa. The curve corresponding to $b'/b = 1.0$ is for a rectangular section, as are the points corresponding to $t/h = 0$ and $t/h = 0.5$.

From the discussion in this section it is seen that Δ and ρ are related, and that for a given Δ and a particular shape it is possible to select a reasonable value for ρ.

7-16 DETERMINATION OF ω, Δ, AND ρ

In the design of a prestressed-concrete beam, the first step is the selection of proper values of ω, ρ, and Δ. The choice of these quantities is discussed in the following paragraphs.

Choice of $\omega = hf'_c/\gamma L^2$ From Eq. (7-8) it can be shown that ϵ decreases with ω and that when ω is less than about 1.50, the required ϵ becomes too large to permit placing of prestressing in the section of the beam. Therefore, in order to have a sufficiently large value for g, the distance between the centroid of steel and the bottom fiber, it is desirable to have a value for ω that is greater than 1.5. Although in spans up to 100 ft the value of ω can be set conveniently in the neighborhood of 3.00, for longer spans it may become impractical to have ω as high as 1.5. Since in a design problem the span length L is known, determination of ω depends upon γ, f'_c, and h.

It should be pointed out that prestressed concrete is not used to its best advantage for large values of ω greater than about 7.0. The large values of ω correspond to short-span beams in which it may not be possible to realize stresses as high as the allowable stresses at the midspan of the beam. Often in short-span beams the stresses at the ends of the beam govern the design.

The current practice in prestressed concrete is based predominantly on ordinary concrete in which the large-size aggregate is gravel or crushed stone. The value of γ in this case is taken as 150 pcf. Lightweight aggregate concrete has also been used extensively with a unit weight of about 100 pcf or less, however.

The strength of concrete most commonly used is $f'_c = 5000$ psi. Precast concrete of much higher strength can be manufactured with no special cost, and is sometimes used.

There is no clear-cut rule in choosing a suitable value for h, the overall depth of the beam. Often the value of h is determined on the basis of clearance or architectural requirements. As a guide in choosing h, one may take L/h in the neighborhood of 25 for short spans and 20 for

WORKING-STRESS DESIGN OF SUPPORTED PRESTRESSED-CONCRETE BEAMS 241

long spans. Any limitation on deflection of the beam will affect the choice of h.

Choice of Δ From Figs. 7-6 through 7-9 it can be seen that the choice of Δ influences two quantities: (1) the concrete area and (2) the eccentricity of prestressing. In dimensionless form these quantities may be presented by R and ϵ, respectively.

It is always desirable to choose Δ such that the area of concrete A will be a minimum or R a maximum. This may be achieved by taking $\Delta = \Delta_e$ if sufficient cover is available. From Fig. 7-6 it can be seen that when ω is less than about 3, sufficient cover is obtained only by reducing ρ to an unreasonably small value.

For small values of ω, sufficient cover may be obtained by taking Δ slightly greater than Δ_e. This will make the section somewhat heavier, but will provide a larger region within which the prestressing steel can be placed.

Hence when $\omega > 3.0$, it is desirable to have $\Delta = \Delta_e$ in order to obtain the least area of concrete. For $\omega < 3.0$, Δ should be increased somewhat to permit placing of the steel.

Choice of $\rho = r^2/h^2$ The practical range of ρ, as shown in Sec. 7-9, is from 0.0833 to about 0.13. For T sections ρ varies between 0.09 to 0.10, while in I sections ρ may approach 0.13. In rectangular sections $\rho = 0.0833$.

It is desirable to have a large value for ρ since an increase in ρ results in a decrease in the cross-sectional area of the beam. From Figs. 7-14 through 7-16 it is seen that the high values of ρ are obtained when t/h is between 0.10 and 0.15. In proportioning a section, t/h should preferably be within this range.

Figures 7-14 through 7-16 also indicate that values of ρ correspond to low values of b'/b. In most cases, however, the lowest practical value of b'/b is in the neighborhood of 0.25. Values of $b'/b < 0.25$ may result in sections which have thin webs.

For reasonably assumed values of k, t/h, and b'/b, the values of ρ and Δ may be obtained from Figs. 7-14 through 7-16. Similarly, for assumed values of k, t/h, and Δ, it is possible to determine ρ and b'/b.

Equations (7-10) and (7-11) may be used to determine a large value for ρ which is consistent with the acceptable range of b'/b, t/h, and Δ. For the assumed value of Δ, $t/h \approx 0.15$, and a small value of b'/b, say 0.25, k may be determined from Eq. (7-11); and for these values of Δ, t/h, b'/b, and k the efficiency ρ may be calculated by Eq. (7-10).

7-17 THE ALLOWABLE STRESS IN PRESTRESSING STEEL

In order to determine the area of the prestressing steel needed once the magnitude of prestressing force is determined, it is necessary to have the allowable stress in steel. The area of steel required can be expressed as follows:

$$A_s = \frac{P_t}{f_{st}}$$

where A_s = required area of prestressing steel
P_t = prestressing force at transfer at midspan
f_{st} = allowable stress in steel at transfer at midspan

The allowable stress in steel is given at the anchorage. According to the ACI code the stress in steel at anchorage immediately after seating has been effected should not exceed $0.70\ f'_s$. In addition, steel stress after losses should not exceed $0.60f'_s$ or $0.80f_{sy}$, whichever is smaller. The quantities f'_s and f_{sy} represent the ultimate strength and nominal yield-point stress of prestressing steel, respectively.

The quantity f_{st} can be calculated approximately by using loss in steel stress (not including friction loss) of 35,000 psi for pretensioning and 25,000 psi for post-tensioning. As discussed in Chap. 6, these quantities approximately represent the complete losses measured from the time immediately after the seating of anchorage has been effected.

According to the ACI code for pretensioned construction the allowable stress f_{st} can be presented in the following three forms, and the lowest of the three should be used in design:

$$f_{st} = \phi(0.70f'_s - 10{,}000)$$
$$f_{st} = \phi(0.60f'_s + 25{,}000)$$
$$f_{st} = \phi(0.80f_{sy} + 25{,}000)$$

The quantity ϕ in pretensioned construction, even with draped strands, is nearly 1.0 and represents the loss due to friction. In most specifications no mention is made of possible loss due to friction in a draped pretensioned beam.

By using the ACI code for post-tensioned construction the allowable stress f_{st} can be presented similarly in the three following forms, and the lowest of the three should be used in design:

$$f_{st} = \phi 0.70 f'_s$$
$$f_{st} = \phi(0.60f'_s + 25{,}000)$$
$$f_{st} = \phi(0.80f_{sy} + 25{,}000)$$

The value of ϕ in post-tensioned beams may vary from 1.00 to 0.85.

WORKING-STRESS DESIGN OF SUPPORTED PRESTRESSED-CONCRETE BEAMS

The ACI code gives the following expression for ϕ:

$$\phi = e^{-(Ks+\mu\alpha)}$$

The quantity ϕ is the ratio P_x/P_0 defined by Eq. (6-6). Table 6-1 gives the values of K and μ for metal sheathing.

7-18 ILLUSTRATIVE PROBLEM 7-2

In order to apply the foregoing to the design of prestressed-concrete beams the following illustrative example is presented.

Design a simply supported beam of 54 ft span subjected to an applied load of 1600 lb per ft. The beam is to be used in a building and is to be pretensioned without non-prestressed reinforcement. Use the 1963 ACI building code. Assume $f'_c = 5000$ psi, $f'_{ci} = 4000$ psi, $\gamma = 150$ pcf, $f'_s = 250,000$ psi, $f_{sy} = 210,000$ psi, and $\eta = 0.85$.

From Table 7-3 the following can be obtained:

$a'_t = 0.038 \qquad c'_t = 0.48$

$c' = 0.45 \qquad a' = 0.085$

For $\eta = 0.85$ and the stress coefficients from Table 7-6, $\Delta_e = 1.02$.

Let us assume that because of clearance requirements the overall depth of the beam, h, cannot exceed 36 in. By assuming $h = 36$ in., ω can be calculated as follows:

$$\omega = \frac{hf'_c}{\gamma L^2} = \frac{3 \times 5 \times 144}{0.15 \times 54 \times 54} = 4.94$$

Since ω is comparatively high, it is possible to provide sufficient cover for the prestressing steel. There is no need to assign a high value for Δ. Let us assume $\Delta = \Delta_e = 1.02$ and consider an idealized I section.

In order to calculate the least-weight area it is necessary to estimate ρ, which depends upon t/h and b'/b. Let us assume $t/h = 0.15$, which corresponds to $t = 5.4$ in., and take $b'/b = 0.3$. By substituting these values for t/h and b'/b and $\Delta = 1.02$ in Eq. (7-11), $k = 0.97$ is obtained, and from Eq. (7-10) $\rho = 0.121$.

The quantity R can now be computed from Eq. (7-6b):

$$R = 8\rho\omega(\eta a_t + c)(\Delta + 1) - (1 - \eta) \qquad (7-6b)$$

where $a_t = a'_t = 0.038 \qquad \omega = 4.94$

$c = c' = 0.45 \qquad \Delta = 1.02$

$\eta = 0.85$

Substituting the above numerical values in Eq. (7-6b), we have

$$R = 8 \times 0.121 \times 4.94(0.85 \times 0.038 + 0.45)2.02 - 0.15 = 4.51$$

Since

$$R = \frac{M_a}{M_g} = \frac{1.6L^2/8}{\gamma A L^2/8} = \frac{1.6}{\gamma A}$$

hence

$$A = \frac{1.6}{\gamma R} = \frac{1.6}{0.15 \times 4.51} = 2.36 \text{ ft}^2 \text{ or } 341 \text{ in.}^2$$

We know

$$A = bh \left[\frac{t}{h}(1 + k) + \frac{b'}{b}\left(1 - 2\frac{t}{h}\right) \right]$$

or

$$340 = 36b[0.167(1 + 0.96) + 0.3(1 - 0.33)] = 19.0b$$
$$b = 17.96 \text{ in.}$$
$$b' = 17.96 \times 0.3 = 5.39 \text{ in.}$$
$$kb = 17.96 \times 0.96 = 17.25 \text{ in.}$$

The sketch in Fig. 7-19a shows the idealized symmetrical I section.

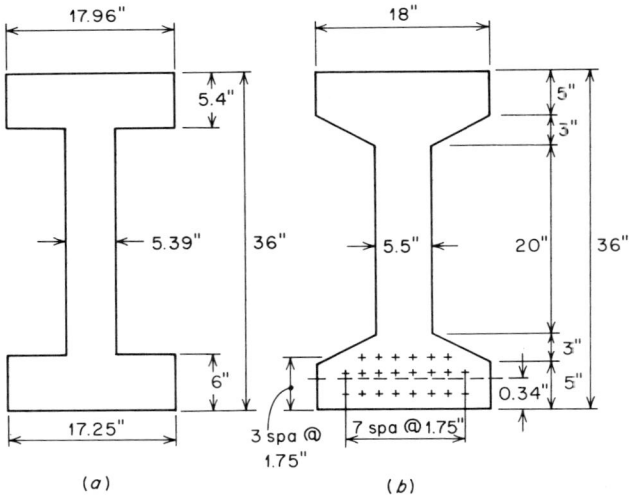

Fig. 7-19 The idealized and actual sections, Illustrative Problem 7-2.

WORKING-STRESS DESIGN OF SUPPORTED PRESTRESSED-CONCRETE BEAMS   245

The prestressing force and the eccentricity may be calculated from Eqs. (7-7a) and (7-8).

$$m = \frac{c_t - \Delta a_t}{1 + \Delta} = \frac{0.48 - 1.02 \times 0.038}{2.02} = 0.218$$

$$P_t = mAf'_c = 0.218 \times 340 \times 5 = 371 \text{ kips}$$

$$\epsilon = \frac{1}{m}\left[\rho(a_t + c_t) + \frac{1}{8\omega}\right]$$

$$= \frac{1}{0.218}\left[0.121(0.038 + 0.48) + \frac{1}{8 \times 4.94}\right] = 0.393$$

$$e = 393 \times 36 = 14.15 \text{ in.}$$
$$g = 18.00 - 14.15 = 3.85 \text{ in.}$$

The design of the idealized section is now complete. It is desirable to have a more practical section which is almost the same as the idealized section. Figure 7-19b shows the actual section adopted. The web thickness is increased to $5\frac{1}{2}$ in. so that sufficient edge distance can be provided for the bent or draped strands in the web if two strands are used in a row. The flanges have been tapered to facilitate the casting of the concrete. The flange width has been increased to 18 in.

The allowable stress in steel is

$$f_{st} = \phi(0.70 f'_s - 10,000)$$

and $f'_s = 250,000$ psi; we may take $\phi \approx 0.95$, which is low for a pretensioned construction.

$$f_{st} = 0.95(0.70 \times 250,000 - 10,000) = 157,000 \text{ psi}$$
$$A_s = \tfrac{371}{157} = 2.37 \text{ in.}^2$$

Use $\tfrac{7}{16}$-in. strand, which has a cross-sectional area of 0.1089 in.^{2}, and

Number of strands $= \tfrac{2.37}{0.1089} = 21.8$, say 22 strands

The arrangement of the strands is also shown in Fig. 7-19b. The center of gravity of the prestressing force can be determined by computing g as follows:

Top row:	$6 \times 5.25 =$	31.5
Middle row:	$8 \times 3.50 =$	28.0
Bottom row:	$8 \times 1.75 =$	14.0
	22 strands	73.5

$$g = \tfrac{73.5}{22} = 3.34 \text{ in.}$$

and
$$e = 18.00 - 3.34 = 14.66 \text{ in.}$$

The properties of the gross section are as follows:

$A = 361$ in.^{2} $\qquad I = 57,095$ in.^{4}

$r^2 = \frac{57,095}{361} = 158.2$ in.^{2} $\qquad \frac{I}{y} = \frac{57,095}{18} = 3172$ in.^{3}

and

$e = 14.66$ in.

$\frac{ey}{r^2} = \frac{14.66 \times 18}{158.2} = 1.668$

$\frac{P_t}{A} = \frac{371}{361} = 1.029$ ksi

The moments and stresses are:

$M_a = \frac{1.6}{8}(54)^2 \times 12 = 7000$ in.-kips $\qquad f_a = \frac{7000}{3172} = 2.207$ ksi

$M_g = \frac{0.376}{8}(54)^2 \times 12 = 1644$ in.-kips $\qquad f_g = \frac{1644}{3172} = 0.518$ ksi

Four requirements

$\frac{P_t}{A}\left(\frac{ey}{r^2} - 1\right) - M_g\frac{y}{I}$
$\qquad = 1.029 \times 0.668 - 0.518 = 0.168 < 0.190$ ksi $\quad$ (7-1)

$\frac{P_t}{A}\left(\frac{ey}{r^2} + 1\right) - M_g\frac{y}{I}$
$\qquad = 1.029 \times 2.668 - 0.518 = 2.227 < 2.400$ ksi $\quad$ (7-2)

$-\eta\frac{P_t}{A}\left(\frac{ey}{r^2} - 1\right) + M_t\frac{y}{I}$
$\qquad = -0.85 \times 1.029 \times 0.668 + 2.725 = 2.142 < 2.250$ ksi $\quad$ (7-3)

$-\eta\frac{P_t}{A}\left(\frac{ey}{r^2} + 1\right) + M_t\frac{y}{I}$
$\qquad = -0.85 \times 1.029 \times 2.668 + 2.725 = 0.392 < 0.425$ ksi $\quad$ (7-4)

Evidently the design is satisfactory.

The above stresses are only approximately correct since they are based upon the gross cross-sectional area of the beam. For a more accurate determination of these stresses the fully transformed section should be used.

In the following paragraphs the stresses based on the fully transformed section are calculated, assuming a modular ratio (the ratio of the modulus of elasticity of steel to that in concrete) of 7.

The section properties are:
$A = 375$ in.^{2} $I = 60{,}070$ in.^{4}
$r^2 = \frac{60{,}070}{375} = 160.2$ in.^{2}
$y_t = 18.56$ in. $y_b = 17.44$ in.
$\frac{I}{y_t} = 3237$ in.^{3} $\frac{I}{y_b} = 3443$ in.^{3}

and
$e = 14.10$ in.
$\frac{ey_t}{r^2} = 1.631$ $\frac{ey_b}{r^2} = 1.535$
$\frac{P_t}{A} = \frac{371}{375} = 0.99$

The moments and stresses are:

$M_g = 1644$ in.-kips $M_a = 7000$ in.-kips
$\frac{M_g y_t}{I} = 0.509$ ksi $\frac{M_g y_b}{I} = 0.477$ ksi
$\frac{M_a y_t}{I} = 2.165$ ksi $\frac{M_a y_b}{I} = 2.030$ ksi
Total 2.674 ksi Total 2.507 ksi

Four requirements

$\frac{P_t}{A}\left(\frac{ey_t}{r^2} - 1\right) - M_g \frac{y_t}{I}$
$= 0.99 \times 0.631 - 0.509 = 0.116 < 0.190$ ksi (7-1)

$\frac{P_t}{A}\left(\frac{ey_b}{r^2} + 1\right) - M_g \frac{y_b}{I}$
$= 0.99 \times 2.535 - 0.477 = 2.033 < 2.400$ ksi (7-2)

$-\eta \frac{P_t}{A}\left(\frac{ey_t}{r^2} - 1\right) + M_t \frac{y_t}{I}$
$= -0.85 \times 0.99 \times 0.631 + 2.674 = 2.153 < 2.250$ ksi (7-3)

$-\eta \frac{P_t}{A}\left(\frac{ey_b}{r^2} + 1\right) + M_t \frac{y_b}{I}$
$= -0.85 \times 0.99 \times 2.535 + 2.507 = 0.377 < 0.425$ ksi (7-4)

It can be seen that the stresses calculated on the basis of the fully transformed section are also within the allowable stresses.

If the assumed value 7 of the modular ratio is correct, the above stresses are more accurate than the ones based on the gross cross-sectional area of the beam. However, since there are inaccuracies associated with the determination of the modular ratio, the above stresses may not be accurate.

In general, the stresses calculated on the basis of the fully transformed section are somewhat lower than those based on the gross section. Therefore, calculation of stresses on the basis of gross cross-sectional area of the beam is usually, but not always, safe.

The ultimate moment in this section on the basis of the stress-strain diagrams for steel and concrete shown in Figs. 3-5 and 3-17, respectively, is 621 in.-kips.

7-19 ILLUSTRATIVE PROBLEM 7-3

Design a girder of 98 ft span that is to be built by using a pretensioned construction. The superimposed dead load consists of a precast slab at 600 plf, and the live load including impact is equivalent to 1500 plf. The overall depth of the girder cannot exceed 66 in. For this problem assume $f'_c = 5000$ psi, $\gamma = 150$ pcf, $f'_s = 250{,}000$ psi, $f_{sy} = 210{,}000$ psi, and design the girder on the basis of the following stress coefficients:

$$a'_t = 0.038 \qquad c'_t = 0.48$$
$$c' = 0.40 \qquad a' = 0$$

The stress coefficients are the ones used in the preceding section to plot the relationship between R and ω for the various criteria. In the following paragraphs these plots will be used directly. Evidently in this case $\Delta_e = 0.94$.

The quantity ω can be computed as follows:

$$\omega = \frac{hf'_c}{\gamma L^2} = \frac{5.5 \times 5 \times 144}{0.15 \times 98 \times 98} = 2.75$$

It is desirable to adopt $\Delta = \Delta_e = 0.94$ in order to make the computed stresses equal to the allowable stresses, which will result in the least area of concrete. However, from Fig. 7-6 it can be seen that for $\omega = 2.75$ sufficient cover is not available for reasonable ρ values. From Fig. 7-7b it is seen that increasing Δ to 1.00 does not increase the cover sufficiently, even if criterion 2 is applied. Figure 7-8b shows that when $\Delta = 1.25$, criterion 2 does provide a cover that is more than sufficient for a reasonable value of ρ. Since increasing Δ also results in an increase in the area of concrete, it is not necessary to increase Δ to as high a value as 1.25. Let us assume $\Delta = 1.20$.

We have

$$\Delta = \frac{y_b}{y_t} = 1.20$$

WORKING-STRESS DESIGN OF SUPPORTED PRESTRESSED-CONCRETE BEAMS

and
$$h = y_b + y_t = 66 \text{ in.}$$
Hence
$$y_t = 30 \text{ in.} \qquad y_b = 36 \text{ in.}$$

Using Fig. 7-14 for an unsymmetrical I section in which the width of the top flange is $1\frac{1}{2}$ times greater than that of the bottom flange, for $\Delta = 1.20$ and $t/h = 0.15$, we have $b'/b = 0.30$ and $\rho = 0.115$. Equation (7-6a) can now be used to calculate R and the required area of concrete:

$$R = 8 \times 0.115 \times 2.75(0.85 \times 0.48)(\tfrac{1}{1.2} + 1) - (1 - 0.15) = 1.74$$

and
$$A = \frac{0.60 + 1.50}{1.74 \times 0.15} \times 144 = 1158 \text{ in.}^2$$

The area of an unsymmetrical I section of the type used here can be presented as follows:

$$A = bh\left[\frac{t}{h}(1 + \tfrac{2}{3}) + \frac{b'}{b}\left(1 - \frac{2t}{h}\right)\right]$$

or
$$1158 = 66b(0.15 \times 1.67 + 0.30 \times 0.70) = 30.35b$$
and
$$b = 38 \text{ in.} \qquad b' = 11.5 \text{ in.}$$
$$kb = 25.5 \text{ in.}$$

Figure 7-20a shows the idealized section. The eccentricity and prestressing force can be calculated for criteria 1 and 2.

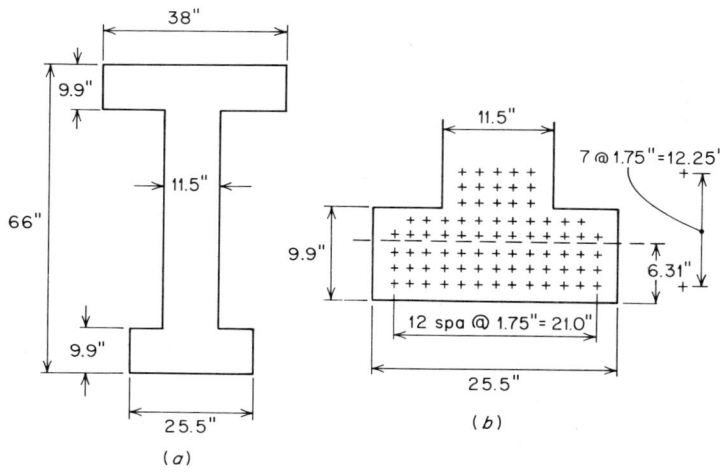

Fig. 7-20 The idealized section, Illustrative Problem 7-3.

Criterion 1 From Eq. (7-7a) we have

$$m = \frac{c_t - \Delta a_t}{1 + \Delta} = \frac{0.48 - 1.20 \times 0.038}{2.20} = 0.197$$

Hence

$$P_t = 0.197 \times 5 \times 1158 = 1142 \text{ kips}$$

and

$$\epsilon = \frac{1}{0.197}\left[0.115(0.48 + 0.038) + \frac{1}{8 \times 2.75}\right] = 0.533$$

Hence

$$e = 0.533 \times 66 = 35.2 \text{ in.}$$

or

$$g = 36.0 - 35.2 = 0.8 \text{ in.}$$

Criterion 2 From Eq. (7-5) the stress coefficient a_t can be calculated for $\Delta = 1.20$:

$$a_t = \frac{1}{\eta}\left(\frac{\eta c'_t + a'}{\Delta} - c'\right) = \frac{1}{0.85}\left(\frac{0.85 \times 0.48}{1.20} - 0.40\right) = -0.071$$

and

$$m = \frac{c_t - \Delta a_t}{1 + \Delta} = \frac{0.48 + 1.20 \times 0.071}{2.20} = 0.256$$

Hence

$$P_t = 0.256 \times 5 \times 1158 = 1480 \text{ kips}$$

and

$$\epsilon = \frac{1}{0.256}\left[0.115(0.48 - 0.07) + \frac{1}{8 \times 2.75}\right] = 0.36$$

Hence

$$e = 0.36 \times 66 = 23.8 \text{ in.}$$

or

$$g = 36.0 - 23.8 = 12.2 \text{ in.}$$

Criterion 1 does not provide sufficient cover and criterion 2 provides too much. If criterion 2 is used, the prestressing force will be unnecessarily increased. In most problems a cover of approximately $0.10h$ gives a reasonable value for g.

In this case let us assume $g = 6.5$ in. or $e = 29.5$ in., which are within the limits set by criteria 1 and 2. In order to calculate the prestressing force for this eccentricity, Eqs. (7-2a) and (7-4a) may be used since these requirements are satisfied exactly even though criteria 1 and 2

WORKING-STRESS DESIGN OF SUPPORTED PRESTRESSED-CONCRETE BEAMS 251

are not used. Let us use Eq. (7-4a):

$$-\eta m \left[\frac{\epsilon \Delta}{\rho(1+\Delta)} + 1\right] + \frac{\Delta(1+R)}{8\rho\omega(1+\Delta)} = a' \qquad (7\text{-}4a)$$

All quantities in the above equation are known except m, which can be computed as follows:

$$-0.85m \left(\frac{\frac{29.5}{66} \times 1.20}{0.115 \times 2.20} + 1\right) + \frac{1.20 \times 2.74}{8 \times 0.115 \times 2.75 \times 2.20} = 0$$

or
$$m = 0.222$$
and
$$P_t = 0.222 \times 5 \times 1158 = 1285 \text{ kips}$$

The variation of eccentricity with prestressing force may be studied for this problem by the method discussed in Sec. 7-8. The variation of $1/P_t$ with e for each of the four requirements is shown in Fig. 7-21. Since requirements (7-2) and (7-4) are satisfied exactly, the region of

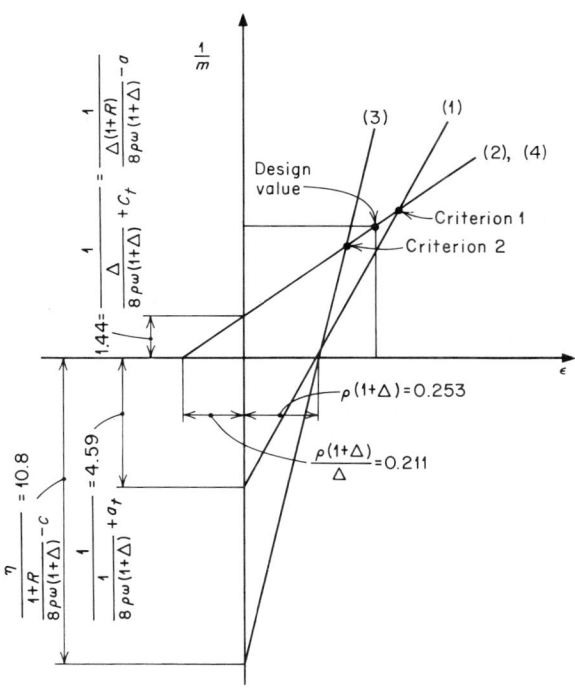

Fig. 7-21 The variation of eccentricity with prestressing force for Illustrative Problem 7-3.

possible solutions becomes a line. An eccentricity of 29.5 in. corresponding to $P_t = 1285$ kips is a possible solution.

The allowable stress in steel at transfer at midspan may be calculated as follows:

$$f_{st} = \phi(0.70 f'_s - 10) = 0.95(0.70 \times 250 - 10) = 157 \text{ ksi}$$

With $\tfrac{7}{16}$-in. strands, the number of strands needed will be

$$1285/(157 \times 0.1089) = 76$$

Figure 7-20b shows the tentative arrangements of the strands in the idealized section. The design on the basis of the idealized section is now complete.

Figure 7-22 shows the actual section. The section properties and the stresses corresponding to the four requirements are shown in Table 7-8. No further explanation seems to be necessary.

It can be shown that the ultimate moment in this section, calculated on the basis of the stress-strain diagrams for steel and concrete shown in Figs. 3-14 and 3-10, respectively, is 102,700 in.-kips.

7-20 STRESSES AT THE ENDS

In the preceding discussions the design of prestressed-concrete beams on the basis of the stresses at midspan of the beam was presented. The four requirements which were developed primarily for the midspan of the beam should also be satisfied at all points along the beam. Simply supported ends, at the centerline of bearing, are particularly important, since there are no stresses at these points caused by the acting loads to counteract the stresses caused by the prestressing force. Generally it

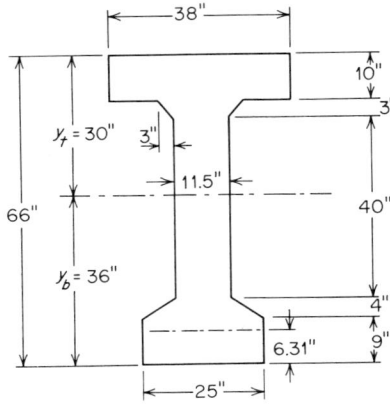

Fig. 7-22 The actual section, Illustrative Problem 7-8.

WORKING-STRESS DESIGN OF SUPPORTED PRESTRESSED-CONCRETE BEAMS

Table 7-8   Section properties and stresses in the actual section

Section properties	Loads and moments	Stresses
$A = 1182.5$ in.^{2}	$W_a = 2.1$ klf	$M_a y_t / I = 1.530$ ksi
$I = 593{,}290$ in.^{4}	$W_g = 1.23$ klf	$M_g y_t / I = 0.897$ ksi
$r^2 = 501.7$ in.^{2}	$P_t = 1290$ kips	$\overline{2.427 \text{ ksi}}$
$I/y_t = 19{,}776$ in.^{3}		$M_a y_b / I = 1.836$ ksi
$I/y_b = 16{,}480$ in.^{3}	$M_a = 30{,}250$ in.-kips	$M_g y_b / I = 1.077$ ksi
$e = 29.69$ in.	$M_g = 17{,}720$ in.-kips	$\overline{2.913 \text{ ksi}}$
$e y_t / r^2 = 1.774$		$P_t / A = 1.08$ ksi
$e y_b / r^2 = 2.130$		

The Four Requirements

$1.08(1.774 - 1) - 0.897 = -0.062$ ksi (compressive)   (7-1)
$1.08(2.130 + 1) - 1.077 = 2.303$ ksi (compressive)   (7-2)
$0.85 \times 1.08(1.774 - 1) - 2.427 = 1.710$ ksi (compressive) (7-3)
$0.85 \times 1.08(2.130 + 1) - 2.913 = 0.013$ ksi (tensile)   (7-4)

is not possible to satisfy the four requirements at simply supported ends without reducing the stresses caused by the prestressing force. Usually this is accomplished by reducing the eccentricity.

After the end eccentricities are obtained, the profile of the center of gravity of steel can be determined and the four requirements checked throughout the span.

The four requirements as applied to a simply supported jacking end, at the centerline of bearing, can be written as follows:

$$\frac{P_t}{\phi A}\left(\frac{e y_t}{r^2} - 1\right) = a_t f'_c \leq a'_t f'_c$$

$$\frac{P_t}{\phi A}\left(\frac{e y_b}{r^2} + 1\right) = c_t f'_c \leq c'_t f'_c$$

$$\eta \frac{P_t}{\phi A}\left(\frac{e y_t}{r^2} - 1\right) = a f'_c \leq a' f'_c$$

$$\eta \frac{P_t}{\phi A}\left(\frac{e y_b}{r^2} + 1\right) = c f'_c \leq c' f'_c$$

In the above expressions A and r are the cross-sectional area and the radius of gyration of the beam section at the centerline of bearing of the end support, respectively. If the beam is designed with end blocks, as is the case for post-tensioned beams, A and r correspond to the section properties of the end block. In pretensioned beams end blocks are sometimes eliminated, in which case A and r will be the same as those at midspan of the beam. The quantity P_t/ϕ is the prestressing force at transfer at the centerline of bearing of the end support. As shown in Sec. 7-17, ϕ is the ratio of the prestressing force at midspan to that at the end of the beam, and it signifies the frictional losses. When there are no frictional losses, $\phi = 1.0$.

These expressions may be restated as follows:

$$e \leq \left(\frac{\phi A}{P_t} a'_t f'_c + 1\right) \frac{r^2}{y_t}$$

$$e \leq \left(\frac{\phi A}{P_t} c'_t f'_c - 1\right) \frac{r^2}{y_b}$$

$$e \leq \left(\frac{\phi A}{\eta P_t} a' f'_c + 1\right) \frac{r^2}{y_t}$$

$$e \leq \left(\frac{\phi A}{\eta P_t} c' f'_c - 1\right) \frac{r^2}{y_b}$$

The lowest of the above four values will give the eccentricity at the centerline of bearing of the end support.

Let us calculate the end eccentricity for Illustrative Problem 7-2 discussed in Sec. 7-18, assuming no end blocks. The properties of the gross section were as follows:

$A = 361$ in.^{2} $I = 57,095$ in.^{4}

$r^2 = 158.2$ in.^{2} $y_t = y_b = 18$ in.

$P_t = 371$ kips

The stress coefficients were

$a'_t = 0.038$ $c'_t = 0.48$

$c' = 0.45$ $a' = 0.085$

and $f'_c = 5000$ psi, $\eta = 0.85$, and $\phi = 0.95$.

Substituting these quantities in the above four expressions for the eccentricity, we have

$$e \leq \left(\frac{0.95 \times 361}{371} \times 0.038 \times 5 + 1\right) \frac{158.2}{18} = 10.33 \text{ in.}$$

$$e \leq \left(\frac{0.95 \times 361}{371} \times 0.48 \times 5 - 1\right) \frac{158.2}{18} = 11.27 \text{ in.}$$

$$e \leq \left(\frac{0.95 \times 361}{0.85 \times 371} \times 0.085 \times 5 + 1\right) \frac{158.2}{18} = 12.89 \text{ in.}$$

$$e \leq \left(\frac{0.95 \times 361}{0.85 \times 371} \times 0.45 \times 5 - 1\right) \frac{158.2}{18} = 12.77 \text{ in.}$$

Evidently the end eccentricity should not exceed 10.33 in.

7-21 THE PROFILE OF THE PRESTRESSING STEEL

The profile of the prestressing steel should be determined in such a way that the four requirements will be satisfied in all points along the beam. Once the eccentricity at midspan and at the ends is known, a reasonable shape for the profile of the center of gravity of prestressing steel may be adopted.[5] In practical problems a chosen number of the strands are bent at one or two points along the span, while the remaining number are carried straight throughout the beam.

Illustrative Problem 7-2 will be used again to show how the profile of the prestressing steel may be determined and stresses checked.

In this problem 22 strands were used with an arrangement that produced an eccentricity of 14.66 in. at midspan. Figure 7-19b shows the arrangement of the strands. In Sec. 7-20 it was shown that in order to limit the stresses at the ends, the end eccentricity should not be greater than 10.33 in.

Let us assume that the six middle strands will be bent at the third points of the span. The position of the six bent strands at the end section—the section at the centerline of bearing of the supports—should be determined in such a way that the eccentricity will not exceed 10.33 in.

It can be shown that the center of gravity of the 16 straight strands is 14.72 in. from the centroidal axis of the section. By assuming the distance between the center of gravity of the six bent strands and the centroidal axis at the end section to be X, the following can be written:

$$(16)(14.72) - 6X = (22)(10.33)$$

or

$$X = \frac{(16)(14.72)}{6} - \frac{(10.33)(22)}{6} = 1.4 \text{ in.}$$

Therefore, the center of gravity of the bent strands at the end section should be at least 1.4 in. above the centroidal axis of the beam. For convenience 1.5 in. has been chosen for this distance.

Figure 7-23a shows the cross section of the beam at midspan—the same as Fig. 7-19b—and Fig. 7-23b shows the cross section at the centerline of bearing of the end support. Figure 7-23c shows the profile of the prestressing steel.

In order to show that the four requirements are met at all points throughout the length of the span, it is convenient to show graphically the variation of stress due to prestressing force at each of the acting loads.

Figure 7-24a shows the variation of the top fiber stress along the length of the span for Illustrative Problem 7-2. The direct compressive stress due to the prestressing force at transfer is shown by a dashed line marked P_t/A, and is plotted below the reference line. The flexural tensile stress due to the prestressing is measured above the dashed line,

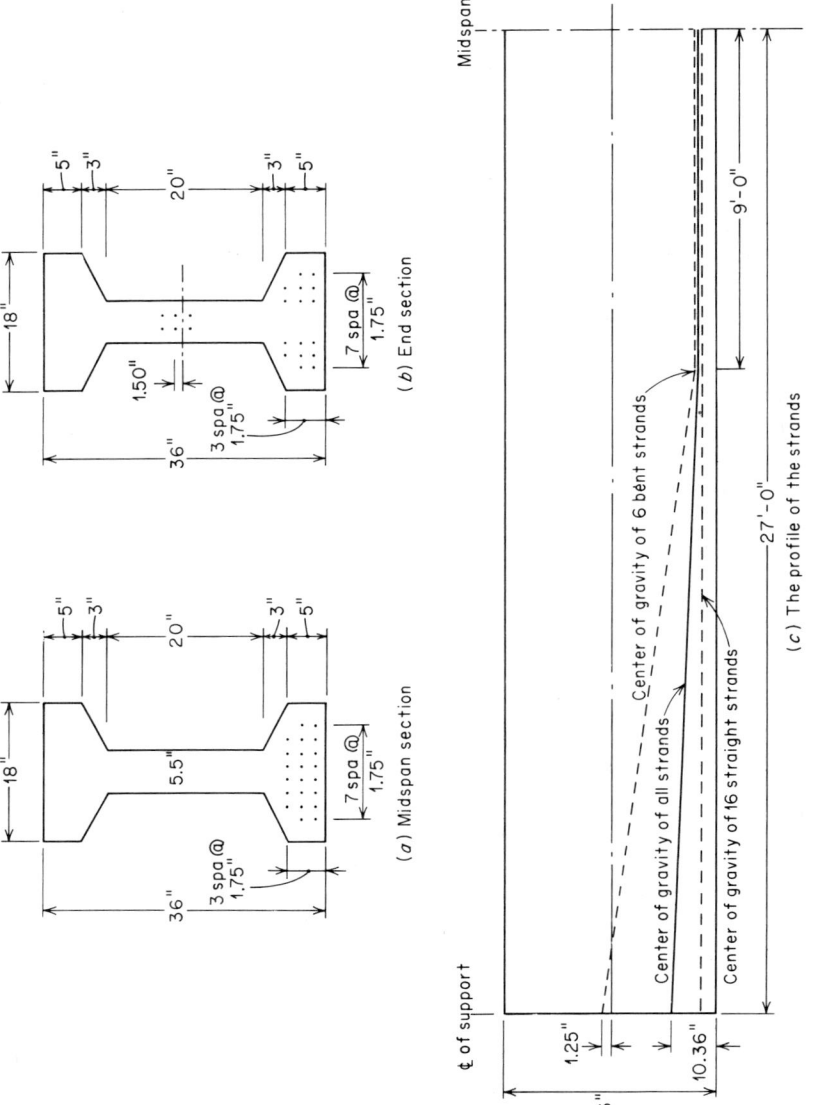

Fig. 7-23 Details of the arrangement of the prestressing steel, Illustrative Problem 7-2.

using this line as a reference. The dashed line above the reference line, consisting of a sloping line and a horizontal line, indicates the total tensile stress due to prestressing force at transfer along the span. The stresses due to prestressing force after losses are shown by solid lines. The parabolic curves plotted above the reference line represent the compressive stresses due to dead load and live load. A study of this presentation indicates that the stresses at the top fiber are mostly within the allowable stresses at all times. The tensile stress at transfer in the region of the third point seems to be slightly more than the allowable stress.

Figure 7-24b represents the variation of the bottom fiber stress for the same problem. Since this figure is similar to Fig. 7-24a, no further

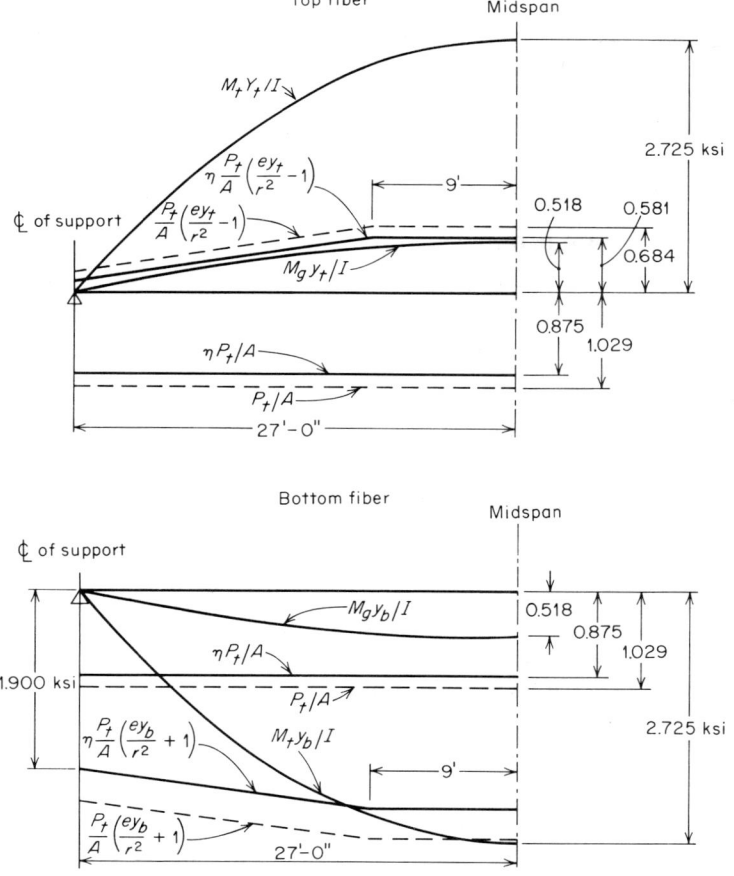

Fig. 7-24 Variation of stresses along the span, Illustrative Problem 7-2.

explanation seems to be needed. It is evident that the stresses at the bottom fiber are within the allowable stresses at all times.

It should be pointed out that in Fig. 7-24 it is assumed that the support shown is at the jacking anchorage or the anchorage closer to the jacking end, where the prestressing force is maximum. In this problem it is assumed that the prestressing force is reduced by 5 percent from the end to the center of the span (in an actual pretensioned beam of the type discussed here the frictional loss may not be as large as 5 percent), and in Fig. 7-24 for convenience the prestressing force is varied linearly from the end to the center of the beam. If instead of the jacking end the anchored end of the beam were considered, the prestressing force would decrease from the center of the beam to the anchor end by about 5 percent.

It should be pointed out that draping the strands is not the only way to reduce the stresses caused by the prestressing force. Sometimes the stress is reduced by destroying the bond of some of the bars in the end region of the beam.

PRESTRESSING WITH SUPERIMPOSED DEAD LOAD

7-22 THE FOUR REQUIREMENTS

In the design method developed in the preceding discussions it is assumed that the only load acting at the time of prestressing is the weight of the beam. This assumption is in agreement with the conventional construction practice. However, occasionally in some special conditions it is possible to prestress a beam when both the weight of the beam and the superimposed dead load are acting. This example is presented here more to show the versatile nature of prestressed concrete than to present a typical problem.

Prestressing with the superimposed dead load is advantageous only in post-tensioned cast-in-place construction. If the falsework is adequate for supporting the additional weight of slab, appreciable saving may be effected in both concrete and steel by applying this method of construction. In the method of design presented in the following paragraphs it is assumed that if the superimposed dead load is a concrete slab, it is not acting compositely with the beam.

In this case the four requirements can be written as follows:

$$\frac{P_t}{A}\left(\frac{ey_t}{r^2} - 1\right) - \frac{(M_g + M_s)y_t}{I} = a_t f'_c \leq a'_t f'_t \qquad (7\text{-}1c)$$

$$\frac{P_t}{A}\left(\frac{ey_b}{r^2} + 1\right) - \frac{(M_g + M_s)y_b}{I} = c_t f'_c \leq c'_t f'_c \qquad (7\text{-}2c)$$

WORKING-STRESS DESIGN OF SUPPORTED PRESTRESSED-CONCRETE BEAMS

$$-\eta \frac{P_t}{A}\left(\frac{ey_t}{r^2} - 1\right) + \frac{(M_g + M_s + M_l)y_t}{I} = cf'_c \leq c'f'_c \qquad (7\text{-}3c)$$

$$-\eta \frac{P_t}{A}\left(\frac{ey_b}{r^2} + 1\right) + \frac{(M_g + M_s + M_l)y_b}{I} = af'_c \leq a'f'_c \qquad (7\text{-}4c)$$

where M_g = moment at midspan due to weight of beam = $W_g L^2/8$, where $W_g = \gamma A$

M_s = moment at midspan due to superimposed dead load = $W_s L^2/8$, where W_s = uniformly distributed superimposed dead load

All other terms in the above equations have been defined in the preceding sections.

Introducing $y_b/y_t = \Delta$, $hf'_c/\gamma L^2 = \omega$, $r^2/h^2 = \rho$, $e/h = \epsilon$, and $P_t/Af'_c = m$ in the above equations and rearranging, we have:

$$m\left[\frac{\epsilon}{\rho(1+\Delta)} - 1\right] - \frac{1 + W_s/W_g}{8\rho\omega(1+\Delta)} = a_t$$

$$m\left[\frac{\epsilon\Delta}{\rho(1+\Delta)} + 1\right] - \frac{\Delta(1 + W_s/W_g)}{8\rho\omega(1+\Delta)} = c_t$$

$$-\eta m\left[\frac{\epsilon}{\rho(1+\Delta)} - 1\right] + \frac{(1 + W_s/W_g)[1 + M_l/(M_g + M_s)]}{8\rho\omega(1+\Delta)} = c$$

$$-\eta m\left[\frac{\epsilon\Delta}{\rho(1+\Delta)} + 1\right] + \frac{\Delta(1 + W_s/W_g)[1 + M_l/(M_g + M_s)]}{8\rho\omega(1+\Delta)} = a$$

A simultaneous solution of the above four equations for Δ, $M_l/(M_g + M_s)$, m, and ϵ yields the following:

$$\Delta = \frac{\eta c_t + a}{\eta a_t + c} \qquad (7\text{-}5)$$

$$\frac{M_l}{M_g + M_s} = \frac{8\rho\omega}{1 + W_s/W_g}[(a + c) + \eta(a_t + c_t)] - (1 - \eta) \qquad (7\text{-}6c)$$

$$m = \frac{c_t - \Delta a_t}{1 + \Delta} \qquad (7\text{-}7a)$$

$$\epsilon = \frac{1}{m}\left[\rho(c_t + a_t) + \frac{1 + W_s/W_g}{8\omega}\right] \qquad (7\text{-}8c)$$

Equation (7-6c) can be written as follows:

$$\left[\frac{8M_l}{\gamma L^2} + (1 - \eta)\frac{W_s}{\gamma}\right]\frac{1}{A} = 8\rho\omega(\eta c_t + a)\left(\frac{1}{\Delta} + 1\right) - (1 - \eta) \qquad (7\text{-}6c)$$

$$\left[\frac{8M_l}{\gamma L^2} + (1 - \eta)\frac{W_s}{\gamma}\right]\frac{1}{A} = 8\rho\omega(\eta a_t + c)(\Delta + 1) - (1 - \eta) \qquad (7\text{-}6d)$$

The above equations indicate that prestressing with superimposed dead load is similar to prestressing with the dead load of the beam only. A study of the above expressions shows that the only difference between the two cases is in term ω.

When the only load acting during prestressing is the weight of the beam—designated as W_g plf here—the expression of ω is $hf'_c/\gamma L^2$.

If superimposed dead load of W_s plf is added during the prestressing, it is only necessary to divide $\omega = hf'_c/\gamma L^2$ by the term $(1 + W_s/W_g)$ to adapt the conventional expressions for use in this case.

It may be concluded that the least-weight design criteria developed in the preceding sections are also applicable to this case.

7-23 ILLUSTRATIVE PROBLEM 7-4

In order to show the difference between prestressing with superimposed dead load and the conventional method of prestressing, Illustrative Problem 7-3 will be considered.

Design a highway bridge girder of 98 ft span that is to be post-tensioned in place. The superimposed dead load is a precast slab which weighs 600 plf. Adequately designed falsework is provided to carry the weight of the beam as well as the weight of the slab. The precast slab is to be placed before the post-tensioning of the girder. The live load including impact is equivalent to 1500 plf. The overall depth of the girder cannot exceed 66 in. For this problem assume $f'_c = 5000$ psi, $\gamma = 150$ pcf, $f_s = 250,000$ psi, and $f_{sy} = 210,000$ psi, and design the girder on the basis of the following stress coefficients:

$$a'_t = 0.038 \qquad c'_t = 0.48$$
$$c' = 0.40 \qquad a' = 0$$

From the expressions developed in Sec. 7-12 it can be shown that for $\eta \approx 0.85$, $\Delta_e = 0.94$.

We know

$$\omega = \frac{hf'_c}{\gamma L^2} = \frac{5.5 \times 5 \times 144}{0.15 \times 98 \times 98} = 2.75$$

Since ω is comparatively small and from Eq. (7-8c) the quantity ϵ is greater in this case than the conventional type of construction, it is unlikely that sufficient cover can be obtained with small Δ values.

Let us assume $\Delta = 1.20$ as in Illustrative Problem 7-3. From Fig. 7-18 (which relates the geometric properties of unsymmetrical I sections for $k = 0.667$) for $\Delta = 1.20$ and $t/h = 0.15$, we have $b'/b = 0.30$ and $\rho = 0.115$.

WORKING-STRESS DESIGN OF SUPPORTED PRESTRESSED-CONCRETE BEAMS

Equation (7-6c) can be used to calculate the least-weight area of concrete:

$$\left[\frac{8M_l}{\gamma L^2} + (1-\eta)\frac{W_s}{\gamma}\right]\frac{1}{A} = 8\rho\omega(\eta c_t + a)\left(\frac{1}{\Delta}+1\right) - (1-\eta) \quad (7\text{-}6c)$$

where

$$\frac{8M_l}{\gamma L^2} = \frac{W_l}{\gamma} = \frac{1.5}{0.15} = 10$$

and

$$(1-\eta)\frac{W_s}{\gamma} = \frac{0.15 \times 0.60}{0.15} = 0.60$$

or

$$\frac{10.6}{A} = 8 \times 0.115 \times 2.75(0.85 \times 0.48)(\tfrac{1}{1.2}+1) - (1-0.85) = 1.74$$

Hence

$$A = \tfrac{10.6}{1.74}\,144 = 878 \text{ in.}^2$$

but

$$A = bh\left[\frac{t}{h}(1+0.667) + \frac{b'}{b}\left(1-\frac{2t}{h}\right)\right]$$

or

$$878 = 66b(0.15 \times 1.667 + 0.3 \times 0.70) = 30.35b$$

Hence

$$b = 29 \text{ in.} \qquad b' = 8.7 \text{ in.}$$
$$kb = 19.3 \text{ in.}$$

Figure 7-25a shows the idealized section. The eccentricity and prestressing force can now be determined from criteria 1 and 2.

Criterion 1 We have

$$m = \frac{c_t - \Delta a_t}{1+\Delta} = \frac{0.48 - 1.20 \times 0.038}{2.20} = 0.197$$

or

$$P_t = 0.197 \times 5 \times 878 = 864 \text{ kips}$$

and

$$\epsilon = \frac{1}{0.197}\left[0.115(0.48+0.038) + \frac{1+\frac{0.6}{0.902}}{8 \times 2.75}\right] = 0.687$$

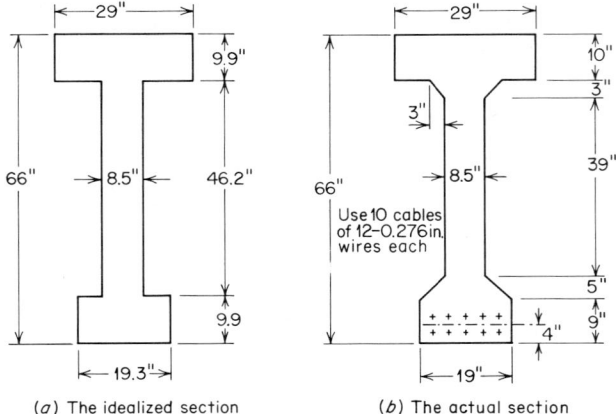

Fig. 7-25 The cross section of the beam, Illustrative Problem 7-4.

Hence

$$e = 0.687 \times 66 = 45.4 \text{ in.}$$
$$g = 36.0 - 45.4 = -9.4 \text{ in.}$$

Criterion 2 From Eq. (7-5) the stress coefficient can be calculated for $\Delta = 1.20$:

$$a_t = \frac{1}{\eta}\left(\frac{\eta c_t' + a'}{\Delta} - c'\right) = \frac{1}{0.85}\left(\frac{0.85 \times 0.48}{1.20} - 0.40\right) = -0.0706$$

and

$$m = \frac{c_t - \Delta a_t}{1 + \Delta} = \frac{0.48 + 1.20 \times 0.0706}{2.20} = 0.256$$

or

$$P_t = 0.256 \times 5 \times 878 = 1122 \text{ kips}$$

and

$$\epsilon = \frac{1}{0.256}\left[0.115(0.48 - 0.07) + \frac{1 + \frac{0.6}{0.9}}{8 + 2.75}\right] = 0.479$$

Hence

$$e = 0.479 \times 66 = 31.60 \text{ in.}$$
$$g = 36.0 - 31.60 = 4.4 \text{ in.}$$

Criterion 1 cannot be used since the center of gravity of prestressing steel falls below the section. Criterion 2, however, can be used. Figure

WORKING-STRESS DESIGN OF SUPPORTED PRESTRESSED-CONCRETE BEAMS

Table 7-9 Section properties and stresses in the actual section

Section properties	Loads and moments	Stresses
$A = 896$ in.^{2}	$P_t = 1140$ kips	$[(M_g + M_s)/I]y_t = 1.463$ ksi
$I = 453{,}160$ in.^{4}	$W_g = 0.935$ klf	
$r^2 = 505$ in.^{2}	$W_s = 0.600$ klf	$[(M_g + M_s)/I]y_b = 1.758$ ksi
$I/y_t = 15{,}105$ in.^{3}	$W_l = 1.500$ klf	
$I/y_b = 12{,}597$ in.^{3}		$M_t y_t/I = 2.895$ ksi
$e = 32$ in.	$M_g = 13{,}480$ in.-kips	
$ey_t/r^2 = 1.90$	$M_s = 8650$ in.-kips	$M_t y_b/I = 3.492$ ksi
$ey_b/r^2 = 2.28$	$M_l = 21{,}600$ in.-kips	
	$M_t = 43{,}730$ in.-kips	$P_t/A = 1.27$ ksi

The Four Requirements

$$1.27(1.90 - 1) - 1.463 = -0.321 \text{ ksi (compressive)} \quad (7\text{-}1c)$$
$$1.27(2.28 + 1) - 1.758 = 2.400 \text{ ksi (compressive)} \quad (7\text{-}2c)$$
$$-0.85 \times 1.27(1.90 - 1) + 2.895 = 1.823 \text{ ksi (compressive)} \quad (7\text{-}3c)$$
$$-0.85 \times 1.27(2.28 + 1) + 3.492 = -0.048 \text{ ksi (compressive)} \quad (7\text{-}4c)$$

7-25b shows the actual section. The section properties and the four requirements are shown in Table 7-9.

By comparing this problem with Illustrative Problem 7-3, it is apparent that by prestressing with superimposed dead load it is possible to reduce the area of concrete.

PROBLEMS

7-1. A prestressed-concrete beam is to span 80 ft. The beam has a symmetrical I section 80 in. deep. Both flanges are 12 in. thick and the web width is $0.3b$, where b is the flange width. The section is constant along the span. Allowable stresses are

$f'_c = 5$ ksi $\quad a'_t = 0.038$
$c'_t = 0.48$ $\quad c' = 0.40$
$a' = 0$

Also
$\eta = 0.85$ $\quad \gamma = 0.15$ kcf

The beam is to carry 4 klf and is to be designed on a least-weight basis.

(a) Find the area of the section.
(b) Find the width of the flanges, b.
(c) Find the maximum prestressing force and the corresponding eccentricity at midspan.
(d) Find the minimum prestressing force and the corresponding eccentricity at midspan.

(e) If g/h at midspan is 0.1, what is the prestressing force required?

(f) If g/h at midspan is 0.1, what is the maximum allowed eccentricity at the end? Friction loss factor is 0.89.

7-2. The section shown in Fig. 7-1a is to be used on a simply supported span of 75 ft length, and is to be subject to a total applied load of 1 klf. Determine the eccentricity and prestressing force. $f'_c = 5000$ psi, $\gamma = 150$ pcf, $a'_t = 0.038$, $c'_t = 0.48$, $c' = 0.45$, $a' = 0.085$.

7-3. The functional requirements for a small warehouse necessitate a layout for the floor as shown in Fig. 7-26. The floor is to be supported by masonry walls and is to be subject to a live load of 125 psf. A precast concrete slab of 6 in. thickness is to constitute the floor.

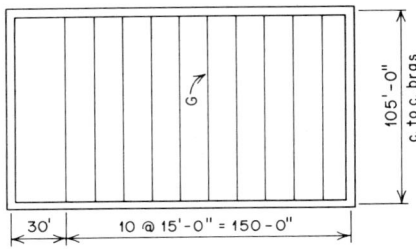

Fig. 7-26 The floor layout, Prob. 7-3.

(a) Design the girder G, assuming a pretensioned precast construction. Show the actual section and the size and spacing of strands.

(b) Design the girder G for a post-tensioned cast-in-place construction, assuming that the prestressing is carried out after the slab is cast. In this case adequate falsework is to be provided to support the weight of the girder and the slab. Only the idealized section is required.

For this problem assume:

$h = 42$ in. $f'_c = 6000$ psi $\gamma = 150$ pcf $\eta = 0.85$

Pretensioned construction:

$a'_t = 0.038$ $c'_t = 0.48$ $c' = 0.45$ $a' = 0.085$

Post-tensioned construction:

$a'_t = 0.038$ $c'_t = 0.44$ $c' = 0.45$ $a' = 0.043$

7-4. It is intended to develop service-load design criteria for a simply supported noncomposite prestressed-concrete beam subjected to upward as well as downward loads. The bending moment at midspan for the downward loads may be taken as M_a and that for the upward loads as αM_a, where α is a number between 0 and 0.50.

(a) Derive expressions for Δ, R, m, and ϵ.

(b) What are the effects of the upward load on the area of the beam, the eccentricity, and the prestressing force?

REFERENCES

1. "Standard Specifications for Highway Bridges," 8th ed., American Association of State Highway Officials, 1961.

2. "Building Code Requirements for Reinforced Concrete (ACI 318-63)," American Concrete Institute, June, 1963.
3. Magnel, G.: "Prestressed Concrete," 3d ed., McGraw-Hill Book Company, New York, 1954.
4. Khachaturian, N., I. Ali, and L. T. Thorpe: Analytical Studies of Relations among Various Design Criteria for Prestressed Concrete, *Univ. Ill. Eng. Expt. Sta. Bull.* 463, 1962.
5. Gurfinkel, G.: Design of Hold-downs in Prestressed Concrete Girders, *Indian Concrete J.*, vol. 41, no. 2, pp. 62–66, February, 1967.

8
Ultimate Design of Simply Supported Prestressed-concrete Beams

8-1 INTRODUCTION

In past design practice, prestressed-concrete beams almost always have been designed and proportioned by working-stress design, discussed in Chap. 7. The provisions of ultimate design have been used to check the flexural strength of a section that has already been designed. In this chapter it is shown that the provisions of ultimate design can be used to proportion a section with a rigorous control of both strength and ductility. The provisions of working-stress design can then be used to check the stresses at transfer, and at service loads in the section so designed. A rational design of a section is considerably simpler by ultimate design than by service-load design.

In the discussions that follow, a simply supported bonded beam is considered, and it is assumed that the strength of the beam is measured by flexure. It is assumed that the only loads acting—in addition to the prestressing force—are the weight of the beam, the superimposed dead load, and the live load.

A brief discussion of the analysis of prestressed-concrete beams at ultimate follows.

8-2 ULTIMATE MOMENT

In Chap. 3 a general method was developed for the analysis of a prestressed-concrete beam for the entire range of loading, including the ultimate. In the following paragraphs this method will be applied specifically to the analysis of the beam for the ultimate load.

In a prestressed-concrete beam, failure can occur when either the prestressing steel or the concrete fails. In most cases, however, the amount of prestressing steel (or non-prestressed steel at the bottom) is large enough to prevent failure of steel before failure of concrete. Hence, practically, the failure of a prestressed-concrete beam occurs when the concrete at the compression zone fails. For estimating the ultimate load, or load corresponding to failure, it is assumed that failure occurs when the strain in the extreme compression fiber in concrete reaches a limiting value ϵ_u. The ultimate load, therefore, may be thought of as a load that by definition corresponds to a strain in concrete at the top fiber, ϵ_u.

The quantity ϵ_u does not necessarily correspond to the complete collapse of the beam. However, a beam with an extreme fiber strain of ϵ_u has deformed beyond usefulness.

The assumptions listed in Sec. 3-7 are all applicable in determination of ultimate moment. These assumptions are listed here again for convenience.

1. The strain distribution in concrete varies linearly with depth.
2. The strain in steel at ultimate can be expressed as follows:

$$\epsilon_{su} = \epsilon_{se} + F_1\epsilon_{ce} + \frac{\epsilon_u}{a}(d - a)F_2 \tag{8-1}$$

   The strain in non-prestressed reinforcement, if any, is equal to strain in concrete at the level of non-prestressed reinforcement.
3. The stress-strain diagrams for all materials are known.
4. Failure occurs when the strain in concrete at the top fiber reaches ϵ_u.
5. The average strain in steel is not greatly different from maximum strain.

In addition to the above assumptions, the tension contributed by concrete may be neglected. The calculations in Chap. 3 show that as the load reaches ultimate the tension contributed by concrete becomes comparatively small until at ultimate it becomes negligible.

On the basis of the above assumptions, a beam with a rectangular section may be analyzed by a simultaneous solution of the following equations:

$$\epsilon_{su} = \epsilon_{se} + F_1\epsilon_{ce} + \frac{\epsilon_u}{a}(d-a)F_2 \qquad (8\text{-}1)$$

$$\epsilon_{s1} = \frac{\epsilon_u}{a}(a - d_1) \qquad (8\text{-}2)$$

$$\epsilon_{s2} = \frac{\epsilon_u}{a}(d_2 - a) \qquad (8\text{-}3)$$

$$\frac{ab}{\epsilon_u}\int_0^{\epsilon_u} f(\epsilon)\, d\epsilon + A_{s1}f_{s1} = A_s f_{su} + A_{s2}f_{s2} \qquad (8\text{-}4)$$

$$M_u = \frac{a^2 b}{\epsilon_u{}^2}\int_0^{\epsilon_u} f(\epsilon)\epsilon\, d\epsilon + A_s f_{su}(d-a)$$
$$\qquad\qquad + A_{s1}f_{s1}(a-d_1) + A_{s2}f_{s2}(d_2-a) \quad (8\text{-}5)$$

$$f_{su} = \phi(\epsilon_{su}) \qquad (8\text{-}6)$$

$$f_{s1} = \phi_{s1}(\epsilon_{s1}) \qquad (8\text{-}7)$$

$$f_{s2} = \phi_{s2}(\epsilon_{s2}) \qquad (8\text{-}8)$$

All the quantities in the above equations were defined in Chap. 3; f_{s1} and f_{s2} are the stresses in the non-prestressed tensile and compressive reinforcements, respectively, at ultimate.

A simultaneous solution of the above equations will result in the unknowns, which in a given beam are ϵ_{su}, f_{su}, ϵ_{s1}, f_{s1}, ϵ_{s2}, f_{s2}, a, and M_u.

All the above equations are applicable to flanged sections if the neutral axis at failure falls in the flange. If the neutral axis falls outside the flange, Eqs. (8-4) and (8-5), the equations of equilibrium of horizontal forces and moments, should be changed to

$$\frac{b'a}{\epsilon_u}\int_0^{\epsilon_u(a-t)/a} f(\epsilon)\, d\epsilon + \frac{ba}{\epsilon_u}\int_{\epsilon_u(a-t)/a}^{\epsilon_u} f(\epsilon)\, d\epsilon$$
$$\qquad\qquad + A_{s1}f_{s1} = A_s f_{su} + A_{s2}f_{s2} \quad (8\text{-}4a)$$

and

$$M_u = \frac{b'a^2}{\epsilon_u{}^2}\int_0^{\epsilon_u(a-t)/a} f(\epsilon)\epsilon\, d\epsilon + \frac{b'a^2}{\epsilon_u{}^2}\int_{\epsilon_u(a-t)/a}^{\epsilon_u} f(\epsilon)\epsilon\, d\epsilon$$
$$\qquad + A_s f_s(d-a) + A_{s1}f_{s1}(a-d_1) + A_{s2}f_{s2}(d_2-a) \quad (8\text{-}5a)$$

The usefulness of the above equations is limited to the case in which the function $f(\epsilon)$ is defined. Ordinarily the above integrals are evaluated numerically.

8-3 ILLUSTRATIVE PROBLEM 8-1

Calculate the ultimate moment for the beam shown in Fig. 8-1. The stress-strain diagrams for steel and concrete are those shown in Figs. 3-16 and 3-17b, respectively. For this problem assume the following:

$$\epsilon_u = 0.003 \qquad \epsilon_{se} = 0.004$$
$$\epsilon_{ce} = 0.0006 \qquad F_1 = F_2 = 1.0$$

Assume that the neutral axis at failure is in the web, and the stress distribution in the compressed zone of concrete is as shown in Fig. 8-1. From Eq. (8-1) we have

$$\epsilon_{su} = 0.004 + 0.0006 + \frac{0.003}{a}(21 - a)$$

or

$$a = \frac{0.063}{\epsilon_{su} - 0.0016}$$

and from Eq. (8-4a) for equilibrium of horizontal forces, we have

$$6 \times 4 \times 20 + \left(\frac{a}{2} - 4\right) 6 \times 5 + \frac{1}{2}\frac{a}{2} \times 6 \times 5 = 2.93 f_{su}$$

or

$$a = 0.13 f_{su} - 16$$

Elimination of a in the above two equations results in the following:

$$f_{su} = \frac{0.484}{\epsilon_{su} - 0.0016} + 123$$

The point of intersection of the above equation with the stress-strain diagram for steel in Fig. 3-16 gives the following values for stress

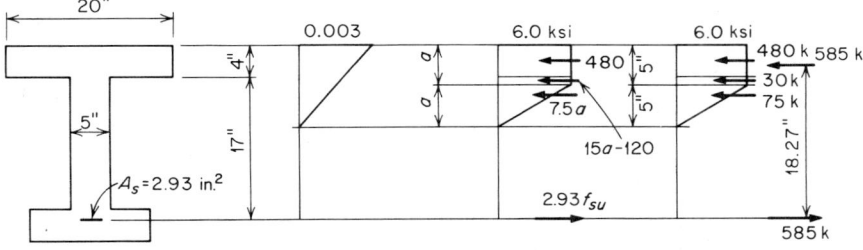

Fig. 8-1 Ultimate strength of a flanged section, Illustrative Problem 8-1.

and strain in steel at ultimate:

$f_{su} = 200$ ksi $\epsilon_{su} = 0.008$

$a = 10$ in.

From Fig. 8-1 we have

$M_u = 585 \times 18.27 = 10{,}690$ in.-kips

8-4 SIMPLIFIED METHODS FOR THE DETERMINATION OF ULTIMATE MOMENT

The procedure for the determination of ultimate moment discussed in the preceding sections is sufficient and is generally applicable to any section. However, since practically the actual stress-strain diagram in concrete in the beam is not known, many simplifications and approximations have been introduced in the determination of ultimate moment. The shape of the stress-strain diagram in concrete is the same as the shape of the stress distribution in the compression zone of a beam since the strain distribution is linear with depth. The area enclosed by the curve that defines the shape of the stress distribution in the compression zone is commonly referred to as the *stress block*. There are many methods for defining the shape of the stress block. A collection and discussion of simple shapes suggested by various engineers is available.[1]

A method which is commonly used for the definition of the shape of the stress block in the beam, and which is of some practical value, is discussed in the following paragraphs.

The simplified methods as a rule are based upon a rectangular section which has only prestressed reinforcement. For a section of this type the ultimate moment may be determined by a simultaneous solution of the following equations:

$$f_{su} = \phi_s(\epsilon_{su}) \tag{8-6}$$

$$\epsilon_{su} = \epsilon_{se} + F_1\epsilon_{ce} + \frac{\epsilon_u}{a}(d - a)F_2 \tag{8-1}$$

$$\frac{ab}{\epsilon_u} \int_0^{\epsilon_u} f(\epsilon)\, d\epsilon = A_s f_{su} \tag{8-4}$$

$$M_u = \frac{a^2 b}{\epsilon_u^2} \int_0^{\epsilon_u} f(\epsilon)\epsilon\, d\epsilon + A_s f_{su}(d - a) \tag{8-5}$$

Since the ultimate moment M_u does not appear in the first three equations, the quantities ϵ_{su}, f_{su}, and a can be determined by a simultaneous solution of these three equations. With these quantities known, M_u can be calculated from Eq. (8-5).

The practical difficulty of the above procedure, as mentioned above,

ULTIMATE DESIGN OF SUPPORTED PRESTRESSED-CONCRETE BEAMS

is that $f(\epsilon)$ is not definitely known for the concrete in the beam. The function $f(\epsilon)$ is the stress in concrete in terms of the strain in concrete. This relation is needed for the calculation of the area of the stress block. The area of the stress block is expressed in the following form:

$$\frac{a}{\epsilon_u} \int_0^{\epsilon_u} f(\epsilon) \, d\epsilon$$

The quantity $f(\epsilon)$ is also needed to determine the moment of the area of the stress block about any point in the section of the beam. The moment of the area of the stress block about the neutral axis is given in the following general form:

$$\frac{a^2}{\epsilon_u^2} \int_0^{\epsilon_u} f(\epsilon) \epsilon \, d\epsilon$$

It can be seen that knowledge of the function $f(\epsilon)$ is essential for the determination of the area of the stress block and its moment about any point in the section.

A method which avoids the function $f(\epsilon)$ and still gives the area of the stress block and its centroid has been developed for rectangular sections through the last several years. The method is referred to as the Stuessi type method after its originator.[2] Many refinements have been developed since the original method first appeared.

In this method the area of the stress block is defined as

$$\frac{a}{\epsilon_u} \int_0^{\epsilon_u} f(\epsilon) \, d\epsilon = k_1 k_3 f_c' a$$

where k_1 = ratio of average to maximum stress in stress block
k_3 = ratio of strength of concrete in beam to that of cylinder
f_c' = cylinder strength of concrete

With the above area of the stress block, the equation of equilibrium of horizontal forces, considering the proper signs, can be expressed as

$$abk_1 k_3 f_c' = A_s f_{su}$$

By introducing $p = A_s/bd$ as the percentage of steel and a as $k_u d$, the above expression can be written as

$$k_u = \frac{p f_{su}}{k_1 k_3 f_c'} \tag{8-9}$$

Equation (8-9) can be used to determine $k_1 k_3$ experimentally, since, for a given beam, p is known and the quantities f_{su}, k_u, and f_c' can be measured. Results of many tests indicate that $k_1 k_3$ has a tendency to decrease with f_c', and various relations have been developed between $k_1 k_3$

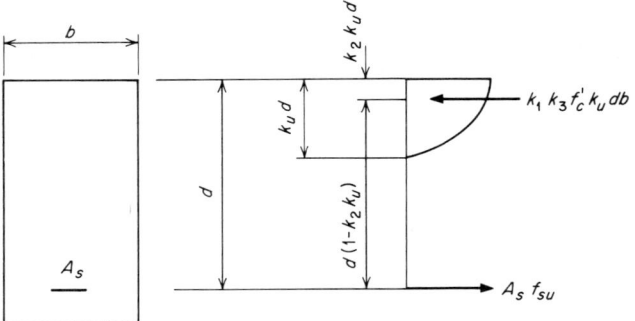

Fig. 8-2 Definition of stress distribution by coefficients k_1, k_2, k_3.

and f'_c.[1,2] The average value of k_1k_3 for $f'_c = 5.0$ ksi is in the neighborhood of 0.7.

The position of centroid in this method is defined by a third ratio k_2. It is assumed that the distance of the centroid of the stress block from the top fiber is k_2k_ud.

Since k_2 is used for the determination of moment and the variation of k_2 has a small effect on moment, an average value may be assigned to k_2. For a rectangular stress block $k_2 = \frac{1}{2}$, and for a triangular stress block $k_2 = \frac{1}{3}$. Since the actual shape of the stress block is between a rectangle and a triangle, k_2 may be taken as the average of $\frac{1}{2}$ and $\frac{1}{3}$, or 0.42. A more accurate evaluation of k_2 indicates that it is in the same neighborhood.[3]

Figure 8-2 shows the stress block as defined by this method.

The equation of equilibrium of horizontal forces, using the above approach and considering the proper signs, can be written as follows:

$$k_1k_3f'_ck_u\,db = A_sf_{su} \tag{8-10}$$

The equation of equilibrium of moments by the above method will be

$$M_u = A_sf_{su}\,d(1 - k_2k_u) \tag{8-11}$$

As before for the determination of M_u, Eqs. (8-6), (8-12), (8-10), and (8-11) can be solved simultaneously:

$$f_{su} = \phi(\epsilon_{su}) \tag{8-6}$$

$$\epsilon_{su} = \epsilon_{se} + F_1\epsilon_{ce} + \frac{\epsilon_u}{k_u}(1 - k_u)F_2 \tag{8-12}$$

$$k_1k_3f'_ck_u\,db = A_sf_{su} \tag{8-10}$$

$$M_u = A_sf_{su}\,d(1 - k_2k_u) \tag{8-11}$$

ULTIMATE DESIGN OF SUPPORTED PRESTRESSED-CONCRETE BEAMS

Equation (8-12) is obtained by substituting $a = k_u d$ in Eq. (8-1). In the above equations all quantities are known or presumed known except ϵ_{su}, f_{su}, k_u, and M_u.

In the simultaneous solution of the above equations, k_u can be eliminated between Eqs. (8-12) and (8-10), resulting in the following expression:

$$f_{su} = \frac{k_1 k_3 f'_c F_2 \epsilon_u}{p(\epsilon_{su} - \epsilon_{se} - F_1 \epsilon_{ce} + F_2 \epsilon_u)} \tag{8-12a}$$

and

$$f_{su} = \phi(\epsilon_{su}) \tag{8-6}$$

In the above two equations, the unknowns are ϵ_{su} and f_{su}. As before, these quantities can be obtained by plotting Eq. (8-12a) on the stress-strain diagram for steel and obtaining the coordinates of the point of intersection. The quantity k_u can be obtained from Eq. (8-10) and M_u from Eq. (8-11).

The ultimate moment for the beam in Illustrative Problem 3-1 can be calculated by the above method. All the values are the same except the stress-strain diagram for concrete—or the shape of the stress block—which is now defined by $k_1 k_3$ and k_2. For the purposes of this problem let us assume the following:

$$k_1 k_3 = 0.70 \qquad k_2 = 0.42$$

In order to have comparable average stresses, the quantity f'_c for this problem, on the basis of the actual stress-strain diagram for concrete, was 4.50 ksi with $\epsilon_u = 0.003$. The following quantities were given or computed in Illustrative Problem 3-1:

$$F_1 = F_2 = 1.0 \qquad \epsilon_{se} = 0.00393$$

$$\epsilon_{ce} = 0.000108 \qquad p = \frac{A_s}{bd} = 0.00282$$

Substitution of the above quantities in Eq. (8-12a) results in

$$f_{su} = \frac{4.77}{\epsilon_{su} - 0.00104}$$

By plotting the above equation on the stress-strain diagram for steel in Fig. 3-16, the coordinates of the point of intersection can be obtained as follows:

$$\epsilon_{su} = 0.0216 \qquad f_{su} = 235 \text{ ksi}$$

and from Eq. (8-9)
$$k_u = \frac{pf_{su}}{k_1 k_3 f'_c} = \frac{0.00282 \times 235}{0.7 \times 6.4} = 0.148$$
or
$$a = k_u d = 1.34 \text{ in.}$$
and
$$M_u = A_s f_{su} d(1 - k_2 k_u) = 0.156 \times 235 \times 9.05(1 - 0.062)$$
$$= 311 \text{ in.-kips}$$

Evidently this result agrees closely with that obtained in Illustrative Problem 3-1 by idealizing the stress-strain diagram for concrete.

It can be seen that the above method for defining the shape of the stress-strain diagram is very useful in the determination of ultimate moment for rectangular sections or beams of constant width. However, for flanged sections or beams of variable width this method is unsuitable.

8-5 ILLUSTRATIVE PROBLEM 8-2

The influence of non-prestressed reinforcement at top and bottom on ultimate curvature and moment can be shown by the following example.

Given is a prestressed-concrete beam of rectangular section with an overall depth of 12 in. and a width of 6 in. The prestressing steel is to be placed 2 in. above the bottom of the section. Hence, the effective depth is to be 10 in.

(a) Determine the area of prestressing steel required so that the stress in steel at failure f_{su} will be 217 ksi.

(b) Using the area of prestressing steel determined in (a), compute the amount of non-prestressed top reinforcement needed to make the curvature at failure $\varphi_u = 160 \times 10^{-5}$ in.$^{-1}$.

(c) Determine the area of the prestressing steel required so that f_{su}, the stress in steel at failure, will be 248 ksi.

(d) Using the area of prestressing steel determined in (c), compute the amount of non-prestressed bottom reinforcement needed to reduce f_{su}, the stress in the prestressing steel, to 217 ksi.

For this problem let us assume

$f'_c = 4.0$ ksi $\epsilon_u = 0.004$
$f_{se} = 120$ ksi $\epsilon_{ce} = 0$
$d_1 = 2$ in. $d_2 = d = 10$ in.
$k_1 k_3 = 0.70$

The stress-strain diagram for steel may be assumed the same as that in Fig. 3-16. The non-prestressed reinforcement is intermediate-grade

ULTIMATE DESIGN OF SUPPORTED PRESTRESSED-CONCRETE BEAMS

steel with a yield-point stress of 50 ksi and a modulus of elasticity of 30,000 ksi. All the compatibility factors may be assumed as unity.

(a) From Fig. 3-16 it can be seen that stress of 217 ksi in prestressed steel corresponds to a strain of 0.01. The quantity k_u may be calculated from Eq. (8-12a).

$$k_u = \frac{\epsilon_u}{\epsilon_{su} - \epsilon_{se} + \epsilon_u} = \frac{0.004}{0.010 - 0.004 + 0.004} = 0.4$$

From equilibrium of horizontal forces, or Eq. (8-10), we have

$$k_1 k_3 f'_c k_u \, db = A_s f_{su}$$

or

$$A_s = \frac{k_1 k_3 f'_c k_u bd}{f_{su}} = \frac{(0.7)(4.0)(0.4)(6)(10)}{217} = 0.31 \text{ in.}^2$$

(b) In order to have a curvature of 160×10^{-5} in.$^{-1}$, $k_u d$ must be equal to the following:

$$k_u d = \frac{0.004}{0.0016} = 2.5 \text{ in.}$$

Hence, the stress in the non-prestressed steel in the top will be

$$f_{s1} = \frac{0.004}{2.5}(0.5)(30{,}000) = 24 \text{ ksi}$$

and the strain in the prestressed steel will be

$$\epsilon_{su} = \frac{0.004}{2.5}(7.5) + 0.004 = 0.016$$

From the stress-strain diagram for prestressed steel we have $f_{su} = 226$ ksi.
From the equilibrium of horizontal forces in the section we have

$$24 A_{s1} + 0.7 \times 4.0 \times 2.5 \times 6 = 226 \times 0.310$$

or

$$A_{s1} = 1.17 \text{ in.}^2$$

(c) From the stress-strain diagram for steel when $f_{su} = 248$ ksi, $\epsilon_{su} = 0.056$; hence, k_u can be calculated as follows:

$$k_u = \frac{0.004}{0.056 - 0.004 + 0.004} = 0.0715$$

and

$$A_s = \frac{k_1 k_3 f'_c k_u bd}{f_{su}} = \frac{0.7 \times 4.0 \times 0.0715 \times 6 \times 10}{248} = 0.049 \text{ in.}^2$$

(d) In this case we have

$$A_{s2} = \frac{0.7 \times 4.0 \times 0.4 \times 6 \times 10 - 217 \times 0.049}{50} = 1.13 \text{ in.}^2$$

The problem can also be solved for a given stress-strain diagram for concrete.

In this problem ϵ_{ce} was assumed to be negligible. The quantity $F_1\epsilon_{ce}$ is small in comparison with ϵ_{se} and ϵ_u, and sometimes it is neglected. Often it is assumed to be equal to ϵ_{ce}.

8-6 APPROXIMATE METHODS FOR THE DETERMINATION OF ULTIMATE MOMENT IN FLANGED SECTIONS

The simplified methods discussed in the preceding sections have been applied to flanged sections. However, the simplifications are not suitable for flanged sections since they result in some inaccuracies.

When $t > k_u d$, the neutral axis at ultimate falls in the flange, and the analysis of the beam is identical with that for rectangular sections. This condition usually occurs when the amount of prestressing steel is small. Figure 8-3 shows a flanged section in which the neutral axis at ultimate falls in the flange.

When $t < k_u d$, the neutral axis at ultimate falls outside the flange, and the equations of equilibrium of horizontal forces and moments should be modified. Figure 8-4 shows a flanged section in which the neutral axis falls in the web.

The equilibrium of horizontal forces can be expressed as

$$k_1 k_3 f'_c k_u \, db' + c_1 f'_c (b - b')t = A_s f_{su} \tag{8-13}$$

The first term of the left side is the force contributed by the web, and the second term is the force contributed by the flange, as shown in Fig. 8-4. The factor c_1 is similar to $k_1 k_3$ for the part of the stress block that develops

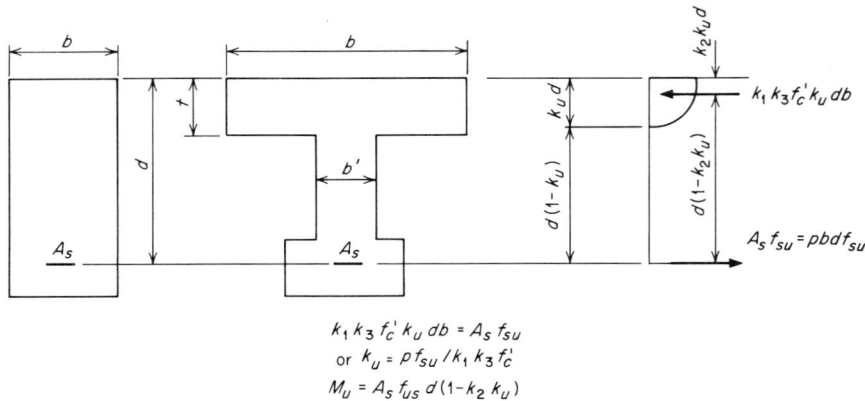

Fig. 8-3 Flanged section in which the neutral axis at ultimate falls in the flange.

ULTIMATE DESIGN OF SUPPORTED PRESTRESSED-CONCRETE BEAMS

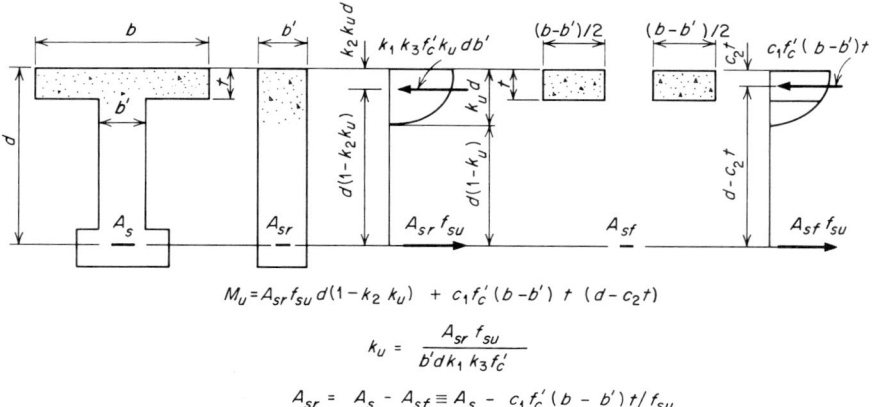

Fig. 8-4 Flanged section in which the neutral axis at ultimate falls outside the flange.

the flange. The quantity c_1 may vary between 0.70 and 1.0 if k_1k_3 is taken as 0.70. When the neutral axis at failure is only slightly below the flange, c_1 is very near 0.70.

The equilibrium of moments results in the following equation:

$$M_u = k_1 k_3 f'_c k_u b' d^2 (1 - k_2 k_u) + c_1 f'_c (b - b') t (d - c_2 t) \qquad (8\text{-}14)$$

where $c_2 t$ is the distance of the centroid, of the part of stress block that develops the flange, to the top fiber.

By introducing A_{sr} as that part of the area of steel which develops the web and A_{sf} as the part of the area of steel that develops the flange, the above expression for moment can also be written

$$M_u = A_{sr} f_{su} d (1 - k_2 k_u) + c_1 f'_c (b - b') t (d - c_2 t) \qquad (8\text{-}15)$$

where

$$A_{sr} = A_s - A_{sf} = A_s - \frac{c_1 f'_c (b - b') t}{f_{su}}$$

As before for the determination of M_u, Eqs. (8-6), (8-12), (8-13), and (8-14) should be solved simultaneously.

$$f_{su} = \phi_s(\epsilon_{su}) \qquad (8\text{-}6)$$

$$\epsilon_{su} = \epsilon_{se} + F_1 \epsilon_{ce} + \frac{\epsilon_u}{k_u}(1 - k_u)F_2 \qquad (8\text{-}12)$$

$$k_1 k_3 f'_c k_u \, db' + c_1 f'_c (b - b') t = A_s f_{su} \qquad (8\text{-}13)$$

$$M_u = k_1 k_3 f'_c k_u b' d^2 (1 - k_2 k_u) + c_1 f'_c (b - b') t (d - c_2 t) \qquad (8\text{-}14)$$

In the above equations all quantities are known or presumed known except ϵ_{su}, f_{su}, k_u, and M_u.

In the simultaneous solution of the above equations, k_u can be eliminated between Eqs. (8-12) and (8-13), resulting in the following expression:

$$f_{su} = \frac{k_1 k_3 f'_c F_2 \epsilon_u (b'/b)}{p(\epsilon_{su} - \epsilon_{se} - F_1 \epsilon_{ce} + F_2 \epsilon_u)} + c_1 \frac{f'_c}{p}\left(1 - \frac{b'}{b}\right)\frac{t}{d} \qquad (8\text{-}16)$$

Coordinates of the point of intersection of the above equation with the stress-strain diagram for steel are f_{su} and ϵ_{su}.

The quantities k_u and M_u can be calculated from Eqs. (8-13) and (8-14).

The ultimate moment for the beam in Illustrative Problem 8-1 can be calculated by this method. Let us assume $k_1 k_3 = 0.7$, $k_2 = 0.42$, $c_1 = 0.85$, $c_2 = 0.50$, and $f'_c = 6.4$ ksi. Substituting $b'/b = 0.25$, $t/d = 0.19$, $p = 0.00697$ in Eq. (8-16), we have

$$f_{su} = \frac{0.481}{\epsilon_{su} - 0.0016} + 111.2$$

A plot of the above equation on the stress-strain diagram for steel in Fig. 3-16 gives the following coordinates for the point of intersection:

$f_{su} = 198$ ksi

$\epsilon_{su} = 0.0071$

and from Eq. (8-13) $k_u = 0.540$ or $k_u d = 11.35$ in.

Ultimate moment can be calculated from Eq. (8-14):

$M_u = 10{,}400$ in.-kips

which is 2.5 percent smaller than the result of Illustrative Problem 8-1.

It can be seen that although c_1 and c_2 are chosen arbitrarily, the results are comparable. The values of c_1 and c_2 adopted here were reasonable, because the neutral axis at ultimate was 11.35 in. from the top fiber. Evidently the above values of c_1 and c_2 would result in a higher estimate of ultimate moment if neutral axis at failure were only slightly below the flange.

8-7 PROVISIONS OF ACI CODE 318-63 FOR THE ULTIMATE FLEXURAL STRENGTH[4,5]

The provisions of the ACI code on ultimate flexural strength are given in two parts. The first part contains the method of determination of ultimate flexural strength, and the second part contains limitations on steel percentage.

ULTIMATE DESIGN OF SUPPORTED PRESTRESSED-CONCRETE BEAMS

It should be emphasized that these provisions are given for design purposes, and their accuracy should be checked with consideration of safety factors recommended in the code.

The text of these two articles of the code is presented below with a slight modification in format.

Ultimate flexural strength* The required ultimate load on a member, determined in accordance with part IV-B of the code, shall not exceed the ultimate flexural strength computed by:

Rectangular sections, or flanged sections in which the neutral axis lies within the flange:†

$$M_u = \phi A_s f_{su} d(1 - 0.59q) \tag{8-17}$$

Flanged sections in which the neutral axis falls outside the flange:‡

$$M_u = \phi A_{sr} f_{su} d \left(1 - \frac{0.59 A_{sr} f_{su}}{b' d f'_c}\right) + 0.85 f'_c (b - b')t(d - 0.5t) \tag{8-18}$$

where

$$A_{sr} = A_s - A_{sf}$$

and

$$A_{sf} = \frac{0.85 f'_c (b - b')t}{f_{su}}$$

Where information for the determination of f_{su} is not available, and provided that f_{se} is not less than $0.5f'_s$, the following approximate values shall be used:

Bonded members:

$$f_{su} = f'_s \left(1 - \frac{0.5 p f'_s}{f'_c}\right) \tag{8-19}$$

Unbonded members:

$$f_{su} = f_{se} + 15{,}000 \tag{8-20}$$

Non-prestressed tensile reinforcement in combination with prestressed steel may be considered to contribute to the tension force in a member at ultimate moment an amount equal to its area times its yield

---

* For flexure ϕ is given as 0.90. The required ultimate moment for sections in which wind and earthquake may be neglected is given as $M_u = 1.5 M_d + 1.8 M_l$, where M_d is the total dead-load moment and M_l is the live-load moment.
† Usually where the flange thickness is more than $1.4 d p f_{su}/f'_c$.
‡ Usually where the flange thickness is less than $1.4 d p f_{su}/f'_c$.

point, provided

$$\frac{pf_{su}}{f'_c} + \frac{p''f_y}{f'_c}$$

does not exceed 0.3.

Limitations on steel percentage

(a) Except as provided in (b), the ratio of prestressing steel used for calculation of M_u shall be such that

$$\frac{pf_{su}}{f'_c}$$

is not more than 0.3.

For flanged sections, p shall be taken as the steel ratio of only that portion of the total tension steel area which is required to develop the compressive strength of the web alone.

(b) When the steel ratio in excess of that specified in (a) is used, the ultimate moment shall be taken as not greater than the following:

Rectangular sections, or flanged sections in which the neutral axis lies within the flange:

$$M_u = \phi(0.25f'_c bd^2) \tag{8-21}$$

Flanged section in which the neutral axis falls outside the flange:

$$M_u = \phi[0.25f'_c b'd^2 + 0.85f'_c(b - b')t(d - 0.5t)] \tag{8-22}$$

(c) The total amount of prestressed and unprestressed reinforcement shall be adequate to develop an ultimate load in flexure at least 1.2 times the cracking load calculated on the basis of a modulus of rupture of $7.5\sqrt{f'_c}$.

In the equations listed above,

ϕ = capacity reduction factor, taken as 0.90

$$q = \frac{pf_{su}}{f'_c}$$

p'' = ratio of unprestressed steel, taken as A_{s2}/bd

It can be seen that the ACI code does not include the effect of non-prestressed reinforcement in calculating the ultimate moment, although its use is permitted.

Equations (8-17) and (8-18) are similar to Eqs. (8-11) and (8-15), respectively. Equation (8-17) may be obtained by substituting $k_u = pf_{su}/k_1k_3f'_c$ and $k_2/k_1k_3 = 0.59$ in Eq. (8-11). Equation (8-18)

ULTIMATE DESIGN OF SUPPORTED PRESTRESSED-CONCRETE BEAMS

may be obtained from Eq. (8-15) by making the following substitutions:

$$k_u = \frac{A_{sr}f_{su}}{b'\,dk_1k_3f'_c} \qquad \frac{k_2}{k_1k_3} = 0.59$$

$$c_1 = 0.85 \qquad\qquad c_2 = 0.50$$

The above set values for c_1 and c_2 result in an overestimate of ultimate moment when the neutral axis at ultimate lies just below the flange.

When the stress-strain diagram for steel is known, the determination of ultimate moment is the same as discussed in Secs. 3-7 and 8-2. When the stress-strain diagram for steel is not known, the code gives Eq. (8-19) for f_{su}, which is intended for bonded beams of rectangular and flanged sections in which the neutral axis falls in the web.

Equation (8-19) gives the stress in steel for bonded rectangular beams approximately but with sufficient accuracy for the present methods and materials used in practice. However, it can be shown that Eq. (8-19) does not apply to the flanged sections, and that it gives higher stresses in steel than actually exist.

Equation (8-20), the approximate expression for f_{su} for unbonded beams, is very much on the safe side; that is, in most cases the actual f_{su} is greater than that obtained by this equation. As discussed in Sec. 3-4, it is difficult to estimate F_2 in an actual unbonded beam; hence an accurate analysis of an unbonded beam is not possible. The small value of f_{su} given by Eq. (8-20) is intended to make the safety factor of an unbonded beam higher than that of an equivalent bonded beam.

According to the code, Eq. (8-17) is applicable when pf_{su}/f'_c is less than 0.3.

If we assume $\epsilon_u \approx 0.0034$, $\epsilon_{se} + \epsilon_{ce} \approx 0.0045$, and $k_1k_3 = 0.7$, this limit is based on $\epsilon_{su} \approx 0.01$. That is, the inequality $pf_{su}/f'_c < 0.3$ is equivalent to $\epsilon_{su} > 0.01$.

From Eq. (8-12), if $\epsilon_{su} > 0.01$, we have

$$\frac{pf_{su}}{f'_c} < \frac{k_1k_3\epsilon_u}{\epsilon_{su} - \epsilon_{se} - \epsilon_{ce} + \epsilon_u} = 0.30$$

Evidently this limit is intended for bonded beams. If this limit is correct for bonded beams, for unbonded beams it should be smaller.

When $pf_{su}/f'_c > 0.30$, the code requires that pf_{su}/f'_c be taken as 0.3. Substituting $pf_{su}/f'_c = 0.3$ in Eq. (8-17), we have

$$M_u = 0.25 f'_c bd^2 \qquad (8\text{-}21)$$

In flanged sections in which the neutral axis at failure falls outside the flange, Eq. (8-18) is applicable when $A_{sr}f_{su}/b'df'_c \leq 0.30$.

When $A_{sr}f_{su}/b'df'_c > 0.30$, the code requires that $A_{sr}f_{su}/b'df'_c$ be taken as 0.3. Substituting $A_{sr}f_{su}/b'df'_c = 0.3$ in Eq. (8-18), we have

$$M_u = 0.25b'd^2 f'_c + 0.85f'_c(b - b')t(d - 0.5t) \tag{8-22}$$

The limitation that $pf_{su}/f'_c + p''f_y/f'_c < 0.30$ is established in a similar manner.

8-8 ANALYSIS OF THE BEAM FOR ULTIMATE DESIGN[6]

In the analysis of prestressed-concrete beams for ultimate design it will be assumed that the section is flanged, the prestressed steel is to be bonded to concrete, and the only non-prestressed reinforcement present is compression steel. Practical sections sometimes contain non-prestressed tensile reinforcement, which increases the flexural strength of the section and reduces its ductility. Non-prestressed tensile reinforcement is not considered here since it does not contribute efficiently to the strength of the beam.

Analysis of the beam for the ultimate design is based upon the following equations:

$$\epsilon_{su} = \epsilon_{se} + F_1 \epsilon_{ce} + \frac{\epsilon_u}{a}(d - a)F_2 \tag{8-1}$$

$$\epsilon_{s1} = \frac{\epsilon_u}{a}(a - d_1) \tag{8-2}$$

$$\frac{ab}{\epsilon_u} \int_0^{\epsilon_u} f(\epsilon)\, d\epsilon + A_{s1} f_{s1} = A_s f_{su} \tag{8-4b}$$

$$M_u = \frac{a^2 b}{\epsilon_u^2} \int_0^{\epsilon_u} f(\epsilon)\epsilon\, d\epsilon + A_{s1} f_{s1}(a - d_1) + A_s f_{su}(d - a) \tag{8-5b}$$

$$f_{su} = \phi(\epsilon_{su}) \tag{8-6}$$

$$f_{s1} = \phi_{s1}(\epsilon_{s1}) \tag{8-7}$$

Equations (8-4b) and (8-5b) are applicable when the neutral axis falls in the flange. If the neutral axis falls below the flange, Eqs. (8-4b) and (8-5b) should be replaced by the following:

$$\frac{b'a}{\epsilon_u} \int_0^{\epsilon_u(a-t)/a} f(\epsilon)\, d\epsilon + \frac{ba}{\epsilon_u} \int_{\epsilon_u(a-t)/a}^{\epsilon_u} f(\epsilon)\, d\epsilon + A_{s1} f_{s1} = A_s f_{su} \tag{8-4c}$$

and

$$M_u = \frac{b'a^2}{\epsilon_u{}^2} \int_0^{\epsilon_u(a-t)/a} f(\epsilon)\epsilon\, d\epsilon + \frac{b'a^2}{\epsilon_u{}^2} \int_{\epsilon_u(a-t)/a}^{\epsilon_u} f(\epsilon)\epsilon\, d\epsilon$$
$$+ A_s f_s(d-a) + A_{s1} f_{s1}(d-d_1) \quad (8\text{-}5c)$$

For convenience in design, Eqs. (8-4b) and (8-5b) will be expressed in dimensionless form as follows:

$$\frac{a/d}{\epsilon_u f'_c} \int_0^{\epsilon_u} f(\epsilon)\, d\epsilon + p'\frac{f_{s1}}{f'_c} = p\frac{f_{su}}{f'_c} \quad (8\text{-}23)$$

$$Q = \frac{M_u}{bd^2 f'_c} = \frac{(a/d)^2}{f'_c \epsilon_u{}^2} \int_0^{\epsilon_u} \epsilon f(\epsilon)\, d\epsilon + p\frac{f_{su}}{f'_c}\left(1 - \frac{a}{d}\right) + p'\frac{f_{s1}}{f'_c}\left(\frac{a}{d} - \frac{d'}{d}\right) \quad (8\text{-}24)$$

where $p = A_s/bd$
$p' = A_{s1}/bd$
f'_c = cylinder strength of concrete at 28 days

Similarly, Eqs. (8-4c) and (8-5c) may be expressed in dimensionless form:

$$\frac{(b'/b)(a/d)}{\epsilon_u f'_c} \int_0^{\epsilon_u} f(\epsilon)\, d\epsilon + \frac{(1 - b'/b)(a/d)}{\epsilon_u f'_c} \int_{\epsilon_u(a-t)/a}^{\epsilon_u} f(\epsilon)\, d\epsilon$$
$$+ p'\frac{f_{s1}}{f'_c} = p\frac{f_{su}}{f'_c} \quad (8\text{-}25)$$

$$Q = \frac{M_u}{bd^2 f'_c} = \frac{(b'/b)(a/d)^2}{f'_c \epsilon_u{}^2} \int_0^{\epsilon_u} \epsilon f(\epsilon)\, d\epsilon$$
$$+ \frac{(1 - b'/b)(a/d)^2}{f'_c \epsilon_u{}^2} \int_{\epsilon_u(a-t)/a}^{\epsilon_u} \epsilon f(\epsilon)\, d\epsilon$$
$$+ p\frac{f_{su}}{f'_c}\left(1 - \frac{a}{d}\right) + p'\frac{f_{s1}}{f'_c}\left(\frac{a}{d} - \frac{d'}{d}\right) \quad (8\text{-}26)$$

8-9 ULTIMATE DESIGN[5,6]

Ultimate design of a prestressed-concrete beam is based upon the ultimate moment and ductility of the section. The section is proportioned in such a way that the ultimate moment is greater than the moment developed under service loads by a prescribed quantity, and that it deforms a certain amount before it fails.

These concepts may be stated in the form of the following requirements:

$$M_u \geq N_d(M_g + M_s) + N_l M_l \quad (8\text{-}27)$$

and

$$\epsilon_{su} \geq \epsilon_{sl} \quad (8\text{-}28)$$

where M_u = flexural strength of beam
N_d = load factor for dead load
M_g = moment due to weight of beam
M_s = moment due to superimposed dead load
N_l = load factor for live load
M_l = moment due to live load
ϵ_{su} = strain in steel at ultimate
ϵ_{sl} = limiting strain in steel

Expression (8-27) states that the required flexural strength of the beam should be at least equal to $N_d(M_g + M_s) + N_l M_l$, which is a requirement for the strength of the beam.

Expression (8-28) states that the ductility of the beam should be large enough that the strain in steel at ultimate is at least equal to a given limiting value designated as ϵ_{sl}. Ductility is usually measured by the curvature at ultimate, which may be defined as follows:

$$\varphi = \frac{\epsilon_u}{a} = \frac{\epsilon_{su} - \epsilon_{se} - \epsilon_{ce} + \epsilon_u}{d}$$

where φ is the curvature of the section. For given values of ϵ_{se}, ϵ_{ce}, ϵ_u, and d, ϵ_{su} may be used as a measure of ductility.

Determination of the area of the beam Expression (8-27) can be written as an equation in the following form:

$$Qbd^2 f'_c = N_d(M_g + M_s) + N_l M_l$$

Substituting $A/h\psi$ for b, where A is the gross cross-sectional area of the beam, h is the overall depth, and ψ is a dimensionless shape factor,

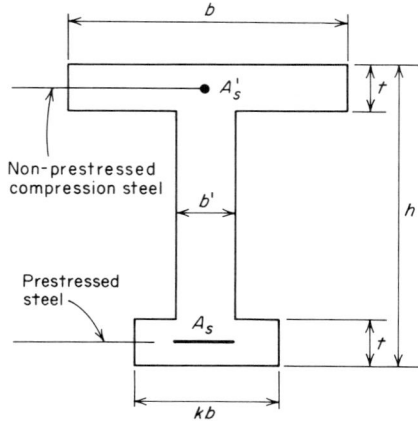

Fig. 8-5 Idealized I section.

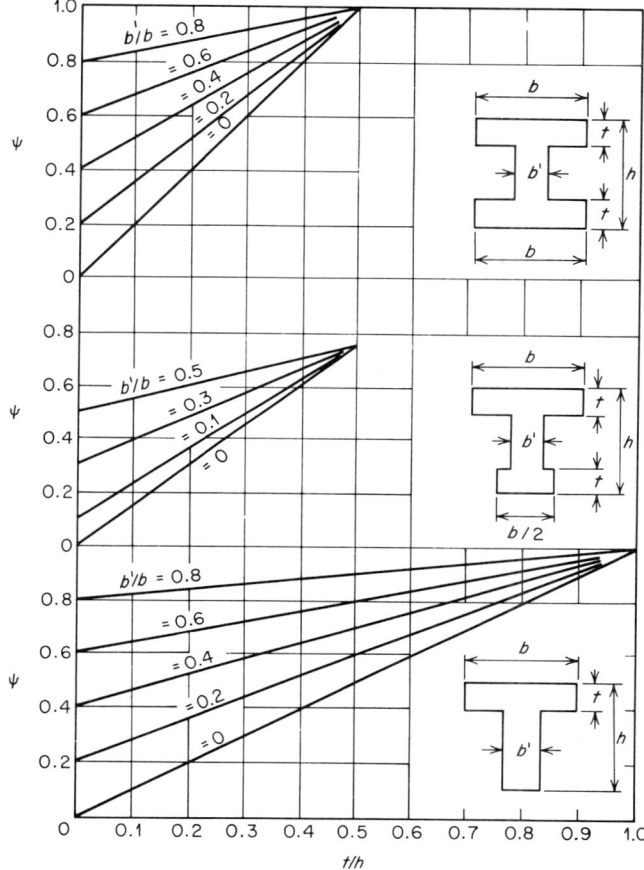

Fig. 8-6 Relationship between ψ and geometric parameters of the section.

we obtain the following:

$$M_g = \frac{\gamma A L^2}{8} = Q \frac{A}{h\psi} d^2 f_c' \frac{1}{N_d} - \frac{N_l}{N_d} M_l - M_s$$

and

$$A = \frac{M_s + (N_l/N_d) M_l}{d^2 f_c' Q/h\psi N_d - \gamma L^2/8} \tag{8-29}$$

where γ is the unit weight of concrete.

For the idealized I section shown in Fig. 8-5, ψ is given by the following expression:

$$\psi = \frac{t}{h}(1+k) + \frac{b'}{b}\left(1 - 2\frac{t}{h}\right) \tag{8-30}$$

The quantity k in the above equation is the ratio of the width of the bottom flange to that of the top flange. Equation (8-30) is plotted in Fig. 8-6 for typical sections.

A study of Eq. (8-29) indicates that for a given design problem, in which the depth and type of concrete are specified, A depends upon ψ and Q only. It can be seen that A decreases with Q and increases with ψ. That is, to decrease the area of the beam it is necessary to increase Q and decrease ψ, or to increase the ratio Q/ψ.

The quantities t/d and b'/b usually decrease with increasing Q/ψ; hence they should be made small, without causing the dimensions of the beam to become unreasonably thin.

From Eq. (8-30) it can be seen that ψ increases and Q/ψ decreases with k. Therefore, a small bottom flange is desirable. However, since the bottom flange of the beam should be large enough to permit the placing of steel, k cannot be reduced indefinitely.

From Eqs. (8-24) and (8-26) it can be seen that Q increases with a/d and hence it is desirable to make a/d as large as possible; however, expression (8-28) for the required ductility sets the upper limit for a/d. Since expression (8-28) sets the required minimum ductility of the beam at a strain in steel equal to ϵ_{sl}, the required maximum a/d consistent with the required ductility can be computed from Eq. (8-1) as follows:

$$\left(\frac{a}{d}\right)_{\max} = \frac{\epsilon_u}{\epsilon_{sl} - \epsilon_{se} - \epsilon_{ce} + \epsilon_u} \tag{8-1a}$$

Equation (8-1a) contains the quantity ϵ_{se}, the strain in steel due to effective prestress. It can be seen that since ϵ_{se} increases with the maximum value of a/d, it should be taken as large as practicable. The practical upper limit for ϵ_{se} for the materials used in pretensioned construction is about 0.005.

It should be pointed out that d/h also influences A, the area of the beam, and from Eq. (8-29) it can be seen that A decreases with d/h. In most practical problems, however, d/h cannot exceed 0.9.

Design procedure In the design method presented here it is assumed that the span length, the acting load, the load factors, the strength, and the unit weight of concrete are given. It is further assumed that the limiting strain in concrete, ϵ_u, the requirement of ductility ϵ_{sl}, and the effective prestrain ϵ_{se}, as well as the stress-strain relations for all materials, are given. Hence, for a selected value of h the calculation of A from Eq. (8-11) means determination of d^2Q/ψ. The quantities d, ψ, and Q may be determined as follows:

ULTIMATE DESIGN OF SUPPORTED PRESTRESSED-CONCRETE BEAMS

1. Assign a reasonable value to d as close to h as the arrangement of strands permits.
2. Assign values to b'/b, t/h, and k, and calculate ψ from Eq. (8-30). These values should be as small as possible.
3. Calculate a/d from Eq. (8-1a), based upon the given values of ϵ_u, ϵ_{se}, and ϵ_{sl}.
4. Calculate $p\,f_{su}/f'_c$ from either Eq. (8-23) or (8-25), whichever applies.
5. Calculate Q from Eq. (8-24) or (8-26), whichever applies.

8-10 ILLUSTRATIVE EXAMPLE 8-3

The following example is presented to illustrate the procedure for the ultimate design of a prestressed-concrete beam and to show the influence of the required ductility on the dimensions of the beam so designed. This problem was solved in Chap. 7 as Illustrative Problem 7-2.

Given is a simply supported beam of 54 ft span subjected to a superimposed dead load of 1.0 klf and a live load of 0.6 klf which produce midspan moments of $M_s = 4370$ in.-kips and $M_l = 2630$ in.-kips, respectively. The load factors are given as $N_d = 1.5$ and $N_l = 1.8$. Design the section (1) for a minimum ductility corresponding to $\epsilon_{sl} = 0.01$ and (2) for a minimum ductility corresponding to $\epsilon_{sl} = 0.02$. Non-prestressed compression steel is not to be used.

The effective prestress or the prestress after losses is given as 145 ksi, which corresponds to a transfer prestress of 170 ksi. The strain due to effective prestress is $\epsilon_{se} = 0.0048$. The quantity ϵ_{ce} is approximated as 0.0005 initially, which may be verified after the section is designed. The limiting strain in concrete is given as $\epsilon_u = 0.003$, unit weight of concrete as $\gamma = 0.15$ kcf, and overall depth as $h = 36$ in. The strength of concrete f'_c is specified as 5 ksi. The stress-strain diagrams for concrete and steel are as shown in Figs. 8-7 and 8-8, respectively.

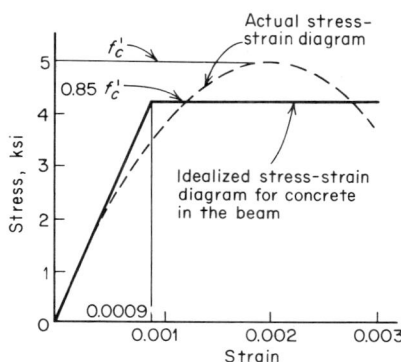

Fig. 8-7 Stress-strain diagram for concrete.

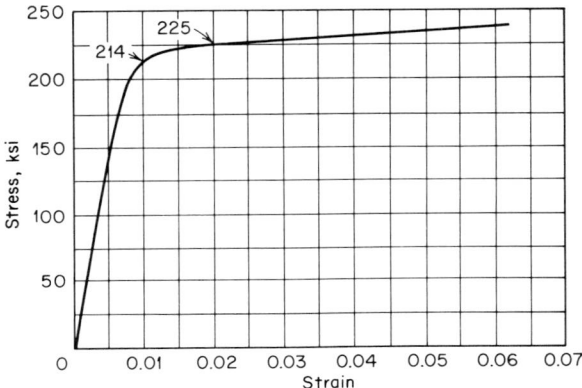

Fig. 8-8 Stress-strain diagram for steel.

a. Section with minimum required ductility corresponding to $\epsilon_{su} = 0.01$

It was shown before that the quantities t/h, b'/b, and k increase with A; hence they should be taken as small as possible. Here they will be taken as $t/h = 1/6$ (or $t = 6$ in.), $b'/b = 0.3$, and $k = 1.0$. Substituting these values in Eq. (8-30) we obtain $\psi = 0.533$. It is further assumed that $d/h = 0.9$, which for $h = 36$ in. yields $d = 32.4$ in. For the given ductility, $\epsilon_{su} = \epsilon_{sl} = 0.01$; from Eq. (8-1a), depth to the neutral axis is $a = 12.64$ in. Since in this case the neutral axis is in the web, Eqs. (8-25) and (8-26) apply.

From Fig. 8-7, $f(\epsilon) = 4722\epsilon$ when $\epsilon \leq 0.0009$, and $f(\epsilon) = 4.25$ when $\epsilon \geq 0.0009$; the quantity pf_{su}/f_c' may be calculated from Eq. (8-25) as follows:

$$p \frac{f_{su}}{f_c'} = \frac{(0.3)(0.390)}{(0.003)(5)} \left(\int_0^{0.0009} 4722\epsilon \, d\epsilon + \int_{0.0009}^{0.003} 4.25 d\epsilon \right)$$
$$+ \frac{(0.70)(0.390)}{(0.003)(5)} \int_{0.00158}^{0.003} 4.25 d\epsilon = 0.085 + 0.11 = 0.195$$

For the above value of pf_{su}/f_c', in a similar way Q is obtained from Eq. (8-26):

$$Q = \frac{(0.3)(0.390)^2}{(5)(0.003)^2} \left(\int_0^{0.0009} 4722\epsilon^2 \, d\epsilon + \int_{0.0009}^{0.003} 4.25\epsilon \, d\epsilon \right)$$
$$+ \frac{(1 - 0.3)(0.390)^2}{(5)(0.003)^2} \int_{0.00158}^{0.003} 4.25\epsilon \, d\epsilon = 0.171$$

The values of pf_{su}/f_c' and Q can also be determined directly, as shown in Fig. 8-9. From Eq. (8-29) the area of the beam is 284 in.² In addition, the following quantities are obtained:

$b = kb = 14.8$ in. $b' = 0.3 \times 14.8 = 4.4$ in.

The stress-strain diagram for steel shown in Fig. 8-8 yields

$f_{su} = 214$ ksi

The amount of prestressing can be found from $pf_{su}/f'_c = 0.195$ to be $p = 0.00455$, from which $A_s = 2.18$ in.², A total of sixteen ½-in. strands are needed. Each ½-in. strand has an area of 0.1438 in.². The final dimensions of the section in this solution are shown in Fig. 8-10a. The bottom flange has been widened to accommodate the reinforcement properly, and it is tapered to facilitate construction. The final width of the top flange is taken the same as that of the bottom flange to maintain the symmetry of the section originally assumed. The properties of the gross section and the transformed section, as well as the stresses at the top and bottom fibers before and after losses for both assumptions, are listed in Table 8-1.

b. Section with minimum required ductility corresponding to $\epsilon_{su} = 0.02$
The ultimate strain in the steel required for this example is large, and is

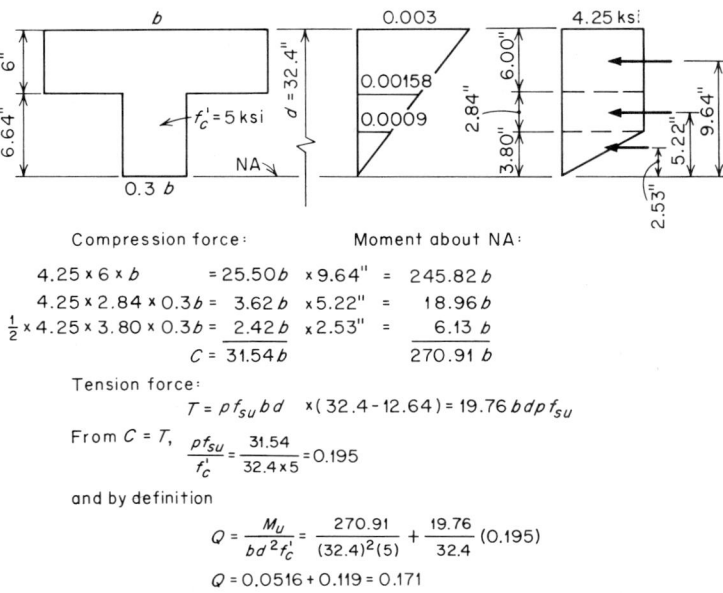

Fig. 8-9 Direct determination of pf_{su}/f'_c and Q.

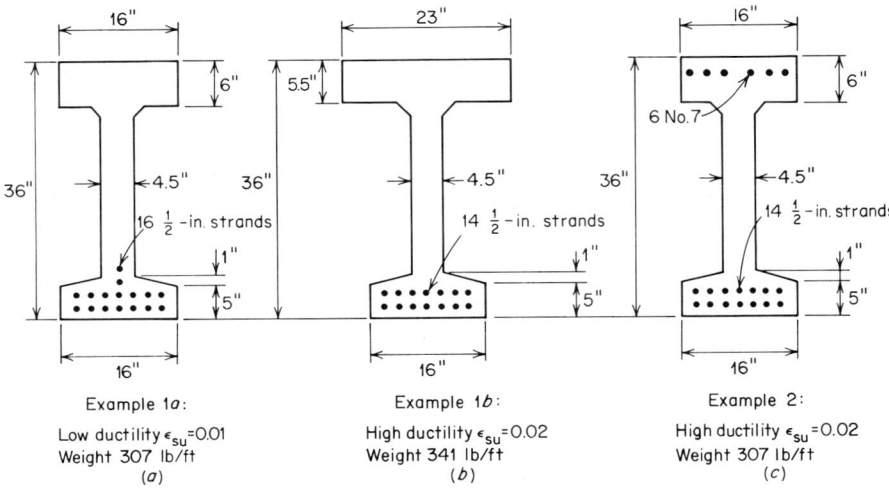

Fig. 8-10 Sections in Illustrative Problems 8-3 and 8-4.

not necessarily used in practice. It has been selected to show that direct design for the largest levels of ductility is possible, and to study how it affects the shape of the section.

All the quantities are the same as in part a of this example except that in this case $\epsilon_{su} = 0.02$, which corresponds to a higher ductility. Since a higher required ductility results in a wider top flange, and the bottom flange need only be large enough to accommodate the reinforcement, k is taken as 0.75 in this case. For a similar reason the quantity b'/b is taken as 0.2. By taking $t = 5.5$ in., $t/h = 0.153$, Eq. (8-30) yields $\psi = 0.407$. As before, $d = 32.4$ in.

From Eq. (8-1a) we obtain $a = 5.49$ in., which places the neutral axis in the flange. For the stress-strain diagram adopted for concrete from Eq. (8-23) we have $pf_{su}/f'_c = 0.122$, and from Eq. (8-24) we obtain $Q = 0.113$. From Eq. (8-29), $A = 336$ in.², from which we can obtain $b = 22.9$ in., $b' = 4.6$ in., $kb = 17.2$ in.

From Fig. 8-8, the stress-strain diagram for steel, $f_{su} = 225$ ksi. From $pf_{su}/f'_c = 0.122$, $p = 0.00271$, from which $A_s = 2.01$ in.². Fourteen $\frac{1}{2}$-in. strands are needed. Figure 8-10b shows the final section of the beam. The dimension of the bottom flange is the minimum required to accommodate the prestressing steel at the required depth. As kb turned out to be larger than necessary, only the required minimum was used, because the bottom flange does not contribute to the strength and ductility of the section. Had the adjustment of the dimensions been

Table 8-1 Section properties and stresses for sections obtained by ultimate design*

Section	A, in.²	y_t, in.	y_b, in.	I, in.⁴	A_s, in.²	A_{s1}, in.²	Weight, lb per ft	Stress before losses (transfer), ksi		Stress after losses, ksi	
								Top (tens)	Bottom (comp)	Top (comp)	Bottom (tens)
Illus. Example 8-3a	294	17.76	18.24	48,080	2.30		307	0.34	3.04	2.37	0.13
$\epsilon_{su} = 0.01$	308	18.43	17.57	51,040				0.27	2.73	2.37	0.14
Illus. Example 8-3b	327	16.20	19.80	55,230	2.01		341	0.23	2.60	1.92	0.37
$\epsilon_{su} = 0.02$	339	16.81	19.19	58,610				0.18	2.36	1.92	0.35
Illus. Example 8-4	294	17.76	18.24	48,080	2.01	3.60	307	0.30	2.66	2.40	0.46
$\epsilon_{su} = 0.02$	328	17.29	18.71	56,270				0.22	2.41	2.02	0.34

* A = area; y_t = distance from centroidal axis to top fiber; y_b = distance from centroidal axis to bottom fiber; I = moment of inertia; A_s = area of prestressed steel; A_{s1} = area of non-prestressed compressive steel; n = modular ratio for both types of steel; f'_c = strength of concrete, f'_{ci} = strength of concrete at transfer; prestress at transfer = 170 ksi; effective prestress after losses = 145 ksi. The underlined quantities correspond to the results obtained on the basis of the transformed section, taking n = 7.

large, recalculation might have been necessary to improve the shape of the section. The properties of the section and the stresses before and after losses for this part are given in Table 8-1.

8-11 ILLUSTRATIVE EXAMPLE 8-4

In order to show that the non-prestressed compressive reinforcement increases the ductility without increasing the area of the section, the following example is presented. It is required to design the section in Illustrative Example 8-3 in such a way that for a ductility corresponding to $\epsilon_{su} = 0.02$, the area of the section will be the same as that for a ductility corresponding to $\epsilon_{su} = 0.01$. The yield-point stress of the compressive reinforcement may be assumed as $f_y = 50$ ksi. The section designed in Illustrative Example 8-3a has a ductility corresponding to $\epsilon_{su} = 0.01$. The problem is to determine how much compressive steel of the given type should be placed so that the ductility of the section will reach that corresponding to $\epsilon_{su} = 0.02$.

The distance of the neutral axis from the top fiber was determined as $a = 5.49$ in. in Illustrative Example 8-3b for the same required ϵ_{su}. Since in this case the neutral axis falls in the flange, Eqs. (8-23) and (8-24) may be used with $Q = 0.171$, as in Illustrative Example 8-3a, and $d' = 2$ in., to write the following independent relations between pf_{su}/f'_c and $p'f'_{su}/f'_c$:

$$0.171 = 0.012 + p\frac{f_{su}}{f'_c}(1 - 0.170) + p'\frac{f_{s1}}{f'_c}(0.170 - 0.062) \quad (8\text{-}23)$$

$$0.122 + p'\frac{f_{s1}}{f'_c} = p\frac{f_{su}}{f'_c} \quad (8\text{-}24)$$

The simultaneous solution of the above equations yields

$$p\frac{f_{su}}{f'_c} = 0.183 \quad \text{and} \quad p'\frac{f_{s1}}{f'_c} = 0.061$$

From Fig. 8-8, the stress-strain diagram for steel, $\epsilon_{su} = 0.02$ corresponds approximately to $f_{su} = 225$ ksi. Therefore, $p = 0.00407$ and $A_s = 1.95$ in.² Fourteen ½-in. strands are needed. Since the strain in compression steel is $0.003 \frac{3.49}{5.49} = 0.0019 > \epsilon_y$, the intermediate-grade steel has yielded and the net f_{s1} is $50 - 4.25 = 45.75$ ksi. Therefore, $p' = 0.00666$, $A_{s1} = 3.20$ in.², and six No. 7 bars of intermediate-grade steel are required. The length of these non-prestressed bars need not be the total span of the beam. Theoretically they are not needed at a section where the required Q is that of the section without the compression reinforcement. The properties of the section and the stresses before and

ULTIMATE DESIGN OF SUPPORTED PRESTRESSED-CONCRETE BEAMS 293

after losses for this example are shown in Table 8-1; Fig. 8-10c shows the beam section.

A reduction in the amount of non-prestressed compression reinforcement is possible with a section having a wider top flange. The parameter $p' f_{s1}/f'_c$ is related to $p f_{su}/f'_c$ by Eq. (8-23). Selection of a smaller value of $p' f_{s1}/f'_c$ would fix $p f_{su}/f'_c$ and permit the determination of the required Q by Eq. (8-24). The area of the section and its final shape can be determined as usual from Eq. (8-29). If the proper values of t/h, b'/b, and k were selected, the new section will present a flange wider than that of Illustrative Example 8-3a, but not as large as that of Illustrative Example 8-3b. Also, the compressive reinforcement required will be smaller than that of Illustrative Example 8-4. This solution would show that, to obtain high ductility, a compromise section can be obtained if some increment of weight is tolerated with a smaller amount of non-prestressed compression steel.

8-12 COMPARISON OF THE THREE SOLUTIONS

It has been shown that ultimate-strength design provides a convenient procedure which leads to well-proportioned sections. The desired ductility and strength were used as the fundamental constraints for proportioning the sections, while the stresses at transfer and under service loads were checked.

An examination of Table 8-1 shows interesting details. The beam of Illustrative Example 8-3a, with a required ductility corresponding to $\epsilon_{su} = 0.01$, required more prestressing steel (two strands) than the beams of Illustrative Examples 8-3b and 8-4, with a required ductility corresponding to $\epsilon_{su} = 0.02$.

For the stress-strain diagram of prestressing steel adopted in these examples, any increase in ductility is accompanied by an increase in stress in steel at ultimate. For the larger ductility considered here, the stress in steel increases at ultimate from 214 to 225 ksi. This increase in steel stress causes a decrease in the required area of prestressing steel.

The beam of Illustrative Example 8-3b shows that by increasing the width of the top flange and thereby adding concrete area to the compression zone, high ductility can be obtained. This, however, increases the weight of the section by 11 percent, but decreases the amount of prestressing steel to 14 strands. The increase in stress in steel at ultimate not only supports the additional weight of the beam but also permits a reduction in the required area of steel. Under the service loads this beam shows, however, a tendency for a larger tensile stress at the bottom fiber due to the smaller amount of prestressing force.

The beam of Illustrative Example 8-4 shows a different way of obtaining high ductility. Six No. 7 intermediate-grade bars are added to the top flange of the low-ductility section of Illustrative Example 8-3a. This increment in compression area raises the neutral axis and increases the lever arm of the resisting couple by approximately 7 percent. In addition, the stress in the steel at ultimate is increased from 214 to 225 ksi, approximately 5 percent. These two factors combined explain the 12 percent reduction in the number of prestressing strands, from 16 to 14, since the required tensile force at ultimate can be obtained with less area of steel at a higher stress and a larger lever arm. The non-prestressed bars also provide additional tensile strength for the top part of the beam at transfer and during handling operations. Furthermore, they have a tendency to reduce the inelastic deflections due to creep.

8-13 PROVISIONS OF ACI CODE 318-63 FOR SHEAR

The provisions of the ACI code for shear concern design of web reinforcement to prevent failure of the beam in the region of combined stresses. For the ultimate design to be meaningful, the beam should be designed to develop its flexural strength. Therefore, provisions for adequate web reinforcement or stirrups are a part of the ultimate design.

In Chap. 4 the bases for the design of web reinforcement were presented in detail. In this section the provisions of ACI Code 318-63 will be discussed briefly.

The ACI code gives the following expression for the required area of web reinforcement:

$$A_v = \frac{V_u - \phi V_c}{\phi d f_y} s \tag{8-31}$$

where A_v = area of web reinforcement
V_u = shear due to specified ultimate load
V_c = shear carried by concrete
s = longitudinal spacing of web reinforcement
d = effective depth
f_y = yield-point stress of web reinforcement
ϕ = capacity reduction factor, taken as 0.85

If $\phi V_c > V_u$, Eq. (8-31) indicates that web reinforcement is not needed. However, the code requires the following minimum:

$$A_v = \frac{A_s f'_s}{80 f_y} \frac{s}{d} \sqrt{\frac{d}{b'}} \tag{8-32}$$

In members of constant overall depth, d in Eqs. (8-31) and (8-32) is specified as the effective depth at the section of maximum moment, and the length of the stirrups at the section under consideration is required to be at least equal to the length of the stirrups at the section of maximum moment.

In sections of varying depth, d is specified as $h(d_m/h_m)$, where d_m and h_m are the effective depth and total depth, respectively, at the section of maximum moment, and h is the total depth at the section under consideration. The stirrups are required to extend into the member a distance d from the compression face.

The shear V_c for normal-weight concrete is given as the lesser of V_{ci} and V_{cw}, introduced in the following discussions,

$$V_{ci} = 0.6 b'd \sqrt{f_c'} + \frac{M_{cr}}{M/V - d/2} + V_d \tag{8-33}$$

but not less than $1.7 b'd \sqrt{f_c'}$,

where $M_{cr} = (I/y)(6\sqrt{f_c'} + f_{pe} - f_d)$
f_{pe} = compressive stress in concrete due to prestress only, after all losses, at extreme fiber of section at which tension stresses are caused by applied loads
f_d = tensile stress due to dead load, at extreme fiber of section at which tension stresses are caused by applied loads

and

$$V_{cw} = b'd(3.5\sqrt{f_c'} + 0.3 f_{pc}) + V_p \tag{8-34}$$

The quantity f_{pc} in the above equation is the compressive stress in concrete, after losses at the centroid of the section, or at the junction of the web and flange when the centroid lies in the flange. The quantity V_p is the vertical component of the effective prestress force.

For lightweight aggregate concrete

$$V_{ci} = 0.1 F_{sp} b'd \sqrt{f_c'} + \frac{M_{cr}}{M/V - d/2} + V_d \tag{8-35}$$

but not less than $0.25 F_{sp} b'd \sqrt{f_c'}$, and

$$V_{cw} = b'd \left[0.5 F_{sp} \sqrt{f_c'} + f_{pc}\left(0.2 + \frac{F_{sp}}{67}\right)\right] + V_p \tag{8-36}$$

where $M_{cr} = (I/y)(0.9 F_{sp} \sqrt{f_c'} + f_{pe} - f_d)$
F_{sp} = ratio of splitting tensile strength to square root of compressive strength

The effective depth d, in Eqs. (8-33) and (8-34), is specified as the distance from the extreme compression fiber to the centroid of the prestressed steel. The value of M/V is specified as that resulting from the position of loads causing maximum moment to occur at the section.

The effective depth in Eqs. (8-34) and (8-36) is specified as the distance from the extreme compression fiber to the centroid of the prestressed steel, or 80 percent of the overall depth of the member, whichever is greater.

The code also permits calculation of V_{cw} as the live load plus dead-load shear corresponding to a principal tensile stress of $4\sqrt{f'_c}$ for normal-weight concrete and $0.6F_{sp}\sqrt{f'_c}$ for lightweight concrete, at the centroidal axis of the section resisting the live load. In flanged sections, if the centroidal axis is not in the web, the principal tensile stress should be determined in the intersection of the flange and the web.

The code requires that when the section $d/2$ from the support is within the transfer length, the actual prestress at the section should be used to calculate V_{cw}. The prestress at the centroid of the section may be assumed to vary linearly from zero at the end face of the beam to a maximum at a distance from the end face equal to the transfer length, assumed to be 50 diameters for strand and 100 diameters for single wire.

Web reinforcement between the face of the support and the section at a distance $d/2$ is to be the same as that required at that section. Shear reinforcement is required for a distance equal to the effective depth d of the member beyond the point theoretically required.

Web reinforcement not less than that determined from Eq. (8-32) is required at all sections, and it is to be spaced not farther apart than three-fourths the depth of the member or 24 in., whichever is the smaller. The code permits omission of web reinforcement when shown by tests that the required flexural strength of the section can be developed.

The code limits the yield strength of the web reinforcement to not more than 60,000 psi.

8-14 ILLUSTRATIVE EXAMPLE 8-5

The following example is presented to illustrate the procedure to determine shear reinforcement in a prestressed-concrete beam by using the provisions of the 1963 building code of the American Concrete Institute. Let us determine the size and spacing of vertical stirrups in the beam of Illustrative Example 8-3.

The beam is 54 ft long between supports and weighs 0.3 klf. It is subjected to uniformly distributed load consisting of 1.0 klf superimposed dead load and 0.6 klf live load. The area of steel strands is 2.30 in.^{2}, at 145 ksi effective prestress. The profile of the center of gravity of the strands is a horizontal line in the middle-third portion of the beam, while hold-down sections at the third points determine an inclined profile for the outer thirds of the beam, as shown in Fig. 8-11. The strength of the concrete, f'_c, is 5 ksi. Other pertinent data can be found in Fig. 8-10a and Table 8-1.

ULTIMATE DESIGN OF SUPPORTED PRESTRESSED-CONCRETE BEAMS

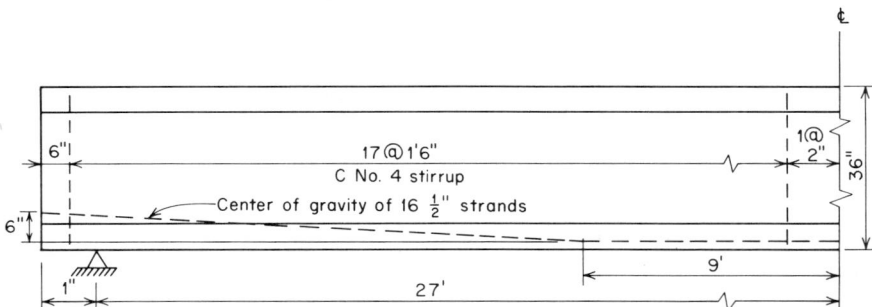

Fig. 8-11 Elevation of the beam of Illustrative Example 8-5.

The required area of shear reinforcement is given by Eq. (8-31) as follows:

$$A_v = \frac{V_u - \phi V_c}{\phi \, d f_y} s \qquad (8\text{-}31)$$

The quantity $V_u - \phi V_c$ must be determined at each section where the area and spacing of web reinforcement are to be computed.

Determination of V_u is based upon the ultimate load which causes flexural failure of the beam, as discussed in Chap. 4. However, since the beam considered here is proportioned on the basis of ultimate design, V_u may be obtained directly from the load factors. The ultimate load on the beam is

$$w_u = (1.5)(0.3 + 1.0) + (1.8)(0.6) = 3.03 \text{ klf}$$

and V_u at a distance x from the left support is given by the following expression:

$$V_u = 3.03 \tfrac{54}{2}\left(1 - \frac{2x}{L}\right) = 81.8\left(1 - \frac{2x}{L}\right)$$

Table 8-2 shows values of V_u at sections with 3 ft intervals. The first section is taken at a distance of $d/2$ from the left support.

The determination of V_c requires the evaluation of V_{ci}, given by Eq. (8-33), and V_{cw}, given by Eq. (8-34). In addition, V_{ci} may not be smaller than $1.7b'd\sqrt{f_c'}$. The code requires that V_c be taken as the lesser of V_{ci} and V_{cw} in each section. The values of ϕV_{ci} and ϕV_{cw} for the beam are given in Table 8-2, and their variation along the span is shown in Fig. 8-12. Other quantities shown in Table 8-2 are d, the distance of the centroid of steel to the top fiber; M_{cr}, the cracking moment; V_d, the shear

Table 8-2 Determination of shear reinforcement for Illustrative Example 8-5

x, ft	x/L	d, in.	M_{cr}, in.-kips	V_d, kips	ϕV_{ci}, kips	V_p, kips	ϕV_{cw}, kips	V_u, kips	$V_u - \phi V_c$, kips	s, in.
1.35	0.025	26.9	6600	33.3	1780	9.3	68.5	77.6	9.1	20
3.0	0.055	27.4	6120	31.2	247	9.3	69.7	72.6	2.9	58
6.0	0.111	28.4	5380	27.3	94.3	9.3	71.9	63.6		
9.0	0.167	29.4	4830	23.4	58.9	9.3	74.1	54.6		
12.0	0.222	30.4	4400	19.5	41.6	9.3	76.3	45.5	3.9	52
15.0	0.278	31.4	4090	15.6	30.9	9.3	78.5	36.3	5.4	38
18.0	0.333	32.4	3930	11.7	23.3	9.3	80.8	27.3	4.0	55
21.0	0.389	32.4	3850	7.8	16.7	0	72.9	18.1	1.4	156
24.0	0.444	32.4	3380	3.9	14.8*	0	72.9	9.2		
27.0	0.500	32.4	3300	0	14.8*	0	72.9	0		

* Minimum value.

due to dead load; and V_p, the vertical component of the prestressing force, all of which are used in the determination of ϕV_{ci} and ϕV_{cw}.

It can be seen in Fig. 8-12 that ϕV_{cw} determines the value of ϕV_c in the portion of the beam close to the supports up to the points where it intersects the descending curve, which shows the variation of ϕV_{ci}. From this point of intersection on, ϕV_{ci} determines the variation of ϕV_c for the

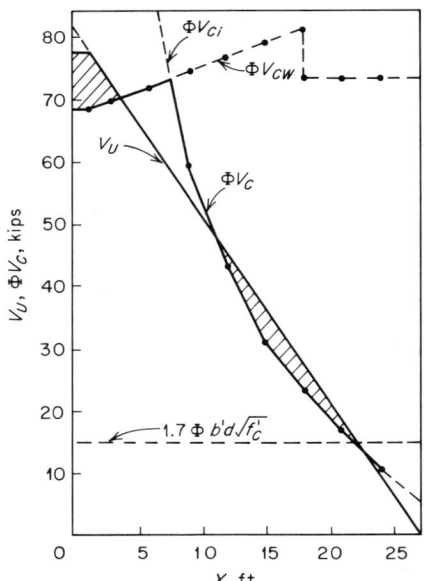

Fig. 8-12 Variation of V_u and ϕV_c for the beam of Illustrative Example 8-5.

central portion of the beam. The shaded portions of Fig. 8-12 represent the regions of the beam where $V_u - \phi V_c$ has a positive sign and for which, therefore, shear reinforcement is necessary.

For practical reasons, such as stiffness and stability during construction of the beam, No. 4 bars are selected for stirrups. The small width of the web requires that one-legged stirrups be used. The necessary spacings at various sections, as determined by Eq. (8-31), are shown in Table 8-2. However, these theoretical spacings do not provide the minimum area of shear reinforcement required by the code.

It is therefore necessary to determine the maximum spacing at which the stirrups can be used. From Eq. (8-32), the maximum spacing can be determined as follows:

$$s = 80 \frac{A_v}{A_s} \frac{f_y}{f'_s} \sqrt{b'd} \qquad (8-32)$$

Substituting $A_v = 0.20$ in.^{2}, $A_s = 2.30$ in.^{2}, $f_y = 40$ ksi, $f'_s = 240$ ksi, $b' = 4.5$ in., and $d = 32.4$ in Eq. (8-32), we have $s = 18$ in. The maximum spacing is smaller than the theoretical spacings shown in Table 8-2. Therefore, wherever shear reinforcement is required along the beam, the spacing for the one-legged No. 4 stirrups should be specified as 18 in.

The provisions of the code extend the stirrup coverage to a distance equal to the effective depth of the member beyond the point theoretically required. The code also requires that stirrups be placed between the face of the support and the section at a distance $d/2$ therefrom. Therefore, in this case, the length of the beam reinforced by stirrups at 18 in., determined by Fig. 8-12, is approximately 25 ft from the face of each support. For practical detailing and construction reasons the unshaded area shown in Fig. 8-12 has been included also.

The remaining central portion of the beam theoretically needs no shear reinforcement. However, the code specifies that stirrups of the same size and type be provided at a maximum spacing given by the smaller of (1) three-fourths of the depth of the member or (2) 24 in. Since three-fourths of the depth is 27 in., one-legged No. 4 stirrups should be used in this portion at 24 in. apart. Figure 8-11 shows an elevation of one-half the beam and the final spacings of the stirrups.

PROBLEMS

8-1. Calculate M_u and φ_u in each of the sections shown in Fig. 8-13. The stress-strain diagrams for steel and concrete are shown in Figs. 3-5 and 3-17b, respectively. The supplementary non-prestressed reinforcement has a yield-point stress of 50,000 psi; the effective prestress is $f_{se} = 120$ ksi.

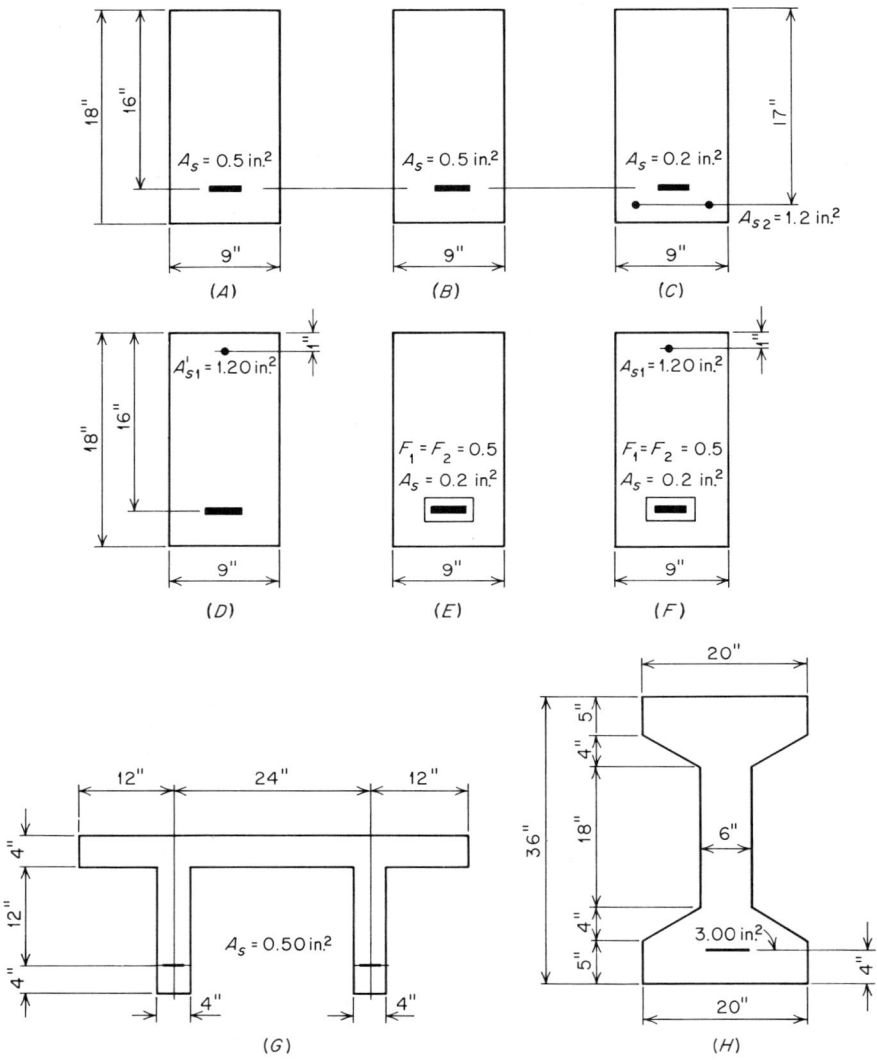

Fig. 8-13 The sections in Prob. 8-1.

8-2. Solve Prob. 8-1 by using the provisions of ACI Code 318-63.

8-3. The section shown in Fig. 8-14 belongs to a precast pretensioned element to be used in the floor construction of a multistory building. The member is simply supported on a 30-ft span. The floors will be finished by adding 2 in. of cast-in-place terrazzo. The design live load is 40 lb per ft². The following information is furnished:

ULTIMATE DESIGN OF SUPPORTED PRESTRESSED-CONCRETE BEAMS   301

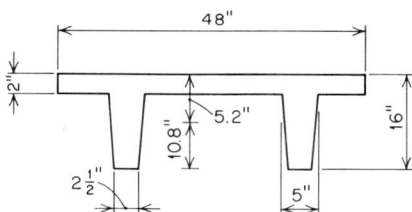

Fig. 8-14 The section in Prob. 8-3.

Material properties:

$f'_c = 6$ ksi $\epsilon_u = 0.004$
$f'_{ci} = 4$ ksi $f_y = 210$ ksi
$f'_s = 240$ ksi $\epsilon_y = 0.01$
$\epsilon'_s = 0.055$

Geometric properties:

$A = 201$ in.² $I = 4920$ in.⁴
$r^2 = 24.5$ in.² $y_t = 5.2$ in.
$y_b = 10.8$ in.

Loads and moments:

$w_g = 0.21$ klf $M_g = 283$ in.-kips
$w_s = 0.10$ klf $M_s = 134$
$wL = 0.16$ klf $M_l = 216$
               $M_t = 633$ in.-kips

Allowable stresses:

$a'_t f'_c = 0.19$ ksi $c'_c f'_c = 2.40$ ksi
$c' f'_c = 2.70$ ksi $a' f'_c = 0.51$ ksi
$\eta = 0.85$ $f_{se} = 160$ ksi

(a) Calculate the minimum number of strands required to comply with service-load design.

(b) Calculate the minimum number of strands required to comply with ACI code provisions for ultimate design. Assume a wide rectangular beam action.

(c) Determine the final number of strands on the basis of both working-stress design and ultimate design. Draw a detail of the section showing placement of the strands, clearance, and cover.

8-4. A prestressed-concrete beam has a section as shown in Fig. 8-15a. The stress-strain diagrams for prestressed steel and concrete are shown in Fig. 8-15b and c, respectively.

(a) Calculate A_s such that the curvature at failure will be 0.0003 in.⁻¹.

(b) Calculate A_s such that the curvature at failure will be 0.0005 in.⁻¹ when $A_s = 2.86$ in.² and $d_1 = 2.0$ in.

(c) Calculate A_s such that at failure $\epsilon_s = 0.057$.

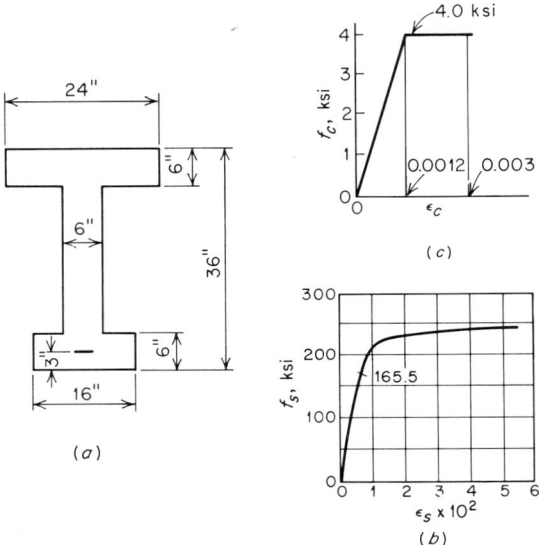

Fig. 8-15 The beam in Prob. 8-4.

For this problem assume $\epsilon_{se} = 0.0048$ and $\epsilon_{ce} = 0.0003$. The non-prestressed compressive steel is intermediate steel with a yield-point stress of 50 ksi.

8-5. The functional requirements for a small warehouse necessitate a layout for the floor, as shown in Fig. 7-26. The floor is to be supported by masonry walls and is to be subject to a live load of 125 psf. Precast concrete slab of 6 in. thickness is to constitute the floor. Design girder G, assuming pretensioned precast construction, and use the provisions of ultimate design. For this problem, assume $h = 68$ in., $\gamma = 150$ pcf, $\epsilon_{se} = 0.0048$, $M_D = 1.50$, $N_l = 2.50$. The stress-strain diagrams for concrete and steel are as shown. The required ductility of the beam corresponds to $\epsilon_{su} = 0.02$.

REFERENCES

1. Hognestad, E.: A Study of Combined Bending and Axial Load in Reinforced Concrete Members, *Univ. Illinois Eng. Expt. Sta. Bull.* 399, 1951.
2. Stuessi, F.: Ueber die Sicherheit des einfach bewehrten Eisenbeton-rechteckbalkens, *Publ., Intern. Assoc. Bridge Structural Eng.*, vol. 1, pp. 487–95, Zürich, April, 1932.
3. Hognestad, E., N. W. Hanson, and D. McHenry: Concrete Stress Distribution in Ultimate Strength Design, *J. Am. Concrete Inst.*, vol. 52, pp. 455–479, December, 1955.
4. "Building Code Requirements for Reinforced Concrete (ACI 318-63)," American Concrete Institute, June, 1963.
5. Gurfinkel, G., and N. Khachaturian: Ultimate Design of Prestressed Concrete Beams, *Univ. Illinois Eng. Expt. Sta. Bull.* 478, April, 1965.
6. Khachaturian, N., and G. Gurfinkel: Ultimate Design of Prestressed Concrete Beams, *Highway Res. Record* 147, December, 1966.

9
Design of Composite Prestressed-concrete Beams

9-1 INTRODUCTION

A composite beam is a structure made of two elements of different material which act as a unit in carrying all or part of the loads.

Composite construction in prestressed concrete usually consists of precast prestressed-concrete beams and cast-in-place reinforced-concrete slab. In practice the precast prestressed-concrete beams are erected first, and subsequently the slab is cast on the top of the beams. After the concrete in the slab is hardened, part of the slab acts as an integral part of the beam in carrying the acting loads. The section made up of the beam section and part of the slab is called the *composite section*.

Precast prestressed-concrete beams are generally cast in a shop or a casting yard, where rigorous control of quality of concrete is possible. Strength of concrete in precast beams is generally high and may reach 7000 psi. On the other hand, since the concrete in the slab is cast in place, it is not economical to apply refined techniques of quality control. Hence the strength of concrete in slab seldom exceeds 4000 psi.

In the type of construction discussed above, the loads that act on the structure before the slab concrete is hardened are carried by the beam alone, and loads that act after the slab concrete is hardened are carried by the composite section. Therefore, the prestressing force and the dead load of the beam and slab are carried by the beam section while the live load is carried by the composite section.

It is possible to build a composite section which can carry all the acting loads. For this purpose it is necessary to carry out the prestressing operation in the field, after the concrete slab is hardened. It is doubtful, however, that it will be economical since it would require shoring for the beam and slab and may involve field post-tensioning. Such a section would be similar to that of a homogeneous beam.

Composite construction results in appreciable saving of material. The actual amount of saving depends on the efficiency of the beam that is to act compositely with the slab.

In order to develop composite action after the slab is hardened, there should be no slip between the slab and the beam at their plane of connection. That is, there should be full shear transfer between the top fiber of the beam and the bottom of the slab. In order to develop the necessary shear stresses or their equivalent, and to prevent slip between these two elements, shear connectors are often provided.

The transference of shear between the slab and the beam is of little consequence in service-load design since there is always sufficient bond to develop the composite section in service loads. However, it is necessary to have sufficient bond between the beam and the slab to develop the flexural strength of the composite section. It is for this purpose that shear connectors are provided. Experimental work indicates that sufficient shear connection can be developed by various means.[1,2]

The shear connectors are provided in the form of shear keys at the top surface of the beam along the span, and vertical ties that are encased in the top fiber of the beam. The keys tend to prevent slip between the slab and the beam, while the vertical ties tend to prevent separation of beam and slab in the direction perpendicular to the contact surfaces. The vertical ties provide some shear connection in form of dowel action; however, they are effective after the shear keys fail, which failure appears unlikely. Often in practice, instead of providing keys, the top of the beam is left unfinished or is artificially roughened. This practice, although not generally followed, seems to be reasonable. In most cases vertical ties are provided by simply extending all web reinforcement into the cast-in-place slab.

As in the design of noncomposite simple beams, composite beams are designed on the basis of the working stress design criteria. The stresses in the beam, due to all possible combinations of loads, are calculated on the basis of the assumption that concrete is an elastic material. These stresses are to be less than or equal to the allowable stresses. Although the slab is an integral part of the composite section, generally the stresses in the beam section govern the design. The stresses at any point in the beam calculated on the basis of the beam section are superposed on the stresses at the same point in the beam calculated on the basis of the composite section.

Design of composite sections, especially in highway bridges, consists in selection of the appropriate standard section, determination of the proper magnitude of the prestressing force, and analysis of the section to ascertain that the stresses are within the allowable limits. Use of the standard sections results in considerable economy since there is no necessity of manufacturing special forms. In highway-bridge practice the AASHO and BPR standard sections are used, often with slight modifications. The selection of standard sections in prestressed concrete is somewhat analogous with the selection of rolled sections in structural steel.

Although in most practical problems the standard sections are used, it is possible that for longer spans it will be necessary to proportion a special section. Furthermore, when a particular section is repeated many times in a structure—as is the case in long structures that have many equal simple spans—proportioning special sections is justified. Because of repetition, the cost of forms becomes small compared with the saving that may be effected if the section is designed for the particular conditions.

The design procedures presented in this part are for simply supported composite beams most commonly used in practice. Generally the most common composite sections have the following features:

1. In addition to the prestressing force the beam is subjected to its own weight, superimposed dead load, and live load—including impact, if any—which act in the same direction.
2. The beam is prestressed in one stage; that is, all the prestressing operation is carried out at one time, and during prestressing the only load acting is the weight of the beam.
3. The centroid of the prestressing force is below the lower kern point of the section.

These assumptions correspond to the most common practice, and they are adopted in order to clarify the fundamental relations that are to be developed. Once these relations are understood, any specialized case can be resolved.

9-2 CONDITIONS OF LOADING

As in a noncomposite simple beam, a composite prestressed-concrete beam is subjected to an infinite variety of conditions of loading. However, from a designer's point of view there are two ideal and limiting conditions in the life of a composite prestressed-concrete beam: the transfer conditions and the final conditions. Transfer conditions correspond to the time when the fabrication of the prestressed beam has just been completed. The prestressing force is at its maximum, and the concrete strength has not yet reached f'_c, the 28-day strength. The final conditions correspond to the time when all the losses in the prestressing force have taken place and the concrete strength has reached f'_c.

The idealized conditions of loading are summarized in Table 9-1. Conditions 1 to 3 correspond to transfer.

Loading condition 1 corresponds to transfer when the beam is subjected to the prestressing force and its own weight. Since beams that are to be used in composite construction are precast, this condition results immediately after the prestressing is complete.

Loading condition 2 may apply to beams in which the slab is cast at transfer or immediately thereafter. Although this condition may be approximated in practice, it is only of academic value.

Loading condition 3 represents a hypothetical condition in which it is assumed that the structure is fully loaded and the live load is carried by

Table 9-1 The idealized conditions of loading

Loading condition	Loads acting*	Effective section	Prestressing force	Strength of concrete
1	$P + G$	Beam section	Maximum	f'_{ci} (Minimum)
2	$P + G + S$	Beam section	Maximum	f'_{ci} (Minimum)
3	$P + G + S + A$	Beam section for P, G, and S; composite section for A	Maximum	f'_{ci} (Minimum)
4	$P + G$	Beam section	Minimum	f'_c (Maximum)
5	$P + G + S$	Beam section	Minimum	f'_c (Maximum)
6	$P + G + S + A$	Beam section for P, G, and S; composite section for A	Minimum	f'_c (Maximum)

* P = prestressing force; G = weight of beam; S = superimposed dead load; A = live load including impact if any and any applied load that comes on the structure after the slab is hardened.

DESIGN OF COMPOSITE PRESTRESSED-CONCRETE BEAMS

the composite section when prestressing force is at its highest. It is unlikely that this condition can be simulated in practice, since before the live load is applied, the concrete in the slab should harden, and during the period that the slab concrete is gaining its strength, the prestressing force decreases.

Conditions 4 to 6 are the final loading conditions.

Loading condition 4 is possible if the precast beam is stored for a long period prior to its use. This condition also applies to a situation in which the precast beams are erected and the work is discontinued for an indefinite period, thus permitting all the losses to take place.

Loading condition 5 corresponds to all beams which have been in place for several years at a time when they are not subjected to live load.

Loading condition 6 represents the fully loaded structure sometime after the construction has been completed.

It can be shown that conditions 1 and 6 are the most important conditions of loading. Condition 1 always governs among the first three conditions, while condition 6 governs among the second three.

As in Chap. 7, the allowable compressive stress in concrete at transfer or before losses is designated as $c'_t f'_c$, and the allowable tensile stress in concrete as $a'_t f'_c$. Similarly, the allowable final compressive stress in concrete is designated as $c' f'_c$, and the allowable final tensile stress as $a' f'_c$. The dimensionless quantities c' and a' are the final stress coefficients that correspond to the allowable compressive and tensile stresses, respectively.

For convenient reference the computed stresses in concrete are also presented as coefficients of f'_c. The computed tensile and compressive stresses in concrete at transfer are designated as $a_t f'_c$ and $c_t f'_c$, respectively. The computed final tensile and compressive stresses are denoted as $a f'_c$ and $c f'_c$, respectively. It is the purpose of service-load design to proportion the section so as to make each computed stress equal to or less than the corresponding allowable stress.

$$a_t \leq a'_t \qquad c_t \leq c'_t$$
$$c \leq c' \qquad a \leq a'$$

Allowable stresses and stress coefficients shown in Tables 7-2 and 7-3 apply to composite construction.

9-3 THE FOUR BASIC REQUIREMENTS

As mentioned in Sec. 9-2, loading condition 1 always governs among the three loading conditions at transfer; that is, if the stresses in the beam are satisfied in loading condition 1, they will automatically be satisfied in loading conditions 2 and 3. This conclusion can be drawn since the allowable concrete stresses at transfer are for the limitation of the stresses in the beam caused by the prestressing force. In loading condition 1 the effect of prestressing is dominant. In this respect there is no difference between a beam that is being manufactured for a noncomposite construction and one that is intended for composite construction.

Similarly, loading condition 6 governs among the three final loading conditions, since the allowable final concrete stresses tend to limit the acting loads, and in loading condition 6 the effect of acting loads dominates.

For each condition of loading the computed stresses at the top and bottom fibers of the beam must be equal to or less than the corresponding allowable concrete stress. On the basis of the preceding discussion there are two governing loading conditions and, therefore, four basic requirements that must be satisfied. These requirements are as follows:

1. For loading condition 1, the tensile stress at the top fiber $(a_t f'_c)$ must be less than or equal to the allowable tensile stress at transfer $(a'_t f'_c)$.
2. For loading condition 1, the compressive stress at the bottom fiber $(c_t f'_c)$ must be less than or equal to the allowable compressive stress at transfer $(c'_t f'_c)$.
3. For loading condition 6, the compressive stress at the top fiber $(c f'_c)$ must be less than or equal to the allowable final compressive stress $(c' f'_c)$.
4. For loading condition 6, the tensile stress at the bottom fiber $(a f'_c)$ must be less than or equal to the allowable final tensile stress $(a' f'_c)$.

The above requirements can be stated algebraically as follows:

$$\frac{P_t}{A}\left(\frac{ey_t}{r^2} - 1\right) - M_g \frac{y_t}{I} = a_t f'_c \leq a'_t f'_c \tag{9-1}$$

$$\frac{P_t}{A}\left(\frac{ey_b}{r^2} + 1\right) - M_g \frac{y_b}{I} = c_t f'_c \leq c'_t f'_c \tag{9-2}$$

$$-\eta \frac{P_t}{A}\left(\frac{ey_t}{r^2} - 1\right) + (M_g + M_s + M_d)\frac{y_t}{I} + M_a \frac{y_{tc}}{I_c} = cf'_c \leq c'f'_c \tag{9-3}$$

$$-\eta \frac{P_t}{A}\left(\frac{ey_b}{r^2} + 1\right) + (M_g + M_s + M_d)\frac{y_b}{I} + M_a \frac{y_{bc}}{I_c} = af'_c \leq a'f'_c \tag{9-4}$$

where M_s = moment due to weight of slab, in.-kips
M_d = moment due to any additional dead load that may act before concrete slab is hardened, in.-kips

M_a = moment due to all loads applied on beam after slab is hardened, in.-kips

y_{tc} = distance from centroidal axis of composite section to top fiber of beam, in.

y_{bc} = distance from centroidal axis of composite section to bottom fiber of beam, in.

I_c = moment of inertia of gross composite section about centroidal axis of composite section, in.^{4}

All other terms have been defined in Chap. 7.

It should be pointed out that in requirements 3 and 4, as presented by expressions (9-2) and (9-4), no consideration is given to the deformation stresses caused by the difference in shrinkage of the bottom fiber of the slab and the top fiber of the beam. Since in the common types of composite construction the beams are precast, a considerable part of the total shrinkage in the beam has already taken place by the time the slab is cast. Although the beam continues to shrink after the slab is cast, the shrinkage in slab starts taking place at a much faster rate immediately after it is in place. This differential shrinkage tends to cause tensile stresses in the slab and in the bottom fiber of the beam, and compressive stresses at the top fiber of the beam.

The most significant effect of the differential shrinkage is the introduction of tensile stresses at the bottom fiber of the beam. Since this effect is in the same direction as that of the applied loads, it results in increased tensile stresses at the bottom fiber. It can be shown that in service-load design the effect of shrinkage is small and may be neglected. However, increased tensile stresses at the bottom fiber of the beam may have a more significant effect upon the cracking resistance of the beam. In this part of the text the shrinkage effect has not been taken into account.

It should be remembered that in developing the above four requirements it has been assumed that the centroid of steel is below the lower kern of the section. If this assumption were not made and the centroid of steel were in the core of the section, the first requirement would automatically be satisfied.

Calculating the section properties of the beam and the composite section on the basis of the gross cross-sectional area is a convenient approximation.

9-4 THE EFFECTIVE SLAB AREA

Most specifications set limitations on the effective width of slab in composite construction. According to the AASHO Standard Specifications for Highway Bridges[3] the effective width of slab shall not exceed the following:

1. One-fourth of the span length of the beam
2. The distance center to center of beams
3. Twelve times the least thickness of slab

In most cases the second limitation governs; that is, the effective width of slab is taken as the distance center to center of beams. It is possible, however, that the first and third limitations will govern when the span length of the beam is small or the spacing of beams is large.

In addition to the above limitations the effectiveness of slab depends upon the quality of concrete in slab and beam. As pointed out in Sec. 9-1, the quality of concrete in the precast prestressed beams is far superior to that of cast-in-place reinforced-concrete slab. Because of the difference in qualities of concrete in beam and slab, in the analysis of composite sections the area of slab should be transformed by the ratio of modulus of elasticity of concrete in the slab to that in the beam.

There is no exact method for the determination of the ratio of the two moduli of elasticity. Modulus of elasticity of concrete is a nonlinear function of strength, and the ratio of modulus of elasticity to strength decreases as the concrete strength increases. ACI Code 318-63 gives the following empirical expression for the modulus of elasticity of concrete:

$$E_c = \gamma^{1.5} 33 \sqrt{f'_c}$$

where γ is the unit weight of concrete in pounds per cubic foot, and f'_c is in pounds per square inch.

The modular ratio of two concretes may be calculated by the above formula. In some cases a modular ratio of 0.6 is used. For a concrete strength of 5000 psi in the beam and 3000 psi in the slab, a modular ratio of 0.6 would seem to be somewhat low. However, it should be pointed out that although precast beams are designed on the basis of 5000 psi concrete, the actual concrete strength is likely to be more than that. On the other hand, cast-in-place concrete with a design strength of 3000 psi is not likely to have an actual strength very much more than the design value.

Considering the uncertainties involved, refinements in the determination of the modular ratio seem unnecessary. Furthermore, a small variation in the modular ratio has comparatively little effect in the design of the beam.

9-5 THE STANDARD SECTIONS

Design of a composite prestressed-concrete beam in most cases is based on standard sections. Standardization of beams in composite construction

DESIGN OF COMPOSITE PRESTRESSED-CONCRETE BEAMS

has introduced appreciable savings in the manufacture of the beam and has helped reduce the design computation time.

The most widespread application of composite construction is in highway bridges. Composite construction can be and has been used in buildings. The standardization of sections for composite construction in buildings, however, is not as well developed as in bridges.

The Bureau of Public Roads and a committee consisting of representatives of the AASHO and PCI each have developed a series of cross sections for prestressed-concrete highway bridges for a wide range of spans and loadings.

Figure 9-1 shows the sections of the four standard AASHO-PCI sections, which are the most common sections used in practice. Table 9-2 shows the section properties of these standard sections.

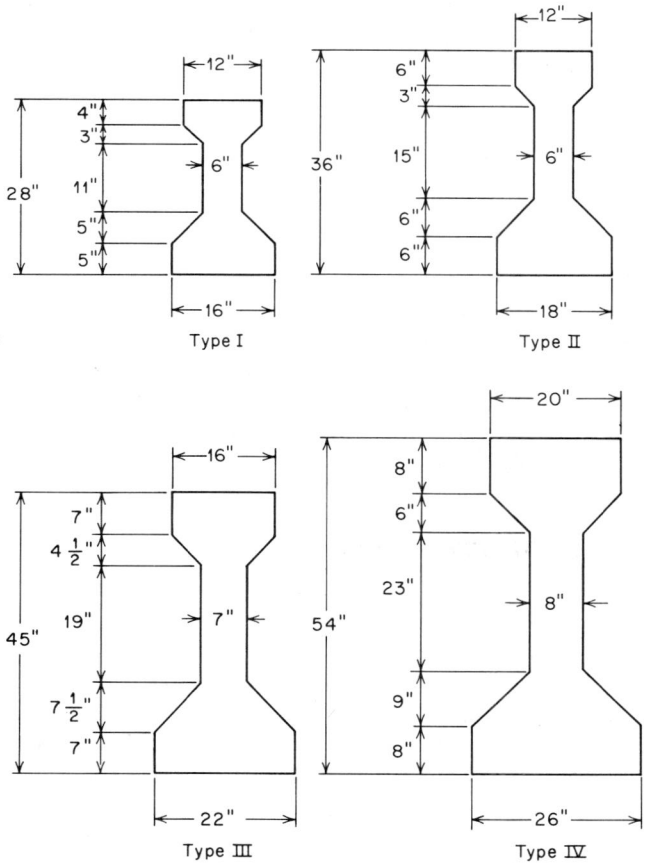

Fig. 9-1 The AASHO-PCI standard sections.

Table 9-2 The properties of AASHO-PCI standard bridge sections for composite construction

Designation	h, in.	A, in.^{2}	y_t, in.	y_b, in.	I, in.^{4}	r^2, in.^{2}	$\dfrac{I}{y_t}$, in.^{3}	$\dfrac{I}{y_b}$, in.^{3}
Type I	28	276	15.41	12.59	22,750	82.4	1476	1,806
Type II	36	369	20.17	15.83	50,980	138.2	2528	3,220
Type III	45	560	24.73	20.27	125,390	223.9	5070	6,190
Type IV	54	789	29.27	24.73	260,730	330.5	8907	10,540

It should be pointed out that composite construction has been applied to multibeam bridges. Standard box sections with shear connectors at the top are placed side by side, and concrete slab is cast on the top.

Design of composite beams by use of a standard section involves the following steps:

1. Choice of the proper section
2. Determination of spacing of beams
3. Determination of the required prestressing force
4. Analysis of the section

In addition to the above steps, the safety factors of the beam should also be checked. However, since this involves calculation of ultimate moment, it will not be discussed here.

Most engineers have design aids which are specifically prepared for the standard sections, and which help to eliminate time-consuming computations. There are also several publications which contain valuable design tables.[4]

9-6 ILLUSTRATIVE PROBLEM 9-1

In order to illustrate the service-load design of a composite prestressed-concrete beam when a standard section is used, the following example is presented:

Select a standard section to be used as an interior stringer for a highway bridge of 60 ft span. Determine the spacing of stringers and the magnitude of the prestressing force. Use H20-S16-44 loading on a 30-ft two-lane roadway with a 7-in.-thick cast-in-place reinforced-concrete slab. The overall depth of the superstructure cannot exceed 4 ft.

The concrete in the precast pretensioned stringers shall have a 28-day strength of 5000 psi and a strength of 4000 psi at the time of cutting of the strands. Use $\tfrac{7}{16}$-in. strands with an ultimate strength $f'_s = 250,000$

DESIGN OF COMPOSITE PRESTRESSED-CONCRETE BEAMS

psi and a yield strength $f_{sy} = 210{,}000$ psi. The 28-day concrete strength in the cast-in-place slab is specified as 3000 psi.

Selection of stringer and spacing Since the overall depth of the structure cannot exceed 4 ft—and this dimension is measured from the bottom of the exterior stringer to the crown of the roadway—it is clear that the overall depth of the stringer cannot exceed 36 in. The slab thickness and crown may amount to about 12 in.

Let us use a standard type II AASHO-PCI section.

The determination of the spacing of the stringers is based upon the required number of stringers of the selected standard shape. It is desirable to use the least number of stringers possible.

The proper number and spacing of the stringers for a given standard section may be determined by trial. However, usually by means of design aids it is possible to determine the proper spacing directly. Let us assume that the spacing of stringers is 5 ft. If this spacing is not adequate, it may be changed. Figure 9-2 shows one-half of the cross section of the superstructure.

From Table 9-2 the section properties of type II standard AASHO-PCI section are:

$A = 369$ in.^{2} $I = 50{,}980$ in.^{4}

$r^2 = \dfrac{I}{A} = 138$ in.^{2} $y_t = 20.17$ in.

$y_b = 15.83$ in. $\dfrac{I}{y_t} = 2528$ in.^{3}

$\dfrac{I}{y_b} = 3220$ in.^{3} $w_g = 0.385$ klf

where w_g is weight per linear foot of the section.

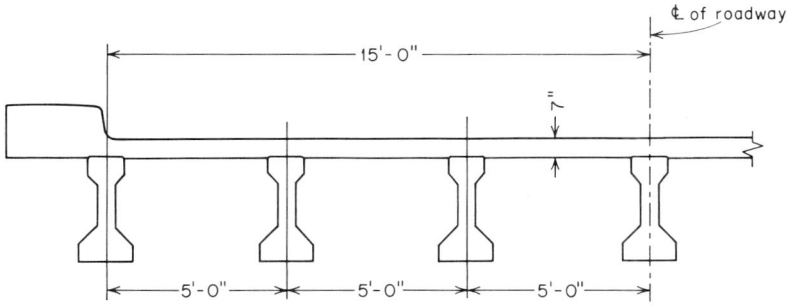

Fig. 9-2 Cross section of the roadway.

The composite section According to the AASHO specifications, the effective slab width is the smallest of the following three quantities:

$60 \times \frac{1 \cdot 2}{4} = 180$ in.

$5 \times 12 = 60$ in.

$7 \times 12 = 84$ in.

Therefore, the effective slab width is the spacing of stringers, which is 60 in. in this problem.

The composite-section properties are calculated by assuming the ratio of modulus of elasticity of concrete in the slab to that in the beam to be 0.6. As pointed out in Sec. 9-4, this value may not be accurate, but is satisfactory for design purposes. It can be shown that a large variation in this modular ratio has a comparatively small effect on the results. Usually increase of the modular ratio causes a decrease in the required prestressing force.

In order to improve the connection between the slab and the stringer, the bottom of the slab is usually placed $\frac{1}{2}$ in. below the top of the stringer. Figure 9-3 shows the composite section for an interior stringer.

Table 9-3 presents a convenient form for the calculation of the properties of the composite section.

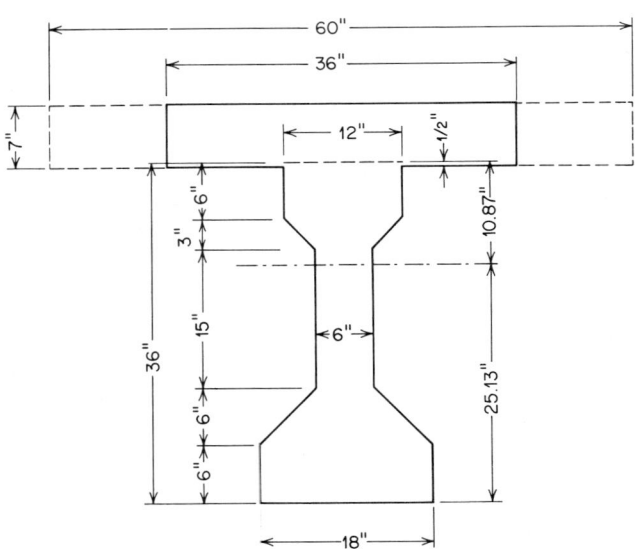

Fig. 9-3 The composite section.

DESIGN OF COMPOSITE PRESTRESSED-CONCRETE BEAMS

Table 9-3 The properties of the composite section

Section	A = area	y	Ay	$\bar{y}$	$\bar{y}^2$	$A\bar{y}^2$	I_0	I_c
Slab	$6.5 \times 36 = 234$	-3.25	-761	14.12	199.4	$46,660$	824	$47,484$
Slab	$0.5 \times 24 = 12$	0.25	3	10.62	112.8	$1,354$		$1,354$
Beam	369	20.17	7443	-9.30	86.5	$31,919$	$50,980$	$82,899$
	615		6685			$79,933$	$51,804$	$131,737$

$y_{tc} = \frac{6685}{615} = 10.87; \bar{y} = y - y_{tc}.$

The composite-section properties are:

$A_c = 615$ in.^{2} $I_c = 131{,}737$ in.^{4}

$y_{tc} = 10.87$ in. $y_{bc} = 25.13$ in.

$\dfrac{I_c}{y_{tc}} = 12{,}120$ in.^{3} $\dfrac{I_c}{y_{bc}} = 5240$ in.^{3}

where A_c is the area of the composite section.

Moments Moments at the midspan of an interior stringer due to the acting loads are as follows:

Beam: $M_g = 0.385 \dfrac{(60)^2}{8} \times 12$
$= 2080$ in.-kips

Slab: $M_s = 0.431 \dfrac{(60)^2}{8} \times 12$
$= 2330$ in.-kips $\Big\}$ 4635 in.-kips

Diaphragm: $M_d = \dfrac{1.25 \times 60}{4} \times 12$
$= 225$ in.-kips

Live load: $M_a = M = \tfrac{1}{2} \times 806.5 \times \tfrac{5}{5}$
$\times 1.27 \times 12 = \dfrac{6145 \text{ in.-kips}}{10{,}780 \text{ in.-kips}}$

In the above calculations it is assumed that one set of diaphragms is provided at the midspan of the bridge. The weight of the diaphragm carried by each interior stringer is conservatively estimated at 1.25 kips. (This figure is based on assuming reinforced-concrete diaphragms that are as deep as the stringer and are about 8 in. wide.)

The impact percentage is calculated from the following expression in the AASHO specifications:

$$I = \frac{50}{L + 125} = \frac{50}{60 + 125} = 27 \text{ percent}$$

The allowable stresses According to the AASHO Specifications for Highway Bridges (see Table 7-2) the allowable stresses in concrete for pretensioned beams without non-prestressed reinforcement are as follows:

At transfer:

Allowable tensile stress = $3 \sqrt{f'_{ci}}$

Allowable compressive stress = $0.60 f'_{ci}$

After losses:

Allowable tensile stress = 0

Allowable compressive stress = $0.40 f'_c$

In this problem since $f'_{ci} = 4000$ psi, the allowable tensile stress at transfer is $3 \sqrt{4000} = 190$ psi, and the allowable compressive stress at transfer is $0.60 \times 4000 = 2400$ psi. The allowable tensile and compressive stresses after losses are zero and $0.4 \times 5000 = 2000$ psi, respectively.

The compressive stresses in the slab cannot exceed $0.40 \times 3000 = 1200$ psi.

Determination of the prestressing force The stress requirements at transfer determine the maximum possible prestressing force. For a reasonably assumed eccentricity, the prestressing force at transfer may be calculated from expressions (9-1) and (9-2) as follows:

$$\frac{P_t}{369}\left(\frac{12.33 \times 20.17}{138} - 1\right) - \frac{2080}{2530} = 0.19 \text{ ksi}$$

$$\frac{P_t}{369}\left(\frac{12.33 \times 15.83}{138} + 1\right) - \frac{2080}{3220} = 2.40 \text{ ksi}$$

In the above expressions the eccentricity is assumed as 12.33 in., which provides sufficient cover for placing the prestressing steel. The quantity g, the distance from the center of gravity of steel to bottom fiber, in this case is taken as 3.5 in. The eccentricity can be expressed as follows:

$$e = y_b - g = 15.83 - 3.50 = 12.33 \text{ in.}$$

DESIGN OF COMPOSITE PRESTRESSED-CONCRETE BEAMS

From the transfer requirements presented above, the values of P_t can be calculated. Solving for P_t from the first expression, we have

$$P_t = \frac{369(0.19 + \frac{2080}{2530})}{\frac{12.33 \times 20.17}{138} - 1} = \frac{369 \times 1.012}{0.80} \equiv 467 \text{ kips}$$

and from the second expression we have

$$P_t = \frac{369(2.40 + \frac{2080}{3220})}{\frac{12.33 \times 15.83}{138} + 1} = \frac{369 \times 3.05}{2.414} \equiv 466 \text{ kips}$$

Evidently P_t cannot exceed 466 kips. In this case requirements (9-1) and (9-2) give almost the same maximum possible value for P_t. Let us use $P_t = 466$.

With $\eta = 0.85$, requirements (9-3) and (9-4) can now be checked:

$$-0.85 \times \frac{466}{369}\left(\frac{12.33 \times 20.17}{138} - 1\right) + \frac{4635}{2530} + \frac{6145}{12{,}120}$$
$$= -0.860 + 1.830 + 0.507 = 1.477 \text{ ksi (compressive)}$$

$$-0.85 \times \frac{466}{369}\left(\frac{12.33 \times 15.83}{138} + 1\right) + \frac{4635}{3220} + \frac{6145}{5240}$$
$$= -2.600 + 1.438 + 1.171 = 0.009 \text{ ksi (tensile)}$$

The above calculations indicate that for a prestressing force of 466 kips, there is a tensile stress of 0.009 ksi at the bottom fiber after losses. Although the specifications require zero stress after losses, in this case the computed tensile stress is too small to require any change in the spacing of the stringers. If this tensile stress were large, it would be necessary to decrease the spacing of the stringers. Let us consider this design satisfactory and take $P_t = 466$ kips.

The stresses in the stringer due to the weight of the beam and the slab (including the diaphragms) after losses are generally small and on the safe side. In this example the dead-load stresses at extreme fibers are:

At top fiber:

$$-0.85 \times \frac{466}{369}\left(\frac{12.33 \times 20.17}{138} - 1\right) + \frac{4635}{2530}$$
$$= -0.860 + 1.830 = 0.970 \text{ ksi (compressive)}$$

At bottom fiber:

$$-0.85 \times \frac{466}{369}\left(\frac{12.33 \times 15.83}{138} + 1\right) + \frac{4635}{3220}$$
$$= -2.600 + 1.438 = -1.162 \text{ ksi (compressive)}$$

The stresses in the slab due to the loads carried by the composite section are small and need not be investigated. Evidently in this case the maximum stress in the slab is $\frac{6140}{7590} \times 0.60 = 0.455$ ksi of compression. The allowable compression stress in the slab concrete is 1.20 ksi.

Determination of number of strands In the AASHO specifications the allowable stress in steel before losses is specified as $0.70f'_s$ and after losses as $0.6f'_s$ or $0.80f_{sy}$, whichever is smaller. The loss of prestress due to all causes is given as 35,000 psi for pretensioned members and 25,000 psi for post-tensioned members. These losses are measured from the time immediately after the seating of anchorage is effected.

In Chap. 7 it was shown that on the basis of the above specification the allowable stress in steel at transfer at midspan can be presented as follows:

$$f_{st} = \phi(0.70f'_s - 10{,}000)$$
$$f_{st} = \phi(0.60f'_s + 25{,}000)$$
$$f_{st} = \phi(0.80f_{sy} + 25{,}000)$$

and the smallest of the above values should be used.

The quantity ϕ is the ratio of the prestress at midspan to that at the jacking anchorage. When the loss due to friction is zero, $\phi = 1.0$. In pretensioned beams with draped strands the magnitude of ϕ may be in the neighborhood of 0.95. It can be shown that in this problem the lowest value of f_{st} is obtained from the first of the above three equations. Assuming $\phi = 0.95$, we have

$$f_{st} = 0.95(0.70 \times 250 - 10) = 157 \text{ ksi}$$

The prestressing force at transfer at midspan in this problem is $P_t = 466$ kips. The number of $\frac{7}{16}$-in. strands needed is

$$\frac{466}{157 \times 0.1089} = 27$$

Figure 9-4 shows the arrangement of the 27 strands. From Fig. 9-4 it is seen that the center of gravity of steel is 3.50 in. from the bottom fiber, as was assumed in calculating P_t.

Therefore, type II AASHO-PCI standard sections spaced at 5-ft centers, with twenty-seven $\frac{7}{16}$-in. strands providing 466 kips of prestressing force in each section, are satisfactory.

DESIGN OF COMPOSITE PRESTRESSED-CONCRETE BEAMS

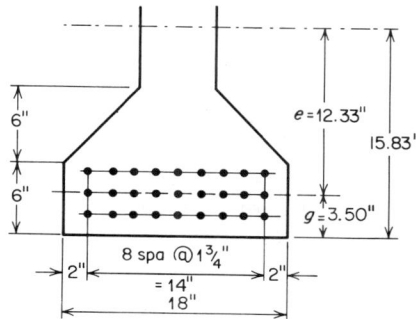

Fig. 9-4 The arrangement of steel.

The effect of the modular ratio In the solution of this design problem the ratio of the modulus of elasticity of concrete in the slab to that in the stringers was assumed as 0.60. It can be shown that increasing this ratio usually causes a decrease in the required prestressing force.

Let us assume the modular ratio of the two concretes to be 0.8 instead of 0.6, and study the effect of the change on this problem. It can be shown that the composite-section properties in this case are

$$A_c = 699 \text{ in.}^2 \qquad I_c = 146{,}280 \text{ in.}^4$$
$$y_{tc} = 9.20 \text{ in.} \qquad y_{bc} = 26.80 \text{ in.}$$
$$\frac{I_c}{y_{tc}} = 15{,}900 \text{ in.}^3 \qquad \frac{I_c}{y_{bc}} = 5460 \text{ in.}^3$$

According to requirements (9-1) and (9-2), P_t cannot exceed 466 kips. If $P_t = 466$ kips and $\eta = 0.85$, requirements (9-3) and (9-4) can be checked on the basis of the above composite-section properties:

$$-0.85 \times \frac{466}{369}\left(\frac{12.33 \times 20.17}{138} - 1\right) + \frac{4635}{2530} + \frac{6145}{15{,}900}$$
$$= -0.860 + 1.830 + 0.386 = 1.356 \text{ ksi (compressive)}$$
$$-0.85 \times \frac{466}{369}\left(\frac{12.33 \times 15.83}{138} + 1\right) + \frac{4635}{3220} + \frac{6140}{5460}$$
$$= -2.600 + 1.438 + 1.126 = -0.036 \text{ ksi (compressive)}$$

It can be seen that by increasing the modular ratio from 0.6 to 0.8 and using $P_t = 466$ kips, the calculations indicate a compressive stress of 0.036 ksi at the bottom fiber after losses. From the composite-section properties it is clear that by increasing the modular ratio both the moment of inertia and the section moduli of the composite section increase. In this case, it is possible to reduce the prestressing force slightly so that the stress at the bottom fiber after losses equals zero. However, this reduction would be too small to change the arrangement of the prestressing steel.

Therefore, one may conclude that the effect of the modular ratio on design is small, and as a rule it is safer to use a smaller value for it.

9-7 DESIGN OF A BEAM FOR COMPOSITE CONSTRUCTION

In the preceding discussions the use of the standard sections for composite construction was described in detail. Although in small structures the use of standard sections results in an economical design, in some cases it is necessary to design the beam that is to act compositely with the cast-in-place slab. This is desirable in structures that are made up of many identical simple spans in which the cost of manufacturing the forms is small in comparison with the saving that may be effected if the beam section is designed efficiently. For unusually long spans it may also become necessary to design the beam section for the particular conditions. In this and subsequent sections the design of the beam section for composite construction will be discussed for a given spacing of beams and a known thickness of slab.

In Sec. 9-3 the four basic requirements for the service-load design of composite construction were developed and presented algebraically in the form of expressions (9-1) through (9-4). These expressions will be rewritten here for convenient reference:

$$\frac{P_t}{A}\left(\frac{ey_t}{r^2} - 1\right) - \frac{M_g y_t}{I} = a_t f'_c \leq a'_t f'_c \tag{9-1}$$

$$\frac{P_t}{A}\left(\frac{ey_b}{r^2} + 1\right) - \frac{M_g y_b}{I} = c_t f'_c \leq c'_t f'_c \tag{9-2}$$

$$-\eta \frac{P_t}{A}\left(\frac{ey_t}{r^2} - 1\right) + (M_g + M_s + M_d)\frac{y_t}{I} + M_a \frac{y_{tc}}{I_c} = cf'_c \leq c'f'_c \tag{9-3}$$

$$-\eta \frac{P_t}{A}\left(\frac{ey_t}{r^2} + 1\right) + (M_g + M_s + M_d)\frac{y_b}{I} + M_a \frac{y_{bc}}{I_c} = af'_c \leq a'f'_c \tag{9-4}$$

All terms in the above expressions were defined in Sec. 9-3.

If a simply supported prismatic beam is assumed, the above four

DESIGN OF COMPOSITE PRESTRESSED-CONCRETE BEAMS

requirements can be written as equations in the following form:

$$\frac{P_t}{Af'_c}\left(\frac{e}{h}\frac{h^2}{r^2}\frac{1}{y_b/y_t+1}-1\right)-\frac{\gamma L^2}{8hf'_c}\frac{1}{y_b/y_t+1}\frac{h^2}{r^2}=a_t$$

$$\frac{P_t}{Af'_c}\left(\frac{e}{h}\frac{h^2}{r^2}\frac{y_b/y_t}{y_b/y_t+1}+1\right)-\frac{\gamma L^2}{8hf'_c}\frac{y_b/y_t}{y_b/y_t+1}\frac{h^2}{r^2}=c_t$$

$$-\eta\frac{P_t}{Af'_c}\left(\frac{e}{h}\frac{h^2}{r^2}\frac{1}{y_b/y_t+1}-1\right)$$

$$+\frac{\gamma L^2}{8hf'_c}\left[1+\frac{A_r}{A}\left(1+\frac{M_d}{M_s}\right)\right]\frac{1}{y_b/y_t+1}\frac{h^2}{r^2}$$

$$+\frac{\gamma L^2}{8hf'_c}\frac{A_r/A}{1+A_{re}/A}\frac{1}{y_{bc}/y_{tc}+1}\frac{M_a}{M_s}\frac{h^2}{r_c^2}=c$$

$$-\eta\frac{P_t}{Af'_c}\left(\frac{e}{h}\frac{h^2}{r^2}\frac{y_b/y_t}{y_b/y_t+1}+1\right)$$

$$+\frac{\gamma L^2}{8hf'_c}\left[1+\frac{A_r}{A}\left(1+\frac{M_d}{M_s}\right)\right]\frac{y_b/y_t}{y_b/y_t+1}\frac{h^2}{r^2}$$

$$+\frac{\gamma L^2}{8hf'_c}\frac{A_r/A}{1+A_{re}/A}\frac{y_{bc}/y_{tc}}{y_{bc}/y_{tc}+1}\frac{M_a}{M_s}\frac{h^2}{r_c^2}=a$$

where A_r = cross-sectional area of slab, taken as St_s, where S is center-to-center spacing of beams and t_s is slab thickness
r_c = radius of gyration of composite section
A_{re} = effective cross-sectional area of slab

The following dimensionless variables have been defined in Chap. 7:

$$\omega = \frac{hf'_c}{\gamma L^2} \qquad \Delta = \frac{y_b}{y_t}$$

$$\rho = \frac{r^2}{h^2} \qquad m = \frac{P_t}{Af'_c}$$

$$\epsilon = \frac{e}{h}$$

The following new variables are introduced:

$$u = \frac{A_r}{A} \qquad \Delta_c = \frac{y_{bc}}{y_{tc}}$$

$$\rho_c = \frac{r_c^2}{h^2} \qquad R_1 = \frac{M_a}{M_s}$$

$$K = \frac{A_{re}}{A_r} \qquad R_2 = \frac{M_d}{M_s}$$

Substituting these dimensionless variables in the above equations, we have

$$m\left[\frac{\epsilon}{\rho(1+\Delta)} - 1\right] - \frac{1}{8\rho\omega(1+\Delta)} = a_t \qquad (9\text{-}1a)$$

$$m\left[\frac{\epsilon\Delta}{\rho(1+\Delta)} + 1\right] - \frac{\Delta}{8\rho\omega(1+\Delta)} = c_t \qquad (9\text{-}2a)$$

$$-\eta m\left[\frac{\epsilon}{\rho(1+\Delta)} + 1\right] + \frac{1 + u + R_2 u}{8\rho\omega(1+\Delta)}$$
$$+ \frac{uR_1}{8\rho_c\omega(1+\Delta_c)(1+Ku)} = c \qquad (9\text{-}3a)$$

$$-\eta m\left[\frac{\epsilon\Delta}{\rho(1+\Delta)} + 1\right] + \frac{\Delta(1 + u + R_2 u)}{8\rho\omega(1+\Delta)}$$
$$+ \frac{\Delta_c u R_1}{8\rho_c\omega(1+\Delta_c)(1+Ku)} = a \qquad (9\text{-}4a)$$

For a given spacing of beams and thickness of slab, the values of R_1, R_2, and K are known in a given problem. Consequently, if the values of a_t, c_t, c, a, and η are known or assumed, there will be eight dimensionless unknown variables in the above four equations, namely, ω, ρ, Δ, u, m, ϵ, Δ_c, and ρ_c. Of these eight variables, ω, ρ, Δ, and u define the section properties of the beam section, and m and ϵ define the magnitude and position of the prestressing force in terms of these section properties. The quantities Δ_c and ρ_c define the properties of the composite section.

The dimensionless variables ω, ρ, Δ, m, and ϵ are the same as those for noncomposite simple beams discussed in Chap. 7. The variables u, Δ_c, and ρ_c are specifically for composite sections.

The quantity u is the ratio of slab area A_r to the area of the beam A. Since the design procedure to be developed here is based upon the assumption that the spacing of beams and the slab thickness are known, u implicitly defines the required cross-sectional area of the beam and is the most significant section property. The practical range of u is from 0.50 to about 2.00. The lower limit defines a section in which the area of the beam is twice as large as the area of the slab. The upper limit is for a section in which the beam area is one-half of the slab area.

The range of variation of ω, ρ, Δ, and u in standard AASHO-PCI sections is shown in Table 9-4. Since ω depends upon the span length of the beam, L, its values for each section are listed in terms of L^2, assuming $f'_c = 5000$ psi and $\gamma = 150$ pcf. Similarly, since $u = A_r/A$ depends upon the spacing of the beam, S, the values of u are listed in terms of S. The quantity u can be expressed as follows:

$$u = \frac{A_r}{A} = \frac{St_s}{A}$$

DESIGN OF COMPOSITE PRESTRESSED-CONCRETE BEAMS

Table 9-4 The dimensionless variables defining the properties of the standard sections

Standard-section designation, AASHO-PCI	$\omega L^2 = \dfrac{h f'_c}{\gamma}$, ft^2*	$\rho = \dfrac{r^2}{h^2}$	$\Delta = \dfrac{y_b}{y_t}$	$\dfrac{u}{s} = \dfrac{t_s}{A}$, ft^{-1}†
Type I	11,200	0.1051	0.816	0.3045
Type II	14,400	0.1066	0.786	0.2277
Type III	18,000	0.1107	0.821	0.1500
Type IV	21,600	0.1135	0.844	0.1064

* $f'_c = 5000$ psi, $\gamma = 150$ pcf.
† $t_s = 7$ in.

where S = spacing of beams
t_s = thickness of slab

Table 9-4 shows the values of $u/S = t_s/A$ for each of the standard sections, assuming $t_s = 7$ in.

A study of the values of ω, ρ, Δ, and u in this table is important in understanding the physical significance of these quantities.

The quantity $\Delta_c = y_{bc}/y_{tc}$ is a measure of the position of the centroidal axis of the composite section. Its practical range of variation is from 2 to 6. The quantity Δ_c is somewhat different from Δ, since y_{tc} is measured from the centroidal axis of the composite section to the top of the beam instead of the top of the slab.

The quantity $\rho_c = r_c^2/h^2$ is a measure of the effective distribution of the area of the composite section. Its range of variation is from 0.11 to 0.18. The range of variation of ρ_c is somewhat higher than that of ρ since in the expression $\rho_c = r_c^2/h^2$ the denominator is in terms of the depth of the beam instead of the overall depth of the composite section.

9-8 RELATIONS AMONG THE DIMENSIONLESS VARIABLES

For design purposes we are primarily interested in Δ, u, m and ϵ. The quantities ω and ρ can be estimated, as discussed in Chap. 7. Since in the method being discussed it is assumed that the spacing of the beams and the slab thickness are known, R_1, R_2, and K may be considered as known quantities.

A simultaneous solution of Eqs. (9-1a) through (9-4a) for Δ, u, m, and ϵ results in the following expressions:

$$\Delta = \frac{8\rho_c\omega(1 + \Delta_c)(1 + Ku)(\eta c_t + a) - \Delta_c R_1 u}{8\rho_c\omega(1 + \Delta_c)(1 + Ku)(\eta a_t + c) - uR_1} \qquad (9\text{-}5)$$

$$u\left[\frac{\rho R_1}{\rho_c(1 + Ku)} + R_2 + 1\right] = 8\rho\omega[\eta(a_t + c_t) + a + c] - (1 - \eta) \qquad (9\text{-}6)$$

$$m = \frac{c_t - \Delta a_t}{1 + \Delta} \qquad (9\text{-}7)$$

$$\epsilon = \frac{1}{m}\left[\rho(a_t + c_t) + \frac{1}{8\omega}\right] \qquad (9\text{-}8)$$

From the geometry of the composite section it can be shown that the following expressions are correct:

$$\Delta_c = \frac{\Delta + Ku(1 + \Delta)(1 + t_s/2h)}{1 - Ku(1 + \Delta)t_s/2h} \qquad (9\text{-}9)$$

$$\rho_c = \frac{\rho + \dfrac{Ku}{12}\left(\dfrac{t}{h}\right)^2}{1 + Ku} + \frac{Ku}{(1 + Ku)^2}\left(\frac{1}{1 + \Delta} + \frac{t_s}{2h}\right)^2 \qquad (9\text{-}10)$$

where t_s is the thickness of the slab.

In the first term on the right-hand side of Eq. (9-10) the quantity $(Ku/12)(t/h)^2$ is small in comparison with ρ and may be ignored:

$$\rho_c = \frac{\rho}{1 + Ku} + \frac{Ku}{(1 + Ku)^2}\left(\frac{1}{1 + \Delta} + \frac{t_s}{2h}\right)^2 \qquad (9\text{-}10a)$$

The derivation of Eqs. (9-9) and (9-10) is presented in Appendix A.

In the discussions that follow, Eq. (9-10a) will be used instead of Eq. (9-10).

Equations (9-5) through (9-8) are equivalent to Eqs. (7-5) through (7-8) for noncomposite beams. When $R_1 = R_2 = 0$, Eqs. (9-5) and (9-6) reduce to Eqs. (7-5) and (7-6), respectively.

The design method for composite sections is similar to that for noncomposite sections.

9-9 DESIGN FOR LEAST AREA OF CONCRETE

The dimensionless quantity $u = A_r/A$ defines the required cross-sectional area of the beam. For given spacing of beams S and thickness of slab t_s, $A_r = St_s$ is a known quantity. Therefore, for a given A_r value an increase in u will result in a decrease in the cross-sectional area A of the beam.

Since R_1, R_2, K, and t_s are known, from Eq. (9-6) it can be seen that

DESIGN OF COMPOSITE PRESTRESSED-CONCRETE BEAMS

for given values of h, ρ, and Δ the quantity u increases with $\eta(a_t + c_t) + a + c$. Hence, in order to obtain the least area of concrete the following must be correct:

$$a_t = a'_t \qquad c_t = c'_t$$
$$c = c' \qquad a = a'$$

that is, all the requirements must be satisfied exactly. However, Eq. (9-5) indicates that Δ must have a specific value in order to satisfy all the requirements exactly. Hence, simultaneous solution of Eqs. (9-5), (9-6), (9-9), and (9-10a) taking $a_t = a'_t$, $c_t = c'_t$, $c = c'$, and $a = a'$, will give the proper values of Δ, u, Δ_c, and ρ_c. The value of Δ thus obtained will correspond to the least-weight beam; however, it will be too small for practical purposes.

It can be shown that for given ω, ρ, and Δ values the following two criteria produce the least-weight design.

Criterion 1 This criterion produces the least area of concrete and lowest corresponding steel area. In this case it is necessary to satisfy requirements (9-1), (9-2), and (9-4) exactly and (9-3) by a margin.

Criterion 2 This criterion produces the least area of concrete and the highest corresponding steel area. In this case it is necessary to satisfy requirements (9-2) to (9-4) exactly and (9-1) by a margin.

Elimination of $(\eta a_t + c)$ between Eqs. (9-5) and (9-6) and substitution of Eqs. (9-9) and (9-10a) for ρ_c and Δ_c, respectively, result in the following expression:

$$\frac{uR_1\rho\left[\Delta + Ku(1 + \Delta)\left(1 + \frac{t_s}{2h}\right)\right]}{\rho(1 + Ku) + Ku\left(\frac{1}{1 + \Delta} + \frac{t_s}{2h}\right)^2}$$
$$+ \Delta(1 + R_2)u = 8\rho\omega\left(1 + \frac{1}{\Delta}\right)(\eta c_t + a) - (1 - \eta) \quad (9\text{-}11)$$

Equation (9-11) is a quadratic equation in terms of u. The quantities ω, t_s/h, ρ, and Δ either are known or can be determined. It can be shown that any increase in h and ρ results in a decrease in the cross-sectional area of the beam. A reduction in Δ has the same effect.

Determination of ω in composite sections does not differ from that of noncomposite sections. Since, in the expression of $\omega = hf'_c/\gamma L^2$, the span length L is known and f'_c and γ are chosen without considering the value of ω, it may be concluded that ω primarily depends on h, the overall depth of the beam section. The factors affecting the determination of h are stiffness and clearance, which have been discussed in Chap. 7 in connection with noncomposite beams.

As shown in Chap. 7, the practical range of ρ is from 0.0833 to about 0.13. It is desirable to have a large value for ρ since from Eq. (9-11) it increases with u and decreases with the cross-sectional area of the beam.

From Eq. (9-11) it can be shown that Δ decreases with u. That is, in order to have a small area, Δ must be reduced. On the other hand, it is not practically possible to have very low values for Δ, since low values of Δ correspond to thin webs and small top flanges. The value of Δ should not be less than 0.7.

The relationship between Δ and ϵ may be studied by Eqs. (9-7) and (9-8):

$$m = \frac{c_t - \Delta a_t}{1 + \Delta} \tag{9-7}$$

$$\epsilon = \frac{1}{m}\left[\rho(a_t + c_t) + \frac{1}{8\omega}\right] = \frac{1 + \Delta}{c_t - \Delta a_t}\left[\rho(a_t + c_t) + \frac{1}{8\omega}\right] \tag{9-8}$$

It can be seen that an increase in Δ results in a decrease in m and an increase in ϵ. Increase of Δ results in an increase in g, permitting the placing of prestressed steel.

Hence, Δ should be small in order to have a small area of concrete, but it must be large enough to permit sufficient cover.

9-10 THE IDEALIZED SECTIONS

In the preceding section it was shown that in composite construction a decrease in Δ results in a decrease in the area of the beam section. Consequently, the beam sections in composite construction should have a heavier bottom flange than top flange.

The actual shapes of sections used in composite prestressed-concrete construction are equivalent to unsymmetrical I sections in which the

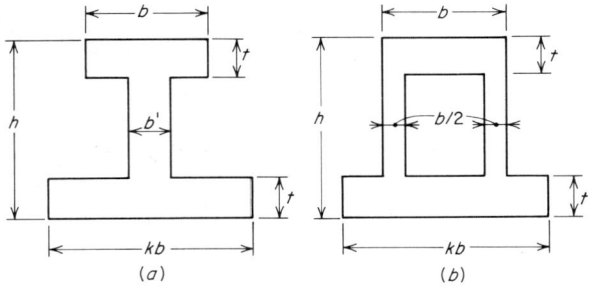

Fig. 9-5 Idealized unsymmetrical I sections in which the bottom flange is wider than the top flange.

DESIGN OF COMPOSITE PRESTRESSED-CONCRETE BEAMS

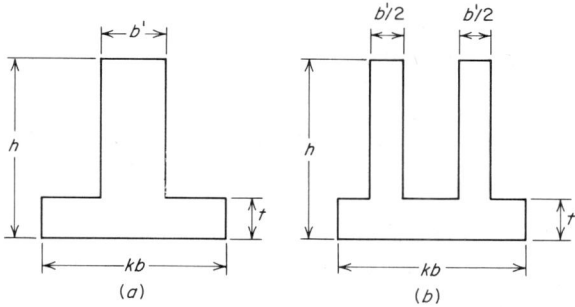

Fig. 9-6 Idealized inverted T sections.

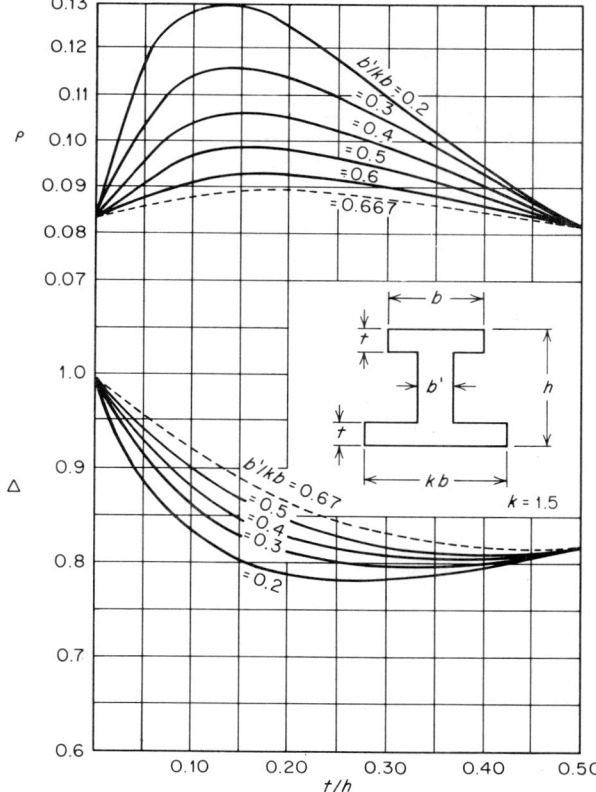

Fig. 9-7 The relations among the geometric properties of an unsymmetrical I section in which the flanges are of equal thickness and the width of the bottom flange is $1\frac{1}{2}$ times greater than that of the top flange.

bottom flange is heavier than the top flange. From the standard sections it is seen that the flanges in the actual sections are tapered. These practical sections can be idealized as sections having web and flanges of uniform thickness.

An inverted T section may also be used in composite construction; however, since the efficiency of this type of section is low, its use will result in a larger area for the beam section.

Figure 9-5a and b shows unsymmetrical I sections, and Fig. 9-6a and b shows inverted T sections. In all these sections $\Delta < 1.0$; flexure about the horizontal axis of these two figures is the same.

For the sections shown in Fig. 9-5a and b Eqs. (7-10) and (7-11) are

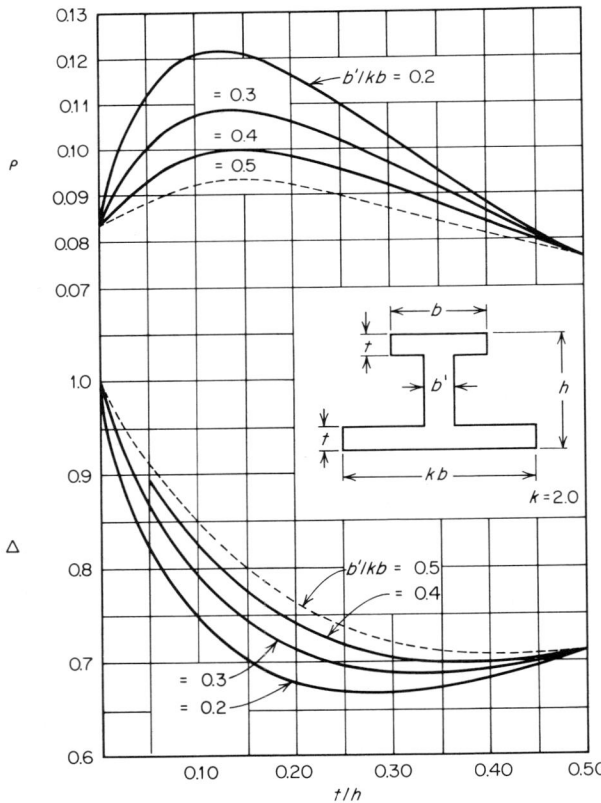

Fig. 9-8 The relations among the geometric properties of an unsymmetrical I section in which the flanges are of equal thickness and the width of the bottom flange is twice greater than that of the top flange.

DESIGN OF COMPOSITE PRESTRESSED-CONCRETE BEAMS

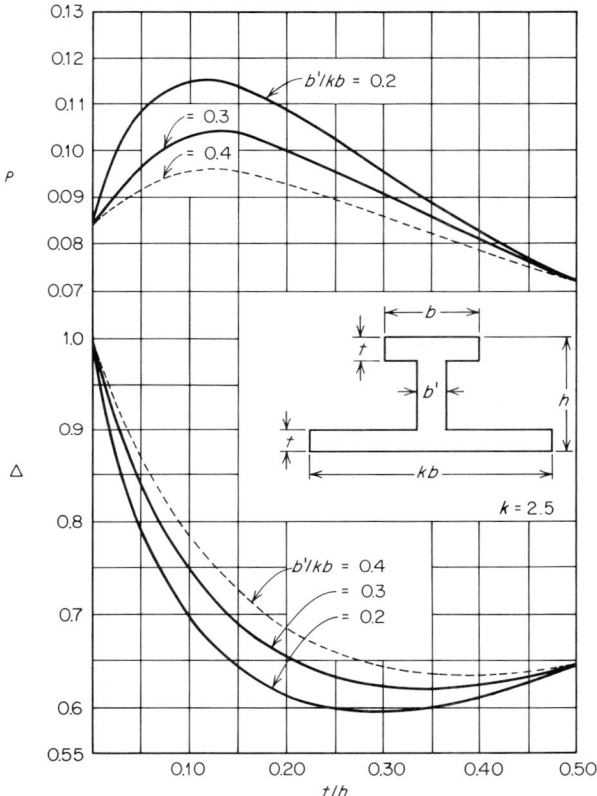

Fig. 9-9 The relations among the geometric properties of an unsymmetrical I section in which the flanges are of equal thickness and the width of the bottom flange is $2\frac{1}{2}$ times greater than that of the top flange.

correct except that Δ is less than unity and k is greater than unity. Similarly, Eqs. (7-12) and (7-13) apply to the sections shown in Fig. 9-6a and b.

Equations (7-10) and (7-11) are plotted in Figs. 9-7 to 9-9 for values of k equal to 1.50, 2.0, and 2.50, corresponding to unsymmetrical I sections in which the bottom flange is 50, 100, and 150 percent wider than the top flange, respectively.

Figure 9-7 is a graphical presentation of Eqs. (7-10) and (7-11) for the specific value of $k = 1.50$. The curves in the top figure represent Eq. (7-10). The quantity ρ is taken as ordinate and t/h is taken as abscissa, and the curves are drawn for six values of b'/kb, namely, 0.2, 0.3,

0.4, 0.5, 0.6, and 0.67. The curve corresponding to $b'/kb = 0.67$ is shown by dashed lines since it corresponds to a T section.

The curves in the bottom of Fig. 9-7 represent Eq. (7-11). In this case Δ is taken as ordinate and t/h is the abscissa, as in the curves above. This equation is plotted for the same values of b'/kb, which are 0.2, 0.3, 0.4, 0.5, 0.6, and 0.67.

These two sets of curves completely define the relations among the geometric properties of an idealized unsymmetrical I section in which $k = 1.50$. By using these curves for a given Δ value and a reasonably selected t/h value, b'/b and ρ can be determined. Similarly, if ρ and t/h are known, consistent values of Δ and b'/b can be obtained.

Figures 9-8 and 9-9 show plots of Eqs. (7-10) and (7-11) for $k = 2.0$ and 2.50, respectively. The curves in these figures are similar to the ones

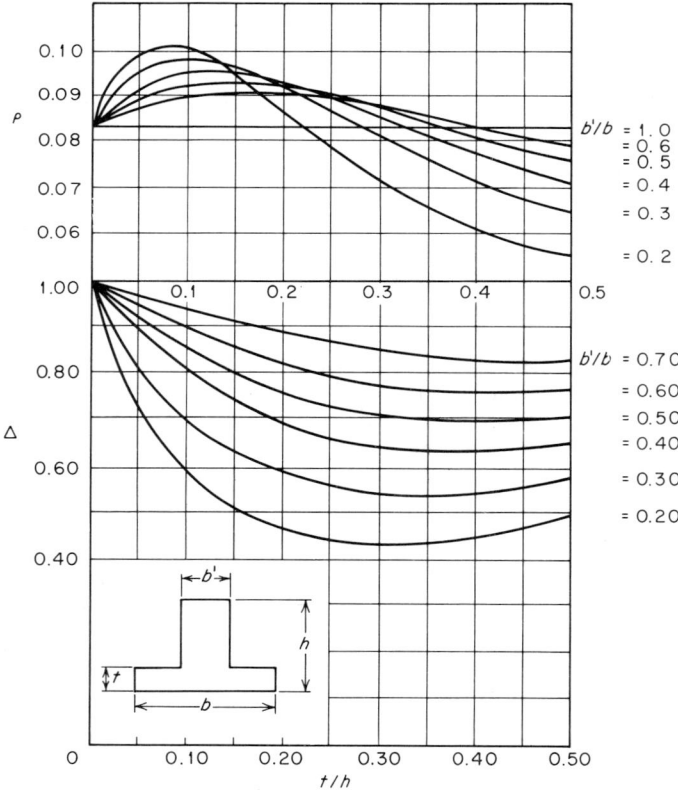

Fig. 9-10 The relations among the geometric properties of an inverted T section.

DESIGN OF COMPOSITE PRESTRESSED-CONCRETE BEAMS   331

in Fig. 9-7. It should be pointed out that the dashed lines in these figures also correspond to T sections.

Figure 9-10 shows similar plots of Eqs. (7-12) and (7-13) for inverted T sections.

9-11 ILLUSTRATIVE PROBLEM 9-2

In order to apply the principles developed in the preceding sections to design of a composite beam, the interior stringer in Illustrative Problem 9-1 will be redesigned. The statement of the problem is the same as that for Illustrative Problem 9-1.

The following stress coefficients will be used:

$a'_t = 0.038 \qquad c'_t = 0.48$
$c' = 0.40 \qquad a' = 0$
$\eta = 0.85$

Solution This problem is the same as Illustrative Problem 9-1 except that instead of a standard section, the beam section is to be proportioned. In order to have a basis for comparing this design with the design in Illustrative Problem 9-1 in which a standard section is used, the spacing of stringers will be taken as 5 ft, the thickness of slab as 7 in., and the overall depth of the beam as 36 in. (the same as in Illustrative Problem 9-1).

The moments at the midspan of an interior stringer due to the acting loads are as follows:

Slab: $M_s = 0.431 \dfrac{(60)^2}{8} \times 12 = 2330$ in.-kips $\Bigg\} = 2555$ in.-kips

Diaphragm: $M_d = \dfrac{1.25 \times 60}{4} \times 12 = 225$ in.-kips

Live load: $M_a = M_l = \tfrac{1}{2} \times 806.5 \times \tfrac{5}{5}$
$\times 1.27 \times 12 = \dfrac{6145 \text{ in.-kips}}{8700 \text{ in.-kips}}$

The above moments are the same as the ones discussed in Sec. 9-6 in connection with Illustrative Problem 9-1.

On the basis of the above quantities we have

$$\omega = \frac{hf'_c}{\gamma L^2} = \frac{3 \times 5 \times 144}{0.15 \times 60 \times 60} = 4.0$$

$$R_1 = \frac{M_a}{M_s} = \frac{6145}{2330} = 2.64$$

$$R_2 = \frac{M_d}{M_s} = \frac{225}{2330} = 0.0965$$

$$\frac{t_s}{h} = \frac{7}{36} = 0.1942$$

and

$$K = \frac{A_{re}}{A_r} = 0.60$$

Assuming $\Delta = 0.75$ and $t/h = 0.15$, for $k = 2.0$ from Fig. 9-5 we have

$$\rho = 0.107$$

and

$$\frac{b'}{kb} = 0.32$$

Since $h = y_b + y_t = 36$ in. and $\Delta = y_b/y_t = 0.75$, it can be shown that $y_b = 15.41$ in. and $y_t = 20.59$ in.

Area of the beam section For least-weight design requirements (9-2) and (9-4) must be satisfied exactly; that is, the following must be correct:

$$c_t = c'_t = 0.48 \qquad a = a' = 0$$

and

$$\eta = 0.85$$

Substituting the above quantities in Eq. (9-11), we have

$$\frac{2.64u[0.75 + 0.6u(1.75)(1 + 0.097)]0.107}{0.107(1 + 0.6u) + 0.6u(\frac{1}{1.75} + 0.097)^2}$$
$$+ 0.175 + 1.0965u = 8 \times 0.107 \times 4.0 \times (1 + \tfrac{1}{0.75})0.85$$
$$\times 0.48 - (1 - 0.85)$$

Simplification of the above expression gives the following equation:

$$u^2 - 0.791u - 0.417 = 0$$

DESIGN OF COMPOSITE PRESTRESSED-CONCRETE BEAMS

It can be shown that $u = 1.153$ is a root of the above equation.

$$u = \frac{A_r}{A} = \frac{60 \times 7}{A} = 1.153$$

$$A = \tfrac{420}{1.153} = 364 \text{ in.}^2$$

The area of an idealized unsymmetrical I section may be expressed as

$$A = kbh \left[\frac{t}{h}\left(1 + \frac{1}{k}\right) + \frac{b'}{bk}\left(1 - 2\frac{t}{h}\right) \right]$$

and for this problem the above expression can be written as

$$364 = 72b(0.15 \times 1.5 + 0.32 \times 0.7) \equiv 32.33b$$

Hence

$b = 11.23$ in.
$kb = 22.46$ in.
$b' = 22.46 \times 0.32 = 7.2$ in.
$t = 0.15 \times 36 = 5.4$ in.

Figure 9-11a shows the idealized section. From Eqs. (9-9) and

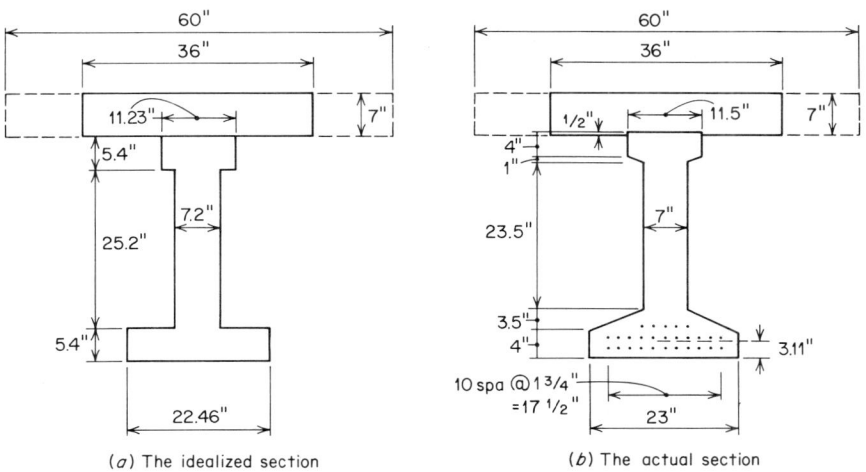

(a) The idealized section (b) The actual section

Fig. 9-11 The idealized and actual composite sections for interior stringers of a highway bridge.

(9-10a) we have

$$\Delta_c = \frac{\Delta + Ku(1 + \Delta)(1 + t_s/2h)}{1 - Ku(1 + \Delta)t_s/2h}$$

$$= \frac{0.75 + 0.6 \times 1.153 \times 1.75 \times 1.097}{1 - 0.6 \times 1.153 \times 1.75 \times 0.097} = 2.36$$

and

$$\rho_c = \frac{\rho}{1 + Ku} + \frac{Ku}{(1 + Ku)^2}\left(\frac{1}{1 + \Delta} + \frac{t_s}{2h}\right)^2$$

$$= \frac{0.107}{1 + 0.6 \times 1.153} + \frac{0.6 \times 1.153}{(1 + 0.6 \times 1.153)^2}\left(\frac{1}{1.75} + 0.097\right)^2$$

$$= 0.171$$

Amount and position of the prestressing force *Criterion 1* This criterion corresponds to the minimum amount of prestressing force and maximum eccentricity. In this case requirements (9-1), (9-2), and (9-4) are satisfied exactly; that is, $a_t = a_t'$, $c_t = c_t'$, and $a = a'$. From Eq. (9-7) we have

$$m = \frac{c_t - \Delta a_t}{1 + \Delta} = \frac{c_t' - \Delta a_t'}{1 + \Delta} = \frac{0.48 - 0.75 \times 0.038}{1.75} = 0.258$$

Hence

$$P_t = Af_c'm = 364 \times 5 \times 0.258 = 470 \text{ kips}$$

and from Eq. (9-8)

$$\epsilon = \frac{1}{m}\left[\rho(a_t + c_t) + \frac{1}{8\omega}\right] = \frac{1}{0.258}(0.107 \times 0.518 + \tfrac{1}{32}) = 0.336$$

Hence

$$e = 0.336 \times 36 = 12.10 \text{ in.}$$

and

$$g = 15.41 - 12.10 = 3.31$$

Criterion 2 This criterion corresponds to the maximum amount of prestressing steel and minimum eccentricity. In this case requirements (9-2) to (9-4) must be satisfied exactly, that is, $c_t = c_t'$, $c = c'$, and $a = a'$. The quantity a_t can be calculated from Eq. (9-5) as follows:

$$\rho = \frac{(\Delta_c - \Delta)(1 + u + R_2u - \eta)}{8\omega(1 + \Delta)[\Delta_c(\eta a_t + c) - (\eta c_t + a)]} \tag{9-5}$$

Solving for a_t from the above equation, we have

$$a_t = \frac{(\Delta_c - \Delta)(1 + u + R_2u - \eta)}{8\rho\omega\Delta_c(1 + \Delta)\eta} + \frac{\eta c_t + a - c\Delta_c}{\eta\Delta_c}$$

DESIGN OF COMPOSITE PRESTRESSED-CONCRETE BEAMS

For criterion 2 in this problem a_t can be calculated as follows:

$$a_t = \frac{(2.36 - 0.75)(1 + 1.153 \times 1.0965 - 0.85)}{8 \times 0.107 \times 4 \times 2.36 \times 1.75 \times 0.85}$$
$$+ \frac{0.85 \times 0.48 - 0.40 \times 2.36}{0.85 \times 2.36} = -0.077$$

From Eq. (9-7) we have

$$m = \frac{c'_t - \Delta a'_t}{1 + \Delta} = \frac{0.48 + 0.75 \times 0.077}{1.75} = 0.3075$$

hence

$$P_t = A f'_c m = 364 \times 5 \times 0.3075 = 560 \text{ kips}$$

and from Eq. (9-8)

$$\epsilon = \frac{1}{0.3075}(0.107 \times 0.403 + \tfrac{1}{32}) = 0.242$$

hence

$$e = 0.242 \times 36 = 8.71 \text{ in.}$$

and

$$g = 15.41 - 8.71 = 6.70 \text{ in.}$$

It appears that criterion 1 can be used since it provides a cover $g = 3.31$ in. On the basis of Eqs. (9-1a) through (9-4a) the four requirements in this case can be written as follows:

$$0.258 \left(\frac{0.336}{0.107 \times 1.75} - 1\right) - \frac{1}{8 \times 0.107 \times 4 \times 1.75} = 0.036$$

$$0.258 \left(\frac{0.336 \times 0.75}{0.107 \times 1.75} + 1\right) - \frac{0.75}{8 \times 0.107 \times 4 \times 1.75} = 0.480$$

$$-0.85 \times 0.258 \left(\frac{0.336}{0.107 \times 1.75} - 1\right) + \frac{(1 + 1.153 \times 1.0965)}{8 \times 0.107 \times 4 \times 1.75}$$
$$+ \frac{1.153 \times 2.64}{8 \times 0.1710 \times 4 \times 3.36 \times 1.694} = 0.302$$

$$-0.85 \times 0.258 \left(\frac{0.336 \times 0.75}{0.107 \times 1.75} + 1\right) + \frac{0.75(1 + 1.153 \times 1.0965)}{8 \times 0.107 \times 4 \times 1.75}$$
$$+ \frac{2.36 \times 1.153 \times 2.64}{8 \times 0.1710 \times 4 \times 3.36 \times 1.694} = 0$$

The above calculations check the design of the idealized section.
The actual beam section, which is similar to the idealized beam section, is shown in Fig. 9-11b.

The section properties of the actual beam section are:

$A = 363$ in.^{2}; $w_g = 0.378$ klf $I = 50{,}166$ in.^{4}
$y_t = 20.79$ in. $y_b = 15.21$ in.
$\dfrac{I}{y_t} = 2415$ in.^{3} $\dfrac{I}{y_b} = 3295$ in.^{3}
$r^2 = 138.2$ in.^{2}
$M_g = 0.378 \dfrac{(60)^2}{8} 12 = 2040$ in.-kips

The section properties of the actual composite section are:

$A_c = 609$ in.^{2} $I_c = 134{,}920$ in.^{2}
$y_{tc} = 11.15$ in. $y_{bc} = 24.85$ in.
$\dfrac{I_c}{y_{tc}} = 12{,}020$ in.^{3} $\dfrac{I_c}{y_{bc}} = 5420$ in.^{3}

The actual prestressing force at transfer can be calculated by using 27 strands and an allowable prestress of 157 ksi at transfer at midspan. In Sec. 9-7 it was shown that this allowable prestress corresponds to a prestress of 175 ksi at the time of pulling at the jacking anchorage.

The actual prestressing force at transfer, therefore, can be computed as follows:

$P_t = 27 \times 0.1089 \times 157 = 461$ kips

Assume the eccentricity of the prestressing force the same as calculated for the idealized section, using criterion 1; that is, assume $e = 12.10$ in.

The four requirements for the actual section can now be checked as follows:

$$\frac{461}{363}\left(\frac{12.10 \times 20.79}{138.2} - 1\right) - \frac{2040}{2415} = 0.196 \text{ psi (tensile)} \quad (9\text{-}1)$$

$$\frac{461}{363}\left(\frac{12.10 \times 15.21}{138.2} + 1\right) - \frac{2040}{3295}$$
$$= 2.396 \text{ psi (compressive)} \quad (9\text{-}2)$$

$$-0.85\frac{461}{363}\left(\frac{12.10 \times 20.79}{138.2} - 1\right) + \frac{4595}{2415} + \frac{6145}{12{,}020}$$
$$= 1.529 \text{ psi (compressive)} \quad (9\text{-}3)$$

$$-0.85\frac{461}{363}\left(\frac{12.10 \times 15.21}{138.2} + 1\right) + \frac{4595}{3295} + \frac{6145}{5420} = 0 \quad (9\text{-}4)$$

Evidently the actual section is satisfactory.

DESIGN OF COMPOSITE PRESTRESSED-CONCRETE BEAMS

A comparison of Figs. 9-3 and 9-9b indicates that both designs give almost identical areas. It should be pointed out that the beam section designed here may be reduced in size if necessary by reducing Δ or increasing ρ and changing t/h or b'/kb ratios. The beam section developed here is only one of many possible sections that may be used in the bridge under consideration.

9-12 ULTIMATE DESIGN

The shear connection between the top of the stringer and the slab should be strong enough that the flexural strength of the composite section is developed. Available test results indicate that if the surface of concrete is kept rough, and sufficient vertical ties are provided, flexural strength of the composite section can be developed.[1]

The strength of a composite section, which consists of the stringer and the slab of known width and thickness, is calculated by assuming that it is a unit all by itself. Strength calculated in this fashion provides only a measure of safety, and should not be confused with the safety of the entire structure for the intended loads.

The flexural strength of the intermediate stringer of Illustrative Example 9-1 can be calculated by assuming that the shear connection between the beam and the slab is sufficiently strong. Figure 9-3 shows the dimensions of the composite section, and Fig. 9-4 shows the details of steel reinforcement. The ultimate moment of this section, if the width of slab is assumed to be 60 in., is 26,400 in.-kips.

Composite sections can also be proportioned by ultimate design.[5] In the following paragraphs this approach is discussed briefly.

In ultimate design the stringer should be designed so that the following two inequalities are satisfied in the composite section:

$$M_{cu} \geq N_d(M_g + M_s + M_d + M_w) + N_l M_l \quad (9\text{-}12)$$

and

$$\epsilon_{su} \geq \epsilon_{sl} \quad (9\text{-}13)$$

The above inequalities are similar to expressions (8-27) and (8-28) corresponding to noncomposite construction. The quantity M_{cu} is the flexural strength of the composite section; ϵ_{su} is the strain in the prestressing steel at failure of the composite section; M_s denotes the moment caused by the weight of that portion of the slab which contributes to the composite section; M_d is the moment due to any additional dead load that may be present before the slab concrete has set; and M_w is the moment due to any dead load (such as wearing surface) that may act on the structure after the slab concrete has set.

In addition, the following inequality should be satisfied in the stringer section:

$$M_u \geq N_d(M_g + M_s + M_d) \tag{9-14}$$

Thus, expressions (9-12) to (9-14) constitute the basis of design.

The condition stated by expression (9-13) can be satisfied easily in a composite section, since a comparatively large area of concrete is available in the compression zone. It can be shown that a considerable compression force may be developed in the slab if the width of slab is large, even though the strength of concrete in the slab is usually in the neighborhood of 3000 psi.

The effective width of slab is usually taken as the center-to-center spacing of stringers, which ranges between 4 and 7 ft. A width of this order of magnitude for the slab provides large compression forces even for high ductilities. On the other hand, the higher the ductility, the smaller are the available compression force and the required area of prestressing steel. Hence, if the ductility is too high, the requirement given by expression (9-12) may no longer be satisfied. Thus, it can be seen that although high ductility is available, expression (9-12) provides an upper limit for it. In practice the highest ductility compatible with expression (9-12) should be used.

The requirement stated by expression (9-13) ensures that the flexural strength of the stringer section is adequate. Where the available area of slab is large, this requirement is automatically satisfied.

Determination of area and prestressing force The discussions in the preceding section may be expressed conveniently in algebraic form for use in design.

It is assumed that the neutral axis of the composite section at failure will fall in the slab. Ignoring the effect of non-prestressed compression reinforcement, if any, and designating S and t_s, respectively, as the width and thickness of the slab, from Eqs. (3-9) and (3-8) the following can be written:

$$M_{cu} = \frac{a^2 S}{\epsilon_u^2} \int_0^{\epsilon_u} \epsilon f(\epsilon) \, d\epsilon + A_s f_{su}(d - a) \tag{9-15}$$

and

$$A_s f_{su} = \frac{aS}{\epsilon_u} \int_0^{\epsilon_u} f(\epsilon) \, d\epsilon \tag{9-16}$$

where

$$a = \frac{\epsilon_u}{\epsilon_{sl} - \epsilon_{se} - \epsilon_{ce} + \epsilon_u} (d + t_s) \tag{9-17}$$

DESIGN OF COMPOSITE PRESTRESSED-CONCRETE BEAMS

The substitution of Eq. (9-15) for M_{cu} and $A\gamma L^2/8$ for M_g in expression (9-12) yields the following for A, the area of the stringer:

$$A \leq \frac{8}{\gamma L^2} \left[\frac{a^2 S}{N_d \epsilon_u{}^2} \int_0^{\epsilon_u} \epsilon f(\epsilon)\, d\epsilon + \frac{A_s f_{su}(d-a)}{N_d} - \frac{N_l}{N_d} M_l - M_s - M_d - M_w \right] \quad (9\text{-}18)$$

Expression (9-18) provides an upper bound for A.

Since expression (9-14) for the stringer section is similar to non-composite design, a lower bound for A may be obtained by expressing (9-14) in the following form:

$$A \geq \frac{M_s + M_d}{d^2 f'_c Q/h\psi N_d - \gamma L^2/8} \quad (9\text{-}19)$$

In practical problems since expression (9-19) seldom governs, it is more convenient to check the flexural strength of the stringer section directly by expression (9-14).

The steps to be taken in the determination of A and A_s may be summarized as follows:

1. On the basis of a ductility greater than or equal to the prescribed ductility, calculate a from Eq. (9-17).
2. Calculate A_s from Eq. (9-16), using the given stress-strain diagram for concrete.
3. Find the upper bound of A from expression (9-18). If A is too large, it means that the ductility may be increased further. On the other hand, if A is unreasonably small, the ductility should be decreased.
4. Determine the proportions of the section and check expression (9-14).

9-13 ILLUSTRATIVE PROBLEM 9-3

Given is a simply supported bridge of 54 ft span consisting of precast prestressed-concrete stringers and cast-in-place slab. The roadway slab is $6\frac{1}{2}$ in. thick, and the structure is to be designed for H20-S16-44 loading. It is anticipated that the structure will have to support a 2-in. wearing surface. The bridge will have one diagram at midspan connecting the stringers, whose weight is equivalent to a concentrated load of 1.25 kips per intermediate stringer. Design an intermediate stringer, assuming a center-to-center spacing of 5 ft and overall depth of 36 in.

The following are given:

For slab: $f'_c = 3000$ psi
For stringer: $f'_c = 5000$ psi
$\epsilon_{se} = 0.0048$
$\epsilon_u = 0.003$
$\gamma = 0.15$ kcf

The quantity ϵ_{ce} may be taken as 0.0005. The stress-strain diagrams for concrete and prestressing steel are shown in Figs. 8-7 and 8-8, respectively.

The moments may be calculated as follows:

$M_s = 0.406 \times \dfrac{(54)^2}{8} \times 12 = 1775$ in.-kips

$M_d = 1.25 \times \dfrac{54}{4} \times 12 = 203$ in.-kips

$M_e = \dfrac{1}{2} \times 699.3 \times \dfrac{5}{5} \times 1.28 \times 12 = 5370$ in.-kips

$M_w = 0.125 \times \dfrac{(54)^2}{8} \times 12 = 546$ in.-kips

Assume $\epsilon_{sl} = 0.031$, and $d + t_s = 33 + 6.5 = 39.5$ in. From Eq. (9-17), $a = 4.12$ in.; from Eq. (9-16), $A_s = 2.33$ in.²; from expression (9-18), the following results:

$A \leq 330$ in.²

To provide the required area for prestressing steel, sixteen ½-in. strands are used. Figure 9-12 shows the stringer section and the arrangement of strands. The web is taken as $5\frac{1}{2}$ in. to provide sufficient room for draping. The top flange is made 12 in. wide in order to provide sufficient area to transmit the shearing stresses which occur between slab and stringer in the composite section.

It can be shown that the section of Fig. 9-12 satisfies expression

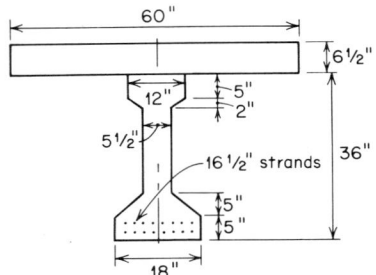

Fig. 9-12 The composite section, Illustrative Problem 9-3.

DESIGN OF COMPOSITE PRESTRESSED-CONCRETE BEAMS 341

Table 9-5 Section properties and stresses in extreme fibers of stringer*

Section	A, in.^{2}	y_t, in.	y_b, in.	I, in.^{4}	Stress before losses (transfer), ksi		Stress after losses, ksi	
					Top (tens)	Bottom (comp)	Top (comp)	Bottom (tens)
Gross:								
Stringer	330.8	20.24	15.76	48,050	−0.23	2.16	1.10	−0.03
Composite	642.8	8.85	27.15	137,960				
Transformed:								
Stringer	344.6	20.78	15.22	50,190	−0.17	1.98	1.14	−0.04
Composite	656.6	9.36	26.64	145,800				

* In the above calculation the ratio of modulus of elasticity of concrete in the slab to that in the stringer is taken as 0.8. The properties of the transformed section are calculated assuming $n = 7$ and $A_s = 2.3$ in.^{2}.

(9-14). Table 9-5 shows a summary of the section properties and stresses in the stringer before and after losses.

In the example presented, the effect of non-prestressed reinforcement in either slab or stringer is not taken into account. The non-prestressed compression reinforcement increases the ductility of the stringer section, but is usually under tensile stress at the failure of the composite section. The longitudinal reinforcement in the slab increases the ductility of the composite section, if placed at the top of the slab.

PROBLEMS

9-1. Solve Illustrative Problems 9-1 and 9-2 for a 75-ft span.

9-2. A type II AASHO-PCI section is manufactured with a concrete strength of $f'_c = 5000$ psi. A concrete slab 7 in. thick and 42 in. wide with $f'_c = 3000$ psi is placed at the top of the section with the proper shear connection. Using the provisions of the ACI Code 318-63 estimate the ultimate moment of the section. The stress-strain diagram for prestressing steel is as shown in Fig. 3-5.

REFERENCES

1. Hanson, N. W.: Precast-Prestressed Concrete Bridges: 2. Horizontal Shear Connections, *J. Res. Develop. Lab.*, vol. 2, no. 2, May, 1960.
2. Evans, R. H., and A. S. Parker: Behavior of Prestressed Concrete Composite Beams, *J. Am. Concrete Inst.*, May, 1955; *Proc.*, vol. 51, pp. 861–878.
3. "Standard Specifications for Highway Bridges," 8th ed., The American Association of State Highway Bridges, 1961.
4. "Reinforced Concrete R/C 34," Portland Cement Association, 1959.
5. Khachaturian, N., and G. Gurfinkel: Ultimate Design of Prestressed Concrete Beams, *Highway Res. Record* 147, pp. 140–156, December, 1966.

10
Analysis and Design of Continuous Prestressed-concrete Beams

10-1 INTRODUCTION

Properly designed continuous prestressed-concrete beams are suitable for longer spans than is usually practical for simple prestressed-concrete beams. In an efficiently designed continuous beam it is possible to have an appreciable saving of material, both concrete and steel, as compared with equivalent simple spans. The efficiency of design depends not only on the proportioning of concrete section and the prestressing force but also on the number of spans and the ratio of span lengths.

Principal advantages of continuous prestressed-concrete beams over simple beams are their stiffness and strength. Continuous beams are stiffer than simple beams; hence it is possible to use a smaller depth for a continuous beam without decreasing its stiffness. Continuous beams can also be designed to have more ductility and strength than is possible in equivalent simple beams.

Continuity provides a versatile means of construction in prestressed concrete, and it can be applied by a multiplicity of methods.[1]

ANALYSIS AND DESIGN OF CONTINUOUS PRESTRESSED-CONCRETE BEAMS

Continuous prestressed-concrete beams can be built as a cast-in-place, post-tensioned construction requiring falsework. It is also possible to avoid the need for falsework by means of a construction technique in which a combination of shop pretensioning and field post-tensioning is used. Precast units may be manufactured with sufficient prestressing to permit their erection without falsework. After the erection of these units is complete, they can be made continuous for live load and superimposed dead load by various means of field post-tensioning.

Figure 10-1a shows a two-span continuous prismatic beam in which the profile of the center of gravity of steel is shown as a curved line. The beam shown in the figure corresponds to a post-tensioned construction, and may be considered as the fundamental type of continuous prestressed-concrete structure. It is seen that in this type of structure for the sake of economy as well as appearance the spans are usually made of equal spans. Figure 10-1b and c shows three- and four-span prismatic continuous beams, respectively.

From Fig. 10-1a through c it can be seen that in order to make the moments due to the prestressing force opposite to those caused by the acting loads, the prestressing force must act at the bottom in the center of each span, while over the supports it must act at the top. In order to satisfy this condition, the profile of the center of gravity of steel has to change its curvature as it passes from a region of positive moment to a

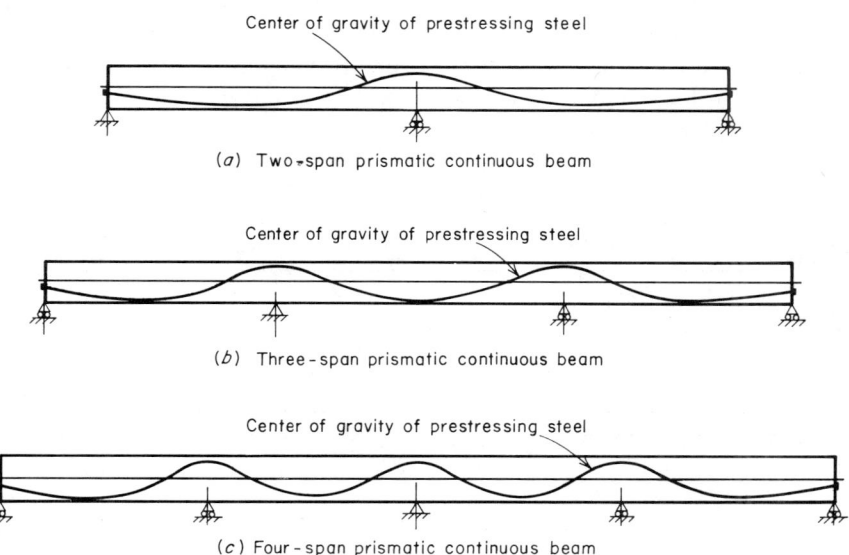

(a) Two-span prismatic continuous beam

(b) Three-span prismatic continuous beam

(c) Four-span prismatic continuous beam

Fig. 10-1 Typical prestressed-concrete prismatic beams.

support. This feature of continuous prestressed-concrete beams results in appreciable friction between the prestressing steel and the sheathing, for which adequate provisions may become necessary.

10-2 LOSS IN THE PRESTRESSING FORCE DUE TO FRICTION

One of the important problems in design and construction of continuous prestressed-concrete beams is the friction between the prestressing steel and the sheathing or the concrete duct through which the steel passes. Under unfavorable conditions it is possible to have such a large friction that a greater part of the prestressing force may be lost between the jacking end and the end support. Evidently under these conditions, continous prestressed-concrete beams would not be practical.

There are several different ways by which the frictional losses may be minimized. Evidently the friction between the prestressing steel and its surrounding material will reduce, (1) by reduction of the length of the prestressing units, (2) by reduction of the curvature of the profile of the prestressing steel, and (3) by reduction of the coefficients of friction. All three ideas may be used to make the loss due to friction small.

Reduction of the length of prestressing units One of the important causes of large frictional losses in continuous prestressed-concrete beams is the great length of the prestressing units when this type of structure is employed. Although in the type of construction shown in Fig. 10-1a through c the prestressing steel can be pulled from both ends, it is still possible to have large losses in the prestressing force due to friction, especially in long spans. The frictional losses may be reduced by using

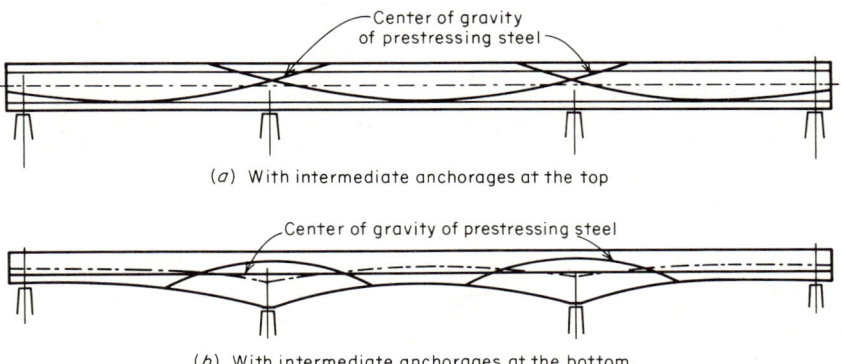

Fig. 10-2 Continuous prestressed-concrete beams with short, discontinuous prestressing elements.

ANALYSIS AND DESIGN OF CONTINUOUS PRESTRESSED-CONCRETE BEAMS

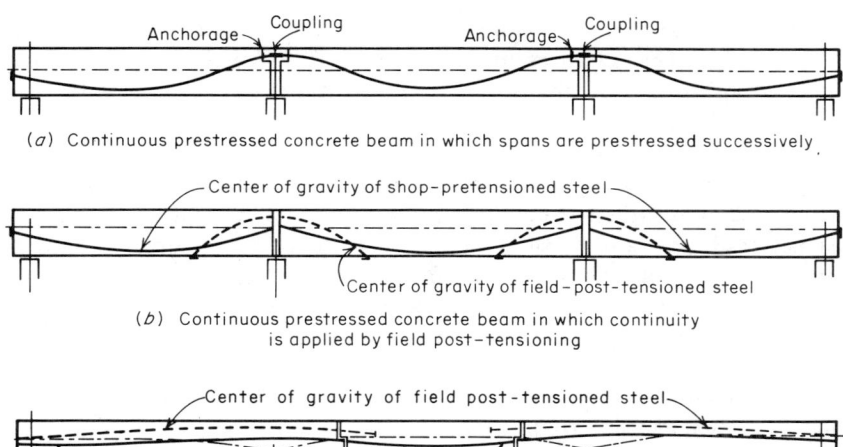

Fig. 10-3 Methods of applying continuity by field post-tensioning.

shorter lengths of discontinuous steel with intermediate anchorages instead of continuous steel with end anchorages only. Figure 10-2a shows such an arrangement in which the prestressing steel is overlapped and intermediate anchorages provided. Although this device reduces the frictional losses, it introduces intermediate anchorages which complicate the construction. Intermediate anchorages are generally subject to high stress concentrations, and adequate reinforcement and additional prestressing are always necessary to prevent cracking. Figure 10-2b shows another arrangement in which the beam is haunched and intermediate anchorages are located at the bottom of the beam.

The length of prestressing units may also be reduced by prestressing one span at a time. In this scheme each span is prestressed and a temporary anchorage provided. Subsequently, by means of coupling, the end of prestressing steel in the next span is joined to the steel in the span already prestressed. This method has been applied successfully by use of high-strength bars.[2] This method is suitable for short spans and has the added advantage of being applicable to either a cast-in-place or a precast beam. Figure 10-3a shows a three-span continuous beam in which this method is applied to obtain continuity.

Figure 10-3b and c shows arrangements to reduce the length of prestressing elements. Precast and pretensioned elements are erected without falsework, and subsequently by field post-tensioning the structure is made continuous for the superimposed dead load and live load. The solid lines in these figures indicate the profile of the shop pretensioned steel, and the dashed lines designate the position of the field post-tensioned steel. This type of construction has been used satisfactorily.[3]

Reduction of the curvature of the profile of prestressing steel Reduction of curvature of prestressing steel causes a reduction in the frictional loss. It is possible to eliminate the curvature entirely by using straight prestressing steel and introducing a small curvature in the beam, as shown in Fig. 10-4a. The curvature in the beam causes the center-of-gravity line of the section to become a curve, thus obtaining the desired eccentricities throughout the structure. The same results may be obtained by haunching the beam or varying its depth, as shown in Fig. 10-4b. The arrangements shown in Fig. 10-4a and b are not entirely satisfactory since it is difficult to control the eccentricity throughout the length of the structure. The arrangement shown in Fig. 10-4a may be objectionable if it is necessary for the top of the girder to be flat. The variable-depth beam shown in Fig. 10-4b may also be objectionable from an architectural or functional point of view.

Figure 10-4c shows an arrangement in which by the slight variation of the depth or haunching it is possible to reduce the curvature of the profile of steel considerably. From a practical point of view this arrangement seems to be satisfactory.

(a) Curved continuous prestressed concrete beam with straight prestressing elements

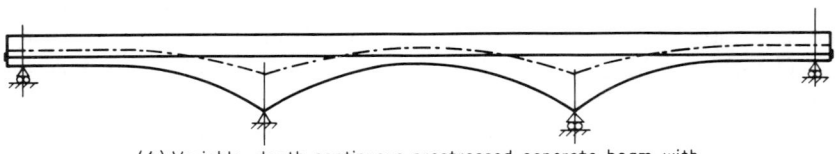
(b) Variable-depth continuous prestressed concrete beam with straight prestressing elements

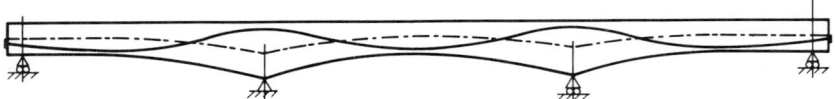
(c) Variable-depth continuous prestressed concrete beam with reduced curvature of the prestressing elements

Fig. 10-4 Curved and variable-depth continuous beams.

Reduction of the coefficient of friction The coefficient of friction between the prestressing steel and its sheathing varies greatly. It depends on the condition of wires at the time of prestressing and the material surrounding the prestressing wires. High frictional coefficients may result when wires with scales and rust are wrapped in softer sheet metal. By the proper choice of materials and care in construction it is possible to reduce the coefficient of friction to a very small quantity.[2]

In addition to the ideas discussed above, friction can also be reduced by exercising extreme care during the construction so that the lateral alignment of the prestressing wires is as nearly perfect as possible. Considerable friction is developed by the lateral bending of the wires, which may cause serious losses in the prestressing force.

10-3 REVERSAL OF SIGN OF THE LIVE-LOAD MOMENT

A limitation which may present itself in continuous prestressed-concrete structures is reversal of sign of the live-load moment, particularly at the region of small dead-load moment. Evidently, prestressed concrete is not suitable when the acting loads change their direction. For example, in simply supported beams if the live load could act in either direction, design of a beam would require small or no eccentricities, resulting in high prestressing force.

The effect of the reversal of sign of moment generally is not important in continuous prestressed-concrete highway bridges since usually the live load is not very high. In other structures, it is possible that in some cases the effect of live load may become important.

In a prismatic, two-span continuous beam such as shown in Fig. 10-1a the dead-load moment is small in a region between one-quarter to about one-third of the span from the center support. In this region, moment due to the prestressing force is negative, and it is possible that the live-load moment may also be negative. In this case the prestressing force is no longer compensating the effects of the acting loads. If the live load is large, the stresses in this region may become too high.

It should be pointed out that the effect of reversal may be minimized by adoption of the proper shape for the profile of the centroid of steel.

10-4 DESIGN OF CONTINUOUS PRESTRESSED-CONCRETE BEAMS

Continuous prestressed-concrete beams, like simple beams, are usually designed on the basis of working-stress design criteria.

The stresses throughout the beam due to all combinations of loads are calculated on the basis of the assumption that concrete is an elastic material. These calculated stresses are to be less than or equal to the allowable stresses.

Elastic analysis is based upon the uncracked section; that is, it is assumed that throughout the continuous beam there are no cracks in the beam under the service loads or, if there are, they do not affect the relative stiffness of the structure.

Since continuous beams are statically indeterminate, design on the basis of strength should depend on the collapse mechanism of the structure. The ultimate design method as presently used is based upon the flexural failure of one section of the structure, and strictly speaking cannot be applied to continuous beams. The limit design concept cannot be used, because the limited ductility of continuous prestressed concrete beams does not provide sufficient deformation to develop the collapse mechanism.

Since design of continuous prestressed-concrete beams depends upon the knowledge of moments caused by the prestressing force, a thorough understanding of the elastic analysis is fundamental to the development of design criteria. In the subsequent sections, therefore, the discussion of the elastic analysis of continuous structures precedes that of the design criteria.

The analysis and design procedures presented in this chapter are for prismatic continuous prestressed-concrete beams.

10-5 ELASTIC ANALYSIS OF CONTINUOUS BEAMS

From the preceding discussion it is clear that there are numerous different types of continuous prestressed-concrete structures. There is no attempt made in this section to present the elastic analysis of each type in detail. The structure considered in this section is assumed to be a prismatic post-tensioned beam in which the profile of the prestressing steel is continuous throughout the length of the structure. Figure 10-1 shows the typical structure considered in this section. The analysis is based upon the following assumptions:

1. During the prestressing operation the supports are restrained against vertical movement.
2. The horizontal component of the tension in the cable is equal to the tension in the cable; that is, the profile of the center of gravity of the prestressing steel is flat.
3. Frictional force is negligible; that is, the prestressing force is the same throughout the length of the structure.
4. The reduction of the cross-sectional area because of cable ducts is negligible.

ANALYSIS AND DESIGN OF CONTINUOUS PRESTRESSED-CONCRETE BEAMS

The elastic analysis of continuous beams for the dead and applied loads is generally well known, and there is no need to discuss it in this section. However, the analysis for the prestressing force and determination of moments due to prestressing force at any point along the beam present certain unusual features which merit some consideration. The following paragraphs contain a brief discussion of elastic analysis of continuous beams.

Figure 10-5 shows that prestressing force produces statically indeterminate moments as well as determinate moments. The statically determinate moments at any point may be obtained by multiplying the

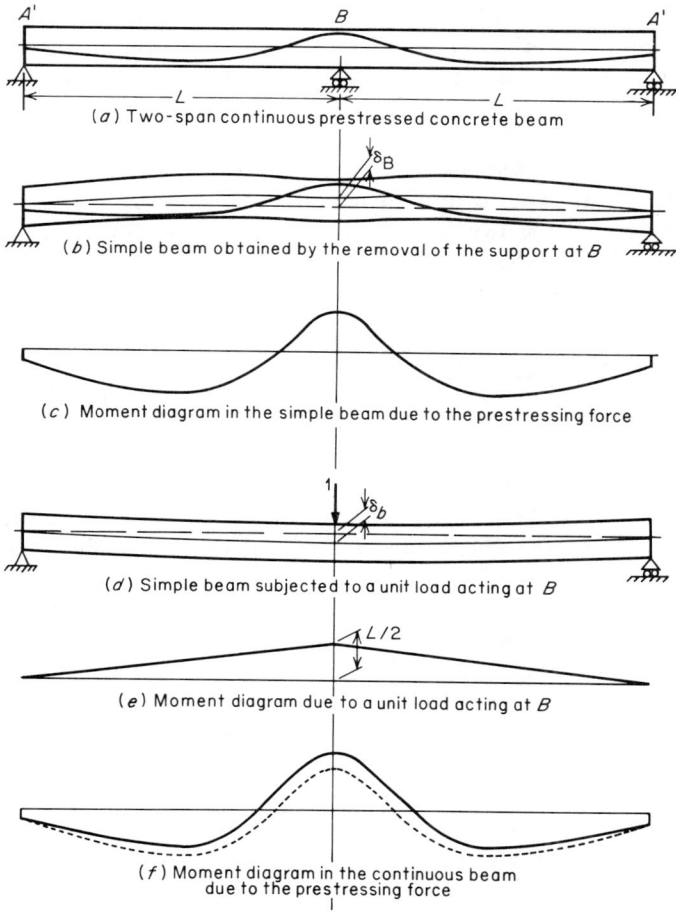

Fig. 10-5 Analysis of a two-span continuous beam.

prestressing force by the eccentricity at that point. The indeterminate moment is a result of the additional restraint caused by the redundant support, since it is assumed that the supports are restrained against vertical movement.

Figure 10-5a shows a two-span continuous prestressed-concrete beam. Removal of the middle support results in a statically determinate structure, as shown in Fig. 10-5b, and the moment diagram is shown in Fig. 10-5c. Because of this moment, point B may deflect upward or downward, depending upon the shape of the moment diagram and the relative magnitude of positive and negative areas under the moment diagram. It is possible that the shape of the moment diagram will be such that no deflection will occur at point B. If point B deflects at all, we may conclude that the prestressing force is introducing an indeterminate reaction at support B which in turn introduces indeterminate moments throughout the span. Figure 10-5b indicates upward deflection of the beam when support B is removed, since the negative area under the moment diagram dominates. The upward deflection at point B is designated as δ_B. Figure 10-5d shows the statically determinate structure with a unit load acting at point B. The resulting moment diagram is shown in Fig. 10-5e. The deflection at point B due to this unit load is designated as δ_b. Evidently the indeterminate reaction due to the prestressing force at the middle support will be

$$\text{Indeterminate reaction at } B = \frac{\delta_B}{\delta_b} = \frac{\int_0^L (Pem/EI)\,dx}{\int_0^L (m^2/EI)\,dx} \qquad (10\text{-}1)$$

where P = prestressing force
$\quad\quad\;\; e$ = eccentricity, that is, distance from center of gravity of steel to center of gravity of beam section, given as function of x
$\quad\quad\; m$ = moment due to unit load acting at point B, given as function of x
$\quad\quad\; L$ = length of one span
$\quad\quad\; E$ = modulus of elasticity, which may be taken as constant quantity
$\quad\quad\;\; I$ = moment of inertia of beam section, which is constant quantity in prismatic beam; it is a function of x in a variable-depth beam
$\quad\quad\; x$ = distance measured from an assumed origin, usually left-end support of structure

The indeterminate moment at point B may be expressed as follows:

$$\text{Indeterminate moment at } B = \frac{\delta_B}{2\delta_b} L$$

ANALYSIS AND DESIGN OF CONTINUOUS PRESTRESSED-CONCRETE BEAMS

The indeterminate-moment diagram will have the same shape as that in Fig. 10-5e except that the moment at point B will be $(\delta_B/2\delta_b)L$ instead of $L/2$.

The moment diagram due to the prestressing force may be obtained by multiplying the ordinates of the moment diagram in Fig. 10-5e by δ_B/δ_b and superposing the result on Fig. 10-5c. The moment diagram due to the prestressing force is shown in Fig. 10-5f by solid lines. The dashed line in this figure indicates the statically determinate moments, the same as that shown in Fig. 10-5c.

The indeterminate moment is positive when δ_B is an upward deflection, and it is negative when δ_B is downward. There are no indeterminate moments when $\delta_B = 0$.

From the above discussion it can be seen that in order to determine the magnitude of the indeterminate moments due to the prestressing force, it is necessary to know δ_B. Determination of δ_B, on the other hand, requires the knowledge of the shape of the statically determinate moment diagram or the shape of the profile of the centroid of the prestressing steel. The position of the center of gravity of steel at any point may be defined by the eccentricity at that point. Eccentricity is the distance from the centroid of the prestressing steel to the center of gravity of the beam section, and its variation defines the shape of the profile of the centroid of steel.

The eccentricity may be defined by several parabolas or by a combination of a parabola and a fourth-degree curve. In the following section, equations are derived for the eccentricity for the end and typical interior spans.

10-6 THE VARIATION OF ECCENTRICITY

As seen in the preceding discussions, eccentricity varies along the beam, and this variation influences the indeterminate moments. In straight prismatic beams since the centroid of the section is straight, the variation of the eccentricity can be presented by an equation defining the position of the centroid of steel. In the following paragraphs the variation of the eccentricity is presented by parabolas and by a combination of a parabola and a fourth-degree curve.

Presentation of eccentricity of parabolas Figure 10-6 represents a typical endspan of a continuous prestressed-concrete beam. The variation of eccentricity, which also defines the position of the centroid of steel, is presented by three parabolas. The origin of the coordinate axes is taken at point 0, the left end of the span. Parabolas AC and CI define the eccentricity from $x = 0$ to $x = \alpha L$ and $x = \alpha L$ to $x = (1 - \alpha_1)L$,

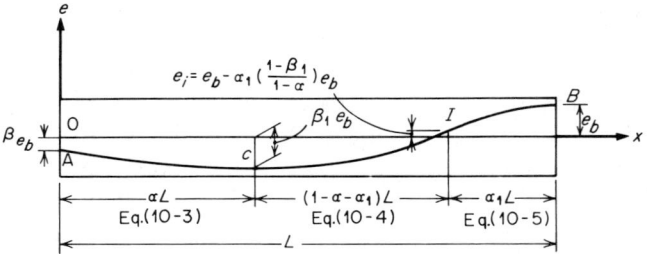

Fig. 10-6 Variation of eccentricity in a typical endspan presented as three parabolas.

respectively, while parabola IB defines the eccentricity from

$$x = (1 - \alpha_1)L$$

to $x = L$. Parabola AC is tangent to parabola CI at point C, and their common tangent is horizontal. Parabolas CI and IB have a common tangent at point I, and the tangent to parabola IB at point B is horizontal.

The eccentricities below the centroidal axis of the beam are taken as *negative*, and those above this axis are taken as *positive*. The eccentricity at point B is taken as e_b, and the eccentricities at points A and C are taken as βe_b and $\beta_1 e_b$, respectively.

It can be shown that in order for parabolas CI and IB to have a common tangent at point I the following expression should be satisfied:

$$e_i = e_b - \alpha_1 \frac{1 - \beta_1}{1 - \alpha} e_b \tag{10-2}$$

where e_i is the eccentricity at point I where parabolas CI and IB have a common tangent. All other terms are discussed above.

The equation for parabola AC, defining the eccentricities from $x = 0$ to $x = \alpha L$, can be written as follows:

$$e = A_0 x^2 + B_0 x + C_0$$

When $x = 0$, $e = \beta e_b$ and it may be concluded that $C_0 = \beta e_b$.

When $x = \alpha L$, $e = \beta_1 e_b$ and $e' = 0$, and the following equations may be obtained:

$$A_0(\alpha L)^2 + B_0(\alpha L) + \beta e_b = \beta_1 e_b$$
$$2A_0(\alpha L) + B_0 = 0$$

The simultaneous solution of the above equations yields the following values for A_0 and B_0:

$$A_0 = -\frac{\beta_1 - \beta}{(\alpha L)^2} e_b$$

$$B_0 = 2\frac{\beta_1 - \beta}{\alpha L} e_b$$

and

$$C_0 = \beta e_b$$

Substituting the above constants in the general equation for a parabola, we have

$$e = -\frac{(\beta_1 - \beta)e_b}{(\alpha L)^2} x^2 + \frac{2(\beta_1 - \beta)e_b}{\alpha L} x + \beta e_b \qquad (10\text{-}3)$$

From $x = \alpha L$ to $x = (1 - \alpha_1)L$ the equation for the eccentricity may be written as follows:

$$e = A_1 x^2 + B_1 x + C_1$$

When $x = \alpha L$, $e = \beta_1 e_b$ and $e' = 0$; and when $x = (1 - \alpha_1)L$, $e = e_i = e_b - \alpha_1[(1 - \beta_1)/(1 - \alpha)]e_b$. These conditions yield the following equation:

$$e = \frac{(1 - \beta_1)e_b}{(1 - \alpha)(1 - \alpha - \alpha_1)L^2} x^2 - \frac{2\alpha(1 - \beta_1)e_b}{(1 - \alpha)(1 - \alpha - \alpha_1)L} x$$
$$+ \frac{\alpha^2(1 - \beta_1)e_b}{(1 - \alpha)(1 - \alpha - \alpha_1)} + \beta_1 e_b \qquad (10\text{-}4)$$

Similarly, it can be shown that from $x = (1 - \alpha_1)L$ to $x = L$ the eccentricity can be expressed as follows:

$$e = -\frac{(1 - \beta_1)e_b}{\alpha_1(1 - \alpha)L^2} x^2 + \frac{2(1 - \beta_1)e_b}{\alpha_1(1 - \alpha)L} x - \frac{(1 - \beta_1)e_b}{\alpha_1(1 - \alpha)} + e_b \qquad (10\text{-}5)$$

Equations can also be developed for the eccentricity in a typical interior span. Figure 10-7 represents a typical interior span of a prestressed-concrete beam in which the variation of the eccentricity is symmetrical about the centerline. If the midspan is assumed to be the origin of the coordinate axes, the equations for the eccentricity can be presented as follows:

From $x = -L/2$ to $x = -(\frac{1}{2} - \alpha_2)L$

$$e = -\frac{2(1 - \beta_2)e_b}{\alpha_2 L^2} x^2 - \frac{2(1 - \beta_2)e_b}{\alpha_2 L} x - \frac{(1 - \beta_2)e_b}{2\alpha_2} + e_b \qquad (10\text{-}6)$$

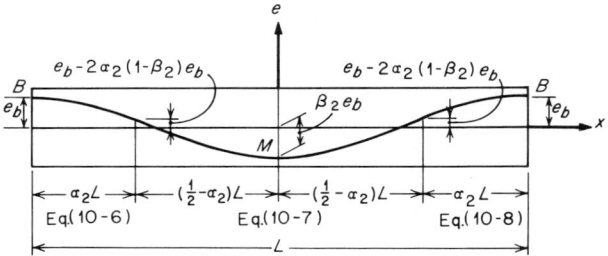

Fig. 10-7 Variation of eccentricity in a typical interior span, presented as three parabolas.

From $x = -(\frac{1}{2} - \alpha_2)L$ to $x = (\frac{1}{2} - \alpha_2)L$

$$e = \frac{2(1-\beta_2)e_b}{(\frac{1}{2} - \alpha_2)L^2} x^2 + \beta_2 e_b \qquad (10\text{-}7)$$

and from $x = (\frac{1}{2} - \alpha_2)L$ to $x = L/2$

$$e = -\frac{2(1-\beta_2)e_b}{\alpha_2 L^2} x^2 + \frac{2(1-\beta_2)e_b}{\alpha_2 L} x - \frac{(1-\beta_2)e_b}{2\alpha_2} + e_b \qquad (10\text{-}8)$$

Presentation of eccentricity by a combination of a parabola and a fourth-degree curve[4] The variation of eccentricity for a typical endspan may be expressed by a combination of a parabola and a fourth-degree curve. Figure 10-8 shows a typical endspan; as before, the origin of the coordinate axes is taken at point 0, the left end of the span. From $x = \alpha L$ to $x = L$ the eccentricity is defined by a fourth-degree parabola. In this case there is one continuous curve from $x = \alpha L$ to L, and point I is the point of inflection of this curve.

The equation for parabola AC is the same as Eq. (10-3).

$$e = -\frac{(\beta_1 - \beta)e_b}{(\alpha L)^2} x^2 + \frac{2(\beta_1 - \beta)e_b}{\alpha L} x + \beta e_b \qquad (10\text{-}3)$$

The equation for the fourth-degree curve defining the eccentricity from $x = \alpha L$ to $x = L$ can be written as follows:

$$e = A_1 x^4 + B_1 x^3 + C_1 x^2 + D_1 x + E_1$$

In order to determine the coefficients of the above fourth-degree equation conveniently, the origin of the coordinate axis will be shifted to point c'. Since the fourth-degree curve shown in Fig. 10-8 is symmetrical about the vertical axis passing through c', $B_1 = D_1 = 0$. Also, when $x = 0$, $e = \beta_1 e_b$, and $E_1 = \beta_1 e_b$.

The values of A_1 and C_1 are obtained from the evaluation of the

ANALYSIS AND DESIGN OF CONTINUOUS PRESTRESSED-CONCRETE BEAMS

preceding conditions, which yield

$$A_1 = -\frac{(1-\beta_1)e_b}{(1-\alpha)^4 L^4}$$

$$C_1 = \frac{2(1-\beta_1)e_b}{(1-\alpha)^2 L^2}$$

Substituting the above coefficients in the general equation for the fourth-degree curve, we have

$$e = -\frac{(1-\beta_1)e_b}{(1-\alpha)^4 L^4} x^4 + \frac{2(1-\beta_1)e_b}{(1-\alpha)^2 L^2} x^2 + \beta_1 e_b$$

It should be pointed out that the above equation is developed by assuming point c' as the origin. Since the origin is taken at point 0, the left end of the structure, it is necessary to modify the above equation by shifting the origin to point 0.

In order to shift the origin from point c' to point 0 we may substitute $x - \alpha L$ for x in the above equation. The equation for the fourth-degree curve defining the eccentricity from $x = \alpha L$ to $x = L$ when the origin is at 0 will be the following:

$$e = -\frac{(1-\beta_1)e_b}{(1-\alpha)^4 L^4} (x - \alpha L)^4 + \frac{2(1-\beta_1)e_b}{(1-\alpha)^2 L^2} (x - \alpha L)^2 + \beta_1 e_b$$

or

$$e = -\frac{(1-\beta_1)e_b}{(1-\alpha)^4 L^4} x^4 + \frac{4\alpha(1-\beta_1)e_b}{(1-\alpha)^4 L^3} x^3$$
$$-\frac{2(2\alpha^2 + 2\alpha - 1)(1-\beta_1)e_b}{(1-\alpha)^4 L^2} x^2 + \frac{4\alpha(2\alpha - 1)(1-\beta_1)e_b}{(1-\alpha)^4 L} x$$
$$+ \frac{\alpha^2(\alpha^2 - 4\alpha + 2)(1-\beta_1)e_b}{(1-\alpha)^4} + \beta_1 e_b \quad (10\text{-}9)$$

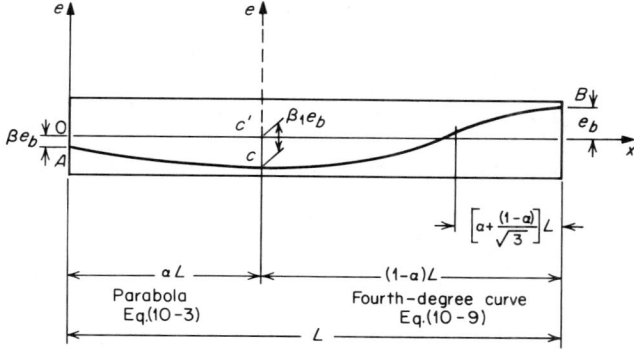

Fig. 10-8 Variation of eccentricity in a typical endspan, presented as a parabola and a fourth-degree curve.

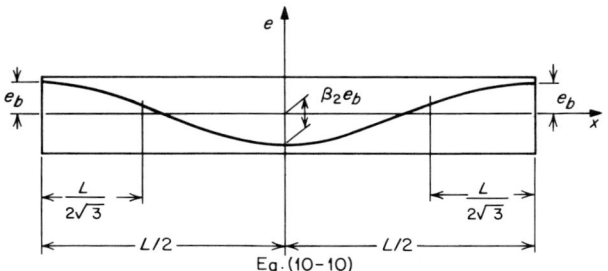

Fig. 10-9 Variation of eccentricity in a typical interior span, presented as a fourth-degree curve.

An equation can also be developed for the eccentricity in a typical interior span. Figure 10-9 represents a typical interior span of a continuous prestressed-concrete beam in which the variation of eccentricity is symmetrical and is defined by a fourth-degree curve.

If the midspan is assumed to be the origin, the eccentricity can be expressed as follows:

$$e = -\frac{16(1-\beta_2)e_b}{L^4}x^4 + \frac{8(1-\beta_2)e_b}{L^2}x^2 + \beta_2 e_b \qquad (10\text{-}10)$$

It should be pointed out that in the fourth-degree curve the position of the point of inflection is known once the value of α is set. The position of the point of inflection in a typical endspan may be determined from Eq. (10-9) by equating the second derivative of e in terms of x to zero:

$$\frac{d^2e}{dx^2} = -\frac{12(1-\beta_1)e_b}{(1-\alpha)^4 L^4}x^2 + \frac{24\alpha(1-\beta_1)e_b}{(1-\alpha)^4 L^3}x - \frac{4(2\alpha^2+2\alpha-1)(1-\beta_1)e_b}{(1-\alpha)^4 L^2} = 0$$

or

$$\left(\frac{x}{L}\right)^2 - 2\alpha\frac{x}{L} + \frac{1}{3}(2\alpha^2 - 2\alpha + 1) = 0$$

It can be shown that the acceptable root in the above equation is the following:

$$\frac{x}{L} = \alpha + \frac{1-\alpha}{\sqrt{3}}$$

that is, the point of inflection is located at a distance $[\alpha + (1-\alpha)/\sqrt{3}]L$ from the left support.

Table 10-1a Expressions for the variation of eccentricity (eccentricity varies parabolically)

Typical endspan   *Typical symmetrical interior span*

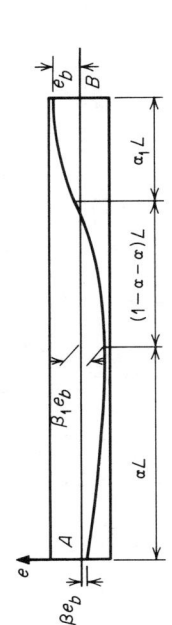

$x = 0$ to $x = \alpha L$:

$$e = -\frac{(\beta_1 - \beta)e_b}{(\alpha L^2)}x^2 + \frac{2(\beta_1 - \beta)e_b}{\alpha L}x + \beta e_b \qquad (10\text{-}3)$$

$x = \alpha L$ to $x = (1 - \alpha_1)L$:

$$e = \frac{(1-\beta_1)e_b}{(1-\alpha)(1-\alpha-\alpha_1)L^2}x^2 - \frac{2\alpha(1-\beta_1)e_b}{(1-\alpha)(1-\alpha-\alpha_1)L}x \\ + \frac{\alpha^2(1-\beta_1)e_b}{(1-\alpha)(1-\alpha-\alpha_1)} + \beta_1 e_b \qquad (10\text{-}4)$$

$x = (1-\alpha_1)L$ to $x = L$:

$$e = -\frac{(1-\beta_1)e_b}{\alpha_1(1-\alpha)L^2}x^2 + \frac{2(1-\beta_1)e_b}{\alpha_1(1-\alpha)L}x - \frac{(1-\beta_1)e_b}{\alpha_1(1-\alpha)} + e_b \qquad (10\text{-}5)$$

$x = -L/2$ to $x = -(\tfrac{1}{2} - \alpha_2)L$:

$$e = -\frac{2(1-\beta_2)e_b}{\alpha_2 L^2}x^2 - \frac{2(1-\beta_2)e_b}{\alpha_2 L}x - \frac{(1-\beta_2)e_b}{2\alpha_2} + e_b \qquad (10\text{-}6)$$

$x = -(\tfrac{1}{2} - \alpha_2)L$ to $x = (\tfrac{1}{2} - \alpha_2)L$:

$$e = \frac{2(1-\beta_2)e_b}{(\tfrac{1}{2} - \alpha_2)L^2}x^2 + \beta_2 e_b \qquad (10\text{-}7)$$

$x = (\tfrac{1}{2} - \alpha_2)L$ to $x = L/2$:

$$e = \frac{2(1-\beta_2)e_b}{\alpha_2 L^2}x^2 + \frac{2(1-\beta_2)e_b}{\alpha_2 L}x - \frac{(1-\beta_2)e_b}{2\alpha_2} + e_b \qquad (10\text{-}8)$$

Table 10-1b Expressions for the variation of eccentricity (eccentricity varies as parabola and fourth-degree curve)

Typical endspan   *Typical symmetrical interior span*

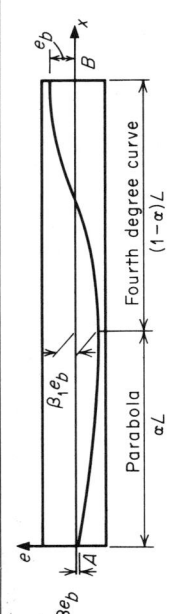

$x = 0$ to $x = \alpha L$:

$$e = -\frac{(\beta_1 - \beta)e_b}{(\alpha L)^2} x^2 + \frac{2(\beta_1 - \beta)e_b}{\alpha L} x + \beta e_b \quad (10\text{-}3)$$

$x = \alpha L$ to $x = L$:

$$e = -\frac{(1-\beta_1)e_b}{(1-\alpha)^4 L^4} x^4 + \frac{4\alpha(1-\beta_1)e_b}{(1-\alpha)^4 L^3} x^3$$
$$-\frac{2(2\alpha^2 + 2\alpha - 1)(1-\beta_1)e_b}{(1-\alpha)^4 L^2} x^2 + \frac{4\alpha(2\alpha-1)(1-\beta_1)e_b}{(1-\alpha)^4 L} x$$
$$+ \frac{\alpha^2(\alpha^2 - 4\alpha + 2)(1-\beta_1)e_b}{(1-\alpha)^4} + \beta_1 e_b \quad (10\text{-}9)$$

$x = -L/2$ to $x = L/2$:

$$e = -\frac{16(1-\beta_2)e_b}{L^4} x^4 + \frac{8(1-\beta_2)e_b}{L^2} x^2 + \beta_2 e_b \quad (10\text{-}10)$$

ANALYSIS AND DESIGN OF CONTINUOUS PRESTRESSED-CONCRETE BEAMS

It can be shown that in a typical interior span the point of inflection is located at a distance $L/(2\sqrt{3})$ from the support.

It may be concluded that when the variation of the eccentricity is presented by the fourth-degree curve, the position of the point of inflection is automatically set. However, when the eccentricity or the centroid of steel is presented by parabolas, the position of the point of inflection only defines the point where one curve ends and another of reverse curvature begins. Therefore, there is an additional degree of freedom when the eccentricity is defined by parabolas.

Both methods have been used in the design of continuous structures. For convenient reference the expressions derived for the variation of eccentricity in the typical end and middle spans are summarized in Tables 10-1a and 10-1b.

10-7 THE FIXED-END MOMENTS

Analysis of the effects of prestressing in a continuous beam can be simple if the method of moment distribution is used. The familiar concepts of fixed-end moments, stiffness, carry-over factors, and moment distribution can be used with advantage to find the indeterminate moments at the supports.

To determine the fixed-end moments due to the prestressing force, at typical end and interior spans, the equations derived for the eccentricity can be used, as follows: Figure 10-10 shows a typical endspan in which the left support A is simply supported and the right support B is fixed. The variation of eccentricity is defined by three parabolas, which are presented by Eqs. (10-3) to (10-5). Introduction of a hinge at B makes the structure statically determinate, and the tangent to the elastic line at B rotates through the angle θ_B, as shown in Fig. 10-10b. The moment diagram of the statically determinate structure is shown in Fig. 10-10c. Application of a unit moment at B in the statically determinate structure causes the tangent to the elastic line to rotate through the angle θ_B. The moment diagram is shown in Fig. 10-10e. Evidently the equation for this moment diagram is $-\dfrac{x}{L}$.

The fixed-end moment at B due to the prestressing force can be expressed as follows:

$$M_{BA} = \frac{\int_0^L Mm\ dx/EI}{\int_0^L (m^2/EI)\ dx} + Pe_b$$

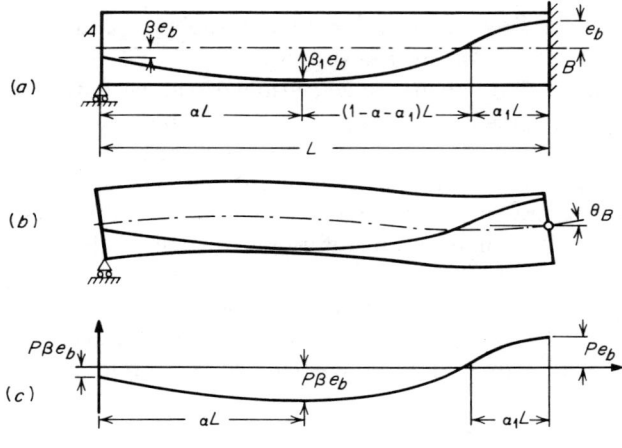

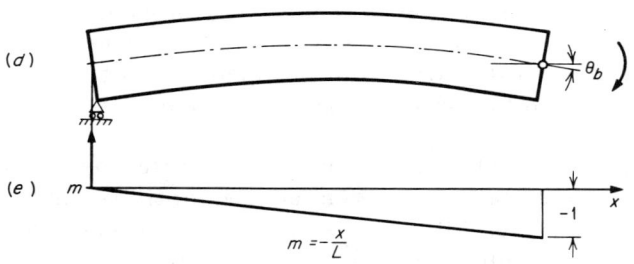

Fig. 10-10 Determination of fixed-end moments caused by the prestressing force.

From Fig. 10-10 it can be seen that M in the above equation is as follows:

$$M = Pe$$

and

$$m = -\frac{x}{L}$$

Substituting the above equations in the expression for the fixed-end moment and considering that in a prismatic beam EI is constant gives the following

$$M_{BA} = -\frac{\int_0^L Pe\,(x/L)\,dx}{\int_0^L (x^2/L^2)\,dx} + Pe_b$$

ANALYSIS AND DESIGN OF CONTINUOUS PRESTRESSED-CONCRETE BEAMS

or

$$M_{BA} = -\frac{3P}{L^2}\int_0^L ex\, dx + Pe_b \tag{10-11}$$

Substituting Eqs. (10-3) to (10-5) for e in the above equation, we have

$$\begin{aligned}M_{BA} =\ & \frac{3P}{L^2}\int_0^{\alpha L}\left[\frac{(\beta_1-\beta)e_b}{(\alpha L)^2}x^2 - \frac{2(\beta_1-\beta)e_b}{\alpha L}x - \beta e_b\right]x\, dx \\ & + \frac{3P}{L^2}\int_{\alpha L}^{(1-\alpha_1)L}\left[-\frac{(1-\beta_1)e_b}{(1-\alpha)(1-\alpha-\alpha_1)L^2}x^2\right. \\ & \left. + \frac{2\alpha(1-\beta_1)e_b}{(1-\alpha)(1-\alpha-\alpha_1)L}x - \frac{\alpha^2(1-\beta_1)e_b}{(1-\alpha)(1-\alpha-\alpha_1)} - \beta_1 e_b\right]x\, dx \\ & + \frac{3P}{L^2}\int_{(1-\alpha_1)L}^{L}\left[\frac{(1-\beta_1)e_b}{\alpha_1(1-\alpha)L^2}x^2 - \frac{2(1-\beta_1)e_b}{\alpha_1(1-\alpha)L}x\right. \\ & \left. + \frac{(1-\beta_1)e_b}{\alpha_1(1-\alpha)} - e_b\right]x\, dx + Pe_b\end{aligned}$$

Integrating and rearranging the above equation, we have

$$M_{BA} = \frac{Pe_b}{4}\{-\beta\alpha^2 - \beta_1[(3-3\alpha_1+\alpha_1^2) + \alpha(2-\alpha_1)] \\ + [(1+\alpha-\alpha_1)^2 - \alpha_1(1-\alpha)]\} \tag{10-12}$$

The expression for the fixed-end moment can be derived similarly for a typical interior span. Figure 10-11 shows a typical interior span with fixed ends in which the profile of the prestressing steel is symmetrical about the midspan. The shape of the profile consists of three parabolas. By introducing hinges at A and B the structure becomes statically determinate, as shown in Fig. 10-11b. Tangents to the elastic line at points A and B rotate through angles θ_A and θ_B, respectively. The moment diagram of the statically determinate structure is shown in Fig. 10-11c. Application of a unit moment at A in the statically determinate structure causes the tangent to the elastic line at points A and B to rotate through the angles θ_a and θ_b, respectively. The moment diagram is shown in Fig. 10-11c.

The following expression can be written for point A:

$$\theta_a M_A + \theta_b M_B = \theta_A$$

where M_A and M_B are the indeterminate moments at A and B. Since by symmetry $M_A = M_B$, we have

$$M_A = M_B = \frac{\theta_A}{\theta_a + \theta_b}$$

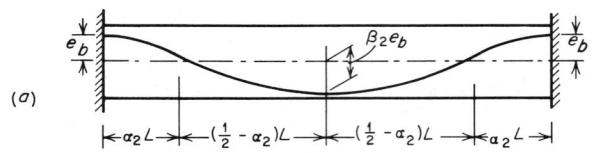

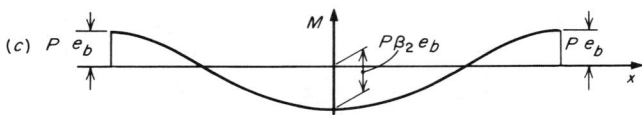
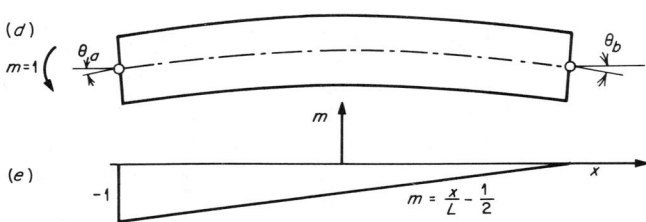

Fig. 10-11 Determination of fixed-end moments caused by the prestressing force.

It can be shown that

$$\theta_a = \frac{L}{3EI} \qquad \theta_b = \frac{L}{6EI}$$

and

$$M_A = M_B = \frac{2EI}{L}\theta_A = \frac{2EI}{L}\int_{-L/2}^{L/2} \frac{Pem}{EI}\,dx$$

where

$$m = \frac{x}{L} - \frac{1}{2}$$

For a prismatic beam the expression for the fixed-end moments will be

$$M_{AB} = M_{BA} = \frac{2P}{L} \int_{-L/2}^{L/2} e\left(\frac{x}{L} - \frac{1}{2}\right) dx + Pe_b \qquad (10\text{-}13)$$

Substituting Eqs. (10-6) to (10-8) for e in the above equation, we have

$$\begin{aligned}M_{AB} = M_{BA} = &\frac{2P}{L} \int_{-L/2}^{-(\frac{1}{2}-\alpha_2)L} \left[-\frac{2(1-\beta_2)e_b}{\alpha_2 L^2} x^2 - \frac{2(1-\beta_2)e_b}{\alpha_2 L} x \right. \\ &\left. - \frac{(1-\beta_2)e_b}{2\alpha_2} + e_b \right]\left(\frac{x}{L} - \frac{1}{2}\right) dx + \frac{2P}{L} \int_{-(\frac{1}{2}-\alpha_2)L}^{(\frac{1}{2}-\alpha_2)L} \left[\frac{2(1-\beta_2)e_b}{(\frac{1}{2}-\alpha_2)L^2} x^2 \right. \\ &\left. + \beta_2 e_b \right]\left(\frac{x}{L} - \frac{1}{2}\right) dx + \frac{2P}{L} \int_{-(\frac{1}{2}-\alpha_2)L}^{L/2} \left[-\frac{2(1-\beta_2)e_b}{\alpha_2 L} x^2 \right. \\ &\left. + \frac{2(1-\beta_2)e_b}{\alpha_2 L} x - \frac{(1-\beta_2)e_b}{2\alpha_2} + e_b \right]\left(\frac{x}{L} - \frac{1}{2}\right) dx + Pe_b \end{aligned}$$

Integration of the above expression gives the following expression for the fixed-end moment:

$$M_{AB} = M_{BA} = \frac{2P}{3}(1 - \alpha_2)(1 - \beta_2)e_b \qquad (10\text{-}14)$$

An expression for the fixed-end moment in a typical endspan can also be developed if the variation of eccentricity is taken as a combination of a parabola and a fourth-degree curve. It can be shown that if Eqs. (10-3) and (10-9) are used for eccentricity, from Eq. (10-11) the following expression may be obtained for the fixed-end moment in a typical endspan:

$$M_{BA} = -\frac{Pe_b}{20}[-5\beta\alpha^2 - \beta_1(3\alpha^2 + 12\alpha + 10) + 4\alpha(2\alpha + 3)] \qquad (10\text{-}15)$$

Similarly, the fixed-end moment in a symmetrical typical interior span, when the eccentricity varies as a fourth-degree curve, can be expressed as follows:

$$M_{BA} = M_{AB} = \frac{8Pe_b}{15}(1 - \beta_2) \qquad (10\text{-}16)$$

For convenient reference the expressions for the fixed-end moments are summarized in Table 10-2.

Table 10-2 Expressions for the fixed-end moments

Beam	Fixed-end moment
	Typical endspan—eccentricity varies parabolically: $$M_{BA} = \frac{Pe_b}{4}\{-\beta\alpha^2 - \beta_1[(3 - 3\alpha_1 + \alpha_1^2) + \alpha(2 - \alpha_1)] + [(1 + \alpha - \alpha_1)^2 - \alpha_1(1 - \alpha)]\} \quad (10\text{-}12)$$
	Typical interior span—eccentricity varies parabolically: $$M_{BA} = M_{AB} = \frac{2Pe_b}{3}(1 - \beta_2)(1 - \alpha_2) \quad (10\text{-}14)$$
	Typical endspan—eccentricity varies as parabola and fourth-degree curve: $$M_{BA} = \frac{Pe_b}{20}[-5\beta\alpha^2 - \beta_1(3\alpha^2 + 12\alpha + 10) + 4\alpha(2\alpha + 3)] \quad (10\text{-}15)$$
	Typical interior span—eccentricity varies as fourth-degree curve: $$M_{BA} = M_{AB} = \frac{8Pe_b}{15}(1 - \beta_2) \quad (10\text{-}16)$$

10-8 GRAPHICAL REPRESENTATION OF FIXED-END MOMENTS

The expressions developed in the preceding section for the fixed-end moments are for the analysis of prismatic continuous structures in which the variation of eccentricity is continuous.

In order to expedite the determination of the fixed-end moments and study the relationship among the variables which affect these moments, the expressions for the fixed-end moments may be plotted.

Figure 10-12a shows plots of Eq. (10-12), which defines the fixed-end moment in a typical endspan when the eccentricity varies as a parabola. The four sets of plots correspond to four values of α_1: 0, 0.10, 0.20, and 0.30. In each plot M_{BA}/Pe_b is taken as ordinate and α as abscissa, and the curves are plotted for $\beta_1 = -0.75, -1.00,$ and -1.25. As shown before, the negative sign of β_1 indicates that the eccentricity at a distance αL from the simple support is negative and opposes the eccentricity at the fixed end. All curves are plotted for $\beta = 0$, that is, for zero eccentricity at the simply supported end. Evidently when $M_{AB}/Pe_b = 1.0$ the indeterminate moment is zero, and Fig. 10-12a shows the combination of variables which result in this condition.

Figure 10-12b shows the plot of Eq. (10-15), which defines the fixed-end moment for a typical endspan when the eccentricity varies as a parabola and a fourth-degree curve. Evidently in this case α and β_1 are the only two variables, and one set of curves is sufficient to define the fixed-end moment.

As before, the line $M_{BA}/Pe_b = 1.0$ corresponds to all combinations of variables which result in zero indeterminate moment.

Figure 10-13 shows the plots of Eqs. (10-14) and (10-16), which define the fixed-end moment in a typical interior span when the eccentricity varies as a parabola and as a fourth-degree curve, respectively.

Equation (10-14) is plotted for three values of α_2: 0, 0.2, and 0.4, taking $(1 - \beta_2)$ as abscissa and M_{BA}/Pe_b as ordinate. This equation corresponds to the parabolic variation of eccentricity, and α_2 defines the position of the point of inflection. Equation (10-16) is shown in the same figure by a thick solid line and corresponds exactly to Eq. (10-14) when $\alpha_2 = 0.2$.

From Fig. 10-12a and b it can be seen that in a typical endspan an increase in the negative eccentricity causes an increase in the fixed-end moment. When the eccentricity varies as a parabola, the closer the point of inflection is to the fixed end, the higher is the fixed-end moment.

In a typical interior span an increase in the negative eccentricity also increases the fixed-end moment; and when the eccentricity varies as a parabola, as the point inflection approaches the ends (as α_2 decreases) the fixed-end moment increases.

$$\frac{M_{BA}}{Pe_b} = -\frac{\beta\alpha^2}{4} - \frac{\beta_1}{4}\left[(3-3\alpha_1+\alpha_1^2)+\alpha(2-\alpha_1)\right] + \frac{1}{4}\left[(1+\alpha-\alpha_1)^2 - \alpha_1(1-\alpha)\right] \quad (10\text{-}12)$$

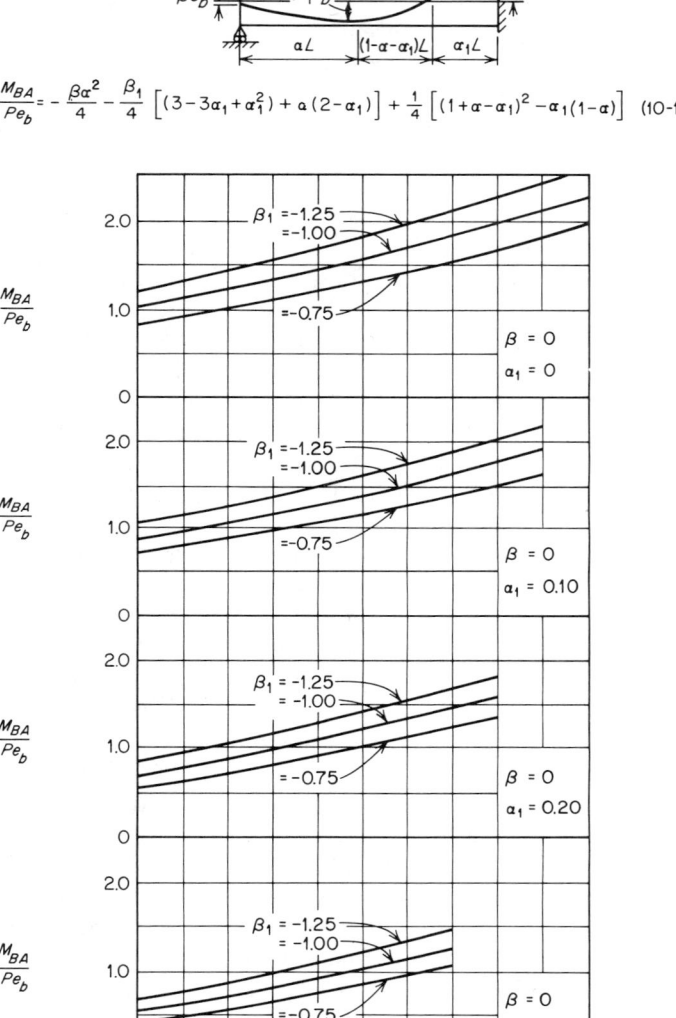

Fig. 10-12(*a*) Variation of M_{BA}/Pe_b with α for a typical endspan. Eccentricity varies parabolically.

ANALYSIS AND DESIGN OF CONTINUOUS PRESTRESSED-CONCRETE BEAMS

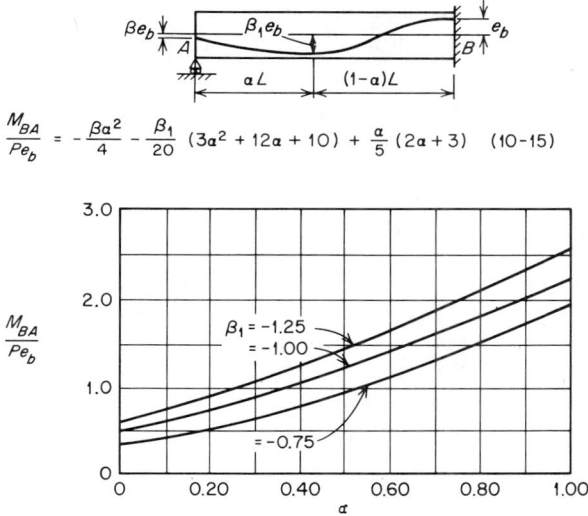

$$\frac{M_{BA}}{Pe_b} = -\frac{\beta a^2}{4} - \frac{\beta_1}{20}(3a^2 + 12a + 10) + \frac{a}{5}(2a + 3) \quad (10\text{-}15)$$

Fig. 10-12(b) Eccentricity varies as a parabola and a fourth-degree curve.

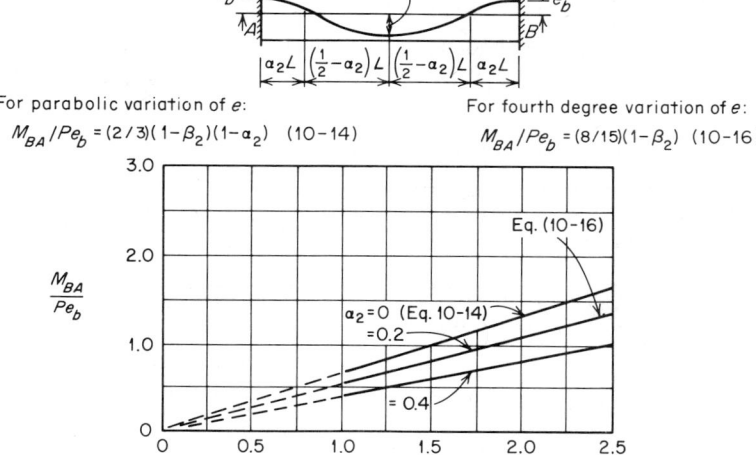

For parabolic variation of e:

$M_{BA}/Pe_b = (2/3)(1-\beta_2)(1-a_2)$ (10-14)

For fourth degree variation of e:

$M_{BA}/Pe_b = (8/15)(1-\beta_2)$ (10-16)

Fig. 10-13 Variation of M_{BA}/Pe_b with $(1 - \beta_2)$ for a typical interior span.

These conclusions are self-evident if one considers variation of eccentricity as that of angle change in the beam.

10-9 ILLUSTRATIVE PROBLEM 10-1

In order to apply the foregoing to the analysis of continuous prestressed-concrete beams the following illustrative problem is presented:

(a) Draw the shear and moment diagrams for the structure shown in Fig. 10-14, assuming that the prestressing force is the only force acting. The profile of the center of gravity of the prestressing steel is made up of sections of parabolas, as shown in the figure.

Since the variation of eccentricity is parabolic, the fixed-end moments may be calculated by Eqs. (10-12) and (10-14).

The fixed-end moment to the left of support B may be calculated by Eq. (10-12) as follows:

$$M_{BA} = M_{B'A'} = \frac{Pe_b}{4}\{-\beta\alpha^2 - \beta_1[(3 - 3\alpha_1 + \alpha_1^2) + \alpha(2 - \alpha_1)] + [(1 + \alpha - \alpha_1)^2 - \alpha_1(1 - \alpha)]\} \quad (10\text{-}12)$$

The following quantities are given (see Fig. 10-14) in this problem:

$$\beta = -\tfrac{1\cdot 0}{2\cdot 5} = -0.40 \qquad \beta_1 = -\tfrac{2\cdot 0}{2\cdot 5} = -0.80$$

$$\alpha = \tfrac{60}{120} = 0.50 \qquad \alpha_1 = \tfrac{18}{120} = 0.15$$

and $e_b = 2.5$ ft.

Substituting the above quantities in Eq. (10-12), we have

$$M_{BA} = \frac{Pe_b}{4}\{0.40(0.50)^2 + 0.80[(3 - 3 \times 0.15 + 0.15^2) + 0.50(2 - 0.15)] + [(1 + 0.50 - 0.15)^2 - 0.15(1 - 0.50)]\} = 1.1614 Pe_b$$

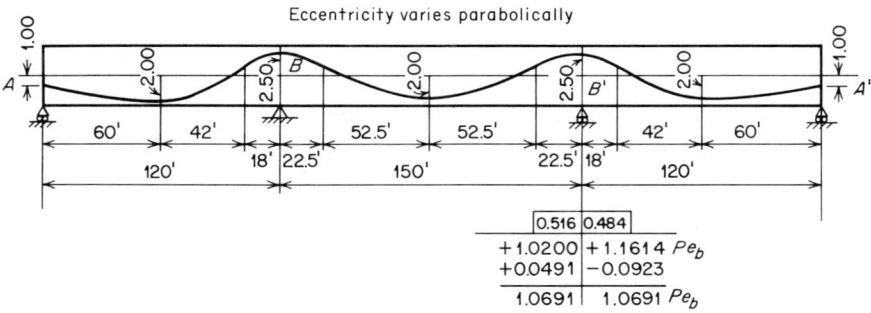

Fig. 10-14 Distribution of moments due to prestressing force. Illustrative Problem 10-1a.

Similarly, the fixed-end moment to the right of support B may be computed from Eq. (10-14) as follows:

$$M_{BB'} = M_{B'B} = \frac{2Pe_b}{3}(1 - \beta_2)(1 - \alpha_2) \tag{10-14}$$

where $\beta_2 = -\frac{2 \cdot 0}{2 \cdot 5} = -0.80$

$\alpha_2 = \frac{22.5}{150} = 0.15$

Hence

$$M_{BB'} = \frac{2Pe_b}{3}(1 + 0.80)(1 - 0.15) = 1.0200 Pe_b$$

The relative stiffnesses and distribution factors may be determined by using the symmetry condition as follows:

Relative stiffness $BA = \dfrac{3 \times 1}{120} = 0.0250$

Relative stiffness $BB' = \dfrac{2 \times 1}{150} = 0.0133$

Distribution factor $BA = \dfrac{0.0250}{0.0383} = 0.653$

Distribution factor $BB' = \dfrac{0.0133}{0.0383} = 0.347$

Figure 10-14 shows the distribution of the fixed-end moments on the basis of the above relative stiffnesses. From this figure it can be seen that the total moment at B due to prestressing force is $1.0704 Pe_b$. Since the statically determinate moment at B is Pe_b, it may be concluded that the statically indeterminate moment at B is positive and is equal to $0.0704 Pe_b$.

Figure 10-15a shows the statically determinate moment diagram for the left half of the structure. Since the figures shown are in terms of P, the prestressing force, the moment diagram is the same as the profile of the center of gravity of the prestressing steel. That is, the moment diagram may be constructed by plotting the eccentricities. Figure 10-15a shows the eccentricities in feet and in terms of P for the tenth points of the spans, and at points of inflection.

Figure 10-15b shows the statically indeterminate moment diagram in terms of P. The ordinate at B is $0.0704 Pe_b = 0.0704 \times 2.50P = 0.1760P$. The ordinates are given at tenth points of the endspan. The indeterminate moment varies linearly at the endspan and is constant at the middle span. Figure 10-15c shows the moment diagram due to the prestressing force which is obtained by the superposition of the statically determinate and indeterminate moments shown in Fig. 10-15a and b, respectively. Figure 10-15d shows the shear diagram due to the prestressing force which may be obtained directly from the moment diagram. It should be pointed out that the numbers shown in Fig. 10-15a through c are in feet and are given in terms of P. Figure 10-15d is in terms of P.

(b) A second example of application of the method is shown in Fig. 10-16, which represents a structure similar to the one analyzed above. In this case the eccentricity varies as a fourth-degree curve throughout the structure, except at the outer halves of the endspans, in which the eccentricity varies parabolically.

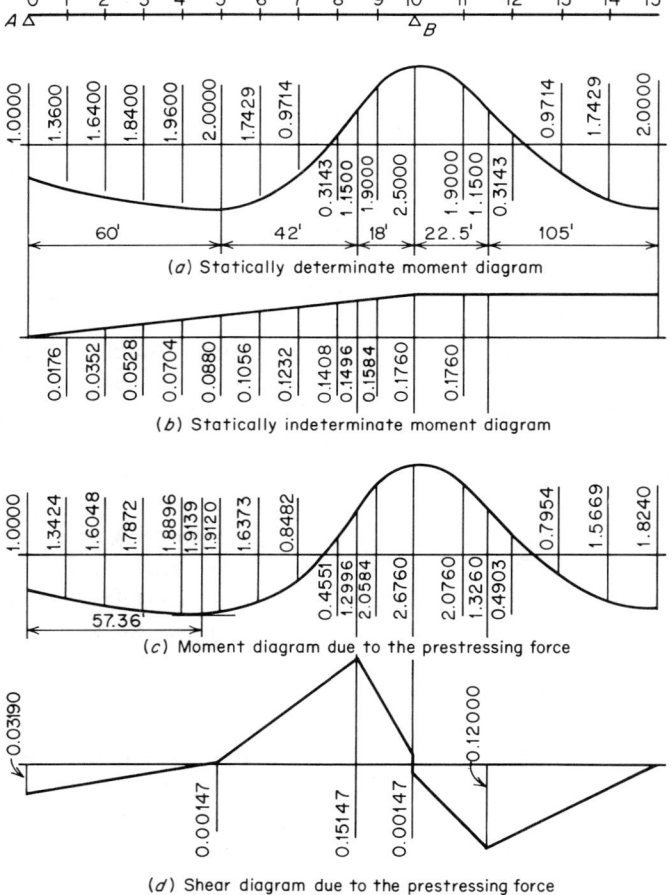

Fig. 10-15 Shear and moment diagrams for the prestressing force (all numbers are coefficients of P, the prestressing force). Illustrative Problem 10-1a.

ANALYSIS AND DESIGN OF CONTINUOUS PRESTRESSED-CONCRETE BEAMS 371

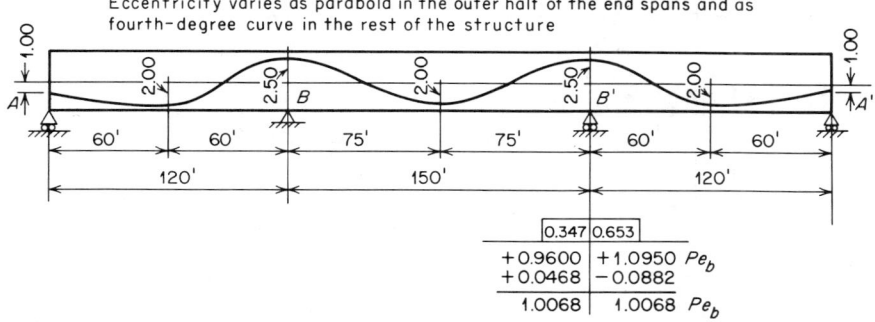

Fig. 10-16 Distribution of moments due to prestressing force. Illustrative Problem 10-1b.

In this case the fixed-end moments to the left and to the right of support B may be calculated from Eqs. (10-15) and (10-16), respectively, as follows:

$$M_{BA} = M_{B'A'} = \frac{Pe_b}{20}[-5\beta\alpha^2 - \beta_1(3\alpha^2 + 12\alpha + 10) + 4\alpha(2\alpha + 3)]$$
$$= \frac{Pe_b}{20}[5 \times 0.40(0.50)^2 - 0.80(3 \times 0.50^2 + 12 \times 0.5 + 10)$$
$$+ 4 \times 0.50(2 \times 0.50 + 3)] = 1.0950 Pe_b$$

and

$$M_{BB'} = M_{B'B} = \frac{8Pe_b}{15}(1 - \beta_2) = \frac{8Pe_b}{15}(1 + 0.8) = 0.9600 Pe_b$$

The distribution factors are the same as in the preceding problem. The total moment due to the prestressing force in this case is $1.0068 Pe_b$.

Figure 10-17 shows the statically determinate and indeterminate moment diagrams as well as the superposition of the two. It should be noted that in this case the shear diagram corresponding to the part of the structure in which the eccentricity varies as a fourth-degree curve is a third-degree curve.

10-10 THE EQUIVALENT-LOAD METHOD

The method used in the preceding sections for the determination of moments caused by the prestressing force in continuous beams may be applied to the analysis of any indeterminate structure with any shape for the variation of eccentricity.

In order to have a better understanding of the analysis of continuous beams for the prestressing forces, it is helpful to determine all the forces

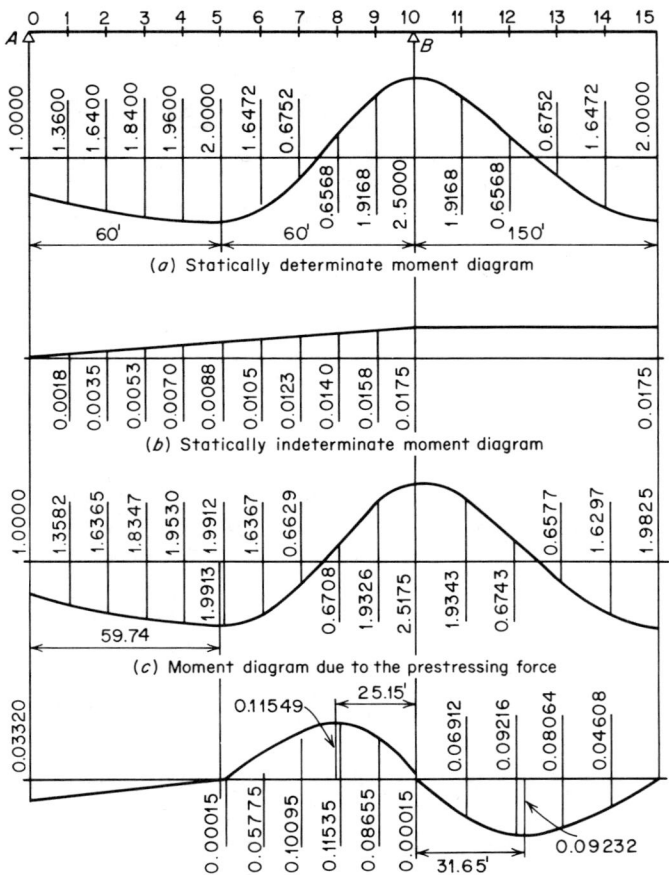

Fig. 10-17 Shear and moment diagrams for the prestressing force (all numbers are coefficients of P, the prestressing force). Illustrative Problem 10-1b.

that the prestressing elements apply on the concrete. Once these forces are determined, the structure may be analyzed for them. This approach will be called the equivalent-load method, and it can be used to derive the expressions for the fixed-end moments obtained in the preceding sections. In the following paragraphs this method will be discussed with some illustrative problems.[5]

The forces that the prestressing elements exert on concrete are the forces at the anchorage, the frictional forces which act on concrete along

ANALYSIS AND DESIGN OF CONTINUOUS PRESTRESSED-CONCRETE BEAMS 373

Fig. 10-18 The forces exerted on concrete by the prestressing elements.

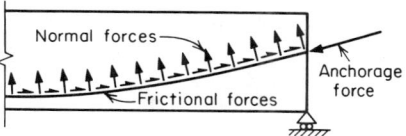

the contact points of steel and concrete, and the forces that are perpendicular to the frictional forces that the steel exerts on concrete. Figure 10-18 shows all the forces that are exerted by prestressing force on concrete. It is assumed that the prestressing force remains constant throughout the length of structure.

In order to study the nature of the lateral forces, the pressure exerted by a cable on a curved surface will be studied.

Let us assume that we have a curved surface as shown in Fig. 10-19a. Cable passing over the surface is being pulled from both ends by a force P, which exerts a distributed pressure on that part of the surface which is in contact with the cable. In order to determine the pressure exerted by the cable let us consider a small section ds of the curved surface, as shown in Fig. 10-19b.

From Fig. 10-19b it can be seen that the following expression is correct:

$$dQ = 2P \sin \frac{d\theta}{2}$$

Since distance ds is small, angle $d\theta$ is also small and we have

$$dQ = 2P \frac{d\theta}{2} = P\, d\theta$$

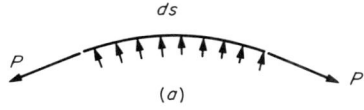

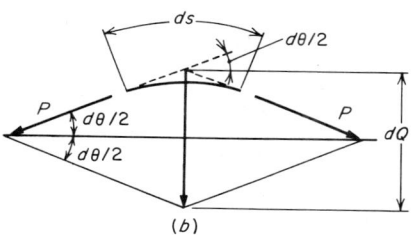

Fig. 10-19 Determination of normal forces caused by a tensioned cable in contact with a curved surface.

The pressure w can, therefore, be presented as follows:

$$w = \frac{dQ}{ds} = P\frac{d\theta}{ds} = \frac{P}{R}$$

where $R = ds/d\theta$ is the radius of curvature of the surface. Evidently for a given curve R, the radius of curvature is a variable quantity, and for a general curve presented by $y = f(x)$, it can be expressed in the following general form:

$$R = \frac{(1 + y'^2)^{\frac{3}{2}}}{y''}$$

and

$$w = \frac{P}{R} = \frac{Py''}{(1 + y'^2)^{\frac{3}{2}}}$$

For flat curves, that is, for curves that have a large radius of curvature, y'^2 in the denominator is small in comparison with unity and may be neglected. Furthermore, for flat curves the distributed loads may be assumed to be vertical. Hence the distributed load may be approximated as follows:

$$w = \frac{P}{R} \approx Py''$$

This approach may be applied to determine the moment due to prestressing force in a simply supported beam.

Figure 10-20a shows a simply supported prestressed-concrete beam in which prestressing steel has a parabolic profile.

The eccentricity can be expressed as

$$e(x) = \frac{4(e_b - e_c)}{L^2} x^2 + e_c$$

The moment due to prestressing force at a distance x from the origin is

$$M = Pe = P\left[\frac{4(e_b - e_c)}{L^2} x^2 + e_c\right] \tag{10-17}$$

In writing the above expression the assumption is made that the horizontal component of the prestressing force is equal to the prestressing force. This assumption is reasonable since the profile of the prestressing force is flat and the cosine of the angle that the tangent to the profile makes with the horizontal is nearly equal to 1.

The same expression may be obtained by using the equivalent-load method.

ANALYSIS AND DESIGN OF CONTINUOUS PRESTRESSED-CONCRETE BEAMS

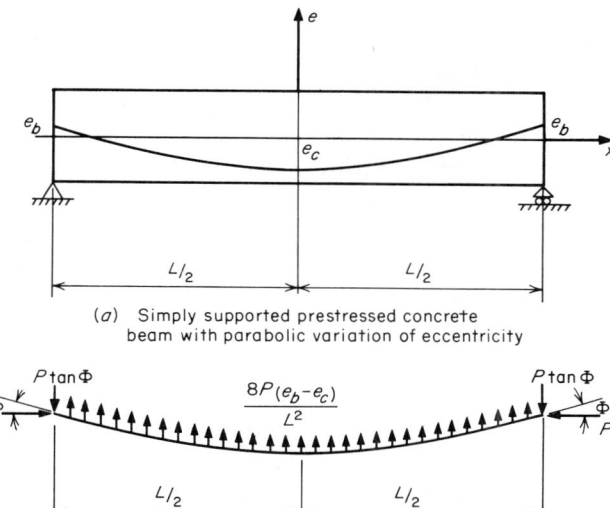

(a) Simply supported prestressed concrete beam with parabolic variation of eccentricity

(b) Free-body diagram of the force acting on concrete

Fig. 10-20 Forces exerted on concrete by the prestressing elements.

Figure 10-20b shows a free-body diagram of all the forces that act on concrete. Since the curve is flat, the distributed forces may be assumed to be vertical, and its magnitude will be

$$w = Pe''(x) = P\frac{8(e_b - e_c)}{L^2}$$

which is a constant quantity.

The horizontal force at the anchorage can be taken as P, and the vertical component will be

$$P \tan \phi = P\frac{8(e_b - e_c)}{L^2}\frac{L}{2} = \frac{4P}{L}(e_b - e_c)$$

The moment at point A may be obtained by taking the moment of all forces to the right of A about A, as follows:

$$M = \frac{-4P}{L}(e_b - e_c)\left(\frac{L}{2} - x\right) + P\frac{8(e_b - e_c)}{2L^2}\left(\frac{L}{2} - x\right)^2 + Pe_b$$

$$M = P\left[\frac{4(e_b - e_c)}{L^2}x^2 + e_c\right] \qquad (10\text{-}17)$$

which is the same expression obtained by multiplying the horizontal component of the prestressing force by the eccentricity.

Continuous structures may also be analyzed by use of this method. Figure 10-21a shows a fixed-end beam representing a typical endspan in which the eccentricity varies as a parabola.

All the forces acting on the concrete are shown in Fig. 10-21b. The distributed load in this case will be uniform since $e''(x)$ is a constant quantity.

From $x = 0$ to $x = \alpha L$, the eccentricity and the uniformly distributed loads are

$$e = -\frac{(\beta_1 - \beta)e_b}{(\alpha L)^2} x^2 + \frac{2(\beta_1 - \beta)e_b}{\alpha L} x + \beta e_b$$

and

$$w = Pe'' = -2P \frac{(\beta_1 - \beta)e_b}{(\alpha L)^2}$$

From $x = \alpha L$ to $x = (1 - \alpha_1)L$

$$e = \frac{(1 - \beta_1)e_b}{(1 - \alpha)(1 - \alpha - \alpha_1)L^2} x^2 - \frac{2\alpha(1 - \beta_1)e_b}{(1 - \alpha)(1 - \alpha - \alpha_1)L} x$$
$$+ \frac{\alpha^2(1 - \beta_1)e_b}{(1 - \alpha)(1 - \alpha - \alpha_1)} + \beta_1 e_b$$

$$w = Pe'' = 2P \frac{(1 - \beta_1)e_b}{(1 - \alpha)(1 - \alpha - \alpha_1)L^2}$$

and from $x = (1 - \alpha_1)L$ to $x = L$

$$e = -\frac{(1 - \beta_1)e_b}{\alpha_1(1 - \alpha)L^2} x^2 + \frac{2(1 - \beta_1)e_b}{\alpha_1(1 - \alpha)L} x - \frac{(1 - \beta_1)e_b}{\alpha_1(1 - \alpha)} + e_b$$

$$w = Pe'' = -2P \frac{(1 - \beta_1)e_b}{\alpha_1(1 - \alpha)L^2}$$

The horizontal force at the simply supported end can be assumed as P, and the vertical force at the simply supported end will be

$$P \tan \phi = 2P \frac{(\beta_1 - \beta)e_b}{\alpha L}$$

Figure 10-21b shows a free-body diagram with all these forces acting on the concrete.

It can be shown that the fixed-end moment at B due to these loads is

$$M_{BA} = \frac{Pe}{4} \{-\beta\alpha^2 - \beta_1[(3 - 3\alpha_1 + \alpha_1^2) + \alpha(2 - \alpha_1)]$$
$$+ [(1 + \alpha - \alpha_1)^2 - \alpha_1(1 - \alpha)]\} \quad (10\text{-}12)$$

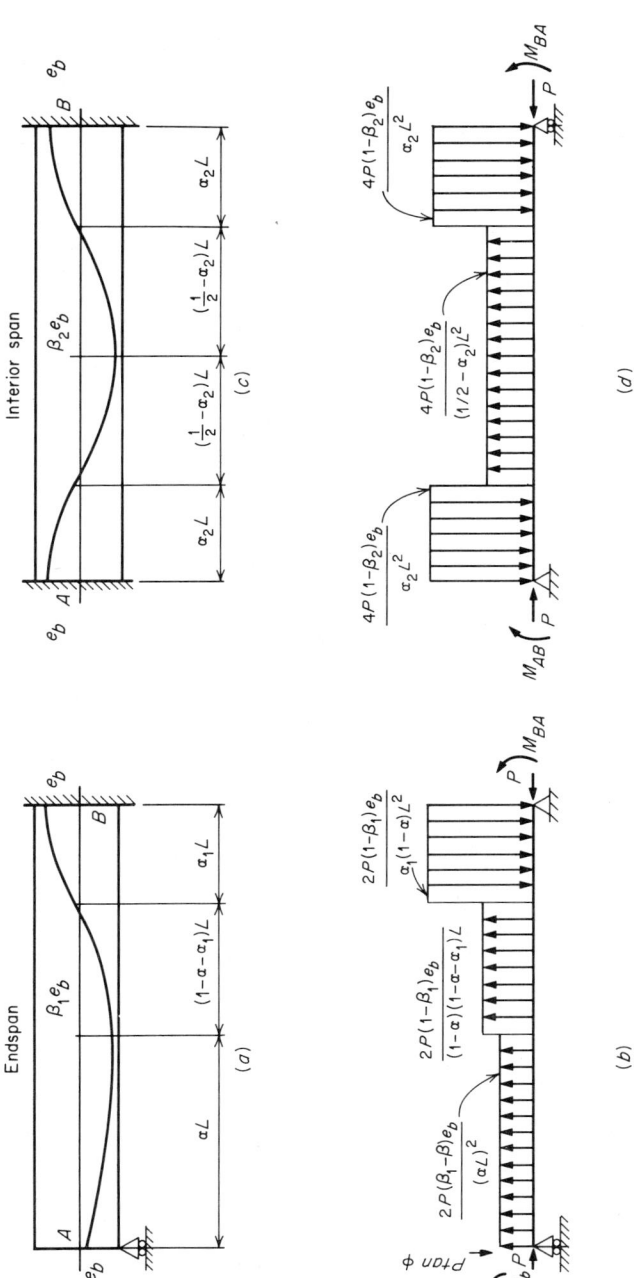

Fig. 10-21 Forces exerted on concrete in typical end and interior spans when eccentricity varies parabolically.

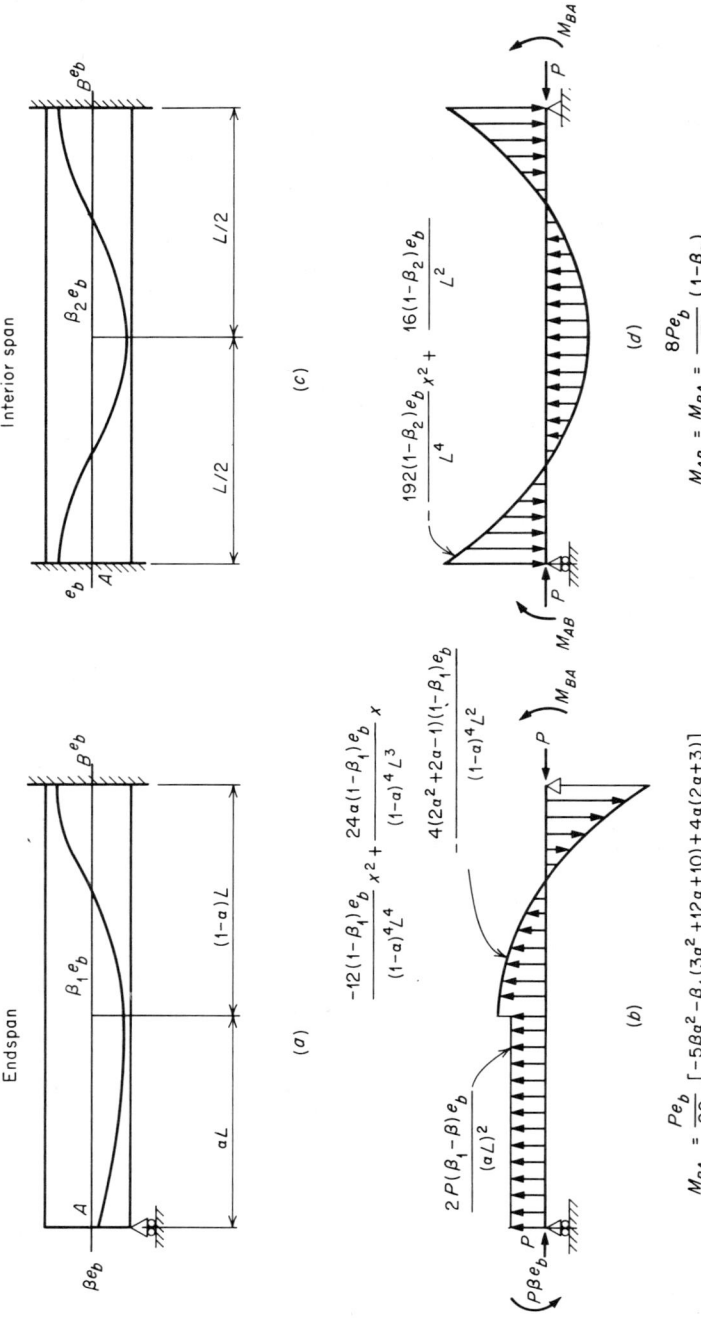

Fig. 10-22 Forces exerted on concrete in typical end and interior spans when eccentricity varies as a parabola and a fourth-degree curve.

ANALYSIS AND DESIGN OF CONTINUOUS PRESTRESSED-CONCRETE BEAMS

Evidently this is the same expression derived in Sec. 10-7. In order to obtain Eq. (10-12) from the loading shown in Fig. 10-21b, convenient use may be made of the expressions for the fixed-end moments for various types of loading shown in Appendix A.

Similarly, Fig. 10-21c shows a typical interior span in which eccentricity varies as a parabola. All the forces acting on concrete are shown in Fig. 10-21d.

From $x = -L/2$ to $x = -(\frac{1}{2} - \alpha_2)L$ the eccentricity and the equivalent uniform load will be

$$e = -\frac{2(1 - \beta_2)e_b}{\alpha_2 L^2} x^2 - \frac{2(1 - \beta_2)e_b}{\alpha_2 L} x - \frac{(1 - \beta_2)e_b}{2\alpha_2} + e_b$$

$$w = Pe'' = \frac{-4P(1 - \beta_2)e_b}{\alpha_2 L^2}$$

Because of symmetry this value of w also expresses the uniformly distributed load from $x = (\frac{1}{2} - \alpha_2)$ to $x = \alpha_2 L$ and from $x = -(\frac{1}{2} - \alpha_2)L$ to $x = (\frac{1}{2} - \alpha_2)L$:

$$e = \frac{2(1 - \beta_2)e_b}{(\frac{1}{2} - \alpha_2)L^2} x^2 + \beta_2 e_b$$

$$w = Pe'' = \frac{4P(1 - \beta_2)e_b}{(\frac{1}{2} - \alpha_2)L^2}$$

Figure 10-21d shows all the forces acting on the concrete, and it can be shown that because of these forces the fixed-end moment at each end will be

$$M = \frac{2Pe_b}{3}(1 - \beta_2)(1 - \alpha_2) \qquad (10\text{-}14)$$

which is the same expression derived in Sec. 10-7.

Figure 10-22 shows the forces exerted on concrete in typical end and interior spans when eccentricity varies as a parabola and a fourth-degree curve. In Fig. 10-22a, which represents a typical endspan, the eccentricity varies parabolically from the left support to a distance αL. The eccentricity varies as a fourth-degree curve from $x = \alpha L$ to the right support. Figure 10-22b shows all the forces acting on the concrete. It should be noted that the load corresponding to the parabolic variation of eccentricity is uniformly distributed and that corresponding to the fourth-degree curve is a parabolic distributed load.

From $x = 0$ to $x = \alpha L$ the variation of eccentricity is parabolic and the distributed load is the same as derived above; that is,

$$w = \frac{-2P(\beta_1 - \beta)e_b}{(\alpha L)^2}$$

From $x = \alpha L$ to $x = L$ the eccentricity and the uniformly distributed load are

$$e = -\frac{(1-\beta_1)e_b}{(1-\alpha)^4 L^4}x^4 + \frac{4\alpha(1-\beta_1)e_b}{(1-\alpha)^4 L^3}x^3$$
$$-\frac{2(2\alpha^2 + 2\alpha - 1)(1-\beta_1)e_b}{(1-\alpha)^4 L^2}x^2 + \frac{4\alpha(2\alpha-1)(1-\beta_1)e_b}{(1-\alpha)^4 L}x$$
$$+\frac{\alpha^2(\alpha^2 - 4\alpha + 2)(1-\beta_1)e_b}{(1-\alpha)^4} + \beta_1 e_b \quad (10\text{-}9)$$

and

$$w = Pe'' = -\frac{12(1-\beta_1)e_b}{(1-\alpha)^4 L^4}x^2 + \frac{24\alpha(1-\beta_1)e_b}{(1-\alpha)^4 L^3}x$$
$$-\frac{4(2\alpha^2 + 2\alpha - 1)(1-\beta_1)e_b}{(1-\alpha)^4 L^2}$$

It can be shown that for the loads shown in Fig. 10-22c the fixed-end moment at B is

$$M_{BA} = \frac{Pe_b}{20}[-5\beta\alpha^2 - \beta_1(3\alpha^2 + 12\alpha + 10) + 4\alpha(2\alpha + 3)] \quad (10\text{-}15)$$

which is the same expression as that derived in Sec. 10-7.

Figure 10-22c and d shows a typical interior span in which eccentricity varies as a fourth-degree curve and the distributed load is parabolic. In this case we have

$$e = -\frac{16(1-\beta_2)e_b}{L^4}x^4 + \frac{8(1-\beta_2)e_b}{L^2}x^2 + \beta_2 e_b \quad (10\text{-}10)$$

and

$$w = Pe'' = -\frac{192(1-\beta_2)e_b}{L^4}x^2 + \frac{16(1-\beta_2)e_b}{L^2}$$

For the loads shown in Fig. 10-22d the fixed-end moments at A and B are

$$M_{AB} = M_{BA} = \frac{8Pe_b}{15}(1-\beta_2) \quad (10\text{-}16)$$

which was also derived in Sec. 10-7.

10-11 NUMERICAL ANALYSIS

The forces exerted by the prestressing elements can be determined from the profile of the centroid of the prestressing steel (which defines the

ANALYSIS AND DESIGN OF CONTINUOUS PRESTRESSED-CONCRETE BEAMS

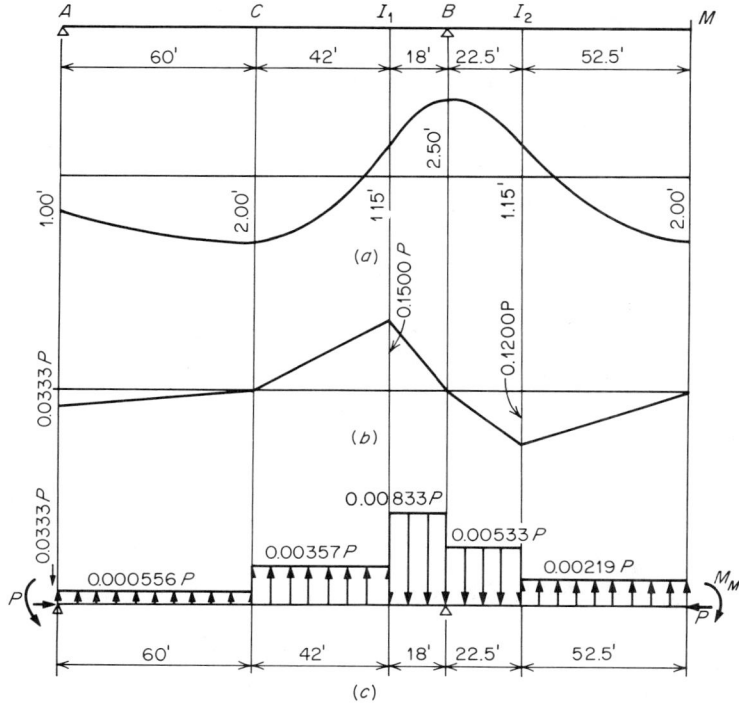

Fig. 10-23 Loads exerted by the prestressing elements on concrete.

variation of the eccentricity) by step-by-step differentiation. In order to show how this procedure works, Illustrative Problem 10-1 (Sec. 10-9) will be considered.

Figure 10-23 shows the left half of a three-span continuous structure in which the eccentricity varies as parabolas. Since the distributed load caused by the prestressing elements is Pe'', it may be assumed that the curve representing the variation of eccentricity is a moment diagram and the required distributed load is the load corresponding to this moment diagram. Figure 10-23a shows the variation of eccentricity throughout the left half of the structure, which is defined by five parabolas. If we assume that this variation of eccentricity is analogous to a moment diagram, the corresponding shear diagram will be as shown in Fig. 10-23b. From A to C the eccentricity, or the analogous moment diagram, is defined by a parabola with a horizontal tangent at point C. Therefore, the shear varies linearly and is zero at point C. The ordinate to the shear diagram at A, V_A may be computed by equating the area under the shear diagram to the change in moment between points A and C as follows:

$$\tfrac{1}{2} V_A (60) = (2.00 - 1.00)P$$

from which $V_A = 0.0333P$.

Similarly, the ordinate to the shear diagram at I_1 is

$$\frac{2(2.00 + 1.15)P}{42} = 0.15P$$

Since the tangent to the moment diagram at B is horizontal, the shear at B will be zero, and the following expression checks the results:

$$\tfrac{1}{2} \times 18 \times 0.15P = (2.50 - 1.15)P = 1.35P$$

The ordinate to the shear diagram at I_2 is

$$\frac{2(2.50 - 1.15)P}{22.5} = 0.12P$$

and at M it is zero.

Figure 10-23c shows the loads resulting from the shear diagram. The calculation of these forces can be seen readily from the figure, and no further explanation seems to be necessary. It should be pointed out that in this case the end eccentricity of 1.00 ft results in an end negative moment of P which should be included among the forces acting on concrete.

It should be noted that in this method there is no need to derive expressions for the variation of eccentricity. This method is particularly suitable when the variation of eccentricity is defined by parabolas.

The analysis of the structure for the loads shown in Fig. 10-23c is routine. The expressions in Appendix A may be used to calculate the fixed-end moments at B. After the fixed-end moments have been obtained, the analysis is identical with Illustrative Problem 10-1.

10-12 ILLUSTRATIVE PROBLEMS 10-2 AND 10-3

In order to show the application of the method of numerical differentiation to the analysis of continuous structures, the following two problems are presented.

Draw the shear and bending-moment diagrams for the structures shown in Figs. 10-24 and 10-25, assuming that the prestressing force is the only force acting.

The complete solutions of these problems are shown in Figs. 10-24 and 10-25, and no further explanation seems to be necessary.

ANALYSIS AND DESIGN OF CONTINUOUS PRESTRESSED-CONCRETE BEAMS

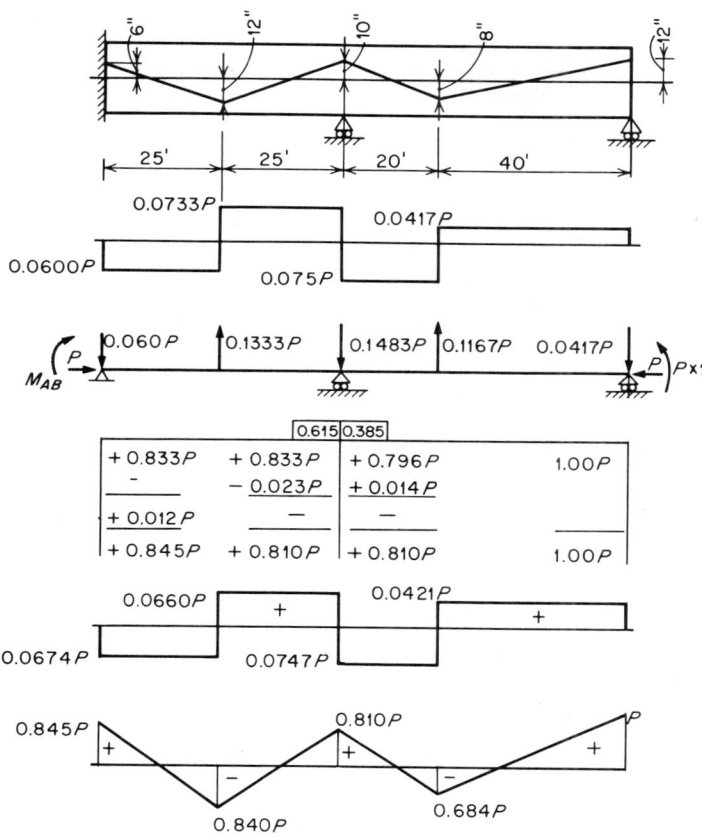

Fig. 10-24 Illustrative Problem 10-2.

10-13 EQUATIONS FOR DESIGN

The design procedure presented here applies to the type of structure discussed in the preceding sections. It is assumed that the center of gravity of the steel is continuous throughout the length of the structure, the supports retain their vertical position during the prestressing, and the frictional loss in the prestressing steel is negligible.

The design may be carried out for a particular section, such as the first interior support, and the stresses checked at other points in the beam. The expressions developed in the following paragraphs are similar to those presented in Chap. 7 of this text for simply supported beams. Expressions are modified to make them applicable to continuous beams.

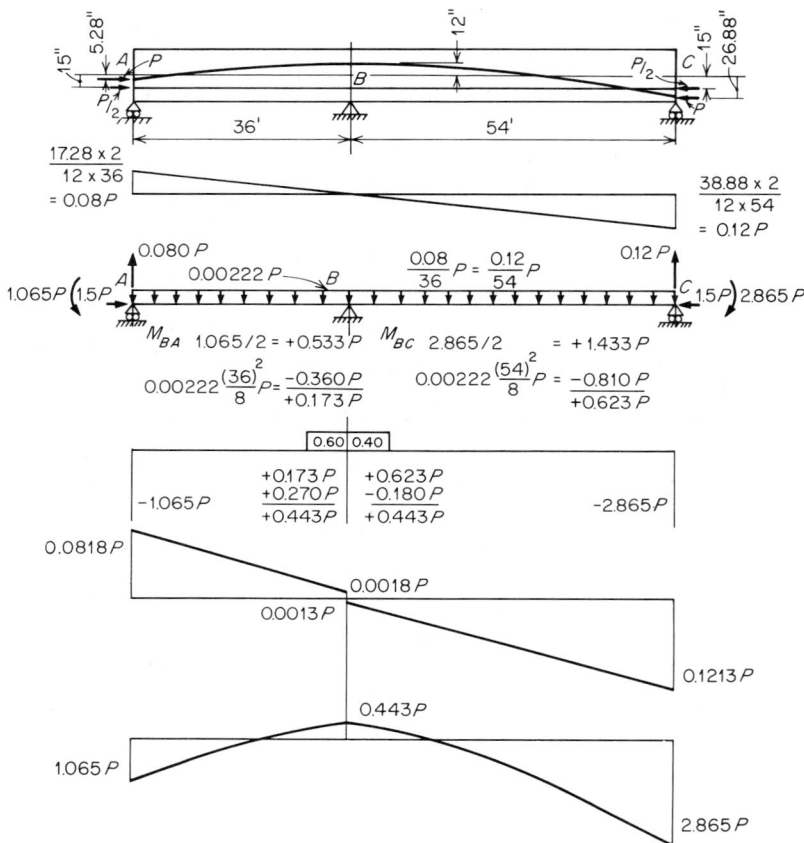

Fig. 10-25 Illustrative Problem 10-3.

The following four requirements should be satisfied at all points in the beam:

$$\frac{P_t}{A}\left(\frac{M_p}{P_t}\frac{y_t}{r^2} - 1\right) - \frac{M_g y_t}{I} = a_t f'_c \leq a'_t f'_c \tag{10-18}$$

$$\frac{P_t}{A}\left(\frac{M_p}{P_t}\frac{y_b}{r^2} + 1\right) - \frac{M_g y_b}{I} = c_t f'_c \leq c'_t f'_c \tag{10-19}$$

$$-\eta \frac{P_t}{A}\left(\frac{M_p}{P_t}\frac{y_t}{r^2} - 1\right) + \frac{M_t y_t}{I} = cf'_c \leq c'f'_c \tag{10-20}$$

$$-\eta \frac{P_t}{A}\left(\frac{M_p}{P_t}\frac{y_b}{r^2} + 1\right) + \frac{M_t y_b}{I} = af'_c \leq a'f'_c \tag{10-21}$$

where P_t = prestressing force at transfer at section under consideration, kips

ANALYSIS AND DESIGN OF CONTINUOUS PRESTRESSED-CONCRETE BEAMS 385

A = gross cross-sectional area of the beam, in.^{2}
M_p = moment due to prestressing force at section under consideration, in.-kips
y_t = distance of top fiber from centroidal axis of gross section, in.
y_b = distance of bottom fiber from centroidal axis of gross section, in.
I = moment of inertia of gross section, in.^{4}
M_g = moment due to weight of beam at section under consideration, in.-kips
M_t = moment due to all loads acting on beam at section under consideration, in.-kips, taken as $M_a + M_g$, where M_a is sum of moments due to superimposed dead load and live load
f_c' = 28-day concrete strength, psi
$a_t f_c', a f_c'$ = computed tensile stress in concrete before and after losses, respectively
$a_t' f_c', a' f_c'$ = allowable tensile stress in concrete before and after losses, respectively
$c_t f_c', c f_c'$ = computed compressive stress in concrete before and after losses, respectively
$c_t' f_c', c' f_c'$ = allowable compressive stress in concrete before and after losses in concrete, respectively
η = effectiveness, taken as P/P_t, where P is prestressing force after losses

It should be pointed out that in expressions (10-18) through (10-21) the term $M_p/P_t = e_1$ represents the effective eccentricity or the distance from the neutral axis of the beam to the *pressure line*.

By introducing h as the overall depth of the beam and noting that

$$M_g = CA\gamma L^2$$

and

$$M_t = M_a + M_g$$

expressions (10-18) through (10-21) can be written in the form of equations as follows:

$$\frac{P_t}{Af_c'}\left(\frac{M_p}{P_t h}\frac{h^2}{r^2}\frac{1}{y_b/y_t + 1} - 1\right) - C\frac{\gamma L^2}{hf_c'}\frac{1}{y_b/y_t + 1}\frac{h^2}{r^2} = a_t$$

$$\frac{P_t}{Af_c'}\left(\frac{M_p}{P_t h}\frac{h^2}{r^2}\frac{y_b/y_t}{y_b/y_t + 1} + 1\right) - C\frac{\gamma L^2}{hf_c'}\frac{y_b/y_t}{y_b/y_t + 1}\frac{h^2}{r^2} = c_t$$

$$-\eta\frac{P_t}{Af_c'}\left(\frac{M_p}{P_t h}\frac{h^2}{r^2}\frac{1}{y_b/y_t + 1} - 1\right) + C\frac{\gamma L^2}{hf_c'}\frac{1 + M_a/M_g}{y_b/y_t + 1}\frac{h^2}{r_2} = c$$

$$-\eta\frac{P_t}{Af_c'}\left(\frac{M_p}{P_t h}\frac{h^2}{r^2}\frac{y_b/y_t}{y_b/y_t + 1} + 1\right) + C\frac{\gamma L^2}{hf_c'}\frac{(y_b/y_t)(1 + M_a/M_g)}{y_b/y_t + 1}\frac{h^2}{r^2} = a$$

It should be pointed out that the quantity C is the coefficient of $A\gamma L^2$, where $A\gamma$ is weight per linear foot of the beam.

The following dimensionless quantities are introduced:

$$\frac{hf'_c}{\gamma L^2} = \omega \qquad \frac{y_b}{y_t} = \Delta$$

$$\frac{r^2}{h^2} = \rho \qquad \frac{M_a}{M_g} = R$$

$$\frac{P_t}{Af'_c} = m \qquad \frac{M_p}{P_t h} = \epsilon_1$$

After substituting these quantities in the above equations and rearranging, the four requirements can be written as follows:

$$m\left[\frac{\epsilon_1}{\rho(1+\Delta)} - 1\right] - \frac{C}{\rho\omega(1+\Delta)} = a_t \tag{10-18a}$$

$$m\left[\frac{\epsilon_1\Delta}{\rho(1+\Delta)} + 1\right] - \frac{C\Delta}{\rho\omega(1+\Delta)} = c_t \tag{10-19a}$$

$$-\eta m\left[\frac{\epsilon_1}{\rho(1+\Delta)} - 1\right] + \frac{C(1+R)}{\rho\omega(1+\Delta)} = c \tag{10-20a}$$

$$-\eta m\left[\frac{\epsilon_1\Delta}{\rho(1+\Delta)} + 1\right] + \frac{C\Delta(1+R)}{\rho\omega(1+\Delta)} = a \tag{10-21a}$$

If the values of a_t, c_t, c, a, and η are known or assumed, there will be six dimensionless variables in the above four equations, namely, ω, Δ, ρ, R, m, and ϵ. These variables and their physical significance are discussed in Chap. 7.

Of these six dimensionless unknowns, the first four define the section properties and the last two define the magnitude and location of the prestressing steel once the section properties of the beam are known.

It can be shown that reasonable values for ω and ρ may be chosen. The quantity ω is an expression for the depth of the beam, and a reasonable value for the depth can be established. The quantity ρ varies in a comparatively narrow range, and an estimate of its value may be made. If ω and ρ are assumed as known quantities, Eqs. (10-18a) through (10-21a) will contain only four dimensionless unknowns, namely, Δ, R, m, and ϵ_1.

A simultaneous solution of Eqs. (10-18a) through (10-21a) yields the following expressions for these four dimensionless unknowns:

$$\Delta = \frac{\eta c_t + a}{\eta a_t + c} \tag{10-22}$$

$$R = \frac{\rho\omega}{C}[(a+c) + \eta(a_t + c_t)] - (1-\eta) \tag{10-23}$$

$$m = \frac{c_t c - a_t a}{(a+c) + \eta(a_t + c_t)} \qquad (10\text{-}24)$$

$$\epsilon_1 = \left[\rho(a_t + c_t) + \frac{C}{\omega}\right]\frac{1}{m} \qquad (10\text{-}25)$$

Equations (10-22) and (10-23) define the section properties of the section under consideration. Equations (10-24) and (10-25) define the magnitude and position of the prestressing force in terms of the section properties.

Elimination of the terms $(\eta a_t + c)$ and $(a + \eta c_t)$ between Eqs. (10-22) and (10-23) results in the following expressions for R, respectively:

$$R = \frac{\rho\omega}{C}(a + \eta c_t)\left(\frac{1}{\Delta} + 1\right) - (1 - \eta) \qquad (10\text{-}23a)$$

$$R = \frac{\rho\omega}{C}(\eta a_t + c)(\Delta + 1) - (1 - \eta) \qquad (10\text{-}23b)$$

The above two equations are alternative forms of Eq. (10-23), and for convenience they may be used instead of Eq. (10-23).

Elimination of a and c_t between Eqs. (10-22) and (10-24) results in the following expressions, respectively:

$$m = \frac{c_t - \Delta a_t}{1 + \Delta} \qquad (10\text{-}24a)$$

$$m = \frac{\Delta c - a}{\eta(1 + \Delta)} \qquad (10\text{-}24b)$$

For convenience Eqs. (10-24a) and (10-24b) may be used instead of Eq. (10-24).

The least-weight design criteria developed in Chap. 7 may be applied to design of continuous beams. These criteria are listed in Table 7-5.

The design of a particular section in a continuous beam is similar to that of a simple beam. Values of ω, Δ, and ρ are selected, and the cross-sectional area of the beam is determined by either Eq. (10-23a) or Eq. (10-23b). From Eqs. (10-24) and (10-25) the magnitude and position of the prestressing force may be determined for each criterion.

10-14 ILLUSTRATIVE PROBLEM 10-4

To show the application of the design method outlined in the preceding section the following design problem is presented:

Design a two-span continuous beam with equal spans of 120 ft each. The live load is 1.4 klf and the superimposed dead load is 1.0 klf. Use the following specifications:

$a'_t = 0.04 \quad c'_t = 0.48$
$c' = 0.40 \quad a' = 0$
$\eta = 0.85$

Assume $f'_c = 5000$ psi, $\gamma = 150$ pcf, $f'_s = 250{,}000$ psi, and $f_{sy} = 210{,}000$ psi.

Figure 10-26a shows the left half of the two-span continuous structure. The moment diagram of the superimposed dead load is shown in Fig. 10-26b, and the curves of maximum positive and negative moments for live load are shown in Fig. 10-26c.

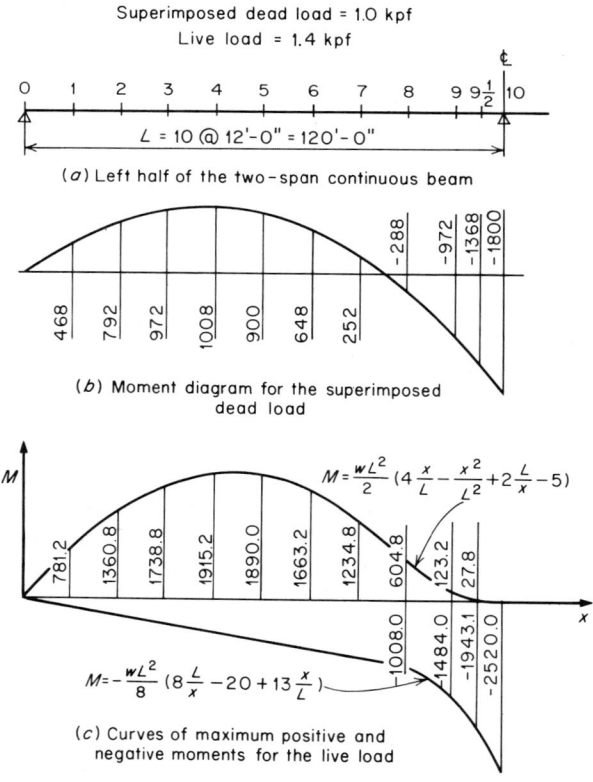

Fig. 10-26 Moment diagram for the superimposed dead load and the curves of maximum moment for live load (all moments are in foot-kips). Illustrative Problem 10-4.

ANALYSIS AND DESIGN OF CONTINUOUS PRESTRESSED-CONCRETE BEAMS

The design criteria presented in the preceding section will be used in designing the section at the interior support.

Let us assume that because of the functional requirements of the structure the overall depth of the beam, h, cannot exceed 75 in. By assuming $h = 75$ in., ω can be calculated as follows:

$$\omega = \frac{hf'_c}{\gamma L^2} = \frac{6.25 \times 5 \times 144}{0.15 \times 120 \times 120} = 2.08$$

For the stress coefficients and the effectiveness specified in this problem we have

$$\Delta_e = \frac{\eta c'_t + a'}{\eta a'_t + c'} = \frac{0.85 \times 0.48}{0.85 \times 0.04 + 0.40} = 0.94$$

Let us assume $\Delta = 1.0$ and consider an idealized symmetrical I section. Since $\Delta > \Delta_e$, in order to obtain the least-weight area it is necessary to satisfy requirements (10-19) and (10-21) exactly. In this case $c_t = c'_t$ and $a = a'$.

It can be shown that in an idealized I section the following expression is correct:

$$\rho = \frac{1 - (1 - b'/b)(1 - 2t/h)^3}{12[1 - (1 - b'/b)(1 - 2t/h)]}$$

where b' = width of web
b = width of top flange
t = thickness of flange

By assuming $t/h = 0.2$, which corresponds to $t = 15.0$ in., and taking $b'/b = 0.45$, the above equation will give $\rho = 0.1096$.

The quantity R can now be computed from Eq. (10-23a):

$$R = \frac{\rho\omega}{C}(\eta c_t + a)\left(\frac{1}{\Delta} + 1\right) - (1 - \eta)$$

where $c_t = c'_t = 0.48$ $\rho = 0.1096$
$a = a' = 0$ $\omega = 2.08$
$\eta = 0.85$ $\Delta = 1.0$

and for a two-span continuous beam $C = 0.125$.

Substituting the above numerical values in Eq. (10-23a), we have

$$R = \frac{M_a}{M_g} = 8 \times 0.1089 \times 2.08(0.85 \times 0.48)2 - 0.15 = 1.34$$

From Fig. 10-26 we have

$$M_a = M_s + M_l = 1800 + 2520 = 4320 \text{ kips}$$

Hence
$$w_g = \frac{8M_g}{L^2} = \frac{8M_a}{RL^2} = \frac{8 \times 4320}{1.34 \times (120)^2} = 1.79 \text{ klf}$$

and
$$A = \tfrac{1.79}{0.15} \times 144 = 1720 \text{ in.}^2$$

but
$$A = bh\left[1 - \left(1 - \frac{b'}{b}\right)\left(1 - 2\frac{t}{h}\right)\right]$$
$$= 75b(1 - 0.55 \times 0.6) = 50.25b = 1720$$

or
$b = 34.2$ in. $t = 15.0$ in.
$b' = 15.4$ in. $h = 75.0$ in.

Figure 10-27a shows the idealized section.

The prestressing force and the eccentricity may be calculated for both criterion 1 and criterion 2. These criteria are applicable since $\Delta > \Delta_e$.

Criterion 1 In this case requirements (10-18), (10-19), and (10-21) are satisfied exactly and $a_t = a'_t$, $c_t = c'_t$, and $a = a'$. From Eq. (10-24a) we have

$$m = \frac{c_t - \Delta a_t}{1 + \Delta} = \frac{c'_t - \Delta a'_t}{1 + \Delta} = \frac{0.48 - 0.04}{2} = 0.22$$

and
$$P_t = mAf'_c = 0.22 \times 1720 \times 5 = 1892 \text{ kips}$$

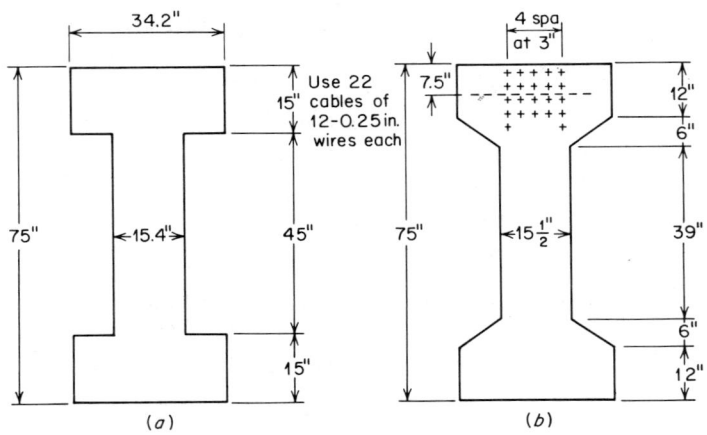

Fig. 10-27 The idealized and actual sections. Illustrative Problem 10-4.

ANALYSIS AND DESIGN OF CONTINUOUS PRESTRESSED-CONCRETE BEAMS

From Eq. (10-25) we have

$$\epsilon_1 = \frac{1}{m}\left[\rho(a_t + c_t) + \frac{C}{\omega}\right]$$

$$= \frac{1}{0.22}\left(0.1096 \times 0.52 + \frac{1}{8 \times 2.08}\right) = 0.532$$

Hence

$$e_1 = 0.532 \times 75 = 39.8 \text{ in.}$$

$$g = \frac{h}{2} - 39.8 = -2.3 \text{ in.}$$

Criterion 2 In this case requirements (10-19) to (10-21) are satisfied exactly, or $c_t = c'_t$, $c = c'$, and $a = a'$. From Eq. (10-24b) we have

$$m = \frac{\Delta c - a}{\eta(1 + \Delta)} = \frac{\Delta c' - a'}{\eta(1 + \Delta)} = \frac{0.40}{0.85 \times 2} = 0.236$$

and

$$P_t = mAf'_c = 0.236 \times 1720 \times 5 = 2030 \text{ kips}$$

From Eq. (10-22) we have

$$a_t = \frac{\eta c_t + a - \Delta c}{\eta \Delta} = \frac{0.85 \times 0.48 - 0.40}{0.85} = 0.009$$

and from Eq. (10-25)

$$\epsilon_1 = \frac{1}{m}\left[\rho(a_t + c_t) + \frac{C}{\omega}\right]$$

$$= \frac{1}{0.236}\left(0.1096 \times 0.489 + \frac{1}{8 \times 2.08}\right) = 0.483$$

Hence

$$e_1 = 0.483 \times 75 = 36.2 \text{ in.}$$

$$g = \frac{h}{2} - 36.2 = 1.3 \text{ in.}$$

The quantity e_1 which is obtained from criteria 1 and 2 represents the eccentricity of the pressure line, that is, the quantity M_p/P_t. In order to determine the eccentricity of the center of gravity of the prestressing force it is necessary to know the profile of the center of gravity of steel.

Let us assume that the eccentricity of the center of gravity of prestressing steel varies as a parabola from the end to midspan and as a fourth-degree curve from midspan to the interior support.

The moment due to prestressing force at the interior support may be obtained from Eq. (10-15) as follows:

$$M_p = \frac{P_t e}{20}[-5\beta\alpha^2 - \beta_1(3\alpha^2 + 12\alpha + 10) + 4\alpha(2\alpha + 3)] \quad (10\text{-}15)$$

Let us assume that in this problem $\alpha = 0.5$, $\beta_1 = -1.0$, and $\beta = 0$. Substituting these quantities in Eq. (10-15) and simplifying, we have

$$M_p = 1.24 P_t e = e_1 P_t$$

Therefore, for criterion 1

$$e = \frac{e_1}{1.24} = \frac{39.8}{1.24} = 32.1 \text{ in.}$$

and for criterion 2

$$e = \frac{e_1}{1.24} = \frac{36.0}{1.24} = 29.00 \text{ in.}$$

Criterion 1 gives the minimum prestressing force of 1892 kips with an eccentricity of 32.1 in. Criterion 2 gives the maximum prestressing force of 2030 kips corresponding to the eccentricity of 29.0 in.

Criterion 1 does not provide a sufficient cover, and probably criterion 2 provides too much. In this case let us assume $e = 30$ in.; we have

$$e_1 = 1.24 \times 30 = 37.2 \text{ in.}$$

and from Eq. (10-21a) the prestressing force corresponding to this eccentricity may be calculated:

$$-\eta m \left[\frac{e\Delta}{\rho(1+\Delta)} + 1\right] + \frac{C\Delta(1+R)}{\rho\omega(1+\Delta)} = a'$$

$$-0.85m\left(\frac{\frac{37.2}{75}}{0.1096 \times 2} + 1\right) + \frac{\frac{1}{8}(1+1.34)}{0.1096 \times 2.08 \times 2} = 0$$

$$m = \frac{0.755}{3.262} = 0.231$$

$$P_t = 0.231 \times 1720 \times 5 = 2000 \text{ kips}$$

The allowable stress in steel at transfer at the first interior support may be computed as follows:

$$f_{st} = 0.70\phi f'_s = 0.70 \times 0.9 \times 250 = 157.5 \text{ ksi}$$

The quantity ϕ in the above expression is the ratio of the prestressing force at the first interior support to that at the jacking end, which in this case is taken as 0.90. By using cables consisting of twelve 0.25-in. wires each, the total number of cables needed will be $2000/(157.5 \times 0.588) \approx 22$ cables.

Figure 10-27b shows the actual section and the arrangements of the cables at the section at the interior support. The properties of the actual section are as follows:

$A = 1747.5$ in.^{2} $I = 1,085,000$ in.^{4}

$r^2 = \dfrac{1,085,000}{1747.5} = 621$ in.^{2} $\dfrac{I}{y} = \dfrac{1,085,000}{1747.5} = 28,940$ in.^{3}

and

$e = 30$ in.

$e_1 = 30 \times 1.24 = 37.2$ in.

$\dfrac{e_1 y}{r^2} = \dfrac{37.2 \times 37.5}{621} = 2.248$

$\dfrac{P_t}{A} = \dfrac{2000}{1747.5} = 1.135$ ksi

The moments and stresses are:

$M_a = 1800 + 2520 = 4320$ ft-kips

$f_a = \dfrac{4320 \times 12}{28,940} = 1.791$ ksi

$M_g = 1.82 \times \dfrac{(120)^2}{8} = 3276$ ft-kips

$f_g = \dfrac{3276 \times 12}{28,940} = 1.359$ ksi

The four requirements at the interior support are:

$1.135 \times 1.248 - 1.359 = 0.057$ ksi

$1.135 \times 3.248 - 1.359 = 2.326$ ksi

$-0.85 \times 1.135 \times 1.248 + 3.150 = 1.948$ ksi

$-0.85 \times 1.135 \times 3.248 + 3.150 = 0.012$ ksi

It should be pointed out that the above stresses are calculated on the basis of the gross cross-sectional properties of the beam. The stresses before losses may be computed more accurately if the net cross-sectional properties of the section are used. The stresses after losses may be computed accurately on the basis of the fully transformed section. These modifications, however, would only slightly change the stresses.

The above calculations indicate that the section at the interior support is adequately designed. Since these requirements should be satisfied at all points along the span, the stresses will be computed at 12 ft intervals along the span.

Figure 10-28 shows the profile of the center of gravity of steel and

394   PRESTRESSED CONCRETE

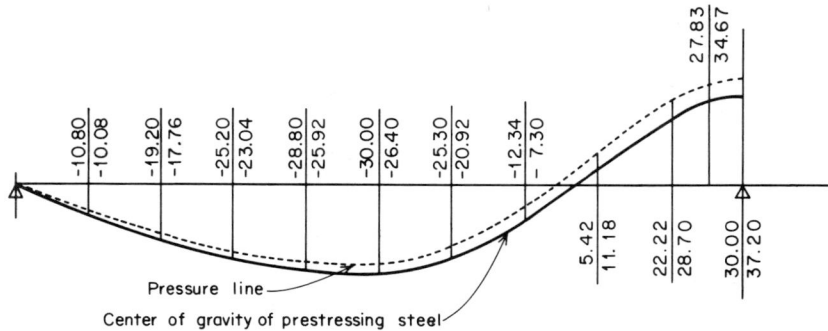

Fig. 10-28 The profile of the center of gravity of steel and the pressure line (all eccentricities are in inches). Illustrative Problem 10-4.

the pressure line as adopted for this problem. The profile of the center of gravity of steel varies as a parabola from the end to midspan and as a fourth-degree curve from midspan to the interior support.

Tables 10-3 and 10-4 show the calculation of stresses along the span, and Fig. 10-29 shows the variation of stress at top and bottom fibers for all combinations of loads.

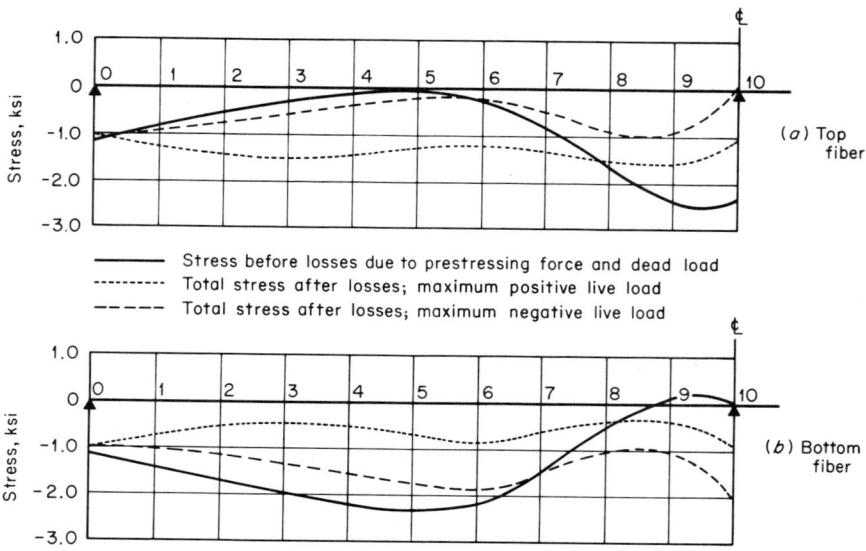

Fig. 10-29 Variation of stress along the span. Illustrative Problem 10-4.

Table 10-3 Stress before losses due to prestressing force and dead load

$P_t = 2000$ kips, $y = 37.5$ in.
$A = 1747.5$ in.^{2} $r^2 = 621$ in.^{2}

Point	e_1, in.	Stress* due to prestressing force, ksi		M_o, ft-kips	Stress* due to dead load, ksi		Total stress, ksi	
		Top fiber $-\dfrac{P_t}{A}\left(\dfrac{e_1 y}{r^2}+1\right)$	Bottom fiber $\dfrac{P_t}{A}\left(\dfrac{e_1 y}{r^2}-1\right)$		Top fiber f_o	Bottom fiber f_o	Top fiber	Bottom fiber
1	−10.08	−0.447	−1.822	851.8	−0.353	0.353	−0.800	−1.469
2	−17.76	0.082	−2.346	1441.4	−0.598	0.598	−0.516	−1.748
3	−23.04	0.445	−2.712	1769.0	−0.732	0.732	−0.287	−1.980
4	−25.92	0.639	−2.910	1834.6	−0.760	0.760	−0.121	−2.150
5	−26.40	0.672	−2.940	1638.0	−0.679	0.679	−0.007	−2.261
6	−20.92	0.297	−2.680	1179.4	−0.489	0.489	−0.192	−2.191
7	−7.30	−0.634	−1.635	458.6	−0.190	0.190	−0.824	−1.445
8	11.18	−1.900	−0.369	−524.2	0.217	−0.217	−1.683	−0.586
9	28.70	−3.098	0.830	−1769.0	0.732	−0.732	−2.366	0.098
$9\tfrac{1}{2}$	34.67	−3.505	1.241	−2489.8	1.031	−1.031	−2.474	0.210
10	37.20	−3.685	1.415	−3276.0	1.357	−1.357	−2.328	0.058

* Negative sign for stress indicates compressive stress.

Table 10-4 Stress after losses due to prestressing force, dead load, superimposed dead load, and live load

$P_t = 2000$ kips, $y = 37.5$ in.
$A = 1747.5$ in.^{2}, $r^2 = 621$ in.^{2}
$\eta = 0.85$

		Stress* due to prestressing force, ksi		Total moment M_t, ft-kips		Stress* due to total moment, ksi				Total stress,* ksi			
		Top fiber $-\eta \dfrac{P_t}{A}\left(\dfrac{e_1 y}{r^2}+1\right)$	Bottom fiber $\eta \dfrac{P_t}{A}\left(\dfrac{e_1 y}{r^2}-1\right)$	Max. positive live load	Max. negative live load	Max. positive live load		Max. negative live load		Max. positive live load		Max. negative live load	
Point	e_1, in.					Top fiber	Bottom fiber	Top fiber	Bottom fiber	Top fiber	Bottom fiber	Top fiber	Bottom fiber
1	−10.08	−0.380	−1.550	2101.0	1193.8	−0.871	0.871	−0.495	0.495	−1.251	−0.679	−0.875	−1.055
2	−17.76	0.069	−1.994	3594.2	1921.4	−1.490	1.490	−0.796	0.796	−1.421	−0.504	−0.727	−1.198
3	−23.04	0.378	−2.305	4479.8	2363.0	−1.858	1.858	−0.980	0.980	−1.480	−0.447	−0.602	−1.325
4	−25.92	0.543	−2.472	4757.8	2338.6	−1.972	1.972	−0.969	0.969	−1.429	−0.500	−0.426	−1.503
5	−26.40	0.571	−2.496	4428.0	1908.0	−1.835	1.835	−0.791	0.791	−1.264	−0.661	−0.220	−1.705
6	−20.92	0.252	−2.278	3490.6	1071.4	−1.449	1.449	−0.445	0.445	−1.197	−0.829	−0.193	−1.833
7	−7.30	−0.539	−1.390	1945.4	−171.4	−0.806	0.806	0.071	−0.071	−1.345	−0.584	−0.468	−1.461
8	11.18	−1.615	−0.314	−207.4	−1820.2	0.086	−0.086	0.755	−0.755	−1.529	−0.400	−0.860	−1.069
9	28.70	−2.632	0.706	−2567.8	−4225.0	1.065	−1.065	1.752	−1.752	−1.567	−0.359	−0.880	−1.046
9½	34.67	−2.980	1.066	−3830.0	−5800.9	1.588	−1.588	2.407	−2.407	−1.392	−0.522	−0.573	−1.341
10	37.20	−3.138	1.202	−5076.0	−7596.0	2.102	−2.102	3.150	−3.150	−1.036	−0.900	0.012	−1.948

* Negative sign for stress indicates compressive stress.

ANALYSIS AND DESIGN OF CONTINUOUS PRESTRESSED-CONCRETE BEAMS   397

It should be pointed out that the stresses shown in Tables 10-3 and 10-4 and their plots in Fig. 10-29 are based upon the prestressing force at the interior support. The prestressing force increases slightly toward the end of the beam since the loss due to friction is less in that direction. If the beam will be prestressed from both ends and the necessary precaution will be made to minimize the friction, the effect of friction will not influence the design.

PROBLEMS

10-1. Draw the shear and bending-moment diagrams for the structures shown in Fig. 10-30a to i. In all cases the prestressing force P is the only force acting.

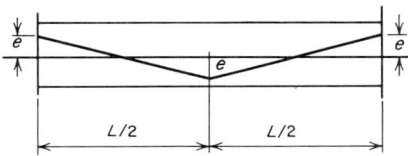

Fig. 10-30(a)

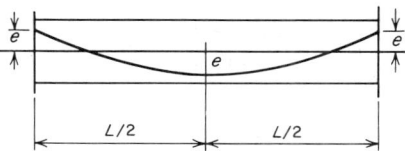

Fig. 10-30(b)

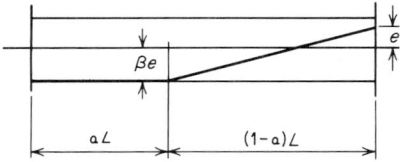

Fig. 10-30(c)

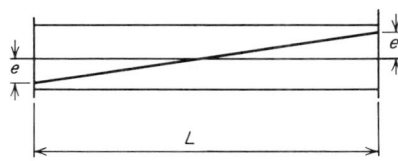

Fig. 10-30(d)

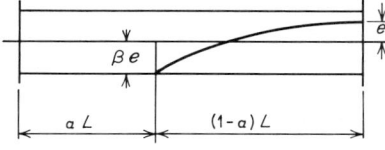

Fig. 10-30(e)

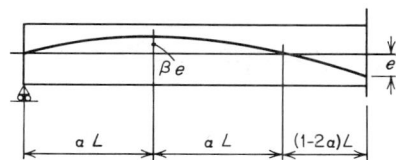

Fig. 10-30(f)

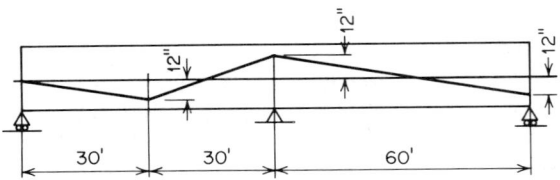

Fig. 10-30(g)

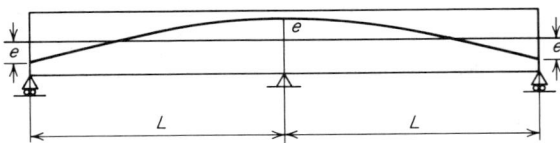

Fig. 10-30(h)

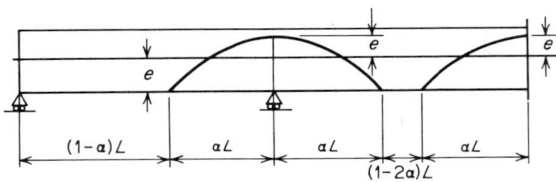

Fig. 10-30(i)

10-2. The structure shown in Fig. 10-31 is to be used in the construction of a warehouse and is to be regarded as a post-tensioned concrete hollow web girder. The members will take only axial stresses, and the loads will be concentrated on the top chord at the individual joints. Freyssinet-type post-tensioning cables (see figure) will provide 77 kips for each 18-wire 0.196-in. cable and 55 kips for each 12-wire 0.196-in. cable. The centroid of the cables in each member coincides with the geometric axis

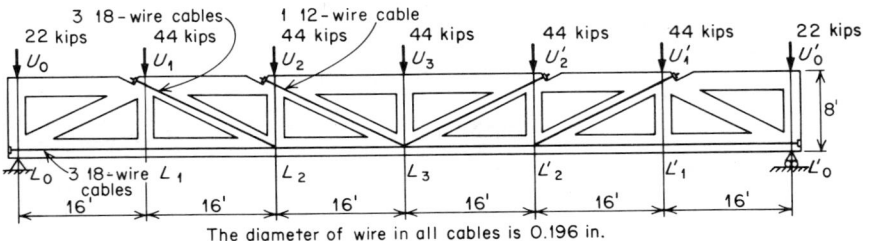

Fig. 10-31 The truss in Problem 10-2.

of the member. Friction is to be neglected. The top chord will have adequate lateral bracing when erected because of the floor structure. The flexure due to self-weight of chords and inclined members is to be neglected. The external loads shown are the total loads on the truss, of which 25 percent correspond to dead load, 25 percent to superimposed dead load, and 50 percent to live load. Determine the member forces due to the following loads:

(a) The prestressing force only

(b) The prestressing force and the weight of the truss

(c) The prestressing force, the weight of the truss, and the superimposed dead load

(d) The prestressing force, the weight of the beam, the superimposed dead load, and the live load

REFERENCES

1. Kaar, P. H., L. B. Kriz, and Eivind Hognestad: Precast-Prestressed Concrete Bridges: 1. Pilot Tests of Continuous Girders, *J. PCA Res. Develop. Lab.*, pp. 21–37, May, 1960.
2. Leonhardt, Fritz: Continuous Prestressed Concrete Beams, *J. Am. Concrete Inst.*, March, 1953; *Proc.* vol. 49, pp. 617–636. See also the Discussion to this paper by D. H. Lee, *J. Am. Concrete Inst.*, December, 1953, pt. 2, pp. 636–1 to 636–3.
3. Koebel, F. E., and Bengt Sonesson: Prestressed Concrete Superstructure for Flood-wrecked International Bridge at Laredo, Texas, *Civil Eng.*, March, 1956, pp. 167–171.
4. Magnel, Gustave: "Prestressed Concrete," pp. 90–95, McGraw-Hill Book Company, New York, and Concrete Publications Limited, London, 1955.
5. Parme, A. L., and G. H. Paris: Designing for Continuity in Prestressed Concrete Structures, *J. Am. Concrete Inst.*, September, 1951, pp. 54–64.
6. Moorman, R. B. B.: Continuous Prestressing, *Trans. Am. Soc. Civil Engrs.*, vol. 121, pp. 814–832, 1956.

11
Prestressed-concrete Columns

11-1 INTRODUCTION

In the preceding chapters prestressed-concrete beams were discussed in detail. Though the most important use of prestressed concrete is in beams, prestressed-concrete columns are used for different situations. Prestressed-concrete piles have been extensively used in foundation work, and slender columns are used in buildings. In this chapter we shall discuss the strength of these members under load.

Figure 11-1a shows a concentrically loaded symmetrically prestressed concrete column. There is no bending moment in an axially loaded column, and its analysis is simple. However, axially loaded columns are not very common in practice since in most cases there is some bending in the column due either to a small eccentricity or to asymmetry in the section. Figure 11-1b shows a column subjected to a load with an eccentricity measured from the middepth of the section. Actually, the eccentricity can be defined as the distance of the load from any conveniently selected reference line in the section of the column.

PRESTRESSED-CONCRETE COLUMNS

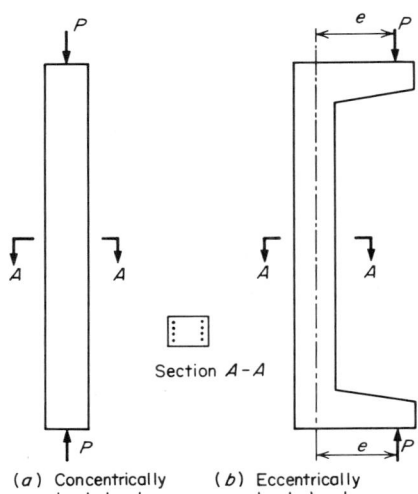

Fig. 11-1 Prestressed-concrete columns.

(*a*) Concentrically loaded column (*b*) Eccentrically loaded column

In the discussions that follow, the strength of a column will be obtained for the entire range of eccentricity in an axially prestressed rectangular section. In all cases the eccentricity of the acting load will be measured from the middepth of the section.

A considerable amount of research effort has been directed toward the study of behavior of prestressed-concrete columns under load. Research work includes experimental work with slender, eccentrically loaded, axially prestressed columns with hinged ends;[1] studies of ultimate strength under combined load and bending moment for standard pile sections;[2] study of effect of partial prestressing on the ultimate strength of eccentrically loaded columns;[3] study of strength of rectangular columns with hinged ends;[4] and studies of the stability of prestressed-concrete long columns.[5,6]

In order to study the strength of column for any type of failure and for any eccentricity of the acting load, certain relationships will be studied.

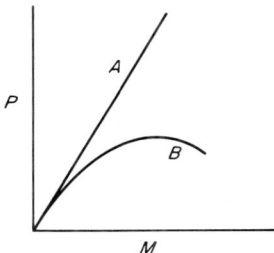

Fig. 11-2 Path to failure—relationship between P and M for the entire range of load.

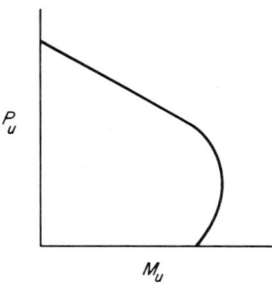

Fig. 11-3 Interaction diagram—relationship between ultimate load and moment for the entire range of eccentricity.

The relationship between the acting load and the bending moment for a given eccentricity when the load is increased from zero to ultimate is very useful since it permits the study of behavior of the column in the entire range of load. This relationship is developed for the critical section of the column, and for convenient reference it is called path to failure of the column.

Figure 11-2 shows typical paths of failure for two typical columns. For short columns this relationship tends to be nearly a line, as shown by curve A. For slender columns this relationship is distinctly nonlinear, as shown by curve B. These curves show that the slender columns are more likely to fail by instability if the peak is reached before concrete crushes.

The relationship between the ultimate load P_u and ultimate moment M_u, called the interaction diagram, is also very useful since it shows the failure load of the column section for any moment-to-load ratio. Figure 11-3 shows a typical interaction diagram from which it is possible to obtain ultimate load and moment that the section can resist.

Figure 11-4 shows the superposition of the two types of curve. Failure occurs when curves A and B, which represent two possible paths to failure of the critical section, intersect the interaction diagram. In a short column described by curve A, the slope of the path to failure is positive, and failure occurs as the load reaches its highest level and the

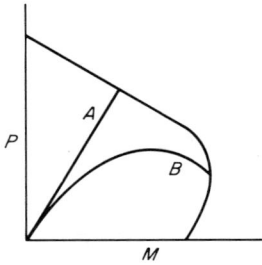

Fig. 11-4 Path to failure and interaction diagram.

concrete crushes. In a long column the maximum load is reached before the curve B describing the path to failure intersects the interaction diagram. In a long column after the maximum load is reached instability begins and at failure the slope of the path to failure is negative.

11-2 DETERMINATION OF ULTIMATE INTERACTION DIAGRAM

Let all possible combinations of axial load P_u and bending moment M_u that create failure in a given section be plotted as points of a graph. Let the ordinates of this graph represent axial loads and the abscissas represent bending moments. The curve that connects all the points in the graph is the ultimate interaction diagram of the section. Any combination of axial load and bending moment which falls inside the area bound by this curve and the coordinate axes can be supported by the section.

The determination of the ultimate interaction diagram requires the following assumptions:

1. The strain distribution in concrete varies linearly with depth in the compression zone.
2. The stress-strain diagrams for concrete and prestressed steel are known.
3. The section fails when the strain in concrete at this extreme fiber reaches ϵ_u or at middepth reaches ϵ_0.

As discussed in Sec. 3-6, ϵ_u, the crushing strain of concrete in bending, is greater than ϵ_0, the crushing strain when the strain is distributed uniformly. In the discussions that follow ϵ_u is taken as 0.003 and ϵ_0 is taken as 0.002.

The various combinations of P and M that create failure in a given prestressed-concrete section can be obtained with these assumptions and the use of relations of equilibrium and compatibility of strains. Let Fig. 11-5 represent the ultimate conditions of a prestressed-concrete column of rectangular section under a given P_u and M_u.

The section has an area of steel A_s equally distributed in two opposite faces. When the steel is prestressed to an initial prestrain ϵ_{se} (after losses), a state of axial prestress in the column is created.

Ultimate conditions in the section can be determined with any given strain distribution that will be in accordance with assumptions 1 and 3. Because of the ultimate strain distribution in the section, as shown in Fig. 11-5, the strain in the prestressing steel in the far side is increased beyond ϵ_{se}, while that of the near side is reduced. The final strains in the steel can be computed for any given strain distribution, and the existing stresses may be determined from the corresponding stress-strain diagram of the material. The forces in the steel, T_s and T'_s, are obtained by multiplying the stresses at ultimate by the corresponding areas of reinforcement in each face. The compression force developed by the concrete, C, can be obtained by calculating the area under the stress distribution and multiplying it by the width of the column.

In order to satisfy the equilibrium of forces, the sum of the internal forces in the section plus the external force P_u must equal zero. Also, the sum of the moments of the internal forces plus M_u, the external moment, must equal zero to satisfy static equilibrium. The pair of values P_u and M_u thus obtained corresponds to a particular combination of axial load and bending moment which causes failure of the section. As mentioned before, if this combination of values is plotted in a graph in which the ordinates represent axial loads and the abscissas represent bending moments, it becomes a point of the interaction diagram.

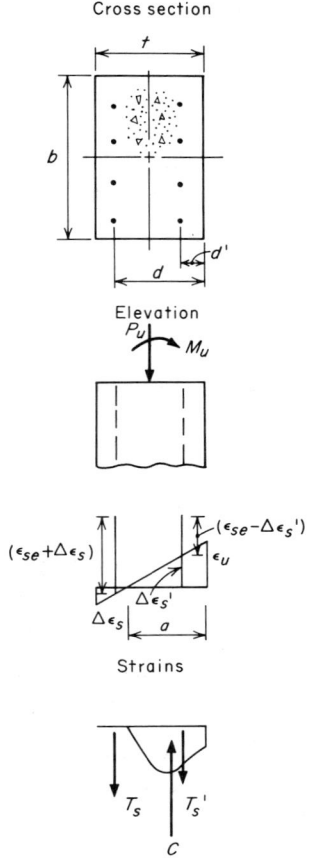

Fig. 11-5 Conditions at ultimate in a prestressed-concrete column.

PRESTRESSED-CONCRETE COLUMNS

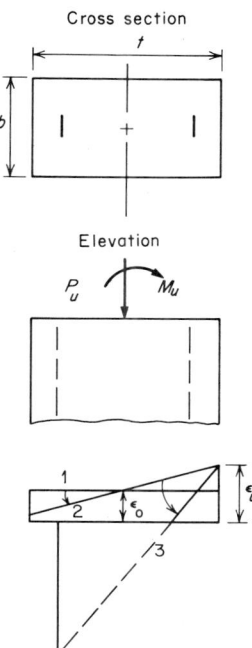

Fig. 11-6 Systematic determination of the interaction diagram using various strain distributions.

A systematic way to determine points of the interaction diagram is illustrated in Fig. 11-6. One may start with a uniform strain distribution over all the section, at a concrete strain ϵ_0 and maximum concrete stress, and determine the maximum axial force that can be supported by the section. This is given by position 1 of the various strain-distribution planes shown in Fig. 11-6. The failure criterion that requires that the middepth strain be equal to ϵ_0 is satisfied by rotating the strain-distribution plane about the middepth section. Any strain distribution contained between planes 1 and 2 satisfies the failure criterion. The top branch of the interaction diagram shown in Fig. 11-7 between points 1 and 2 was obtained from the strain-distribution planes bound by planes 1 and 2 of Fig. 11-6. The top branch is characterized by failures due to large axial loads and small bending moments. The rest of the interaction diagram, between points 2 and 3 of Fig. 11-6, may be obtained by satisfying the criterion that failure of the section occurs when the crushing strain ϵ_u is attained in the extreme end fiber. Strain distributions that meet this criterion can be obtained through rotations about the end fiber, between the limits set by planes 2 and 3. The latter corresponds to failure of the section due to bending moment alone, such as in the conven-

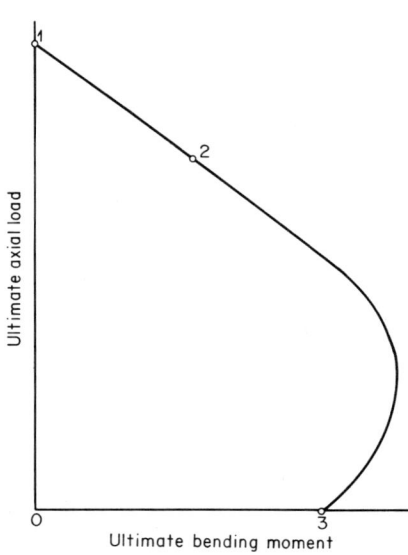

Fig. 11-7 Typical ultimate interaction diagram for a prestressed-concrete column.

tional case of a beam section, and cannot be predicted a priori. It can be determined, however, by using the general method of Sec. 8-2.

The determination of the interaction diagram of an actual section is given in the following illustrative example.

11-3 ILLUSTRATIVE EXAMPLE 11-1

A prestressed-concrete column with a 16- by 10-in. rectangular cross section is reinforced with eight $\tfrac{1}{2}$-in. strands, four in each wide face. The strength of the concrete is $f'_c = 6000$ psi. The strain at which maximum stress occurs, ϵ_0, and the crushing strain ϵ_u are assumed equal to 0.002 and 0.003, respectively. The stress-strain diagram for concrete is described by k_1, k_2, and k_3. For this problem assume $k_1 = 0.75$, $k_2 = 0.42$, and $k_3 = 0.85$. The stress-strain diagram in steel is shown in Fig. 11-8.

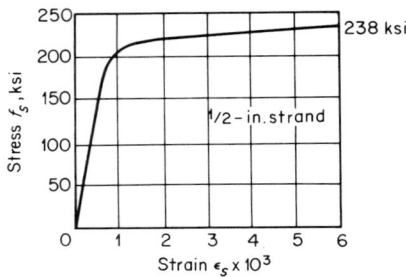

Fig. 11-8 Stress-strain diagram of prestressing steel.

PRESTRESSED-CONCRETE COLUMNS

Two levels of initial prestrain in the prestressing steel, ϵ_{se}, are considered. In the first case, full prestressing to $\epsilon_{se} = 0.005$ is applied, while in the second case the section has no prestressing at all, $\epsilon_{se} = 0$. The object of the problem is to determine the ultimate interaction diagram for the given section in both cases and to discuss how it is influenced by the level of initial strain in the prestressing steel.

The determination of three points of the ultimate interaction diagram for the section subjected to an initial prestrain $\epsilon_{se} = 0.005$ in the prestressing steel is shown in Figs. 11-9 to 11-11. The three points correspond, respectively, to a uniform strain distribution ϵ_u in the section, to a strain distribution given by ϵ_u at one end and zero at the other end of the section, and finally to the condition of failure created in the absence of

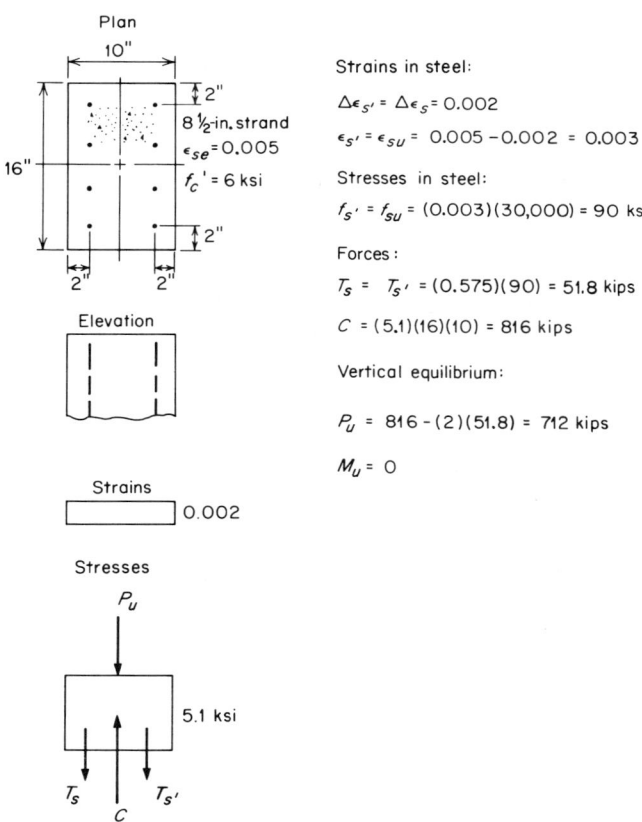

Fig. 11-9 Determination of ultimate axial load for a prestressed column.

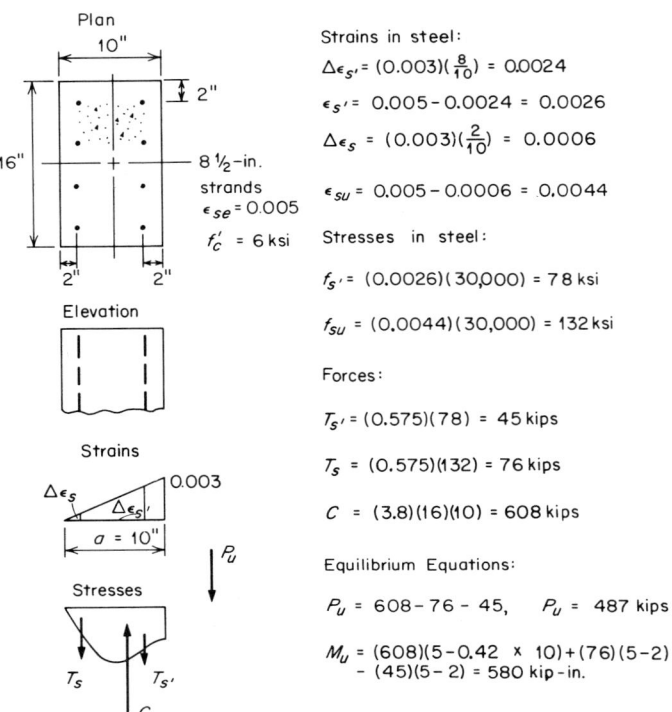

Fig. 11-10 Determination of ultimate moment and axial load for zero-tensile strain in a prestressed column.

axial load by bending moment alone. The numerical process necessary to obtain the values of P_u and M_u in each case is given in the corresponding figure. It appears that no additional explanation is necessary in the text.

The ultimate interaction diagram is the solid curve $\epsilon_{se} = 0.005$ shown in Fig. 11-12. The three points that were previously determined are labeled 1, 2, 3, respectively. More points are necessary if the diagram is to have any reasonable accuracy. This is especially true between points 2 and 3, where the curve has particularly pronounced curvatures. Additional points can be easily and systematically determined by the reader. For this purpose it may be helpful to observe that the concrete stress block is fully developed for any point in the lower branch of the interaction diagram, and the values of k_1, k_2 and k_3 are valid between points 2 and 3. However, between points 1 and 2, the stress block is not fully developed and these values of the coefficients cannot be used logically. To

PRESTRESSED-CONCRETE COLUMNS

determine the compression force developed by the concrete and its corresponding moment when only the general stress-strain diagram of concrete is available, a numerical-integration method such as that discussed in Sec. 3-16 may be necessary. If, however, a stress-strain diagram for concrete, such as that developed by Hognestad (Chap. 3, Ref. 3) and shown in Fig. 3-10c, is used, then expressions exist which simplify the

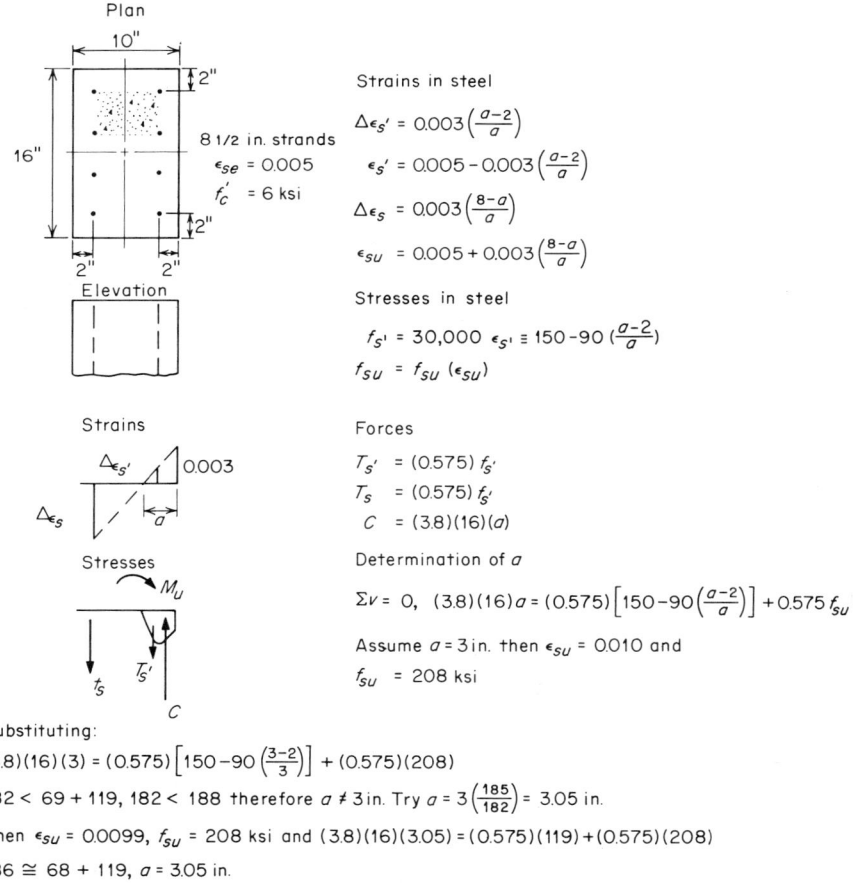

Strains in steel

$$\Delta\epsilon_s' = 0.003\left(\frac{a-2}{a}\right)$$

$$\epsilon_s' = 0.005 - 0.003\left(\frac{a-2}{a}\right)$$

$$\Delta\epsilon_s = 0.003\left(\frac{8-a}{a}\right)$$

$$\epsilon_{su} = 0.005 + 0.003\left(\frac{8-a}{a}\right)$$

Stresses in steel

$$f_{s'} = 30{,}000 \ \epsilon_{s'} \equiv 150 - 90\left(\frac{a-2}{a}\right)$$

$$f_{su} = f_{su}(\epsilon_{su})$$

Forces

$$T_{s'} = (0.575)f_{s'}$$
$$T_s = (0.575)f_{s'}$$
$$C = (3.8)(16)(a)$$

Determination of a

$$\Sigma v = 0, \quad (3.8)(16)a = (0.575)\left[150 - 90\left(\frac{a-2}{a}\right)\right] + 0.575 f_{su}$$

Assume $a = 3$ in. then $\epsilon_{su} = 0.010$ and
$f_{su} = 208$ ksi

Substituting:

$(3.8)(16)(3) = (0.575)\left[150 - 90\left(\frac{3-2}{3}\right)\right] + (0.575)(208)$

$182 < 69 + 119$, $182 < 188$ therefore $a \neq 3$ in. Try $a = 3\left(\frac{185}{182}\right) = 3.05$ in.

Then $\epsilon_{su} = 0.0099$, $f_{su} = 208$ ksi and $(3.8)(16)(3.05) = (0.575)(119) + (0.575)(208)$
$186 \cong 68 + 119$, $a = 3.05$ in.

Equilibrium Equations

$P_u = 0$

$M_u = (186)(5 - 0.42 \times 3.05) - (68)(5-2) + (119)(5-2) = 844$ kip-in.

Fig. 11-11 Determination of ultimate moment in a prestressed column.

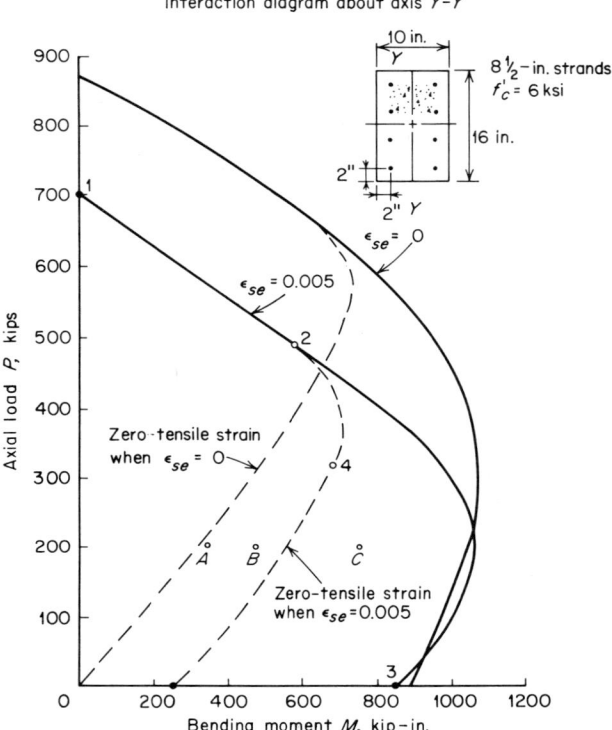

Fig. 11-12 Effect of initial prestrain ϵ_{se} on the interaction diagram of a prestressed-concrete column.

required determination. In most practical cases, however, there is no need for the determination of additional points between 1 and 2, as it can be shown that a straight line is an accurate approximation to the actual curve.

The determination of the ultimate interaction diagram for the case where ϵ_{se}, the initial prestrain in the prestressing steel, is zero is not given, but the interaction diagram is shown in Fig. 11-12. The section behaves like a conventional reinforced-concrete section, except that it is reinforced with high-strength steel which lacks a definite yield point. This accounts for the flat curvature of the diagram, as shown in Fig. 11-12, and for the absence of a sharp "balanced point," a definite characteristic of the interaction diagram of conventionally reinforced concrete sections.

A comparison between both diagrams is of interest. As may have

PRESTRESSED-CONCRETE COLUMNS

been expected, initial prestrain in the steel reduces the capacity of the section to resist external axial load. This, of course, is due to the action of the internal compression force imposed by the prestressing steel on the concrete. In the lower region of the diagram, as shown in Fig. 11-9, both sections attain approximately the same value of maximum bending moment although under different axial loads. The same is true for the value of bending moment under zero axial load, which is somewhat smaller for the axially prestressed column. The statement can be made that the flexural strength of axially prestressed concrete sections is independent of ϵ_{se}, the amount of initial prestrain in the prestressing steel. The ductility of the section, however, as measured by ultimate curvature, is greatly influenced by ϵ_{se}. The variation of ultimate axial load with curvature in the section is shown in Fig. 11-13 for $\epsilon_{se} = 0.005$ and $\epsilon_{se} = 0$. It is readily noticeable that the non-prestressed section is more ductile, that is, the ultimate curvature is larger, at any value of ultimate axial load. When failure occurs in the section under the action of bending moment only, the difference in curvature is approximately 40 percent in favor of the non-prestressed section.

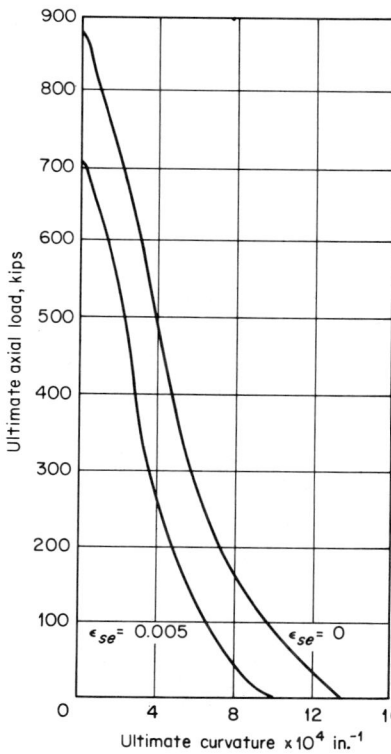

Fig. 11-13 Effect of initial prestrain ϵ_{se} on the ductility of a prestressed-concrete column.

11-4 DETERMINATION OF ZERO-TENSILE-STRAIN INTERACTION DIAGRAM

Full understanding of the behavior under service loads of prestressed-concrete columns, especially the long, slender ones, requires more information than that provided by the ultimate interaction diagrams of the section. For the usual case of long prestressed columns subjected to the combined action of axial load and bending moment, the secondary effects due to the action of the axial load on the deflection configuration can be substantial. The product of the axial load and the transverse deflection at each section creates secondary moments that when added to the primary bending moment may contribute significantly to the final deflection configuration of the column.

If the column remains uncracked throughout its length, the transverse deflections are smaller than if cracking occurs. This is due to the fact that the moment of inertia of an uncracked section is larger than that of a cracked section. The column section may be designed in such a way as to prevent cracking of the section under service-load conditions, thereby ensuring the maximum stiffness of the member.

In order to determine whether the section of a column will be cracked or not under a given axial load and bending moment, a cracking interaction diagram is necessary. Such a diagram would constitute the envelope of all possible combinations of axial load and bending moment which create a maximum tensile stress in the section lower than the cracking strength of the concrete. Since the tensile strength of the concrete in flexure is difficult to estimate accurately, and the contribution of concrete to tension is negligible, it is reasonable and conservative to neglect the tensile strength of the concrete. Instead of the cracking interaction diagram, then, a more realistic and useful concept would be that of a zero-tensile-strain interaction diagram. The latter may be defined as the envelope of all possible combinations of axial load and bending moment for which tensile strains are not created in the section. Obviously, the cracking and the zero-tensile-strain interaction diagrams coincide for a theoretical material with no tensile strength.

The systematic determination of the zero-tensile-strain interaction diagram follows the same principles used in the determination of the ultimate interaction diagram. The various strain distributions are shown in Fig. 11-14. The process is the same as before between positions 1 and 2; that is, both interaction diagrams have the same top branch. Beyond position 2, the rotation of the strain-distribution planes occurs about the near end fiber at the level of the crushing strain as before, except that limit position 3 is attained in this case when the strain at the far end fiber equals zero. The last portion of the diagram is obtained by a rotation of the strain distribution planes about the far end fiber at the

PRESTRESSED-CONCRETE COLUMNS 413

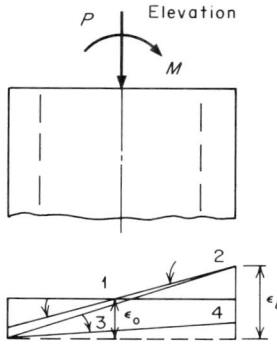

Fig. 11-14 Systematic determination of the zero-tensile-strain interaction diagram using various strain distributions.

level of zero strain until limit position 4 is reached, as shown in Fig. 11-14, for which the section is subjected to bending moment only with no axial load.

The zero-tensile-strain interaction diagrams for the prestressed and the non-prestressed concrete column of the preceding example have been determined and plotted in Fig. 11-12. The fact that the stress block in the concrete is not fully developed when the maximum strain in the section is smaller than ϵ_u prevents the use of Stuessi's expressions to determine points in this interaction diagram beyond plane 3 of Fig. 11-11. However, if Hognestad's stress-strain diagram is used, as shown in Fig. 3-10c, the following simple expressions can be used to determine C, the compression force developed by the concrete, and M_c, its moment about the centerline of the section:

$$C = (0.85 f'_c bt) \frac{\epsilon_4}{\epsilon_0} \left(1 - \tfrac{1}{3} \frac{\epsilon_4}{\epsilon_0}\right)$$

$$M_c = \tfrac{1}{6}(0.85 f'_c bt^2) \frac{\epsilon_4}{\epsilon_0} \left(1 - \tfrac{1}{2} \frac{\epsilon_4}{\epsilon_0}\right)$$

where all symbols have been defined before with the exception of ϵ_4, which is the strain in the concrete at the near end fiber. The preceding expressions are valid for the range $0 \leq \epsilon_4 \leq \epsilon_0$ when the strain in the concrete at the far end is zero. The use of the preceding expressions is illustrated in Fig. 11-15 to determine a point of the zero-tensile-strain interaction diagram.

An observation of Fig. 11-12 indicates that the prestressed section has a zero-tensile-strain interaction diagram that contains that of the non-prestressed section. Under loading conditions such as given by point 4 in Fig. 11-12, the non-prestressed section may be cracked while the pre-

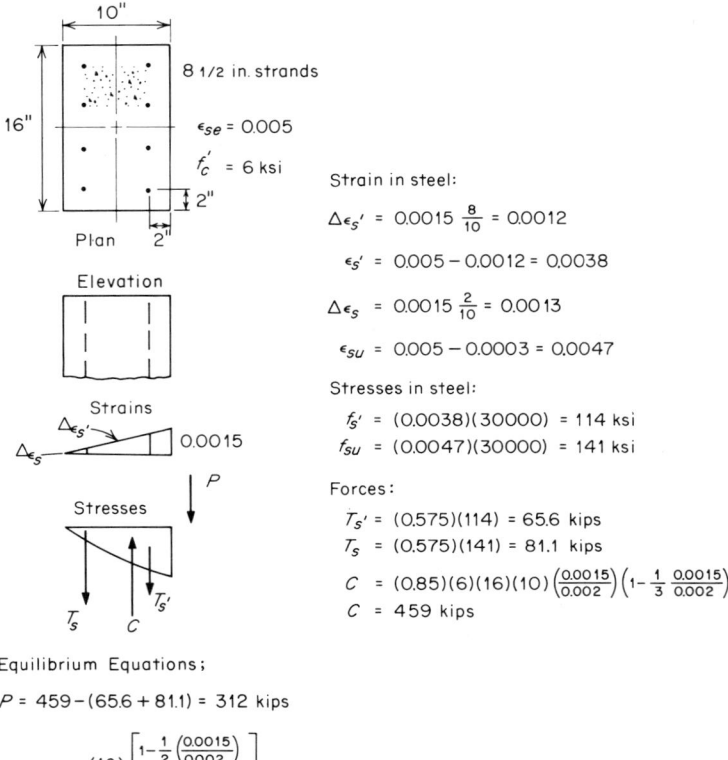

Fig. 11-15 Determination of a point in the zero-tensile-stress interaction diagram.

stressed section is not. Because of the reduction in stiffness of a cracked section, larger curvatures and deflections occur under given bending moments in a cracked column. The integrated effect on the behavior of a long column can be significant, as secondary effects may add substantially to the final deflection configuration. The following example is presented to illustrate this point fully.

11-5 ILLUSTRATIVE EXAMPLE 11-2

The member shown in Fig. 11-16 is a typical column in a given building. The column supports the end reaction $P = 200$ kips of a precast girder at

PRESTRESSED-CONCRETE COLUMNS

its top and is subjected to the action of a wind load estimated at 20 psf. The columns are spaced at 20 ft centers. The rectangular section of the column and material properties are identical to those of Illustrative Example 11-1. The wide face of the column is parallel to the axis of the wall. Bending of the column due to wind, therefore, takes place about the weak axis of the section. The column may be assumed as restrained at its top and bottom sections against lateral displacement but free to rotate at each end. Study the behavior of the column (a) for the case in which the initial prestrain in the prestressing steel is $\epsilon_{se} = 0.005$ and (b) for the case in which the initial prestrain is zero.

(a) Behavior of prestressed column with initial prestrain $\epsilon_{se} = 0.005$
The worst combination of axial load and bending moment will occur at midspan under the action of the wind. It will be equal to

$$P = 200 \text{ kips} \qquad M = \tfrac{1}{8}(0.4)(24)^2(12) = 346 \text{ kip-in.}$$

and may be represented in the interaction diagram of Fig. 11-12 as point A. It can be seen readily that no tensile strains, and therefore no cracks, occur in the midspan section under the values of P and M.

Because of the flexibility of the member and the combined action of the transverse wind load and the axial load, an additional moment $\Delta M = P\delta$ exists, where δ is the horizontal deflection at midheight. With

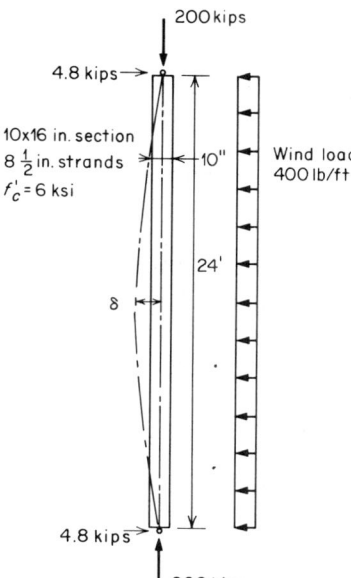

Fig. 11-16 Column subjected to combined axial load and bending moment.

the assumption that the material behaves elastically, and because of the fact that no cracks occur in the section which may change the value of the moment of inertia along the column, the following expression[7] can be used to evaluate the midspan deflection:

$$\delta = \left(\frac{5}{384}\frac{qL^4}{EI}\right)\frac{12(2\sec u - 2 - u^2)}{5u^4}$$

where $u = \frac{L}{2}\sqrt{\frac{P}{EI}}$ and q, L, E, and I are the conventional symbols for transverse load, free span, modulus of elasticity of the column, and moment of inertia of the section, respectively.

In the preceding expression the factor $\frac{5}{384}(qL^4/EI)$ evaluates the midspan deflection when the member is subjected to a longitudinal load $P = 0$. When $P > 0$, δ is amplified by the second factor in the expression. In the limit, when $u = \pi/2$ under the critical load $P = \pi^2(EI/L^2)$ the quantity δ theoretically reaches an infinite value and for all practical purposes the column fails due to elastic instability.

The modulus of elasticity of the column may be assumed conservatively as equal to E_c, the modulus of elasticity of the concrete of the column. The American Concrete Institute recommends the following expression to evaluate the tangent modulus of elasticity of concrete in pounds per square inch:

$$E_c = w^{1.5}(33)\sqrt{f'_c}$$

where w is the weight of concrete in pounds per cubic foot. If we assume that $w = 150$ pcf and $f'_c = 6000$ psi, the preceding expression yields

$$E_c = (150)^{1.5}(33)\sqrt{6000} = 4{,}660{,}000 \text{ psi}$$

The required value of the moment of inertia can be estimated conservatively as that of the uncracked section of the column. The small contribution of the prestressing steel can be safely neglected. For the given case in which flexure takes place about the weak axis of the section the corresponding moment of inertia is given by

$$I = \tfrac{1}{12}(16)(10^3) = 1330 \text{ in.}^4$$

The value of u can then be determined as follows:

$$u = \frac{(24)(12)}{2}\sqrt{\frac{200}{(4660)(1330)}} = 0.82$$

which, when substituted in the expression for the midspan deflection, yields

$$\delta = \left[\frac{5}{384} \frac{(\frac{0.4}{12})(24 \times 12)^4}{(4660)(1330)}\right] \frac{12(2 \sec 0.82 - 2 - 0.82^2)}{(5)(0.82)^4}$$
$$= (0.48)(1.38) = 0.66 \text{ in.}$$

The additional bending moment is

$$\Delta M = (200)(0.66) = 132 \text{ kip-in.}$$

and the final combination of maximum moment and axial load at midspan is the following:

$$P = 200 \text{ kips} \qquad M = 346 + 132 = 478 \text{ kip-in.}$$

Point B in Fig. 11-12 represents the preceding actual conditions at the midspan section. It still corresponds to a section with no tensile strains. The assumption of an uncracked column is therefore justified, and point B may be considered an accurate representation of the critical section of the column under service loads. The interaction diagram shows that a minimum factor of safety of 2.2 exists against failure of the column due to increase of the wind load. Therefore, the prestressed column may be considered a satisfactory solution.

(b) Behavior of non-prestressed column With no initial prestrain in the steel, the member behaves as a conventional reinforced-concrete column subjected to axial load and bending moment. It is likely, then, that the member will be cracked at some sections because of different factors such as shrinkage, erection stresses, or temporary overloading due to various reasons. The conditions at the midspan section under the action of wind and axial load, as represented by point A in Fig. 11-12, are also conducive to possible cracking. The secondary action of the axial load on the deflected configuration created by the transverse load adds to the cracking of the column.

It is difficult to estimate the moment of inertia of the cracked section at midspan, as it depends on the final moment at the section, which in turn depends on the moment of inertia. The final configuration of the column can be determined accurately by numerical analysis.[8,9] This treatment, however, is beyond the scope of this text.

A good estimate of the final conditions at the midspan section can be obtained if a reasonable assumption is made on the average value of the moment of inertia of the column. If the section were cracked to the level of the tensile steel, i.e., for a distance of 2 in. into the far end side of the column, the moment of inertia of the section would be

$$I = (\tfrac{1}{12})(16)(8)^3 = 686 \text{ in.}^4$$

To justify taking the preceding value of I as the moment of inertia of the column, let us consider the following argument. If a linear stress distribution is assumed in the section, the maximum stress in the compression face of the section to develop a compression force of 200 kips is $200/(\tfrac{1}{2})(8)(16) = 3.1$ ksi. The moment in the section would be $(200)(5 - \tfrac{1}{3} \times 8) = 467$ kip-in. These values are close to the values of P and M at the uncracked midspan section of the prestressed column, as determined in part a of this problem. It is true that the final value of moment at midspan in the non-prestressed column will be greater because of the increased flexibility of the cracked column. However, cracking will be less severe at sections away from midspan and will disappear toward the ends of the column, a fact which may compensate for the more severe cracking in the center portion of the column. It may be possible, therefore, to take the moment of inertia of the non-prestressed column conservatively as $I = 682$ in.^{4}. With $E_c = 4660$ ksi as before, the corresponding value of u for this case is

$$u = \frac{(24)(12)}{2} \sqrt{\frac{200}{(4660)(682)}} = 1.14$$

The midspan deflection under the combined action of axial load and transverse wind load can be obtained as follows:

$$\delta = \left[\frac{5}{384} \frac{(\tfrac{0.4}{12})(24 \times 12)^4}{(4660)(682)} \right] \frac{12(2 \sec 1.14 - 2 - 1.14^2)}{(5)(1.14)^4}$$

$$= (0.94)(2.20) = 2.07 \text{ in.}$$

The additional bending moment is then

$$\Delta M = (200)(2.07) = 414 \text{ kip-in.}$$

and the final combination of maximum moment and axial load at midspan is

$P = 200$ kips $M = 346 + 414 = 760$ kip-in.

Point C of Fig. 11-8 represents the preceding conditions. It can be seen readily how uncomfortably close is point C to the limits of the interaction curve. The safety of the column under these conditions may be estimated as smaller than 1.4 for failure due to increase in bending moment. In view of the crude nature of the assumptions made to arrive at point C, and the small factor of safety obtained, it seems that the non-prestressed column is not a satisfactory solution at all and should not be used.

It can be seen that the prestressed column will have a better behavior under the conditions of this problem than the non-prestressed column. This is true in long columns, subjected to relatively small axial loads and to bending moments which may increase with increasing deflections. The major effect of prestressing consists in guaranteeing an uncracked member with elastic behavior under service loads, which will be subjected to minimum additional deformations due to secondary effects. The latter may be of such importance that, even when the interaction diagram for the prestressed section is contained by the interaction diagram of the non-prestressed section, the behavior under load of the long prestressed column is better than that of the non-prestressed column.

PROBLEMS

11-1. Determine the ultimate interaction diagram of the section of Illustrative Example 11-1 for the following cases: (a) When the initial prestrain in the prestressing steel is $\epsilon_{se} = 0.0025$; (b) when the strength of the concrete is reduced to $f'_c = 5000$ psi and the initial prestrain is $\epsilon_{se} = 0.005$; (c) same as before except $\epsilon_{se} = 0$.

11-2. Using the results of parts (b) and (c) of the preceding problem, study the effects of reducing by 1000 psi the strength of the concrete in the column of Illustrative Problem 11-2.

REFERENCES

1. Aroni, Samuel: Slender Prestressed Concrete Columns, *J. Structural Div. ASCE, Proc. Paper* 5886, vol. 94, no. ST4, pp. 875–904, April, 1968.
2. Zia, Paul, and E. C. Guillermo: Combined Bending and Axial Load in Prestressed Concrete Columns, *J. Prestressed Concrete Inst.*, June, 1967, pp. 52–59.
3. Lin, T. Y., and T. R. Lakhwara: Ultimate Strength of Eccentrically Loaded Partially Prestressed Columns, *J. Prestressed Concrete Inst.*, June, 1966, pp. 37–49.
4. Zia, Paul, and F. L. Moreadith: Ultimate Load Capacity of Prestressed Concrete Columns, *J. Am. Concrete Inst.*, July, 1966; *Proc.*, vol. 63, no. 7, pp. 767–788.
5. Breckenridge, R. A.: Study of the Characteristics of Prestressed Concrete Columns, *Univ. Southern Calif., Los Angeles, Eng. Center Rept.* 18-6, April, 1953.
6. Hromadik, J. J.: Column Strength of Long Piles, *J. Am. Concrete Inst.*, June, 1962; *Proc.*, vol. 59, no. 6, pp. 757–788.
7. Timoshenko, S. P., and J. M. Gere: "Theory of Elastic Stability," 2d ed., McGraw-Hill Book Company, New York, 1961.
8. Newmark, Nathan M.: Numerical Procedure for Computing Deflections, Moments and Buckling Loads, *Trans. ASCE*, vol. 108, 1943.
9. Gurfinkel, G., and A. R. Robinson: Determination of Strain Distribution and Curvature in a Reinforced Concrete Section Subjected to Bending Moment and Longitudinal Load, *J. Am. Concrete Inst.*, July, 1967; *Proc.*, vol. 64, no. 7, pp. 398–403.

appendix A

A-1 DERIVATION OF EQS. (7-5) THROUGH (7-8)

The following equations are developed in the text:

$$m\left[\frac{\epsilon}{\rho(1+\Delta)} - 1\right] - \frac{1}{8\rho\omega(1+\Delta)} = a_t \qquad (7\text{-}1a)$$

$$m\left[\frac{\epsilon\Delta}{\rho(1+\Delta)} + 1\right] - \frac{\Delta}{8\rho\omega(1+\Delta)} = c_t \qquad (7\text{-}2a)$$

$$-\eta m\left[\frac{\epsilon}{\rho(1+\Delta)} - 1\right] + \frac{1+R}{8\rho\omega(1+\Delta)} = c \qquad (7\text{-}3a)$$

$$-\eta m\left[\frac{\epsilon\Delta}{\rho(1+\Delta)} + 1\right] + \frac{\Delta(1+R)}{8\rho\omega(1+\Delta)} = a \qquad (7\text{-}4a)$$

Derivation of Eq. (7-5) Dividing Eq. (7-2a) by Δ and subtracting it from Eq. (7-1a), we have

$$-m\left(1 + \frac{1}{\Delta}\right) = \frac{a_t - c_t}{\Delta}$$

Similarly, dividing Eq. (7-4a) by Δ and subtracting it from Eq. (7-3a), we have

$$\eta m \left(1 + \frac{1}{\Delta}\right) = \frac{c - a}{\Delta}$$

Dividing the first of the above equations by the second and rearranging give the following:

$$\Delta = \frac{\eta c_t + a}{\eta a_t + c} \tag{7-5}$$

Derivation of Eq. (7-6) Multiplying Eqs. (7-1a) and (7-2a) by η and adding them to Eqs. (7-3a) and (7-4a), we have

$$\frac{-\eta + 1 + R}{8\rho\omega} = \eta(a_t + c_t) + a + c$$

or

$$R = 8\rho\omega \left[\eta(a_t + c_t) + a + c\right] - (1 - \eta) \tag{7-6}$$

Derivation of Eq. (7-7) Dividing Eq. (7-2a) by Δ and subtracting it from Eq. (7-1a), we have

$$-m\left(1 + \frac{1}{\Delta}\right) = \frac{a_t - c_t}{\Delta}$$

or

$$m = \frac{c_t - \Delta a_t}{1 + \Delta} \tag{7-7a}$$

Similarly, dividing Eq. (7-4a) by Δ and subtracting it from Eq. (7-3a), we have

$$\eta m \left(1 + \frac{1}{\Delta}\right) = \frac{c - a}{\Delta}$$

or

$$m = \frac{\Delta c - a}{\eta(1 + \Delta)} \tag{7-7b}$$

Substituting Eq. (7-5) for Δ in either Eq. (7-7a) or Eq. (7-7b) gives the following:

$$m = \frac{c_t c - a_t a}{(a + c) + \eta(a_t + c_t)} \tag{7-7}$$

Derivation of Eq. (7-8) Adding Eqs. (7-1a) and (7-2a), we have

$$m \frac{\epsilon}{\rho} = \frac{1}{8\rho\omega} + a_t + c_t$$

APPENDIX A

or

$$\epsilon = \frac{1}{m}\left[\rho(a_t + c_t) + \frac{1}{8\omega}\right] = \frac{(a+c) + \eta(a_t + c_t)}{c_t c - a_t a}\left[\rho(a_t + c_t) + \frac{1}{8\omega}\right] \quad (7\text{-}8)$$

A-2 DERIVATION OF EQS. (7-10) THROUGH (7-14)

Derivation of Eq. (7-11) The cross-sectional area of the idealized unsymmetrical I section shown in Fig. A-1 can be presented as follows:

$$A = bt + kbt + (h - 2t)b'$$

or

$$A = bh\left[\frac{t}{h}(1+k) + \frac{b'}{b}\left(1 - \frac{2t}{h}\right)\right]$$

and

$$y_b = \frac{bt}{A}\left(h - \frac{t}{2}\right) + \frac{b'h}{2A}(h - 2t) + \frac{kbt^2}{2A}$$
$$= \frac{bh}{A}\left[\frac{t^2}{2h}(k-1) + \frac{b'}{b}h\left(\frac{1}{2} - \frac{t}{h}\right) + t\right]$$

or

$$y_b = \frac{\frac{t^2}{2h}(k-1) + \frac{b'}{b}h\left(\frac{1}{2} - \frac{t}{h}\right) + t}{\frac{t}{h}(k+1) + \frac{b'}{b}\left(1 - \frac{2t}{h}\right)}$$

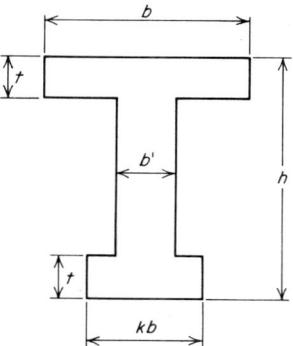

Fig. A-1 An idealized unsymmetrical I section.

Substituting the above expression in $\Delta = y_b/y_t \equiv y_b/(h - y_b)$ and rearranging, we have

$$\Delta = \frac{\frac{b'}{b}\left(1 - \frac{2t}{h}\right) + \frac{2t}{h} + \frac{t^2}{h^2}(k-1)}{\frac{b'}{b}\left(1 - \frac{2t}{h}\right) + 2k\frac{t}{h} - \frac{t^2}{h^2}(k-1)} \quad (7\text{-}11)$$

The moment of inertia of an unsymmetrical I section can be presented as follows:

$$I = (kb - b')\frac{t^3}{3} + (h - t)^3 \frac{b'}{3} + \tfrac{1}{12}bt^3 + bt\left(h - \frac{t}{2}\right)^2 - Ay_b^2$$

and

$$\rho = \frac{I}{Ah^2} = \frac{hb}{3A}\left[\left(k - \frac{b'}{b}\right)\left(\frac{t}{h}\right)^3 + \left(1 - \frac{t}{h}\right)^3 \frac{b'}{b}\right.$$
$$\left. + \frac{1}{4}\left(\frac{t}{h}\right)^3 + \frac{3t}{h}\left(1 - \frac{t}{2h}\right)^2\right] - \left(\frac{y_b}{h}\right)^2$$

Substituting the expressions for A and y_b in the above equation and rearranging, we have

$$\rho = \frac{\left(1 + k - \frac{2b'}{b}\right)\left(\frac{t}{h}\right)^3 + \frac{b'}{b} + \frac{3t}{h}\left(1 - \frac{t}{h}\right)\left(1 - \frac{b'}{b}\right)}{3\left[\left(1 + k - \frac{2b'}{b}\right)\frac{t}{h} + \frac{b'}{b}\right]} - \left(\frac{\Delta}{1 + \Delta}\right)^2 \tag{7-10}$$

Symmetrical I sections By substituting $k = 1.0$ and $\Delta = 1.0$ in Eq. (7-10), the efficiency of a symmetrical I section, shown in Fig. A-2, may be obtained.

$$\rho = \frac{1 - (1 - b'/b)(1 - 2t/h)^3}{12[1 - (1 - b'/b)(1 - 2t/h)]} \tag{7-14}$$

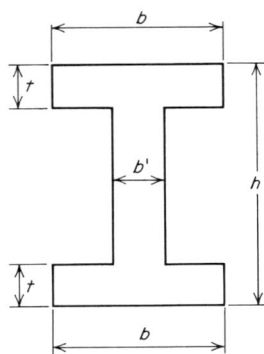

Fig. A-2 An idealized symmetrical I section.

APPENDIX A

T sections Figure A-3 shows an idealized T section. Substituting $k = b'/b$ in Eqs. (7-10) and (7-11) and rearranging, we have

$$\rho = \frac{\left(1 - \frac{b'}{b}\right)\left(\frac{t}{h}\right)^3 + \frac{b'}{b} + \frac{3t}{h}\left(1 - \frac{t}{h}\right)\left(1 - \frac{b'}{b}\right)}{3\left[\left(1 - \frac{b'}{b}\right)\frac{t}{h} + \frac{b'}{b}\right]} - \left(\frac{\Delta}{1 + \Delta}\right)^2$$

(7-12)

$$\Delta = \frac{\frac{b'}{b}\left(1 - \frac{t}{h}\right)^2 + \frac{2t}{h}\left(1 - \frac{t}{2h}\right)}{\frac{b'}{b}\left(1 - \frac{t^2}{h^2}\right) + \frac{t^2}{h^2}}$$

(7-13)

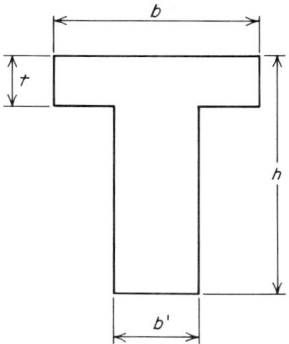

Fig. A-3 An idealized T section.

A-3 DERIVATION OF EQ. (9-9)

The cross-sectional area of an idealized composite section is shown in Fig. A-4. The centroidal axis of the composite section may be defined

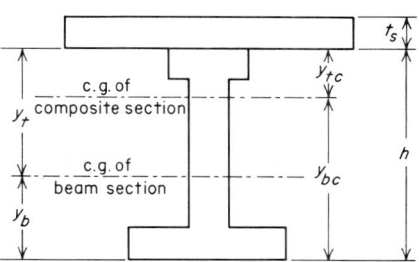

Fig. A-4 An idealized composite section.

as follows:

$$y_{bc} = \frac{Ay_b + A_{re}(h + t_s/2)}{A + A_{re}}$$

and

$$y_{tc} = \frac{Ay_t - A_{re}t_s/2}{A + A_{re}}$$

where A = area of beam section
$A_{re} = KSt_s$ = effective area of slab

Hence

$$\Delta_c = \frac{y_{bc}}{y_{tc}} = \frac{Ay_b + A_{re}h(1 + t_s/2h)}{Ay_t - A_{re}h\, t_s/2h}$$

or

$$\Delta_c = \frac{\dfrac{y_b}{y_t} + \dfrac{A_{re}}{A}\left(1 + \dfrac{t_s}{2h}\right)\left(1 + \dfrac{y_b}{y_t}\right)}{1 - \dfrac{A_{re}}{A}\dfrac{t_s}{2h}\left(1 + \dfrac{y_b}{y_t}\right)}$$

and

$$\Delta_c = \frac{\Delta + Ku(1 + \Delta)(1 + t_s/2h)}{1 - Ku(1 + \Delta)t_s/2h} \tag{9-9}$$

A-4 DERIVATION OF EQ. (9-10)

The quantity ρ_c has been defined in the text as follows:

$$\rho_c = \frac{I_c}{(A + A_{re})h^2}$$

or

$$\rho_c = \frac{I/A}{(1 + A_{re}/A)h^2} + \frac{A_{re}/A}{12(1 + A_{re}/A)}\left(\frac{t_s}{h}\right)^2$$
$$+ \frac{1}{1 + A_{re}/A}\left(\frac{y_t - y_{tc}}{h}\right)^2 + \frac{A_{re}/A}{1 + A_{re}/A}\left(\frac{y_{tc} + t_s/2}{h}\right)^2$$

The above expression may be written as follows:

$$\rho_c = \frac{I/A}{(1 + A_{re}/A)h^2} + \frac{A_{re}/A}{12(1 + A_{re}/A)}\left(\frac{t_s}{h}\right)^2$$
$$+ \frac{(A_{re}/A)^2}{(1 + A_{re}/A)^3}\left(\frac{1}{1 + y_b/y_t} + \frac{t_s}{2h}\right)^2$$
$$+ \frac{A_{re}/A}{(1 + A_{re}/A)^3}\left(\frac{1}{1 + y_b/y_t} + \frac{t_s}{2h}\right)^2$$

or

$$\rho_c = \frac{\rho + \dfrac{Ku}{12}\left(\dfrac{t_s}{h}\right)^2}{1 + Ku} + \frac{Ku}{(1 + Ku)^2}\left(\frac{1}{1 + \Delta} + \frac{t_s}{2h}\right)^2 \tag{9-10}$$

A-5 FIXED-END MOMENTS

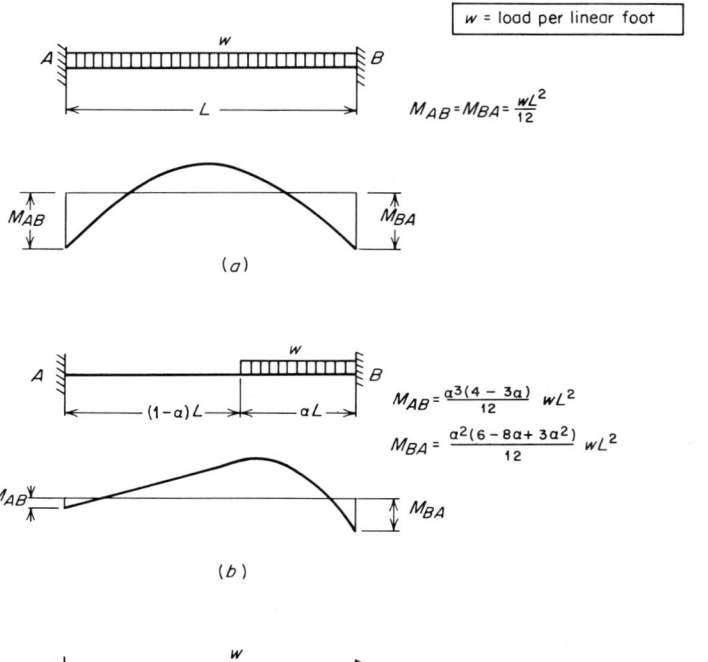

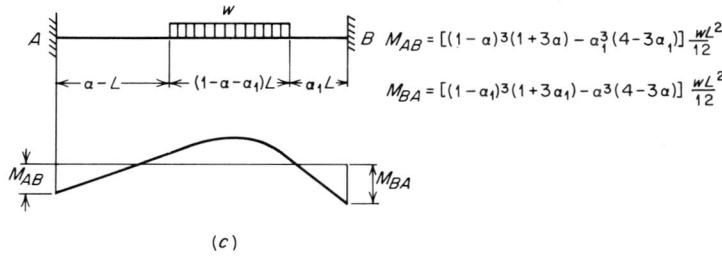

Fig. A-5 Fixed-end moments for various types of loading.

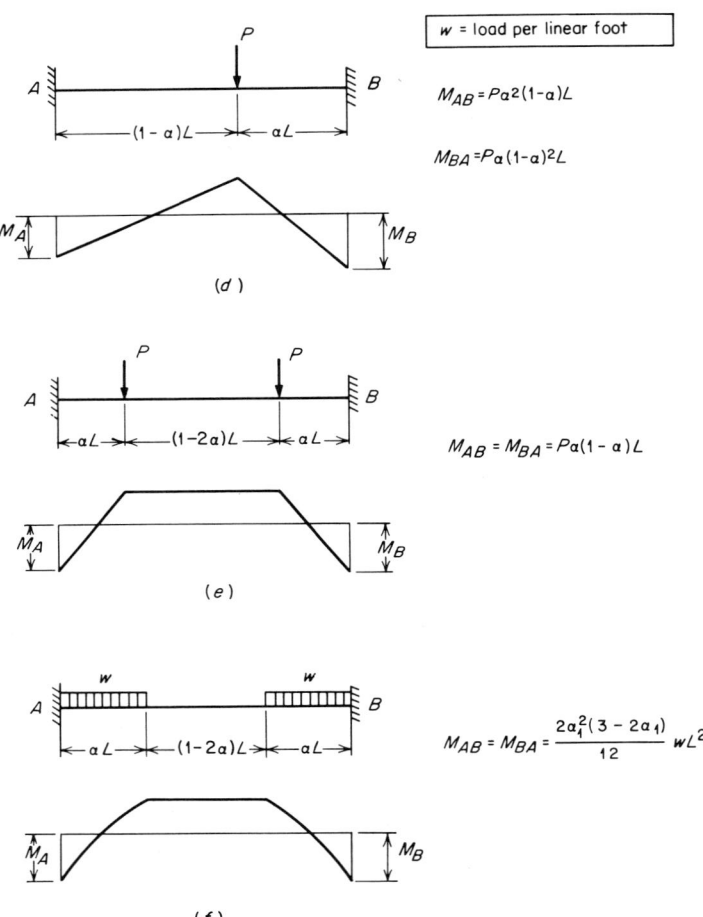

w = load per linear foot

$M_{AB} = P\alpha^2(1-\alpha)L$

$M_{BA} = P\alpha(1-\alpha)^2 L$

$M_{AB} = M_{BA} = P\alpha(1-\alpha)L$

$M_{AB} = M_{BA} = \dfrac{2\alpha_1^2(3-2\alpha_1)}{12} wL^2$

Fig. A-5 (*Continued*)

APPENDIX A 429

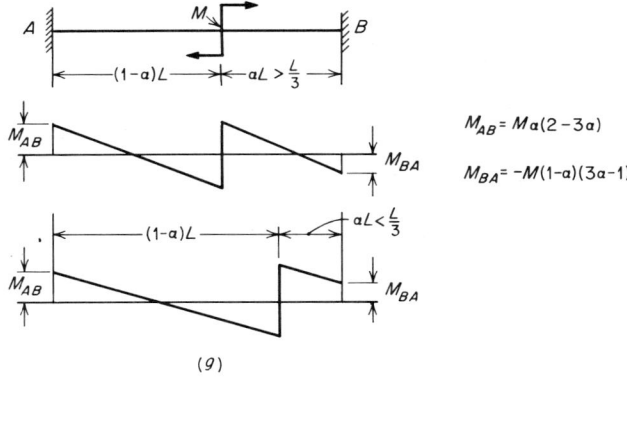

$M_{AB} = Ma(2-3a)$

$M_{BA} = -M(1-a)(3a-1)$

(g)

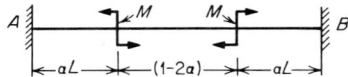

$M_{AB} = M_{-BA} = M(1-2a)$

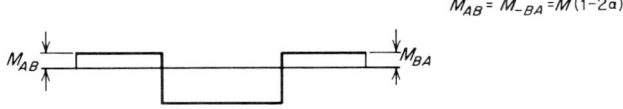

(h)

Fig. A-5 (*Continued*)

appendix B
Physical and Geometric Properties of Prestressing Systems

B-1 FREYSSINET SYSTEM

CABLE CHARACTERISTICS

Cable Size	12/.196	18/.196	12/.276
Nominal Steel Area (in²)	0.362	0.543	0.718
Ultimate Strength (lbs)	90,500	136,000	170,000
Max. Tensioning Load (80% Ultimate)	72,500	109,000	136,000
Max. Design Load – see graph	No. 1	No. 2	No. 3
Cable Weight – sheath not included (lbs/ft)	1.23	1.85	2.45
Recommended hole diameter	1⅝ in.	1½ in.	1½ in.

APPENDIX B

ANCHORAGE DIMENSIONS

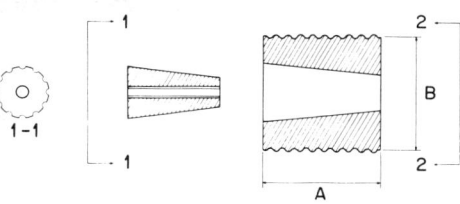
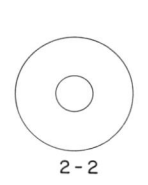

DIMENSIONS OF CONES		
	A	B
12/.196	4"	3⅞"
18/.196	4⅞"	4⅞"
12/.276	4⅞"	4⅞"

ASSEMBLED VIEW

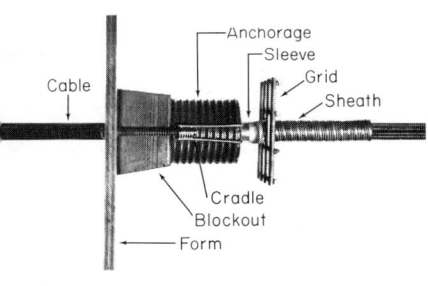

JACK CLEARANCES

ANCHORAGE TYPE	A	B	C	D	E
12/.196	11"	4"	23"	5"	6"
18/.196 & 12/.276	17"	5½"	36"	7"	11"

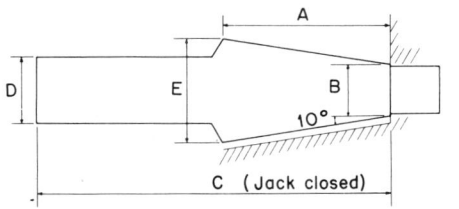

END ZONE DETAILS

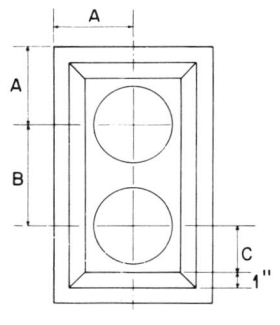
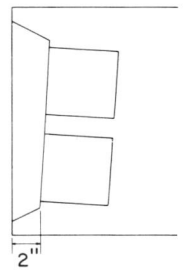

	A	B	C
12/.196	4"	5"	2½"
18/.196	5"	6½"	3"
12/.276	5½"	6½"	3"

TABLE 1   TENDON CHARACTERISTICS

QUALITY	ASTM GRADE				TYPE 270K			
Ultimate Strength of One Strand	36,000 Lb.				41,300 Lb.			
Nominal Steel Area of One Strand	.1438 In.^{2}				.1531 In.^{2}			
Number of Strands	6	8	9	12	6	8	9	12
Nominal Steel Area (In.^{2})	0.86	1.15	1.29	1.73	0.92	1.22	1.38	1.84
Ultimate Tendon Strength (Lb.)	216,000	288,000	324,000	432,000	247,800	330,400	371,700	495,600
Maximum Initial* Tensioning Load (Lb.) (75% of Ultimate)	162,000	216,000	243,000	324,000	185,850	247,800	278,775	371,700
Tendon Weight (Lb./Ft.) (without enclosure)	2.96	3.95	4.45	5.93	3.15	4.20	4.73	6.30
Recommended Hole I.D. (In.)	1-7/8	2-1/4	2-1/4	2-5/8	1-7/8	2-1/4	2-1/4	2-5/8

APPENDIX B

END BLOCK DETAILS

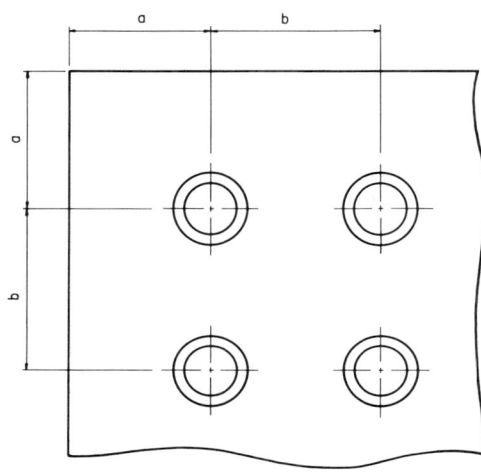

TENDON TYPE	Minimum Edge Distance (In.) a	Minimum Center Distance (In.) b
Internal Anchorage		
6 strands	7	9
8 strands	9	10
9 strands	9	10
12 strands	10	12
External Anchorage		
6 strands	6-1/2	9
8 strands	7	9-1/2
9 strands	7	9-1/2
12 strands	8	10-1/2

Note: Assume f'c = 4,000 psi.
In cases where edge and/or center distances are less please consult our engineering staff for recommendations.

ANCHORAGE RECESS

● It is normal to recess anchorages into pockets. Typical details are illustrated below for internal and external anchorages with bearing plates.

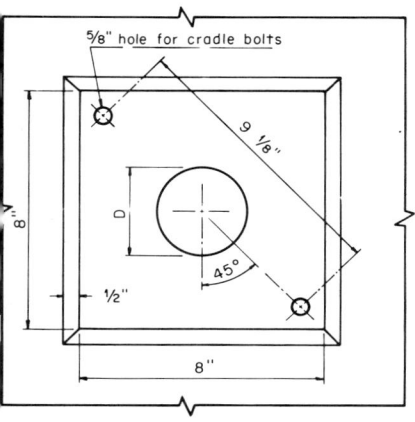

Note: Dimension "D" - hole for projecting tendon = 3" ±

Internal blockout form

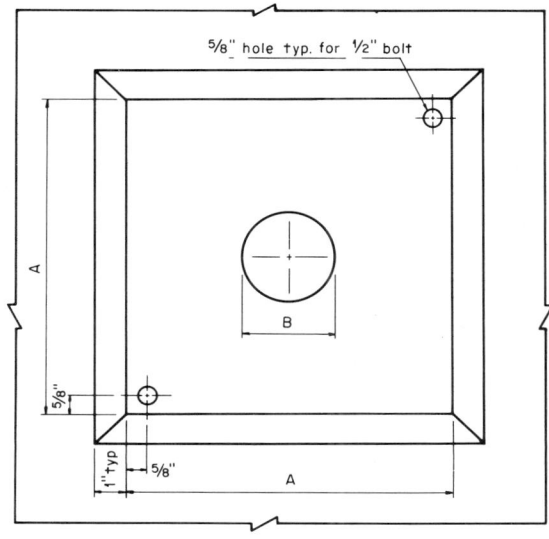

Note: Dimension "A" to suit proper bearing plate
Dimension "B" - hole for projecting tendon = 3" ±

External blockout form

PRESTRESSED CONCRETE

SPIRAL

Tendon	P (In.)	No. Turns
6 strands	2	5
8 strands	1-3/4	6
9 strands	1-3/4	6
12 strands	1-1/2	7

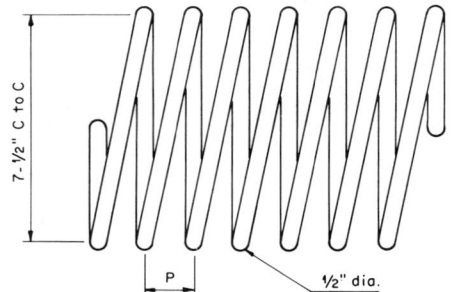

GRID

Tendon	Bar Size
6 strands	No. 3
8 strands	No. 3
9 strands	No. 4
12 strands	No. 4

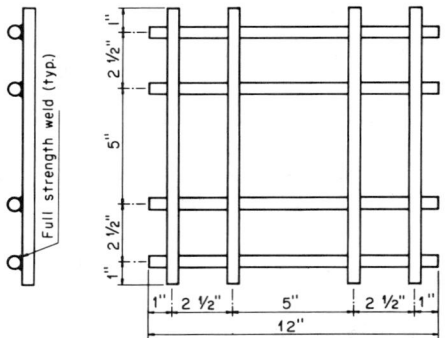

Note: In narrow members, the grid may be shortened in one direction and lengthened in the other direction provided the amount of reinforcing is not reduced. In all cases however, the 5" center dimension must be maintained due to clearance for the connection sleeve.

BEARING PLATE

Tendon	Bearing Plate Dimensions (In.)			
	$a^* = b^*$	c^{**}	d	e
6/500K	9	1	4-5/8	5-3/4
8/500K	9-1/2	1-1/2	4-3/8	5-3/4
9/500K	9-1/2	1-1/2	4-3/8	5-3/4
12/500K	10-1/2	1-3/4	4-3/8	5-3/4

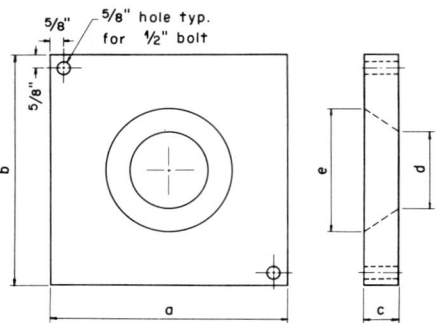

- Bearing stresses under plates do not exceed 3,000 psi under final load. These dimensions can vary provided same bearing area is obtained. Minimum dimension is 9".
- Minimum thickness.

  Note: Large hole in bearing plate should be flame cut to fit cone.

APPENDIX B

B-2 STRESSTEEL SYSTEM
Design Properties of Stressteel Bars

Nominal Bar Size ⌀"	Nominal Weight Pounds Lin/Ft.	Nominal Area Sq. Inches	Ultimate Strength Guaranteed Minimum		Recommended Initial Tensioning Load—0.7 f's†		Maximum Recommended Final Design Load—0.6 f's†	
			REGULAR 145 ksi	SPECIAL 160 ksi	REGULAR 101.5 ksi	SPECIAL 112 ksi	REGULAR 87 ksi	SPECIAL 96 ksi
			(All units in values of 1000 pounds)					
¾	1.50	.442	64	71	45	50	39	42
⅞	2.04	.601	87	96	61	67	52	58
1	2.67	.785	114	126	80	88	68	75
1⅛	3.38	.994	144	159	101	111	87	95
1¼	4.17	1.227	178	196	125	137	107	118
1⅜	5.05	1.485	215	238	151	166	129	143

*Design properties indicated are in accordance with ACI Building Code 318-63, Sections 2606 and 2607. Temporary jacking stresses up to 0.8f's are permitted to overcome losses due to tendon friction, anchorage seating and elastic shortening. Losses due to creep, shrinkage and steel relaxation should be deducted from the recommended initial tensioning load to obtain actual final design load. Actual final design load, after losses are accounted for, may be less than 0.6f's.

Bar Anchorage Detailing Information

Wedges and Nuts

WEDGE
W—

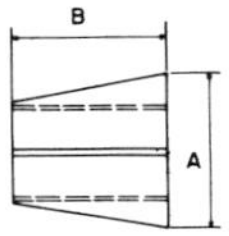

NUT
N—

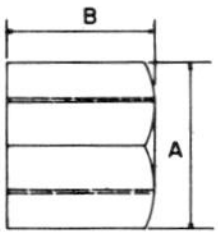

GRIP NUT
G—

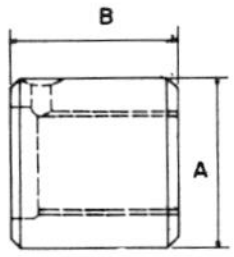

BAR Ø"	WEDGES Part No.	A	B	WT. # Ea.	NUTS* Part No.	A	B	WT. # Ea.	GRIP NUTS Part No.	A	B	WT. # Ea.
3/4	W6	1 3/8	1 1/4	.2	N6	1 3/8	1 1/4	.5	G6	1 7/8	1 3/4	1.0
7/8	W7	1 3/4	1 1/2	.3	N7	1 5/8	1 7/16	.7	G7	2 1/16	2	1.5
1	W8	1 3/4	1 1/2	.5	N8	1 7/8	1 5/8	1.0	G8	2 1/4	2 1/4	2.5
1 1/8	W9	2	1 3/4	.7	N9	2 1/8	1 13/16	1.5	G9	2 1/2	2 1/2	2.7
1 1/4	W10	2 1/4	2	.8	N10	2 1/4	2	2.0	G10	2 3/4	2 5/8	3.2
1 3/8	W11	2 3/8	2 3/16	1.1	N11	2 3/8	2	2.0	G11	3	2 7/8	4.2

All data subject to revision as new developments or improvements are made. All dimensions in inches

*3/16" washer shipped with each nut.

Couplers

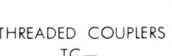

THREADED COUPLERS
TC—

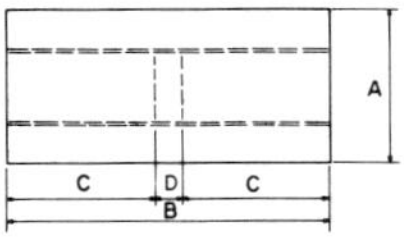

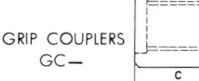

GRIP COUPLERS
GC—

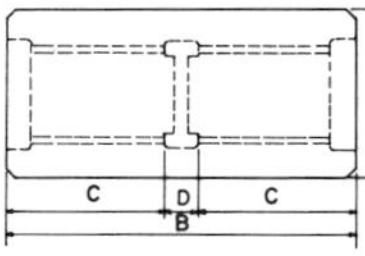

BAR Ø"	THREADED COUPLERS Part No.	A	B	C	D	I.D." Coupler Shield	WT. # Ea.	GRIP COUPLERS Part No.	A	B	C	D	I.D." Coupler Shield	WT. # Ea.
3/4	TC6	1 1/4	3	1 3/8	1/4	1 1/2	1.1	GC6	1 7/8	3 3/8	1 7/16	3/8	2 1/4	2.0
7/8	TC7	1 1/2	3 1/2	1 5/8	1/4	2	1.8	GC7	2 1/8	3 7/8	1 3/4	3/8	2 1/4	3.0
1	TC8	1 3/4	3 3/4	1 3/4	1/4	2	1.9	GC8	2 1/4	4 1/4	1 15/16	3/8	2 3/4	3.5
1 1/8	TC9	2	4 1/4	1 7/8	1/2	2 1/2	2.5	GC9	2 1/2	4 5/8	2 1/8	3/8	3	4.9
1 1/4	TC10	2 1/4	4 3/4	2 1/8	1/2	2 1/2	3.7	GC10	2 3/4	5 1/8	2 3/8	3/8	3	6.2
1 3/8	TC11	2 1/4	5	2 1/4	1/2	2 1/2	3.5	GC11	3	5 5/8	2 5/8	3/8	3 1/2	8.1

All data subject to revision as new developments or improvements are made. All dimensions in inches

Plates

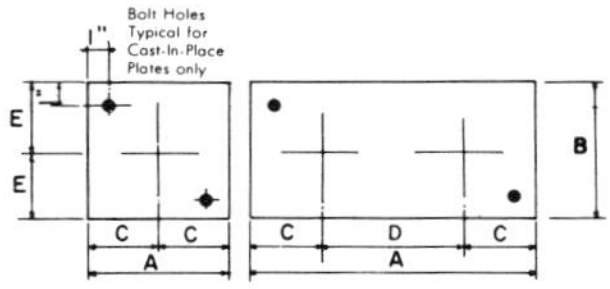

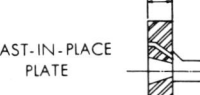

CAST-IN-PLACE PLATE

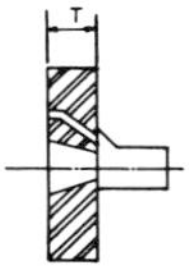

DRY PLACED PLATE

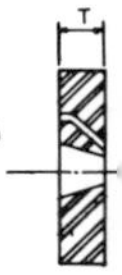

BAR Ø"	Part No. WP.-TP. or P.*	No. of Holes	DIMENSIONS IN INCHES A	B	C	D	E	T	WT. LBS. EA.
3/4	6	1	4	4	2	—	2	1	4.5
7/8	7	1	5	4 1/2	2 1/2	—	2 1/4	1 1/2	9.5
1	8	1	5 1/2	5	2 3/4	—	2 1/2	1 1/2	11.9
1 1/8	9	1	6	6	3	—	3	1 3/4	17.8
1 1/4	10	1	7	6	3 1/2	—	3	1 3/4	20.8
1 3/8	11	1	7 1/2	7	3 3/4	—	3 1/2	2	29.7
2 @ 1	8-2	2	11	5	3	5	2 1/2	1 1/2	23.4
2 @ 1 1/8	9-2	2	11 1/2	6	3 1/4	5	3	1 3/4	32.2
2 @ 1 1/4	10-2	2	12	7	3 1/2	5	3 1/2	1 3/4	41.6
2 @ 1 3/8	11-2	2	14 1/2	7	4 1/4	6	3 1/2	2	57.5

*Precede part number by using appropriate letter designation: WP=Wedge Plate; TP=Threaded Plate; EP=Plate with Drilled Hole; Hole 1/4" larger than bar Ø.

NOTES:

1. Standard Plate sizes shown, when used with regular grade bars, conform to allowable bearing stresses of ACI Building Code 318-63 Section 2605, assuming that f'c=5000 psi and Ab'/Ab=1.0. For lower strength concrete a larger bearing area should be provided. For higher strength concrete, or whe Ab'/Ab exceeds 1.0, smaller plates may be used. Consult o Engineering Department for recommendations on plate desig
2. Standard Plates are fabricated from AISI C-1040 Steel.
3. Plates to anchor a larger number of bars, plates to anch bars of different diameters, or plates of non-rectangular sha may be designed to meet the specific needs of the engineer.

APPENDIX B

Strand Anchorage Detailing Information

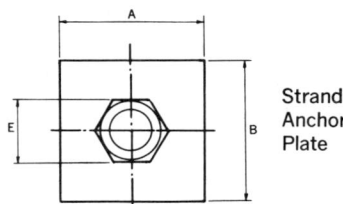

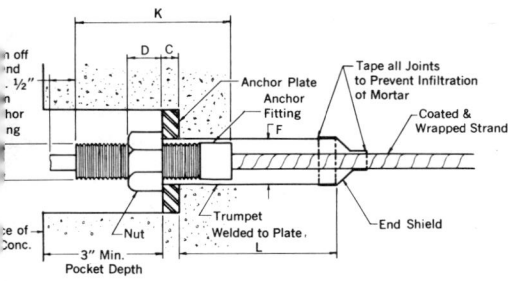

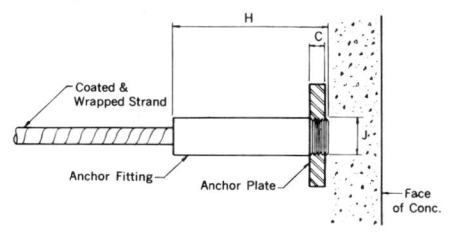

Jacking End Anchorage — Monostrand

Non-Jacking End Anchorage — Monostrand

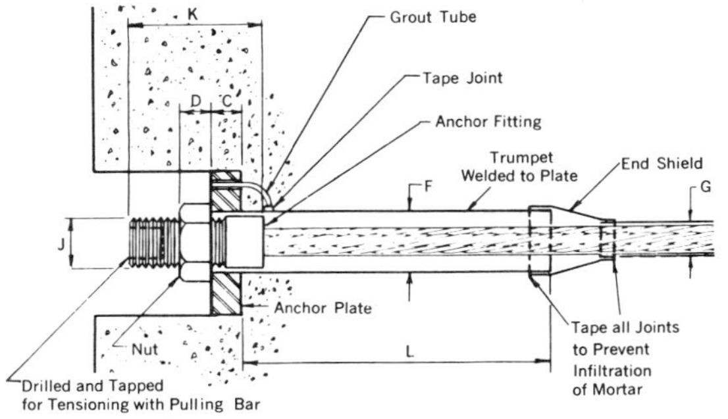

Jacking End Anchorage — Multistrand

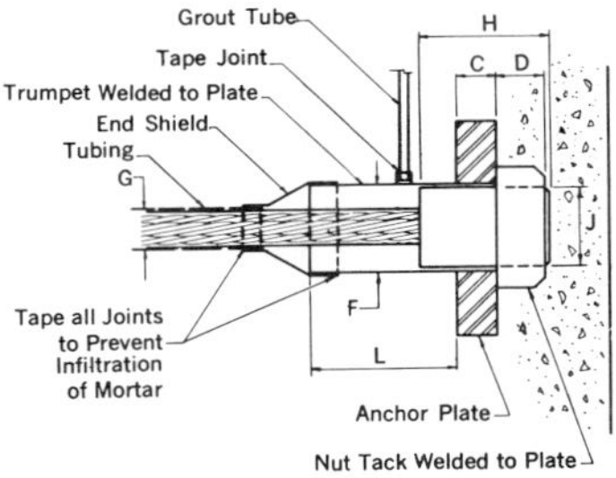

Non-Jacking End Anchorage — Multistrand

Design and Detailing Information

Tendon Type	Ult. Load Kips	No. & Size of Strands	Wgt. - Lb. 100-ft.	Max. O.L. (0.8 f's) Kips	Initial Load (0.7 f's) Kips	Final Load (0.6 f's) Kips	DIMENSIONS IN INCHES										
							A*	B*	C	D	E	F	G	H	J	L + Δe	K
S1-5	41.3	1x½	54	33.0	28.9	24.8	3½	3½	¾	1¼	2	1½	¾	5	1¼	3	5
S1-6	54.0	1x.6	74	43.2	37.8	32.4	4	4	¾	1¼	2	1½	⅞	5	1¼	3	5
S3-5	123.9	3x½	162	99.0	86.7	74.4	6	6	1	1-9/32	3⅜	2½	1⅜	5	2¼	14	7
S3-6	162.0	3x.6	222	129.6	113.4	97.2	6	7	1	1-9/32	3⅜	2½	1½	5	2¼	14	7
S4-6	216.0	4x.6	296	172.8	151.2	129.6	7	8	1	1-17/32	3¾	3	1⅝	5	2⅝	14	8
S5-6	270.0	5x.6	370	216.0	189.0	162.0	8	8½	1	1-17/32	3¾	3	1¾	5	2⅝	14	8
S6-5	247.8	6x½	324	198.0	173.4	148.8	8	8½	1	1-25/32	4½	3½	1¾	5	3⅛	14	8
S6-6	324.0	6x.6	444	259.2	226.8	194.4	9	9	1¼	1-25/32	4½	3½	1⅞	5	3⅛	14	8
S7-5	289.1	7x½	378	231.0	202.3	173.6	9	9	1¼	1-25/32	4½	3½	1¾	5	3⅛	14	8
S7-6	378.0	7x.6	518	302.4	264.6	226.8	9	10½	1¼	1-25/32	4½	3½	2	5	3⅛	14	8
S12-5	495.6	12x½	648	396.5	346.9	297.4	12	12	1½	2-9/16	6⅝	5	2½	5	4½	14	9
S19-5	784.7	19x½	1026	627.8	549.3	470.8	14	14½	1¾	2-9/16	6⅝	5	3	22	4½	16	22
**S28-5	1156.4	28x½	1508	924.8	809.2	694.0	17½	17½	2½	2-13/16	8⅜	6	3½	9½	5½	14	12

* Standard plate sizes shown conform to allowable bearing stresses of ACI Building Code 318-63, Section 2605, assuming that f'c=5,000 psi and Ab'/Ab=1.0. For other values of f'c and/or Ab'/Ab, plate sizes should be calculated.

** This tendon and larger ones are available for special applications. Consult your Stressteel representative.

All data subject to revision as new developments or improvements are made.

Couplers

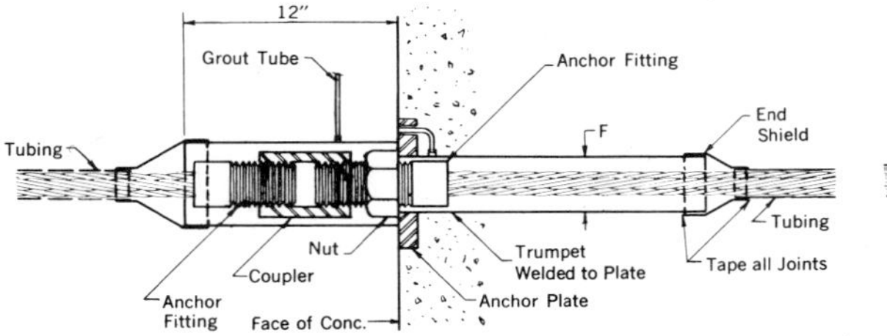

Intermediate End Anchorage with Coupling for Extending Tendon

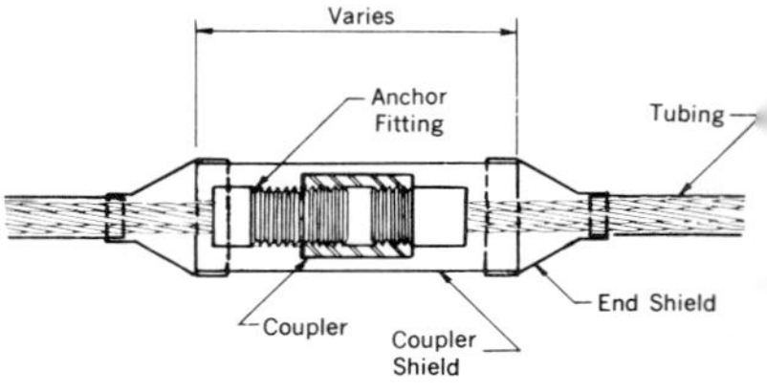

Typical Coupler Arrangement

APPENDIX B

B-3 ROEBLING SYSTEM

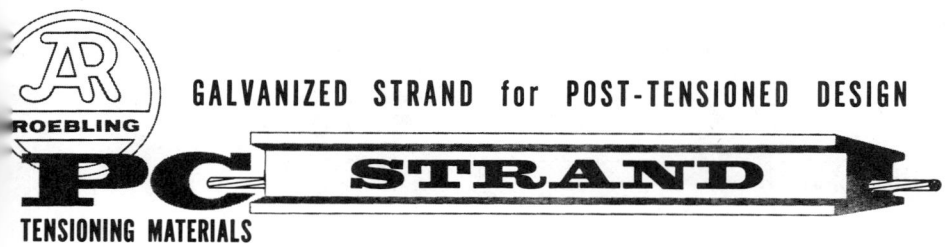

GALVANIZED STRAND for POST-TENSIONED DESIGN

DIAMETER (INCHES)	WEIGHT PER FOOT (POUNDS)	AREA (SQUARE INCHES)	MINIMUM GUARANTEED ULTIMATE STRENGTH (POUNDS)	RECOMMENDED FINAL DESIGN LOAD (POUNDS)
0.600	0.737	0.215	46,000	26,000
0.835	1.412	0.409	86,000	49,000
1	2.00	0.577	122,000	69,000
1 1/8	2.61	0.751	156,000	90,000
1 1/4	3.22	0.931	192,000	112,000
1 3/8	3.89	1.12	232,000	134,000
1 1/2	4.70	1.36	276,000	163,000
1 9/16	5.11	1.48	300,000	177,000
1 5/8	5.52	1.60	324,000	192,000
1 11/16	5.98	1.73	352,000	208,000

ROEBLING

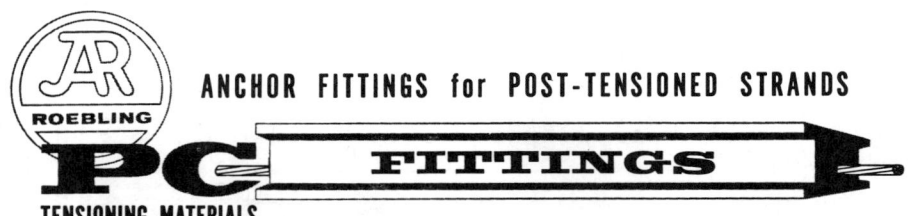

ANCHOR FITTINGS for POST-TENSIONED STRANDS

PC FITTINGS

TENSIONING MATERIALS

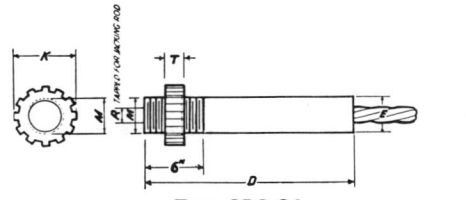

Type SDS 34

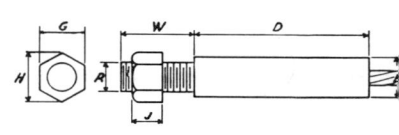

Type SDS 35

STRAND DIAMETER	MEASUREMENTS IN INCHES										TOTAL WEIGHT (POUNDS)	
	D	W	E	M	R	G	H	J	K	T	Type SDS34	Type SDS35
.600	9½	8	1¹¹⁄₁₆	-------	1¼–12N	1¹³⁄₁₆	2¹⁄₁₆	1³⁄₁₆	——	——	——	9¼
.835	12½	10	2¼	12N	1⅝–12N	2¼	2⁹⁄₁₆	1⁵⁄₁₆	3¾	1	16½	21
1	13	11	2¾	8N	2 – 8N	3	3⁷⁄₁₆	1¾	4⅜	1¼	24¾	32½
1⅛	16½	11	3	8N	2¼– 8N	3	3⁷⁄₁₆	1¾	4¾	1½	39½	50

For fittings Type SDS35, standard studs having dimension W, shown above, are carried in stock.

Other stud lengths must be fabricated to order. All SDS 34 and 35 fittings are proofloaded to a stress in excess of the recommended design stress after being attached to the strand.

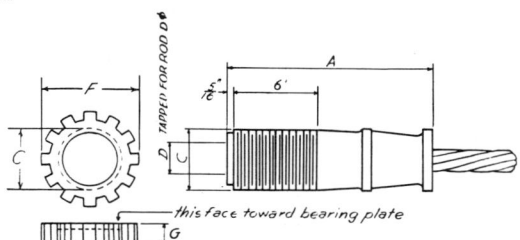

Type SS 2

STRAND DIAMETER	MEASUREMENTS IN INCHES					TOTAL WEIGHT (POUNDS)
	A	C	D	F	G	
1	10	3½–4N	2 –4½NC	5⅜	1¼	29
1⅛	10¾	4 –4N	2¼–4½NC	6	1⅜	38
1¼	11⅜	4⅜–4N	2½–4NC	6½	1⅝	45
1⅜	11¾	4⅞–4N	3–4NC	7	1⅝	54
1½	12⅜	5¼–4N	3–4NC	7⅝	1¾	74
1⁹⁄₁₆	12¾	5½–4N	3½–4NC	7⅞	1¾	77
1⅝	13	5⅝–4N	3½–4NC	8	1¾	84
1¹¹⁄₁₆	13¼	5⅞–4N	3½–4NC	8⅜	1¾	88

Fittings Type SS–2 can also be supplied with permanent studs and no external threads when necessary.

ROEBLING

PC

TENSIONING MATERIALS

DETAILS for Post-Tensioned Strands

BEARING PLATE ASSEMBLY

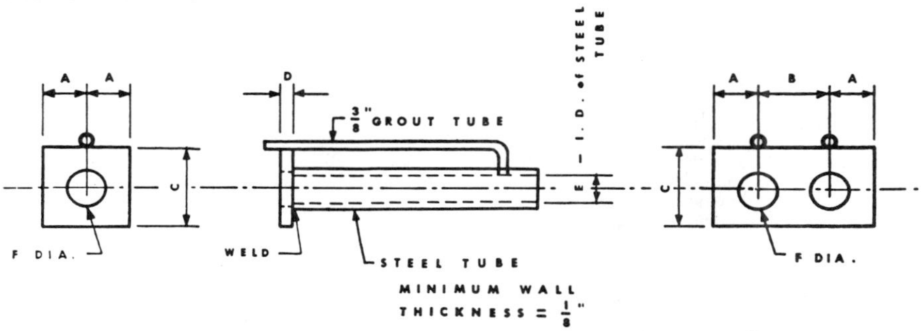

Type of Fitting	Strand Diam. (Inches)	A*	B*	C*	D	E		F
						Min.	Max.	
SDS 34	.835	3	5¼	6	⅞	2⅜	2½	E + 1/16
	1	3	5½	6	⅞	2⅞	3	E + 1/16
	1⅛	3½	6¼	7	1	3⅛	3¼	E + 1/16
SDS 35	.600	2	3¼	4	⅝	1¾	1⅞	1½
	.835	3	4½	6	⅞	2⅜	2½	1⅞
	1	3	4⅞	6	⅞	2⅞	3	2¼
	1⅛	3½	5½	7	1	3⅛	3¼	2½

Type of Fitting	Strand Diam. (Inches)	A*	B*	C*	D	E		F
						Min.	Max.	
SS2	1	4	7	8	⅞	3¾	3⅞	E + 1/16
	1⅛	4¼	7½	8½	1	4¼	4⅜	E + 1/16
	1¼	4½	8	9	1¼	4⅝	4¾	E + 1/16
	1⅜	4¾	8½	9½	1¼	5⅛	5¼	E + 1/16
	1½	5⅜	9½	10¾	1⅜	5½	5⅝	E + 1/16
	1 9/16	5⅞	10¼	11¾	1⅜	5¾	5⅞	E + 1/16
	1⅝	5⅞	10¼	11¾	1½	5⅞	6	E + 1/16
	1 11/16	6	10½	12	1½	6⅛	6¼	E + 1/16

*These are standard minimum dimensions. They can be reduced slightly by using special jack bases.
Bearing pressure under above plates does not exceed 2500 p.s.i.

B-4 BBRV SYSTEM

Type B Anchor Type designation				B 32	B 64	B 100	B 138
Steel wires per anchor, max. number		5 mm dia.		14	28	44	—
		6 mm dia.		10	20	32	44
		7 mm dia.		8	16	24	34
Permanent load in metric tons		max.	t	32,3	64,6	99,5	137,4
Short-time load on being over-stressed, metric tons		max.	t	36	72	110	150
Anchor head	B1: thread diameter	D_A	mm	75	100	115	130
	standard length	L_A	mm	40	60	80	90
	[1] special length	L_A	mm	60	80	100	120
Lock nut	B2: outside diameter	D_M	mm	105	135	155	180
	height	H_M	mm	22	30	40	50
Bearing plate	B3: side length	S_P	mm	140	180	220	260
	thickness	D_P	mm	14	16	20	25
End-trumpet	B4: outside diameter	D_T	mm	87	112	128	148
	connection diameter	D_I	mm	33,5	41,5	57,5	62,5
	[2] basic length approx.	L_C	mm	120	155	170	190
	[2] standard lengths	L_T	mm	210	260	345	345
			mm	345	450	450	450
			mm	510	570	570	570
	length of cone	L_K	mm	110	130	150	170
Spiral	B5: outside diameter	D_S	mm	140	180	220	260
	length (about)	L_S	mm	250	250	250	250
Assembly rod B11: min. length			mm	500	500	500	500
Temp. pullrod B21: min. length			mm	350	400	400	450
	thread dia.	D_H	mm	42	52	52	62
[3] Installation length, standard		a	mm	110	130	150	170
[3] Installation length, standard		b	mm	170	200	230	255

[1] for very long prestressing cables with uncertain frictional conditions.
[2] L_T depends on the displacement of the anchor head B1; $L_T \sim L_C$ + displacement.
[3] a and b can be reduced considerably on taking all the determinative factors into account.

APPENDIX B

Type J Anchor	Type designation				J 32	J 64	J 100	J 138
Steel wires per anchor, max. number		5 mm dia.			14	28	44	—
		6 mm dia.			10	20	32	44
		7 mm dia.			8	16	24	34
Permanent load in metric tons		max.		t	32,3	64,6	99,5	137,4
Short-time load on being over-stressed, metric tons		max.		t	36	72	110	150
Anchor head	J 1: outside diameter		D_A	mm	74	95	110	125
	length		L_A	mm	35	45	60	70
End-trumpet	J 2: outside diameter		D_T	mm	82	107	122	142
	connection diameter		D_I	mm	33,5	41,5	57,5	62,5
	[1] standard length		L_T	mm	205	205	255	255
				mm	340	340	340	340
				mm	505	505	505	505
	length of cone		L_K	mm	140	218	250	263
Spiral	J 3: outside diameter		D_S	mm	140	180	220	260
	[2] min. length		L_S	mm	300	300	300	300
Assembly rod J 11: min. length				mm	500	500	500	500
Bearing plate J 21: side lengths			S_P	mm	150	200	220	260
Temp. pullrod J 23: min. length				mm	350	400	400	450
	thread dia.		D_H	mm	42	52	52	62
Side length of wooden insert plate			a	mm	200	250	270	300

[1] L_T depends on the displacement of the anchor head J 1; $L_T \sim L_A + 30 +$ displacement.
[2] with sufficient concrete coverage and additional reinforcement for the end-trumpet J 2! Otherwise $L_S = L_T + L_K$.

Type C Anchor	Type designation			Types 6 mm	C 130	C 170	—
				Types 7 mm	C 125	C 170	C 220
Steel wires per anchor, max. number		6 mm dia.			42	55	—
		7 mm dia.			31	42	55
Permanent load in metric tons		max.		t	130,6	171	222,2
Short-time load on being over-stressed, metric tons		max.		t	140	190	245
Basic element	C 1: thread diameter		D_G	mm	80	90	98
Pull sleeve	C 2: outside diameter		D_A	mm	118	130	144
	min. length		L_A	mm	104	118	138
Lock nut	C 3: outside diameter		D_M	mm	162	178	198
	height		H_M	mm	39	45	53
Bearing plate	C 4: outside diameter		D_P	mm	290	320	360
	height		H_P	mm	51	58	67
End-trumpet	C 5: outside diameter		D_T	mm	133	147	162
	connection diameter		D_I	mm	63	68	76,5
	[1] standard lengths		L_T	mm	160	160	170
				mm	290	290	300
				mm	420	420	430
	length of cone		L_K	mm	185	210	230
Spiral	C 6: outside diameter		D_S	mm	290	320	360
	length		L_S	mm	250	250	250
Assembly rod C 11: min. length				mm	500	500	500
Temp. sleeve short C 23: length			L_Z	mm	104	118	138
Temp. sleeve long:	length $2 L_Z + 4$ mm			mm	212	240	280
Pullrod:	thread diameter		D_H	mm	62	72	82
[2] Installation length, standard			a	mm	155	170	190
[2] Installation length, standard			b	mm	220	230	260

[1] L_T depends on the displacement of the pull sleeve C 2; $L_T \sim$ displacement $+ 20$.
[2] a and b can be reduced considerably on taking all the determinative factors into account.

Type S Anchor Type designation			S 32	S 64	S 100	S 138
Steel wires per anchor, max. number	5 mm dia.		14	28	44	—
	6 mm dia.		10	20	32	44
	7 mm dia.		8	16	24	34
Bearing plate, square Sq 1	B/L	mm	—	—	220/220	260/260
Free wire length	S	mm	—	—	550	650
Venting pipe S 3, length	E	mm	—	—	850	950
Bearing plate, rectangular Sr 1	B/L	mm	120/150	150/220	160/300	180/360
Free wire length	S	mm	500	550	550	650
Venting pipe S 3, length	E	mm	800	850	850	950
Bearing plate, long Sl 1	B/L	mm	60/300	80/400	80/560	120/560
Free wire length	S	mm	500	600	700	750
Venting pipe S 3, length	E	mm	800	900	1000	1050
Connection diameter	D_I	mm	33,5	41,5	57,5	62,5

Type E Anchor Type designation	Types 6 mm		E 130	E 170	–
	Types 7 mm		E 125	E 170	E 220
Steel wire per anchor, max. number	6 mm dia.		42	55	—
	7 mm dia.		31	42	55
Bearing plate E 2: outside diameter	D_P	mm	235	270	300
height	H_P	mm	52	60	68
End-trumpet E 4: length	L_T	mm	215	245	265
connection diameter	D_I	mm	62,5	68	76,5
Assembly tube E 11: min. length		mm	130	130	130
Length of anchorage	L_V	mm	288	326	354

APPENDIX B

B-5 RYERSON SYSTEM

TENDON FORCE CALCULATOR

When the location and magnitude of the prestressing force requirement for a structural member has been developed, Ryerson can, at your option, select the appropriate type of anchorage and tendon size. And we will provide placing drawings.

Wire is 0.25 inches nominal diameter (Area = 0.04909 square inches) complying with ASTM-A421-58T specification. Minimum ultimate strength of wire is 240,000 P.S.I.

NUMBER OF WIRES	CROSS SECTIONAL AREA Square inches	EFFECTIVE PRESTRESSING FORCE Max. in Kips	TEMPORARY OVERSTRESSING FORCE Max. in Kips
1	.04909	7	9.4
6	.29454	42.5	56.5
8	.39272	56.5	75.4
12	.58908	84.8	113.1
14	.68726	99	132
18	.88362	127.2	169.7
20	.98180	141.4	188.5
24	1.17816	169.7	226.2
27	1.32541	190.9	254.5
28	1.37452	197.9	263.9
30	1.47270	212	282.8
36	1.76724	254.5	339.3
38	1.86542	268.6	358.7
40	1.96360	282.8	377
46	2.25814	325.2	433.6
54	2.65082	381.7	509
72	3.53448	509	678.6
90	4.4181	636.2	848.3

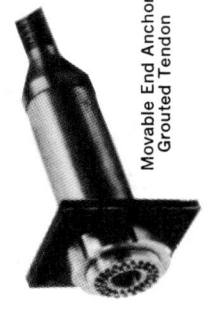

Movable End Anchor Grouted Tendon

Fixed End Anchor Grouted Tendon

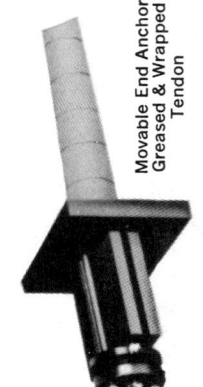

Movable End Anchor Greased & Wrapped Tendon

Fixed End Anchor Greased & Wrapped Tendon

B-6 PRESCON SYSTEM

TYPE C (COUPLED) TENDON SIZE CHART

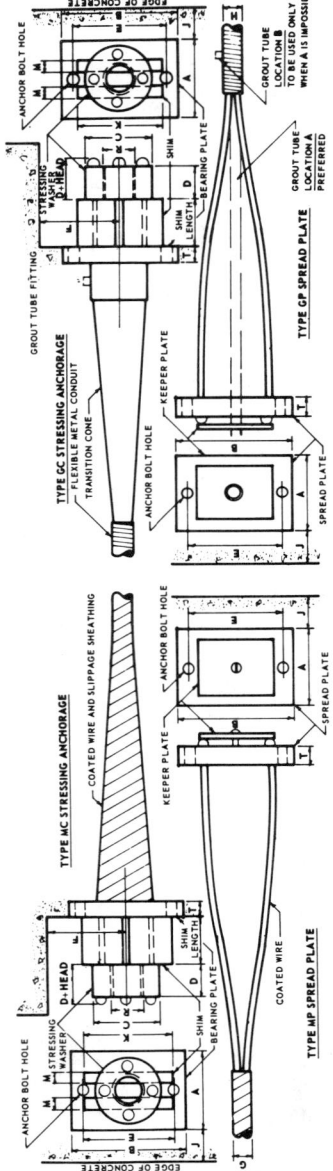

Prescon Prestressing Tendons are described by the number of wires, sheathing (coated or grouted) and type of end anchorage.

EXAMPLES:

6MCP
6 wires in tendon
M - coated and wrapped
C - one stressing end
P - one spread plate end

10GCS
10 wires in tendon
G - grouted
C - one coupled stressing end
S - one standard stressing end

1. Anchor bolt hole diameter:
   $\frac{1}{2}''$ - 2 thru 14 wire
   $\frac{5}{8}''$ - 15 wire and above
2. O.D. of conduit $\frac{1}{8}''$ larger than H.
3. Clear dimension required on one side to insert shims, measured in the direction of long dimension of plate.
4. If either J or f'_c is less than the value shown in the table, bearing plate must be increased in size to prevent excess bearing stress in concrete.
5. Add "D PLUS Head" shim length and cover to determine pocket depth.
6. Concrete strength at time of stressing.

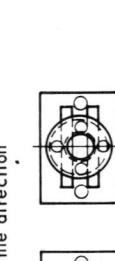
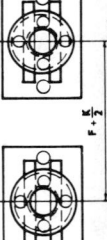

$F + \frac{K}{2}$

NO. WIRES 0.250" DIA.	PRE-STRESS FORCE KIPS 0.6 f's	BEARING PLATE				SPREAD PLATE				WASHER			NO. WIRES 0.250" DIA.	SHIM		TENDON		CLEAR		CONCRETE		D PLUS HEAD	NO. WIRES 0.250" DIA.
		Width A	Length B	Thick. T	Hole Spacing E[1]	Width A	Length B	Thick. T	Hole Spacing E[1]	Rod Size R	Dia. C	Thick. D		Width K	Thick. M	Dia. Coated G	I.D. Cond. H[2]	F[3]	J[4]	28 Day Strength f'c [4]	Stressing Strength f'ci [6]	[5]	
2	14.1	3	4½	⅝	3	3	4	⅝	3	1	2½	1¼	2	4	*⅜	½	1	5¾	¾	3000	2250	1½	2
3	21.2		4½		3		4		3				3		¼	½	1		¾	3000	2250		3
4	28.3		4½		3½		4½		3½				4		*⅜	⅝	1¼		1¼	4000	3000		4
5	35.3	3	5½	⅝	4½	3	5½	⅝	4½	1	2½		5	4	¼	¾		5¾					5
6	42.4	4	6	¾	5	4	6	¾	5	1¼	3		6	5	⅜			6					6
7	49.5		6				6						7	5	⅜			6					7
8	56.6		7				7						8	5	⅜	¾	1¼	7					8
9	63.6		7½		5		7½		5	1¼		1¼	9	6	½	1	1½	7				1½	9
10	70.7	4	8½	¾	6	4	8½	¾	6	1½	3	1½	10			1		7				1¾	10
11	77.8	5	8	1	6	5	8	1	6		3½		11			1⅛		7¼	1¼				11
12	84.8	5			6	5			6				12					7¼	1½				12
13	91.9	6			7	6			7				13				1½	7¼					13
14	98.9		8				8			1½	3½	1½	14				1¾	8¼				1¾	14
15	106.0		8½				8½			1¾	4	1⅞	15			1⅛		8½				2	15
16	113.1		9½				9½					1⅞	16			1¼						2⅛	16
17	120.2		9½		7		9½		7				17			1¼	1¾						17
18	127.2		10		9		10		9				18			1¼	2						18
19	134.3		11				11						19			1⅜		8½					19
20	141.4		11				11			1¾	4	1⅞	20	6	½			9¾				2⅛	20
21	148.5		11½				11½			2	4¾	2	21	8	⅝							2¼	21
22	155.5		11½				11½						22			1⅜							22
23	162.6		12				12						23			1½							23
24	169.7	6	12	1		6	12	1			4¾		24			1½		9¾		4000	3000		24
25	176.8	7	11½	1¼		7	11½	1¼			5¼		25			1⅝		11		5000	3750		25
26	183.8		11½				11½			2		2	26									2¼	26
27	190.9		12				12			2¼		2¼	27									2½	27
28	197.9		12		9		12		9				28			1⅝							28
29	205.0	7	12½		10	7	12½		10				29		⅝	1¾	2						29
30	212.0	8	12		10	8	12		10				30		¾		2¼						30
31	219.2		12		10		12		10				31										31
32	226.2		13		11½		13		11½				32			1¾							32
33	233.3		13				13						33			1⅞							33
34	240.4		14				14			2¼	5¼	2¼	34			1⅞		11				2½	34
35	247.5	8	14	1¼		8	14	1¼		2½	6	2¾	35			1⅞		11½				3	35
36	254.5	9	13	1½		9	13	1½					36			2							36
37	261.6		13		11½		13		11½				37			2							37
38	268.7		15		13		15		13				38			2	2¼						38
39	275.7												39			2¼	2½						39
40	282.8										6		40					11½					40
41	289.9										6½		41					11¾					41
42	296.9										6½		42			2¼		11¾					42
43	304.0	9	15	1½	13	9	15	1½	13	2½	6½	2¾	43	8	¾	2½	2½	11¾	1½	5000	3750	3	43

*One shim only

TYPE D (DONUT) TENDON SIZE CHART

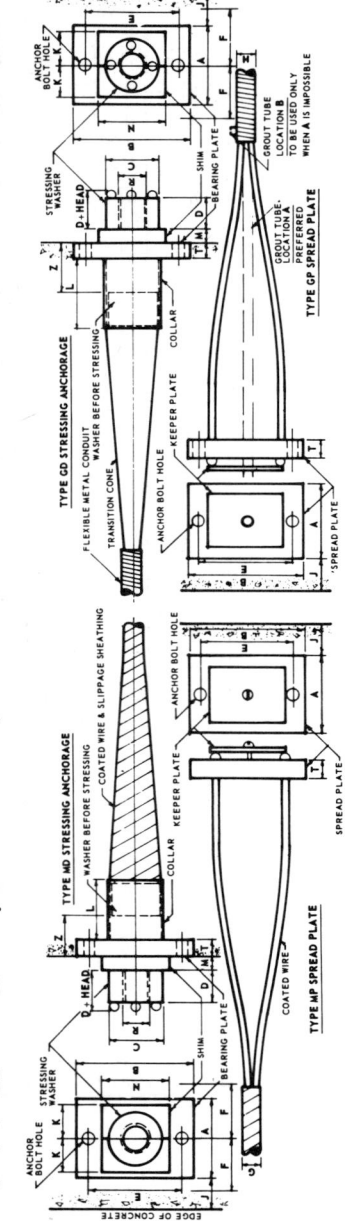

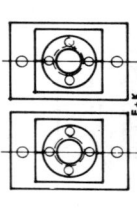

MINIMUM DISTANCE BETWEEN TENDONS

Prescon Prestressing Tendons are described by the number of wires, sheathing (coated or grouted) and the type of end anchorage.

EXAMPLES:
6MDP
 6 wires in tendon
 M- coated & wrapped
 D- one stressing end
 P- one spread plate end
10GDD
 10 wires in tendon
 G- grouted
 D- two stressing ends
 D- two stressing ends

1. Anchor bolt hole diameter:
 $\frac{1}{2}$"- 2 thru 14 wire
 $\frac{5}{8}$"- 15 wire & above
2. O.D. of conduit $\frac{1}{8}$" larger than H.
3. Clear dimension required on each side to insert shims, measured in the direction of the short dimension of plate.
4. If either J or f'_c is less than the value shown in the table, bearing plate must be increased in size to prevent excess bearing stress in concrete.
5. Add "D PLUS Head" to cover to determine pocket depth.
6. Concrete strength at time of stressing.
7. Z = Elongation minus D+M value in table.
8. L = Elongation minus T+M value in table.

APPENDIX B

NO. WIRES 0.250" DIA.	PRE-STRESS FORCE KIPS 0.6 f's	BEARING PLATE				SPREAD PLATE				WASHER				NO. WIRES 0.250" DIA.	SHIM			TENDON		CLEAR		CONCRETE		D + HEAD & SHIM	Z (SEE NOTE 7)	L (SEE NOTE 8)		NO. WIRES 0.250" DIA.
		Width	Length	Thick.	Hole Spcg.	Width	Length	Thick.	Hole Spcg.	Rod Size	Dia.	Thick.		Width	Thick.	Length	Dia. Coated	I.D. Cond.			28 Day Strength	Stressing Strength			Coated	Grout	
		A	B	T	E1	A	B	T	E1	R	C	D		K	M	N	G	H2	F3	J	f'c 4	f'ci 6	D5				
2	14.1	4½	5	1	4	3	4	⅝	3	1	2½	1¼	2	2	⅝	4	½	1	3¼	¾	3000	2250	2¼	1⅞	⅝	*-⅜	2
3	21.2	↑	↑	↑	↑		4		3	↑	↑	↑	3	↑	↑	↑	½	1	↑	¾	3000	2250	↑	↑	↑	↑	3
4	28.3						4½		3½				4				⅝	1¼		1¼	4000	3000					4
5	35.3	4½	5	↓	4	3	5½	⅝	4½	1	2½	↓	5	2	⅝	4	¾	↑	3¼	↑	↑	↑		1⅞	⅝	*-⅜	5
6	42.4	5	6½		5	4	6	¾	5	1¼	3		6	2½	¾	5	↑		4					2	¾	*-¼	6
7	49.5	↑	6½		↑		6			↑	↑		7											↑	↑	↑	7
8	56.6	↓	6½				7						8				¾	1¼									8
9	63.6	5	7½				7½		5	1¼		1¼	9				1	1¼						2¼	2		9
10	70.7	6	7			4	8½	¾	6	1½		1½	10				1	↑						2½	2¼		10
11	77.8	↑	7½		5	5	8	1	6	↑	3	↑	11				1⅛		4	1¼				↑	↑		11
12	84.8		8		7	5			6		3½		12				↓		4¼	1½							12
13	91.9		8½			6	↓		7	↓	3½	↓	13				1½		4¼	↑				↓	↓		13
14	98.9		9		↑	↑	8			1½	3½	1½	14	2½	¾	5	1¾		4¼					2½	2¼	¾	*-¼	14
15	106.0		10				8½			1¾	4	1⅞	15	3	1	6	1⅛	↑	5					3¼	2⅞	1	0	15
16	113.1		11	↑	↑		9½		↑				16	↑	↑	↑	1¼							↑	1	0	16
17	120.2	6	11	↑	7		9½		7				17				1¼	1¾							1	0	17
18	127.2	7	10	1¼	9		10		9				18				1¼	2							1¼	¼	18
19	134.3	7	11				11						19				1⅜	↑							↑	↑	19
20	141.4	7					11			1¾	4	1⅞	20				↑		5					2⅞			20
21	148.5	8					11½			2	4¾	2	21				↑		5½					3			21
22	155.5	↑	11				11½			↑	↑	↑	22				1⅜		↑					↑			22
23	162.6		12				12						23				1½										23
24	169.7	8	12	1¼		6	12	1					24				1½				4000	3000			1¼	¼	24
25	176.8	8½	10	1½		7	11½	1¼					25				1⅝				5000	3750			1½	½	25
26	183.8	9	↑	↑		↑	11½	↑		2	↑	2	26	↓		↓	↑		↑		↑	↑	3¼	3	↑	↑	26
27	190.9	9					12			2¼		2¼	27	3		6			5½				3½	3¼			27
28	197.9	9½					12		9	↑	4¾	↑	28	4		8	1⅝		6½				↑	↑			28
29	205.0	9½				7	12½		10		5¼		29				1¾	2	7								29
30	212.0	10	10			8	12		10		↑		30				↑	2¼									30
31	219.2		10½				12		10				31				↑	↑									31
32	226.2		10½		↓		13		11½				32				1¾										32
33	233.3		11		9		13		↑				33				1⅞										33
34	240.4		11½		10	↓	14	↓		2¼		2¼	34				1⅞						3½	3¼			34
35	247.5		11½		↑	8	14	1¼		2½		2½	35				1⅞						3¾	3½			35
36	254.5		12			9	13	1½		↑	5¼	↑	36				2						↑	↑			36
37	261.6		12½			↑	13	↑	11½		5½		37			↓	2										37
38	268.7		13				15		13		↑		38				2	2¼									38
39	275.7		13				↑		↑	↓		↓	39				2¼	2½									39
40	282.8		13½							5½		2½	40				↑	↑						3¾	3½			40
41	289.9		14							6	2¾		41				↑	↑						4	3¾			41
42	296.9		14	↓	↓	↓	↓	↓	↓	↑	2¾	↓	42				2½	↑	↓	↓				4	3¾	↓	↓	42
43	304.0	10	14½	1½	10	9	15	1½	13	2½	6	2¾	43	4	1	8	2½	2½	7	1½	5000	3750	4	3¾	1½	½	43

*Example: 2⅝" − (−⅜") = 2⅝" + ⅜" = 3"

TYPE S (STANDARD) TENDON SIZE CHART

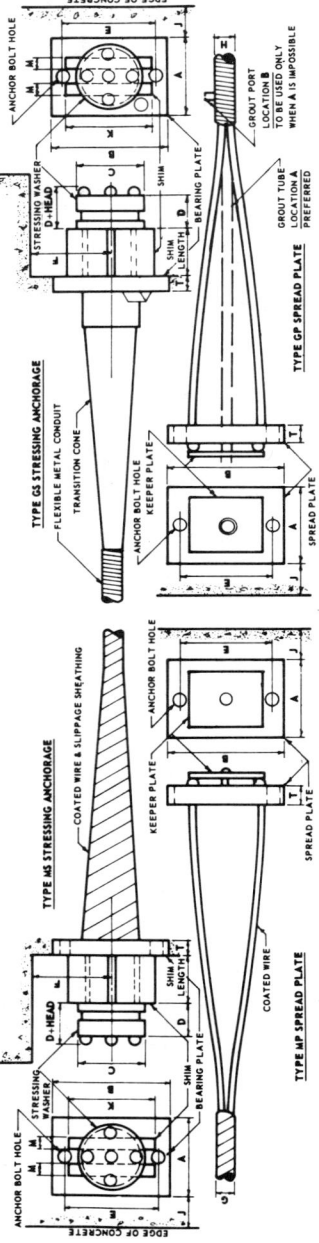

Prescon Prestressing Tendons are described by the number of wires, sheathing (coated or grouted) and type of end anchorage.

EXAMPLES:
6MSP
  6 wires in tendon
  M - coated and wrapped
  S - one stressing end
  P - one spread plate end
10GSS
  10 wires in tendon
  G - grouted
  S - two stressing ends

1. Anchor bolt hole diameter:
   $1/2"$ - 2 thru 14 wire
   $5/8"$ - 15 wire and above
2. O.D. of conduit $1/8"$ larger than H.
3. Clear dimension required on one side to insert shims, measured in the direction of long dimension of plate.
4. If either J or f'_c is less than the value shown in table, bearing plate must be increased in size to prevent excess bearing stress in concrete.
5. Add "D PLUS Head" shim length and cover to determine pocket depth.
6. Concrete strength at time of stressing.

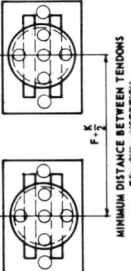

$F = \frac{K}{2}$

MINIMUM DISTANCE BETWEEN TENDONS

APPENDIX B

NO. WIRES 0.250 DIA.	PRE-STRESS FORCE KIPS 0.6 f's	BEARING PLATE				SPREAD PLATE				WASHER		NO. WIRES 0.250 DIA.	SHIM		TENDON		CLEAR		CONCRETE		D PLUS HEAD	NO. WIRES 0.250 DIA.
		Width	Length	Thick.	Hole Spcg.	Width	Length	Thick.	Hole Spcg.	Dia.	Thick.		Width	Thick.	Dia. Coated	I.D. Cond.			28 Day Strength	Stressing Strength		
		A	B	T	E1	A	B	T	E1	C	D		K	M	G	H2	F3	J4	f'c4	f'ci6	5	
2	14.1	3	4	⅝	3	3	4	⅝	3	2	1¼	2	2½	*⅜	½	1	5¾	¾	3000	2250	1½	2
3	21.2	↑	4	↑	3	↑	4	↑	3	↑	↑	3	4	¼	½	1	↑	¾	3000	2250	↑	3
4	28.3	↓	4½	↓	3½	↓	4½	↓	3½	↓	↓	4	4	¼	⅝	1¼	↓	1¼	4000	3000	↓	4
5	35.3	3	5½	⅝	4½	3	5½	⅝	4½	2		5	4	¼	¾		5¾					5
6	42.4	4	6	¾	5	4	6	¾	5	2½		6	5	⅜	↑	↑	6					6
7	49.5	↑	6	↑	↑	↑	6	↑	↑	2½		7	↑	↑			6					7
8	56.6		7				7			2½		8			¾	1¼	7					8
9	63.6	↓	7½	↓	5	↓	7½	↓	5	3		9			1	1½	7					9
10	70.7	4	8½	¾	6	4	8½	¾	6	↑		10			1		7					10
11	77.8	5	8	1	6	5	8	1	6			11	↓	↓	1⅛		7¼	1¼				11
12	84.8	5	↑	↑	6	5	↑	↑	6	3		12	5	⅜	↑		7¼	1½				12
13	91.9	6	↓		7	6	↓		7	3½		13	6	½		1½	7¼	↑				13
14	98.9	↑	8	↑	↑	↑	8	↑	↑	↑		14	↑	↑		1¾	8¼					14
15	106.0		8½				8½					15			1⅛		8½					15
16	113.1		9½	↓			9½	↓		↓	↓	16	↓	↓	1¼		↑	↑			↑	16
17	120.2		9½		7		9½		7		1¼	17			1¼	1¾					1½	17
18	127.2		10		9		10		9		1½	18			1¼	2	↓				1¾	18
19	134.3		11		↑		11		↑		1½	19	↓	↓	1⅜	↑	8½				1¾	19
20	141.4		11				11			3½	1½	20	6	½	↑		9¾				1¾	20
21	148.5		11½				11½			4	1⅞	21	8	⅝			↑				2⅛	21
22	155.5		11½				11½			↑	↑	22	↑	↑	1⅜						↑	22
23	162.6	↓	12	↓		↓	12	↓				23			1½							23
24	169.7	6	12	1		6	12	1				24			1½		9¾		4000	3000		24
25	176.8	7	11½	1¼		7	11½	1¼				25			1⅝		11		5000	3750		25
26	183.8	↑	11½	↑		↑	11½	↑				26	↑	↑	↑		↑		↑	↑	↑	26
27	190.9		12				12					27										27
28	197.9		12		9		12		9	↓	↓	28			1⅝							28
29	205.0	7	12½		10	7	12½		10	4	1⅞	29		⅝	1¾	2					2⅛	29
30	212.0	8	12		10	8	12		10	4¾	2¼	30		¾		2¼					2½	30
31	219.2	↑	12		10	↑	12		10			31	↓	↓								31
32	226.2		13		11½		13		11½			32		1¾								32
33	233.3		13		↑		13		↑			33		1⅞								33
34	240.4	↓	14	↓		↓	14	↓				34		1⅞		11						34
35	247.5	8	14	1¼		8	14	1¼		4¾		35		1⅞		11½						35
36	254.5	9	13	1½	↓	9	13	1½	↓	5¼		36		2		↑						36
37	261.6	↑	13	↑	11½	↑	13	↑	11½			37		2								37
38	268.7		15		13		15		13			38		2		2¼						38
39	275.7		↑		↑		↑		↑			39		2¼		2½						39
40	282.8											40		↑			11½					40
41	289.9											41					11¾					41
42	296.9	↓	↓	↓	↓	↓	↓	↓	↓	↓	↓	42		2¼		11¾					↓	42
43	304.0	9	15	1½	13	9	15	1½	13	5¼	2¼	43	8	¾	2½	2½	11¾	1½	5000	3750	2½	43

*One shim only

B-7 CCL SYSTEM

DATA

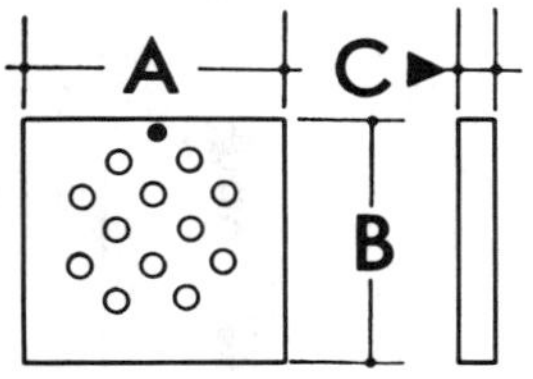

anchor plates

Plaque d'ancrages
Ankerplattes
Placa de Anclajes

No. of WIRES Nombre de fils Drahtezahl No. de Alambres	INCHES A	INCHES B	INCHES C	M/M A	M/M B	M/M C
12	$4\frac{1}{2}$	$4\frac{1}{2}$	$\frac{3}{4}$	114	114	19
8	$2\frac{3}{4}$	5	$\frac{1}{2}$	70	127	13
4	$2\frac{1}{4}$	$2\frac{1}{4}$	$\frac{3}{8}$	57	57	10

TABLE OF PRESTRESSING FORCES design detail

Forces de Precontrainte

Spannkrafte

Fuerzas de Pretensado

No. OF WIRES Nombre de fils Drahtezahl No. de Alambres	LBS—LIVRES—PFUND—LIBRAS: INITIAL P_I 0.70 F_s	LBS—LIVRES—PFUND—LIBRAS: FINAL PF 0.80 P_I	KILOS: Valeur Initiale P_I Anfangswert P_I Inicial 0.70 F_s	KILOS: Valeur Finale P_I Endvert P_I Final 0.80 P_I
12	112,500	90,000	51,000	40,800
8	75,000	60,000	34,000	27,200
4	37,500	30,000	17,000	13,600

A TYPICAL DETAIL AT centreline of span

- Vue de detail typique a l'axe de la portee
- Typisches detail in Balkenmitte
- Seccion en el centro del vano

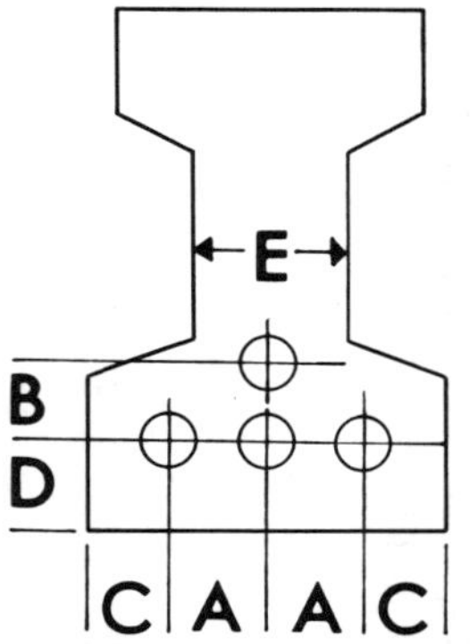

A	B	C	D	E	Size of Duct
INS.	INS.	INS.	INS.	INS.	INS.
$2\frac{3}{4}$	$2\frac{1}{2}$	$2\frac{1}{4}$	$2\frac{1}{4}$	$4\frac{1}{2}$	$1\frac{5}{8}$
$2\frac{1}{2}$	$2\frac{1}{4}$	$2\frac{1}{4}$	$2\frac{1}{4}$	$4\frac{1}{4}$	$1\frac{1}{2}$
$1\frac{3}{4}$	$1\frac{1}{2}$	2	2	4	$\frac{3}{4}$
or $2\frac{1}{2}$	or $2\frac{1}{4}$	or $2\frac{1}{4}$	or $2\frac{1}{4}$	or $4\frac{1}{2}$	or $1\frac{1}{2}$

Diametre de l'alveole—Spannkanaldurchmesser
Tamano del Conducto

m/m	m/m	m/m	m/m	m/m	m/m
70	65	60	60	115	42
65	60	60	60	115	40
45	40	50	50	100	20
or 65	or 60	or 60	or 60	or 115	or 40

APPENDIX B

DETAILING DATA
FOR THE SPIRAL WIRE SYSTEM

2 No. meshes of ⅜" dia bars at 3" c/c both ways
2 armatures en treillis, barres o 12 mm, 7.5 cm entre-axes, des deux cotes
2 Bewehrungsnetze, o 12 mm Stabe, 7.5 cm Mittenabstande, in beiden Richtungen
Emparrillado de o ⅜" Separados 3"

⅜" Dia. Stirrups at 3" Centres
Etriers o 12 mm, 6.5 cm entre-axis
Bugel o 12 mm, 6.5 cm Mittenabstande
Estribos de o ⅜" Cada 3"

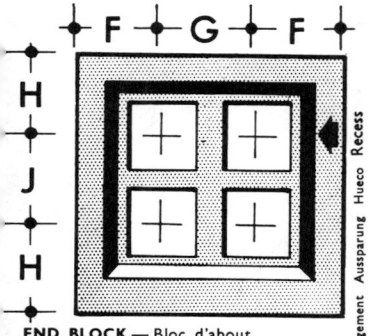

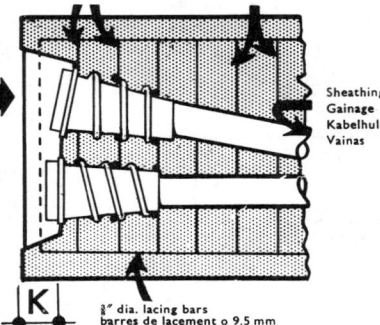

Sheathing
Gainage
Kabelhulle
Vainas

DONNÉES D'ÉPURE
Pour le système
d'ancrage hélicoïdal
de fils

ANGABEN
für Spiralanker
für draht

DATOS
INFORMATIVOS
Para el sistema
espiral de alambre

END BLOCK — Bloc d'about
Balkenende — Bloque de Culata

Logement Aussparung Hueco Recess

⅜" dia. lacing bars
barres de lacement o 9.5 mm
Bindestabe o 12 mm
Redondos de ⅜"

MINIMUM DIMENSION IN INCHES
Cotes Minimales en Pouces/en mm   Mindestmasse in Zoll/in mm   Dimensiones Minimas en Pulgadas/en mm

	F	G	H	J	K
12 WIRE	4½	6	4½	6	3
8 WIRE	3½	4	4¾	6½	3
4 WIRE	3¼	3½	3¼	3½	3

Cotes Minimales en Millimètres   Mindestmasse in Millimeter   Dimensiones Minimas en Milimetros

	F	G	H	J	K
12 Fils / 12 Draht / 12 Alambres	115	150	115	150	75
8 Fils / 8 Draht / 8 Alambres	90	100	120	165	75
4 Fils / 4 Draht / 4 Alambres	85	90	85	90	75

NOTE

Please note: The above figures are recommended minimum dimensions. Where possible the centres should be increased above the dimensions shown. We will be pleased to advise on any problems concerning the detailing of "end blocks".

Veuillez noter que les chiffres précités sont les cotes minimales recommandées. Là où c'est possible il faut augmenter les entre-axes à une cote plus forte que celle indiquée. Nous serons heureux de vous conseiller sur tout problème afférent au dimensionnement des "blocs d'about".

Zur Beachtung: Obige Zahlen stellen die empfehlenswerten Mindestmasse dar. Nach Möglichkeit sind die Abstände grösser zu wählen. Mit Bezug auf die Gestaltung der Balkenende sind wir mit unserem Rat gern zur Verfügung.

Observacion: Las cifras anteriores se aconsejada como dimensiones minimas. Cuando ello sea posible, las distancias entre centros deben su aumentadas. Con mucho gusto averoraremos en cualquien problema relativo a los bloques de anclaje o culata.

CCL CABCO SYSTEMS

SINGLE STRAND SYSTEMS
SYSTEME A TORON UNIQUE
EINZELLITZENSYSTEME
SISTEMA A TRECCE SINGOLO
SISTEMA DE UN SOLO CORDON

						CCL SPIRAL / SPIRALE / SPIRAL / SPIRALE / ESPIRAL		GB TUBE / TUBE / ROHR / TUBO / TUBO		
TENDON	TENDON	SPANNGLIED	CAVO	TENDON	ins.	1.125	1.125	0.7	0.6	0.5
					mm.	29	29	18	15	13
INITIAL PRESTRESSING FORCE	FORCE TENSION INITIALE	ANFANGS- SPANNKRAFT	FORZA INIZIALE DIE PRE- COMPRESSIONE	FUERZA DE PRETEN- SADO INICIAL	kips	130	130	58	35.7	25.9
					tonnes	59	59	26.5	16.2	11.8
ANCHORAGE / ANCRAGE / VERANKERUNG / ANCORAGGIO / ANCLAJE	SIZE OF SQUARE FLANGE	SURFACE DE LA BRIDE CARRE	GRÖSSE DER VIERECKIGEN FLANSCHE	GRANDEZZA DELLA FLANGIA QUADRATA	DIMENSIONES DE LA BRIDA CUADRADA	ins.	4.5 sq.	4.5 sq.	3.5 sq.	2.5 sq.	2.25 sq.
						mm.	114 sq.	114 sq.	89 sq.	64 sq.	57 sq.
	LENGTH	LONGUEUR	LÄNGE	LUNGHEZZA	LONGITUD	ins.	7	6	5	3.25	3
						mm.	178	153	127	83	76
	TYPE	TYPE	TYPE	TIPO	TIPO		S6	T7	T1	S2	S5
DIAMETER OF DUCT	DIAMETRE DE LA GAINE	SPANNKANAL- DURCHMESSER	DIAMETRO DELLA GUAINA	DIAMETRO DEL CONDUCTO		ins.	1.5	1.5	1.25	1.0	0.75
						mm.	39	39	33	25	18
DIMENSIONS					A	ins.	2.5	2.5	2.25	2.0	1.75
						mm.	65	65	60	50	45
					B	ins.	2.25	2.25	2.0	1.75	1.5
						mm.	60	60	50	45	40
					C	ins.	2.5	2.5	2.5	2.25	2.0
						mm.	65	65	65	60	50
					D	ins.	2.5	2.5	2.5	2.25	2.0
						mm.	65	65	65	60	50
					E	ins.	5.0	5.0	5.0	4.5	4.0
						mm.	130	130	130	120	100
					F	ins.	4.25	4.25	3.75	3.25	3.25
						mm.	110	110	95	85	85
					G	ins.	6.0	6.0	5.0	3.5	3.75
						mm.	150	150	130	90	95
					H	ins.	4.25	4.25	3.75	3.25	3.5
						mm.	110	110	95	85	90
					J	ins.	6.0	6.0	5.0	3.5	4.5
						mm.	150	150	130	90	115
MINIMUM DEPTH OF RECESS	MINIMUN DE PROFUNDEUR DE LA CAVITE	AUSSPARUNG DER MINIMUM VERTIEFUNG	PROFONDITA MINIMA DEL L'ALVEOLE	PROFUNDIDAD MINIMA DE LA CAVIDAD		ins.	2.0	5.25	4.0	3.5	3.0
						mm.	50	135	100	90	75

APPENDIX B

MULTI-STRAND SYSTEMS
SYSTEME A TORONS MULTIPLES
MEHRFACH-LITZENSYSTEME
SISTEMA A TRECCE MULTIPLI
SISTEMA DE CORDONES MULTIPLES

7/0.7	4/0.7	12/0.6	7/0.6	4/0.6	12/0.5	7/0.5	4/0.5
7/18	4/18	12/15	7/15	4/15	12/13	7/13	4/13
406	232	428	250	143	311	181	104
184.5	105.5	194.5	113.5	65	141.5	82.5	47.5
8.5 sq.	7.0 sq.	9.0 sq.	7.0 sq.	6.0 sq.	8.5 sq.	6.0 sq.	4.75 sq.
216 sq.	178 sq.	229 sq.	178 sq.	152 sq.	216 sq.	152 sq.	121 sq.
13	10	13	10	10	13	10	8
330	254	330	254	254	330	254	203
T6	T5	T8	T5	T4	T6	T4	S4
3.0	2.0	3.25	2.5	2.0	3.0	2.0	1.625
75	51	81	63	51	75	51	42
4.0	3.0	4.25	3.5	3.0	4.0	3.0	2.75
100	75	110	90	75	100	75	70
3.75	2.75	4.0	3.25	2.75	3.75	2.75	2.5
95	70	100	85	70	95	70	65
3.25	2.75	3.5	3.0	2.75	3.25	2.75	2.5
85	70	90	75	70	85	70	65
3.25	2.75	3.5	3.0	2.75	3.25	2.75	2.5
85	70	90	75	70	85	70	65
6.5	5.5	7.0	6.0	5.5	6.5	5.5	5.0
170	140	180	150	140	170	140	130
7.0	5.5	7.5	5.5	5.0	7.0	5.0	4.5
180	140	190	140	130	180	130	115
10.0	8.5	10.5	8.5	7.5	10.0	7.5	6.0
225	215	265	215	190	255	190	150
7.0	5.5	7.5	5.5	5.0	7.0	5.0	4.5
180	140	190	140	130	180	130	115
10.0	8.5	10.5	8.5	7.5	10.0	7.5	6.0
225	215	265	215	190	255	190	150
5.5	5.0	5.25	5.0	4.75	5.0	4.25	4.0
140	130	135	130	120	130	110	100

SPIDER PROFILERS — Spider Profilers required only for sharp or reversed curves

Radius of Duct	Centres
Straight—400 ft.	0—6 ft.
400—200 ft.	6 ft.—4 ft. 6 in.
200—100 ft.	4 ft. 6 in.—3 ft.
100—30 ft.	3 ft.—2 ft.
Under 30 ft.	1 ft. 6 in.

ECARTEURS — Les écarteurs ne sont nécessaires que si le parcours du câble comporte de faibles rayons de courbure

Rayon de la Gaine	Centres
Droit —120 m.	0—180 cm.
120 m. — 60 m.	180 cm.—140 cm.
60 m. — 30 m.	140 cm.— 90 cm.
30 m. — 10 m.	90 cm.— 60 cm.
moindre 10 m.	45 cm.

ABSTANDHALTER — Die Abstandhälter sind nur für sehr starke oder Gegenkrümmungen erforderlich

Radius der Spannkanal	Mittenabstände
Gerade —120 m.	0—180 cm.
120 m. — 60 m.	180 cm.—140 cm.
60 m. — 30 m.	140 cm.— 90 cm.
30 m. — 10 m.	90 cm.— 60 cm.
Weniger als—10 m.	45 cm.

DISTANZIATORE — Gli spaziatori sono usati solamente nelle curve acute o inverse

Distanze	Centres
Diritto —120 m.	0—180 cm.
120 m. — 60 m.	180 cm.—140 cm.
60 m. — 30 m.	140 cm.— 90 cm.
30 m. — 10 m.	90 cm.— 60 cm.
Meno di 10 m.	45 cm.

ESPACIADORES — Los espaciadores solo son necesarios con curvaturas muy fuertes o inflexiones

Radio del Conducto	Separacion
Mas de —120 m.	0—180 cm.
120 m. — 60 m.	180 cm.—140 cm.
60 m. — 30 m.	140 cm.— 90 cm.
30 m. — 10 m.	90 cm.— 60 cm.
Menos de 10 m.	45 cm.

Index

AASHO-BPR standard sections, 305, 309, 312
AASHO specifications, 222, 224, 310, 314, 316
Abeles, P. W., 15
ACI Building Code 1963, 222, 242, 243, 277, 278, 294, 310
Ali, I., 265
Allowable stress:
  in concrete, 201
  in end stirrups, 169
  in steel, 242
Analysis, 60
  of continuous beams, 348
  illustrative example, 368
Anchorage, 156
  transverse reinforcement, 166
Anchorage-zone stresses, 163
Appleton, J. H., 124
Approximate methods for calculating ultimate moment, 277
  provisions of ACI code, 278
Aroni, S., 419

Basic requirements:
  beams with superimposed dead load, 258
  composite design, 307
  continuous beams, 384
  noncomposite design, 204
BBRV system, 23, 30–34
Behavior of prestressed-concrete beams, influence of the variables, 107
Billet, D. F., 124
Bond, 156
  flexural, 161
  strands, 160
  stress-transfer, 156
  wires, 157
Breckenridge, R. A., 419
Bridge stringer:
  selection and spacing, 313, 331
  ultimate design, 339

Brittle behavior, 43
Bursting stresses, 165

CCL system, 23, 37
Christodoulides, S. P., 172
Columns, 400
  behavior of non-prestressed, 417
  behavior of prestressed, 415
Combined stresses, behavior of beams in the region of, 126
Compatibility equations, 62
Compatibility factor, 47–49, 53, 60
Composite sections, 303, 314
Concrete:
  modulus of rupture, 59
  stress-strain diagram: beams, 56–58
    cylinders, 56–58
  tensile strength, 59
Conditions of loading, 199, 200
Continuous beams, 342
Corley, W. G., 196
Cracking, 48, 49, 132, 133
  behavior of beam before cracking, 128
  inclined, 136
  longitudinal ends, 163
Creep, 6, 12, 180, 181
  on long-time deflections, effect of, 184, 187
Curing, 21
Curvature:
  of beam, 70
  of cable, 373
  reduction of, 346

Davis, R. E., 196
Deflections:
  calculation of, 85
  long-time, 184
    calculation of, 189
Depth factor ω, 213
  determination of, 240

Design:
    composite construction, 320
    continuous beams, 347, 383
    for economy, 216
    illustrative examples, 208, 243, 248, 260, 287, 292, 312, 331, 339, 387
    least-weight, 212, 324
    problems in, 206
    ultimate, 283, 337
    using standard sections, 207, 312
    working-stress, 197
Dill, R. E., 6, 15
Dimensionless variables, 212
    relations among, 215, 223, 323
Ductile behavior, 43
Ductility, 283
    minimum required, 288

Eccentricity:
    determination of, 207
    typical endspan, 352
    typical interior span, 352–353
    variation as a fourth-degree curve and parabola, 354, 358
    variation as parabolas, 351, 357
    variation with prestressing force, 210
Eccentricity ratio ϵ, 214
Effective slab area, 309
Effectiveness, 182, 205
Efficiency ρ, 213
    choice of, 241
    determination of, 240
Equations:
    of compatibility, 62
    of equilibrium, 61
Equivalent-load method, 371
    illustrative examples, 382
Evans, R. H., 341

Failure:
    path to, 402
    shear-compression, 135, 137
    web-distress, 135, 140
Fixed-end moments, 359, 364
    endspan, 361
    intermediate span, 362–363
    variation with the shape of profile, 365
Flanged sections, 112–114, 268, 269
Flexural behavior, 42
    ductile and brittle behavior, 43

Flexural behavior:
    flexural strength, 44
    ultimate moment, 44
Forms:
    steel, 21
    wood, 21
Freyssinet, E., 6, 15, 23
Freyssinet system, 6, 22, 23, 179
Friction, 174
    coefficients, 175
    loss in continuous beams, 344
    loss of prestress, 174
    reduction of coefficient of, 347
Frictional stresses, 175

Gere, J. M., 419
Gergely, P., 172
Grouting, 22
Guillermo, E. C., 419
Gurfinkel, G., 265, 302, 341, 419
Guyon, Y., 166, 172
Guys, 5

Hanson, N. W., 124, 172, 302, 341
Hewett, W. H., 6
High-strength steel bars, 22, 25
Hognestad, E., 124, 302, 399, 409, 413
Hold-down device, 19, 20
Hromadik, J. J., 419

Idealized loading conditions, composite sections, 306
Idealized sections, 232
    composite, 326
    I sections, 239
    T sections, 238
    unsymmetrical I sections, 234–237
Inclined cracks, 136
Indeterminate moment, 350
Indeterminate reaction, 350
Inflection point, 352, 356
Iyengar, K. T. S. R., 172

Jack:
    hydraulic, 19
    short-stroke, 19
Jackson, P. A., 5
Janney, J. R., 172
Jensen, V. P., 125

INDEX

Kaar, P. H., 172, 399
Kern point, 198
Khachaturian, N., 155, 265, 302, 341
King-post truss, 4
Koebel, F. E., 399
Krahl, N. W., 155
Kriz, L. B., 399

LaFraugh, R. W., 172
Lakhwara, T. R., 419
Least area of concrete, 217
Least-weight design:
  of composite sections, 320
  of noncomposite sections, 212
Leonhardt, F., 399
Limitations on steel percentage, 280
Limiting value of strain, 267
Lin, T. Y., 419
Loading:
  conditions of, 199, 306
  idealized conditions of, 200, 306
Long-time deflections, 173, 184, 189
  with prestressing alone, 184
  rate of creep method, 189
  superposition method, 189
  with transverse load, 187
Long-time effects, 180
Losses of prestress, 173
  due to anchorage set, 179
  due to elastic shortening, 177
  due to friction, 174, 344
Lund, J. G. F., 6

MacGregor, J. G., 155
McHenry, D., 124, 125, 196, 302
Magnel, G., 166, 172
Magura, D. D., 196
Marshall, W. T., 172
Mass, M. A., 172
Mattock, A. H., 172
Modular ratio, 310
  effect of, 319
Modulus of elasticity for concrete, 310
Mohr's circle, 129, 130
Moment-curvature relationship, 72
  cracking load, 77
  prestressing force only, 72
  ultimate, 83
  zero strain, 76
Moment-deflection relationship, 90

Moment ratio R, 214
Moorman, R. B. B., 399
Moreadith, F. L., 419

Newmark, N. M., 419
Non-prestressed steel, 52, 55, 56
Numerical analysis of continuous beams, 380

Olsen, S. E., 155

Paris, G. H., 399
Parker, A. S., 341
Parme, A. L., 399
Path to failure, 402
Post-tensioning systems, 16, 17, 22, 430
Prescon system, 23, 30, 34, 36, 37
Prestress, effective, 46, 47
Prestressing with superimposed dead load, 258
Prestressing bed, 19
Prestressing force, 2
  determination of, 207, 316
Prestressing steel, 53, 54
  the allowable stress, 242
  bars, 53
  the profile of, 255
  strands, 54
  summary of types, 54
  wires, 53, 54
Prestressing systems, 17
Pretensioning, 16, 17
Pretensioning systems, 17
Principal stresses, 127–129
Profile of prestressing steel, 255

Queen-post truss, 5
Quick-release grip, 18

Ramaley, D., 125
Ransome, E. L., 15
Raphael, J. M., 196
Region of combined stresses, 43
  behavior after cracking, 132
  behavior before cracking, 128
Region of pure moment, 43
Reinforced concrete, 64
Reinforcement ratio m, 214
Relaxation, 180, 182

Relaxation:
  on long-time deflections, effect of, 184, 187
Robinson, A. R., 419
Roebling system, 23, 29
Ryerson system, 23, 30, 34, 35

Saurbrey, A., 15
Schorer, H., 180, 196
Segmental beams, 222
Shape factor Δ, 214
  choice of, 241
  determination of, 240
Shear, 126, 127
  at inclined cracking: composite sections, 139
    noncomposite sections, 137
  provisions of ACI code, 294
Shear-compression failure, 133, 135
Shear connectors, 304
Shear reinforcement, 296
Shear span, 126
Shrinkage, 12, 180
  on long-time deflections, 184, 187
Siess, C. P., 125, 155, 172, 196
Single-element beams, 222
Sonesson, B., 399
Sozen, M. A., 125, 155, 172, 196
Spalling stresses, 165
Split-cone wedge, 18
Stages of loading, 45
  prestressing force only, 45, 46
  ultimate, 48–50
  zero strain in concrete, 46, 47
Standard sections, 207, 310
Steiner, G. R., 5
Straight-line theory, 64
Strain distribution with depth, 44
Strands, 17–19
  determination of number, 209, 245, 252, 283, 290, 292, 318
Strength of beam, 43
Stress block, 270
Stress coefficients, 201
  summary of, 203
Stress trajectories, 127–130
Stress-strain diagrams:
  concrete, 56–58
    approximate, 67
  non-prestressed steel, 55
  prestressing steel, 53, 54

Stress-strain relationship, 44
Stresses:
  at the ends, 252
  summary of allowable, 202
Stressteel system, 22, 25
  SEEE, 27
Stuessi, F., 271, 302

Thorpe, L. T., 265
Timoshenko, S. P., 419
Transfer, 199
Transfer length, 157, 159
Transverse end reinforcement, 166, 169
Troxell, G. D., 196

Ultimate design, 267, 282, 283
  composite section, 337
  examples of, 287, 292, 339
Ultimate interaction diagram, 402
  determination of, 403
  effect of initial prestrain, 410
  illustrative example, 406
Ultimate moment, 267
  simplified methods: for determination of, 270
    in flanged sections, 276
Unbonded beam, illustrative problem, 105

Vertical ties, 304
Vertical upright, 17, 18

Warwaruk, J., 125
Web-distress failure, 134
  composite sections, 141
  cracking noncomposite sections, 140
  shear at inclined, 135
Web reinforcement, 128, 143, 144
Welsh, W. A., 172
Whipple, Squire, 3, 15
Whipple truss, 3
Whitney, C. S., 125
Wobble coefficients, 175
Wobble effect, 174

Zero-tensile-strain diagram, 412
Zia, P., 419
Zwoyer, E. M., 155

This book was set in Modern by The Maple Press Company, and printed on permanent paper and bound by The Maple Press Company. The designer was Richard Paul Kluga; the drawings were done by J. & R. Technical Services, Inc. The editors were B. J. Clark and David Damstra. Paul Poss supervised the production.

Date Due

WITHDRAWN